Nonlinear & Complex Systems

Nonlinear & Complex Systems

Editor:

Olga Moreira, Ph.D.

ArclerPress

www.arclerpress.com

Nonlinear & Complex Systems

Editor: Olga Moreira, Ph.D.

708 3rd Avenue, 6th Floor
New York
NY 10017
United States of America
www.arclerpress.com

ISBN: 978-1-68094-433-4

Library of Congress Control Number: 2016951368

About the Editor

Olga Moreira, Ph.D.

Olga Moreira obtained her Ph.D. in Astrophysics from the University of Liege (Belgium) in 2010, her BSc. in Physics and Applied Mathematics from the University of Porto (Portugal). Her post-graduate travels and international collaborations with the European Space Agency (ESA) and European Southern Observatory (ESO) led to great personal and professional growth as a scientist. Currently, she is working as an independent researcher, technical writer, and editor in the fields of Mathematics, Physics, Astronomy and Astrophysics.

Contents

Section II: Perturbation Theory and Homotopy Perturbation Methods

List of Contributors

Henk W. Broer
Johann Bernoulli Institute for Mathematics and Computer Science, University of Groningen, The Netherlands

Marwan Alquran
Department of Mathematics and Statistics, Jordan University of Science and Technology, Irbid 22110, Jordan

Kamel Al-Khaled
Department of Mathematics and Statistics, Jordan University of Science and Technology, Irbid 22110, Jordan

Simon A. Neild
Faculty of Engineering, University of Bristol, Bristol BS8 1TR, UK

Alan R. Champneys
Faculty of Engineering, University of Bristol, Bristol BS8 1TR, UK

David J. Wagg
Department of Mechanical Engineering, University of Sheffield, Sheffield S1 3JD, UK

Thomas L. Hill
Faculty of Engineering, University of Bristol, Bristol BS8 1TR, UK

Andrea Cammarano
Faculty of Engineering, University of Bristol, Bristol BS8 1TR, UK

Huixian Cai
Department of Mathematics, South China University of Technology, Guangzhou 510640, China

Chaohong Pan
Department of Mathematics, South China University of Technology, Guangzhou 510640, China

Zhengrong Liu
Department of Mathematics, South China University of Technology, Guangzhou 510640, China

Daniele Avitabile
Centre for Mathematical Medicine and Biology, School of Mathematical Sciences, University of Nottingham, University Park, Nottingham NG7 2RD, UK

Piotr Słowi´nski
Department of Mathematics, College of Engineering, Mathematics and Physical Sciences, University of Exeter, Exeter, Devon EX4 4QF, UK

Benoit Bardy
EuroMov, Montpellier University, 700 Avenue du Pic Saint-Loup, 34090 Montpellier, France
Institut Universitaire de France, Paris, France

Krasimira Tsaneva-Atanasova
Department of Mathematics, College of Engineering, Mathematics and Physical Sciences, University of Exeter, Exeter, Devon EX4 4QF, UK
EPSRC Centre for Predictive Modelling in Healthcare, University of Exeter, Exeter, Devon EX4 4QF, UK

Mehmet Pakdemirli
Applied Mathematics and Computation Center Celal Bayar University

Süleyman Cengizci
Institute of Applied Mathematics, Middle East Technical University, 06800 Ankara, Turkey

Mehmet Tarık Atay
Department of Mechanical Engineering, Abdullah Gül University, 38039 Kayseri, Turkey

Aytekin Eryılmaz
Department of Mathematics, Nevşehir Hacı Bektaş Veli University, 50300 Nevşehir, Turkey.

U. Filobello Nino
Facultad de Instrumentación Electrónica, Universidad Veracruzana, Circuito Gonzalo Aguirre Beltrán S/N, 91000 Xalapa, Veracruz, Mexico

H. Vazquez Leal
Facultad de Instrumentación Electrónica, Universidad Veracruzana, Circuito Gonzalo Aguirre Beltrán S/N, 91000 Xalapa, Veracruz, Mexico

M. M. Rashidi
Shanghai Key Lab of Vehicle Aerodynamics and Vehicle Thermal Management Systems, Tongji University, 4800 Cao An Rd., Jiading, Shanghai 201804, China
ENN-Tongji Clean Energy Institute of Advanced Studies, Shanghai, China

H. M. Sedighi
Department of Mechanical Engineering, Shahid Chamran University, Ahvaz, Iran

A. Perez Sesma
Facultad de Instrumentación Electrónica, Universidad Veracruzana, Circuito Gonzalo Aguirre Beltrán S/N, 91000 Xalapa, Veracruz, Mexico

M. Sandoval Hernandez
Doctorado en Ciencia, Cultura y Tecnología, Universidad de Xalapa, Km 2 Carretera Xalapa Veracruz, 91190 Xalapa, Veracruz, Mexico

A. Sarmiento Reyes
National Institute for Astrophysics, Optics and Electronics, Luis Enrique Erro #1, Sta. María Tonantzintla, 72840 Puebla, Mexico

A. D. Contreras Hernandez
Facultad de Instrumentación Electrónica, Universidad Veracruzana, Circuito Gonzalo Aguirre Beltrán S/N, 91000 Xalapa, Veracruz, Mexico

D. Pereyra Diaz
Facultad de Instrumentación Electrónica, Universidad Veracruzana, Circuito Gonzalo Aguirre Beltrán S/N, 91000 Xalapa, Veracruz, Mexico

C. Hoyos Reyes
Facultad de Instrumentación Electrónica, Universidad Veracruzana, Circuito Gonzalo Aguirre Beltrán S/N, 91000 Xalapa, Veracruz, Mexico

V. M. Jimenez Fernandez
Facultad de Instrumentación Electrónica, Universidad Veracruzana, Circuito Gonzalo Aguirre Beltrán S/N, 91000 Xalapa, Veracruz, Mexico

J. Huerta Chua
Facultad de Ingeniería Electrónica y Comunicaciones, Universidad Veracruzana, Venustiano Carranza S/N, Col. Revolución, 93390 Poza Rica, Veracruz, Mexico

F. Castro Gonzalez
Facultad de Instrumentación Electrónica, Universidad Veracruzana, Circuito Gonzalo Aguirre Beltrán S/N, 91000 Xalapa, Veracruz, Mexico

J. R. Laguna Camacho
Facultad de Ingeniería Mecánica Eléctrica, Universidad Veracruzana, Venustiano Carranza S/N, Col. Revolución, 93390 Poza Rica, Veracruz, Mexico.

Md. Abdur Razzak
Department of Mathematics, Rajshahi University of Engineering and Technology (RUET), Kazla, Rajshahi 6204, Bangladesh

Md. Shamsul Alam
Department of Mathematics, Rajshahi University of Engineering and Technology (RUET), Kazla, Rajshahi 6204, Bangladesh

S. Narayanamoorthy
Department of MathematicsBharathiar University

S. P. Sathiyapriya
Department of MathematicsBharathiar University

Heng-you Lan
Institute of Nonlinear Science and Engineering Computing, Sichuan University of Science & Engineering, Zigong, Sichuan 643000, P.R. China
Key Laboratory of Higher Education of Sichuan Province for Enterprise Informationalization and Internet of Things, Zigong, Sichuan 643000, P.R. China

Yi-shun Cui
College of Materials and Chemical Engineering, Sichuan University of Sciences & Engineering, Zigong, Sichuan 643000, P.R. China.

Jinbo Lu
School of Energy Science and Engineering, University of Electronic Science and Technology of China, Chengdu, Sichuan 611131, China

Xiaorong Hou
School of Energy Science and Engineering, University of Electronic Science and Technology of China, Chengdu, Sichuan 611131, China

Min Luo
School of Energy Science and Engineering, University of Electronic Science and Technology of China, Chengdu, Sichuan 611131, China

Ji Chul Kim
Department of Psychological Sciences, University of Connecticut, Storrs, CT, USA

Edward W. Large
Department of Psychological Sciences, University of Connecticut, Storrs, CT, USA

Junling Sun
School of Mathematics and Information ScienceHenan Polytechnic University

Xuefeng Han
School of Mathematics and Information ScienceHenan Polytechnic University

Yi Li
State Key Laboratory of Information Photonics and Optical Communications, Beijing University of Posts and Telecommunications, Beijing 100876, China
School of Science, Beijing University of Posts and Telecommunications, Beijing 100876, China

Wen-rui Shan
State Key Laboratory of Information Photonics and Optical Communications, Beijing University of Posts and Telecommunications, Beijing 100876, China
School of Science, Beijing University of Posts and Telecommunications, Beijing 100876, China

Tianping Shuai
State Key Laboratory of Information Photonics and Optical Communications, Beijing University of Posts and Telecommunications, Beijing 100876, China
School of Science, Beijing University of Posts and Telecommunications, Beijing 100876, China

Ke Rao
State Key Laboratory of Information Photonics and Optical Communications, Beijing University of Posts and Telecommunications, Beijing 100876, China
School of Science, Beijing University of Posts and Telecommunications, Beijing 100876, China

L. Borkowski
Division of Dynamics, Lodz University of Technology, Stefanowskiego 1/15, 90-924 Lodz, Poland

A. Stefanski
Division of Dynamics, Lodz University of Technology, Stefanowskiego 1/15, 90-924 Lodz, Poland

Cheng-Hsiung Yang
Graduate Institute of Automation and Control, National Taiwan University of Science and Technology, No. 43, Section 4, Keelung Road, Taipei 106, Taiwan

Cheng-Lin Wu
Graduate Institute of Automation and Control, National Taiwan University of Science and Technology, No. 43, Section 4, Keelung Road, Taipei 106, Taiwan

Mihaela Neamțu
West University of Timisoara, 4 Bulevardul Vasile Pârvan, 300115 Timisoara, Romania

Anamaria Lițoiu
West University of Timisoara, 4 Bulevardul Vasile Pârvan, 300115 Timisoara, Romania

Petru C. Strain
West University of Timisoara, 4 Bulevardul Vasile Pârvan, 300115 Timisoara, Romania

Preface

This book aims to provide an overview of the recent research devoted to nonlinear and chaotic systems by discussing the modeling of different phenomena. It will encompass methods and techniques from the analysis of nonlinear ordinary differential equations, perturbation theory, bifurcation analysis and chaos synchronization. It contains 19 chapters, organized into 3 sections:

- Section 1 - Resonance Phenomena, Nonlinear Vibrations and Waves.
- Section 2 - Perturbation Theory and Homotopy Perturbation Methods.
- Section 3 - Bifurcation Analysis and Chaos Synchronization.

Section 1 is centered on the analysis of nonlinear ordinary differential equations. It devoted to the study of resonance, oscillations and waves phenomena of from the point of view of dynamical systems close to resonances or bifurcation points. This will include review approximate methods for strongly nonlinear differential equations with oscillations (e.g. Lindstedt-Poincare method and Krylov-Bogoliubov first approximate method) and introduce the normal forms theory for simplifying nonlinear dynamical systems. It will consider examples of coupled nonlinear oscillators, including Drinfeld-Sokolov system and Haken–Kelso–Bunz model (rhythmic movement patterns in human bimanual coordination). Section 2 is focused on the perturbation methods, singular perturbation problem, homotopy perturbation method, variational method, and successive complementary expansion method (a combination of the homotopy and singular perturbation methods). It will include examples of these methods applied to boundary problems, nonlinear large-amplitude oscillations, and nonlinear integral-differential equations. Section 3 is first focused bifurcation analysis, in particular, the Hopf bifurcation of nonlinear dynamic systems. It will include the examples of the Hopf bifurcation as applied to neural networks and Boussinesq systems. Next, it will investigate a higher-order nonlinear Schrödinger equation with non-Kerr term by using the bifurcation theory method as well as the dynamics of a ring of seven unidirectionally coupled nonlinear Duffing oscillators using Fast Fourier Transform analysis presented in the form of a bifurcation graph. The second subject covered in this section is the chaos synchronization, which occurs when two or more dissipative chaotic systems are coupled. The first example will be an adaptive control scheme developed to analyze the generalized adaptive chaos synchronization with uncertain chaotic behavior parameters between two identical chaotic dynamic systems. The second example is the synchronization of a four-dimensional Lorenz-Stenflo system modeled by an electronic circuit. The last example investigates the chaos synchronization of a three-dimensional autonomous nonlinear system (called the general T-system)

Editor
Olga Moreira, Ph.D.

INTRODUCTION

Systems are entities composed of well-defined components. When integrated the components act together as to form a functioning whole with dynamical behaviors and responses to the environment Systems can be embedded into other functional entities as components. Identifying a system or a hierarchy of systems requires a certain level of abstraction and simplification. One delineates simple from complex systems.

Simple Systems have few components and their behavior is in all respects fully understandable and predictable. An example would be a solid ball falling under the action of gravity through air. This Simple System consists of the ball, the air, and the gravitational force. Here we usually assume a single ball, constant acceleration of gravity, a viscous drag on the ball, and Newton's laws. When making these assumptions, we arrive at very useful answers. We did, however, neglect many aspects. If, for example, we would ask how the behavior changes when we go from one ball, to two, to three, or even more balls that fall close to each other, our "Simple System" assumption fails. It is not sufficient to generalize from one ball's behavior to many. Instead we need to consider the interaction of the balls through their self-generated vortices. This makes the latter example a topic for Complex Systems Research.

Complex Systems may have many components (elements or spatio-temporal fields) that collaborate to create a functioning whole. Thereby the function creates itself, i.e., it "comes about" by the dynamical interaction of the components without an intervening regulatory body. One speaks of "Self-Organization" or also of "Emergence". Important is that the word "complex" is not to be confused with the word "complicated". Let's consider, for example, an "architectural building complex" that serves a multitude of functionalities. As we all know it is not complicated, but shows its true functional complexity when we use it.

Nonlinear system is a system in which the output is not directly proportional to the input. Nonlinear problems are of interest to engineers, physicists and mathematicians and many other scientists because most systems are inherently nonlinear in nature. Nonlinear systems may appear chaotic, unpredictable or counterintuitive, contrasting with the much simpler linear systems.

Typically, the behavior of a nonlinear system is described in mathematics by a nonlinear system of equations, which is a set of simultaneous equations in which the unknowns (or the unknown functions in the case of differential equations) appear as variables of a polynomial of degree higher than one or in the argument of a function which is not a polynomial of degree one. In other words, in a nonlinear system of equations, the equation(s) to be solved cannot be written as a linear combination of the unknown variables or functions that appear in them. It does not matter if nonlinear known functions appear in the equations. In particular, a differential equation is linear if it is linear in terms of the unknown function and its derivatives, even if nonlinear in terms of the other variables appearing in it.

As nonlinear equations are difficult to solve, nonlinear systems are commonly approximated by linear equations (linearization). This works well up to some accuracy and some range for the input values, but some interesting phenomena such as solutions, chaos and singularities are hidden by linearization. It follows that some aspects of the behavior of a nonlinear system appear commonly to be counterintuitive, unpredictable or even chaotic. Although such chaotic behavior may resemble random behavior, it is absolutely not random.

For example, some aspects of the weather are seen to be chaotic, where simple changes in one part of the system produce complex effects throughout. This nonlinearity is one of the reasons why accurate long-term forecasts are impossible with current technology.

A well-known, simple example for a complex system where elements interact is a double pendulum that moves in the gravitational field [see the Science Express experiment]. While a single pendulum shows only two simple types of behavior -- swinging motion at low speeds and rotation in a fixed direction at high speeds, two coupled pendula show chaotic dynamics -- they switch in an apparently erratic manner between phases of rotation and of oscillations. Let's consider also another well-known example for a complex system where spatio-temporal fields like temperature, fluid velocity, and humidity collaborate, namely cloud patterns. They emerge all by themselves. All that is required is the thermal energy deposited by the Sun on Earth's surface, the heat radiated from the cloud top into space, the Earth's gravitational field, and the moist atmosphere. The dynamics and structure of the cloud patterns are sustained by the perpetual flux of energy from the Sun to the Earth and from the Earth's surface into space. It is evident that Complex Systems span the whole spectrum from life sciences and medicine, physics, chemistry and engineering, to social, economic, and cognitive sciences. Research in Complex Systems requires a truly interdisciplinary approach that crosses traditional disciplines. In the following we will discuss in detail further examples that include "Self-Organization and Pattern-Formation", "Complex Technical Systems", "Turbulence" and "Collective Phenomena". These examples only give a small glimpse of the field. Nevertheless they capture well the challenges and opportunities that lie ahead of us.

Complex Systems Research

The key feature of complex systems is that the cooperative interactions of the individual components determine the emergent functionalities, which individually do not exist Complex systems need energy to sustain their dynamical and structural behavior. Little changes in one component can have far reaching consequences for the system as a whole. Natural complex systems often show a high level of robustness due to redundancy in their components and interactions. Complex subsystems often combine to create new levels of functionalities. One of the main features of complex systems is that the behavior observed in different systems can be described by the same principles. For example, the aggregation of social amoebae, the catalytic gas reaction on a platinum surface and the cardiac fibrillation that leads to sudden death are all well described by the same theory.

It is the topic of Complex Systems Research to indendfy and to understand the fundamental and general principles of complex systems. Theories need to be further developed that at the same time are sufficiently abstract and detailed as to apply to the wide range of classes of complex systems. They must capture the interaction of different temporal or spatial scales, the interplay between the individual history of a system and the universal features, the self-organized coordination of different elements or parts, the emergence of collective phenomena on the basis of local and nonlocal interactions, and so on. Complex systems research, therefore, encompasses the interaction between general principles and methods on the one hand and the detailed investigation of concrete complex systems on the other. Identifying, understanding, and controlling the organizational, structural, and dynamical principles of complex systems and utilizing them for the benefit of human kind will be the challenge for the next decade and beyond.

Representative Examples

(A) Self-Organization and Pattern-Formation:

The past two and a half decades have seen major progress in the theory of pattern-forming, self-organizing systems. An example of a pattern-forming process that determines the material properties of metallic alloys, as used in turbine plates of airplanes, is given in Fig. 1.

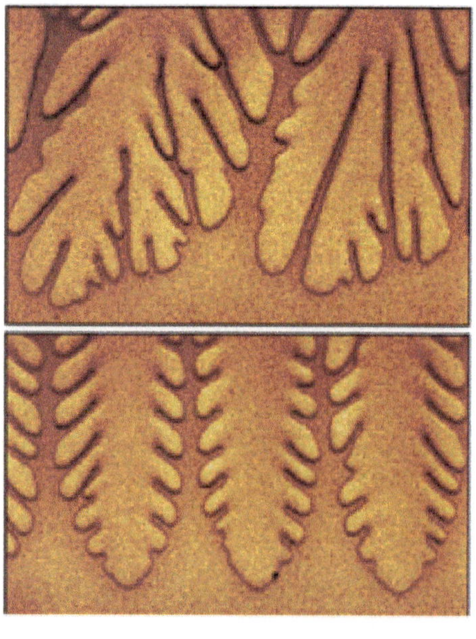

Figure 1: Dendritic single crystal growth of a model alloy between two plates in a temperature gradient. The selected solidification pattern depends on the direction of the growth with respect to the crystalline axes. Dendritic structures like these determine the mechanical properties of super alloy turbine plates and of metallic alloy casts. In single crystal growth for semiconductor technology instabilities like the one shown need to be avoided.

Numerical simulation and advanced experimental tools now allow highly resolved quantitative investigations of the spatio-temporal evolution, not only of spatially extended systems, but also of highly coupled networks. This is well illustrated by the example of epidemic outbreaks and the subsequent geographic spreading of the infectious disease. For diseases that are transmitted from person-to-person, the spreading is caused by human mobility. Pathogens can be transported by air travel and generate new nuclei of infection on long distances. The resulting patterns rather resemble a fractal than a spreading front in a recent empirical and theoretical study this statistics has been explored irrespective of the means of transportation by using banknotes as a proxy. Concurrently to Physics, Chemistry and Material Science, Biology and Medicine are progressing rapidly. We are gathering an ever-increasing amount of quantitative data. In the life science this includes patterns of gene expression and protein interaction in time and space during biological function and development New imaging technologies (high speed imaging, nano-scale fluorescent imaging of proteins, ultrashort x-ray imaging, cryo-electron microscopy etc.) give us fundamental data into the complexity and pattern formation from the molecular and intracellular scale, to the multi-cellular scale of whole organs and organisms. Experimental tools begin to interface with, manipulate, and control the dynamics of biological cells with a spatiotemporal precision that matches the microscopic organization. For example, on the level of single neurons, in vitro fast multi-site uncaging of neurotransmitter applied to dendritic arbors can mimic the complex distributed patterns of synaptic inputs received in vivo. Genetically encoded light-gated ion channels and pumps now allow the control of precise temporal patterning of ion channels. These emerging techniques are driving a revolutionary change. This allows for the first time to quantitatively address many long-standing questions about the mechanisms of pattern-formation and self-organization from Physics, Chemistry and Material Science, to Biology and Medicine. Progress requires seamless collaboration across all disciplines and with it a new breed of scientists that are knowledgeable both in the field of complex systems and their specific field of scholarship.

1. Complexity of Heart and Mind

Among the organs of our body, the function of heart and brain are unique in that their operation emerges from the collective dynamics of millions of strongly interacting cells well organized in their geometrical structure and connectivity. In the heart muscle the propagation of a nonlinear wave pulse, the cardiac action potential, controls the contraction. Usually the propagation is well organized both in space and time and the heart functions as an efficient biological pump. During cardiac fibrillation, synchronous contraction is disrupted by vortex-like rotating waves of electrical activity resulting in spatio-temporal-chaotic excitation patterns (see Fig. 2).

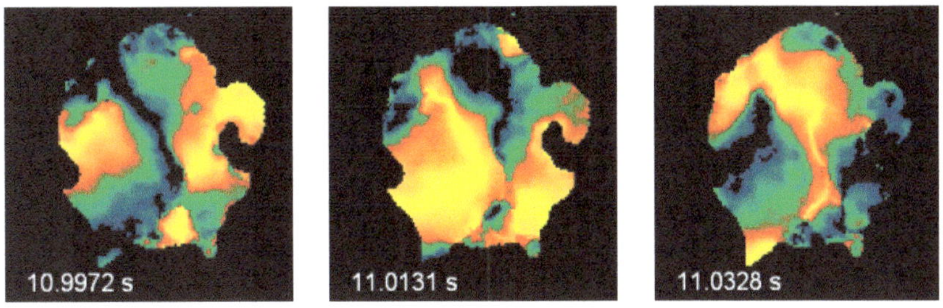

Figure 2: Cardiac fibrillation: spatial-temporal chaotic electric excitation pattern on the surface of heart tissue (field of view 6 x 6 cm^2). Colour code: black = resting tissue, yellow = excited tissue. The missing synchronization leads to the failure of the heart as a pump and to sudden death.

This electro-mechanical malfunction of the heart can rapidly evolve into Sudden Cardiac Death, which causes an estimated 738.000 deaths per year in the European Community alone. Yet the physical mechanisms underlying the dynamics and control continue to pose a fundamental scientific riddle. Similarly, in the brain the propagation of a nonlinear wave pulse, namely the neural action potential, is at the basis of the computational and memory power of the brain, i.e., what determines the workings of our 'minds'. Here, however, due to the high degree of interconnectivity and topological complexity of the neuronal network, the coordinated activity of millions of interacting nerve cells is more complex. The field of network dynamical systems is largely driven by the steady stream of novel dynamical phenomena that result from the interplay of local nonlinear dynamics and complex network structure. Also, the real time processing of input interacts with the slower process of learning through activity dependent changes of synaptic weights. The brain, however, would be only incompletely understood when just viewed as a complex dynamical system. For example, the processing of hierarchically structured sequences (sentences) can be described at the neural systems level as an input driven initial phase of local structure building which is immediately followed by a second phase of building hierarchical dependencies. This specific ability to extract rules underlying hierarchical dependencies appears to differentiate humans from non-human primates. Therefore, understanding the operation of the mind also requires describing and analyzing its emergent information processing functions. To achieve this, many aspects of neural computation have been successfully formulated as problems of pattern-formation, statistical inference and optimal decision-making, phrasing them in the mathematical language of statistical physics.

2. Cellular Organization of Life: From Intracellular Dynamics to the Emergence of Patterns and Shapes

Higher organisms develop by the collective organization of many cells. These cells multiply by repeated divisions and create widely varying morphologies. The biological cells themselves are complex systems that utilize highly evolved genetic regulatory schemes to create structures that can dynamically maintain themselves or contribute to a higher multi-cellular organism. These regulatory schemes in turn integrate complementary types of information, the internally stored genetic and epigenetic one and the externally provided and variable information from the environment or other cells via specific signals. On the single cell as well as the multi-cellular level it remains a fundamental challenge to understand the principles underlying the developmental process that lead to the formation of complex patterns and morphologies. The collective behaviors of many cells are guided by their physical properties and active behaviors. Approaches from statistical physics and soft matter physics play therefore an increasing role for the understanding of multi-cellular systems. An example of current study is the two-dimensional tissue, the so-called epithela. Physical descriptions of the dynamics of cell packing reveal that the tissue as a whole behaves as an active form of matter, being dynamically remodeled and exhibiting dynamic behaviors as a result of cell division and cell death.

Figure 3: Tissues are organized in cellular compartments separated by sharp interfaces. In the fly wing two compartments are apparent by a GFP labeling of cells (left, courtesy of Christian Dahmann). Simulation of a compartment boundary generated by increased cell bond tension (right, courtesy of Frank Julicher).

The combination of cellular signaling for example by morphogens with cell division and cell flows suggests that novel mechanisms for pattern-formation are at work. A key process is the control of the cell division axis, which determines cell rearrangements and its emergent behavior.

(B) Complex Technical Systems

In the past significant progress has been made to increase the productivity, selectivity, and sustainability of chemical and biotechnological production processes. But in the next two decades, further scientific and technological breakthroughs in process systems engineering are needed in order to organize the transition from fossil fuels and petro-chemical feed stocks towards renewable energies and raw materials, thereby closing CO_2 cycles by integrating biological substances. To accomplish this challenging task, theoretical and experimental approaches need to be developed for the analysis and design of chemical and biochemical production systems with focus on their inherent multi-level structure.

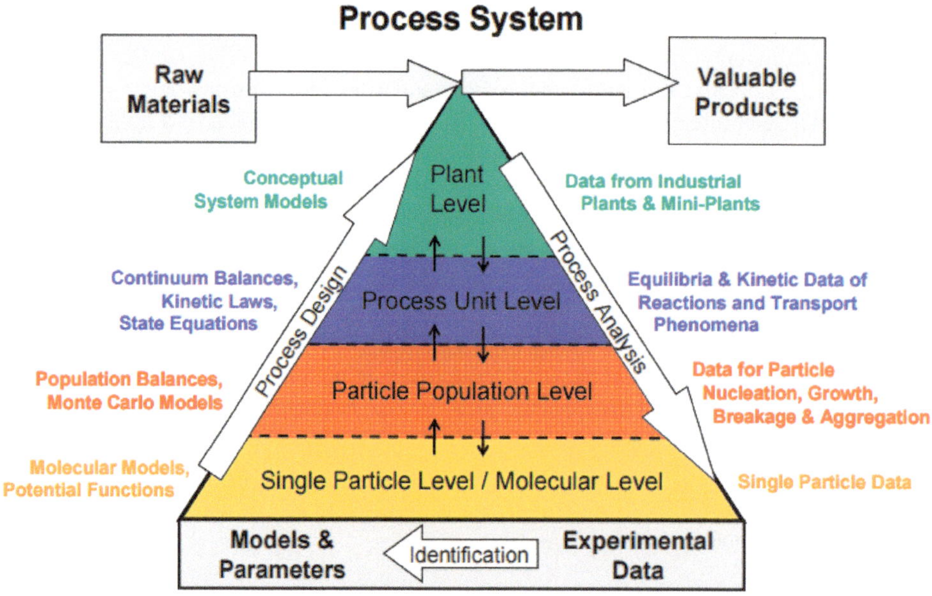

Figure 4: Multilevel structure of chemical and biochemical process systems.

(C) Turbulence

Turbulence manifests itself in myriad ways across the physical, chemical, engineering and even biological sciences. It occurs whenever fluid viscous forces are small compared to the dominant driving forces of the flow; in practice this includes most macroscopic natural and technological flows. The mixing and transport properties critically impact the fundamental behavior of phenomena found in astrophysics, in fusion and plasma physics, in earth, solar and planetary sciences, in process engineering, and in the ecological sciences. Turbulence is a fundamental topic in fluid and dynamical systems research and among others, raises basic questions in both applied mathematics and informatics. The subject of turbulence unites students of the terrestrial biosphere studying the exchange of heat, momentum and matter within the forest canopy, with physicists exploring strategies for confining plasmas, with mathematicians developing fundamental theories and numerical methods, or computational scientists designing methods for structuring computations across hundreds of thousands of computational elements.

Key challenges:

- Turbulence is key to the spread of pollutants, aerosols and bio-agents in the biosphere. Unclear is how turbulence influences particle dynamics as function of particle density, size, and concentration, and how particles at higher concentration affect the dynamics of the flow
- The lack of understanding of the coupled turbulence/aerosol dynamics and of convective currents underlies a persistent uncertainty in the prediction of the earth's future climate. A better understanding is needed to evaluate pressures on environment and climate from anthropogenic and natural emissions and to render more precise the predictions on the energy balance due to phase changes that are highly coupled to the dynamics of drop formation in turbulent clouds and the concomitant effect on radiation balance.
- The transfer of energy due to thermal convection determines the structure of planets and stars. A better understanding of high Rayleigh-number turbulent convection and of turbulent boundary layers is needed.
- In wind energy gain is maximal for the highest wind speeds, i.e., strongest turbulence. Current technology does not incorporate well the effect of turbulent velocity fluctuations on the design of wind-turbines. A better knowledge of turbulence and its interaction with wind farms at the large scale and the propeller turbulence boundary layer interaction at the small scale is needed to advance the technology.
- The future of fusion reactors relies on an efficient confinement of the plasma. Instabilities and turbulence disturb this confinement. A better understanding of turbulence and its control is needed to bring fusion from the laboratory to the power plant.
- The development of more efficient and environmentally friendly air, ground, and water transportation relies to a large part on the development of quieter, cleaner and more efficient propulsion systems. Progress requires a better knowledge of turbulent fluid flows in turbines and engines.
- Any process of energy efficient combustion and burning requires turbulent mixing of the combustibles. Better fundamental stochastic models for turbulent mixing are needed to allow for the development of more efficient and cleaner technologies.
- Turbulence determines the environment of plankton, which is the main food source for most higher developed ocean life. Currently, our understanding of the effects of turbulence on the biology of plankton is rudimentary.
- Transport of aerosols and sand plays an important part in erosion and desertification of land and forest environments. An increased understanding of the impact of turbulence in boundary layers and aerosol transport is needed for better management and risk assessment of biological resources.
- The clustering of matter in the beginning universe, as well as the initial stages of planet formation is attributed to turbulence. Detailed mechanisms although proposed are not clear due to a lack of knowledge of particle transport in turbulence.

(D) Collective Phenomena far from Equilibrium

A particularly well"defined and easily accessible model system for the study of collective phenomena far from thermal equilibrium is a granular gas. In the recent past, it has been shown that phase transitions in wet granular gases can be well understood on the basis of just a few characteristic parameters. This demonstrates similarities with the universality in equilibrium phase transitions, although in the case of a granular system, large granular temperature gradients are present in steady state. In simulations of driven dense granular gases a Kolmogorov-type instability accompanied by periodic undulations and, at stronger drive, chaotic behavior has been observed suggesting similarities between granular gases and fluids.

SECTION I - RESONANCE PHENOMENA, NONLINEAR VIBRATIONS AND WAVES

CHAPTER 1 RESONANCE AND FRACTAL GEOMETRY

Henk W. Broer

Keywords: ResonanceResonance tongue, Subharmonic bifurcation, Covering space, Cantor set, Fractal geometry, Devil's staircase, Lyapunov diagram

WHAT IS RESONANCE?

A heuristic definition of resonance considers a dynamical system, usually depending on parameters, with several oscillatory subsystems having a rational ratio of frequencies and a resulting combined and compatible motion that may be amplified as well. Often the latter motion is also periodic, but it can be more complicated as will be shown below. We shall take a rather eclectic point of view, discussing several examples first. Later we shall turn to a number of universal cases, these are context-free models that occur generically in any system of sufficiently high-dimensional state and parameter space.

Among the examples are the famous problem of Huygens's synchronizing clocks and that of the Botafumeiro in the Cathedral of Santiago de Compostela, but also we briefly touch on tidal resonances in the planetary system. As universal models we shall deal with the Hopf–Neĭmark–Sacker bifurcation and the Hopf saddle-node bifurcation for mappings. The latter two examples form 'next cases' in the development of generic bifurcation theory. The term universal refers to the context independence of their occurrence: in any with certain generic specifications these bifurcations occur in a persistent way. We witness an increase in complexity in the sense that in the parameter space the resonant phenomena correspond to an open & dense subset union of tongues, while in the complement of this a nowhere dense set of positive measure exists, corresponding to multi- or quasi-periodic dynamics. This nowhere dense set has a fractal geometry in a sense that will be explained later. From the above it follows that this global array of resonance tongues and fractal geometry has a universal character. As we shall see, both locally and globally Singularity Theory can give organizing principles. It should be noted at once that next to periodic and quasi-periodic dynamics also forms of chaotic dynamics will show up.

REMARK

In many cases the resonant bifurcations are repeated at ever smaller scales inside the tongues, leading to an infinite regress. Then we have to extend the notion of open & dense to residualand that of nowhere dense to meagre. Here a residual set contains a countable intersection of open & dense sets, while a meagre set is a countable union of nowhere dense sets. One sometimes also speaks in terms of G_δ- or F_σ-sets, respectively [73].

Periodically Driven Oscillators

Many of the resonant phenomena of interest to us are modelled by periodically driven or coupled oscillators. To fix thoughts we now present two examples, namely the harmonic and the Duffing oscillator subjected to periodic forcing.

The Driven Harmonic Oscillator

One of the simplest class-room examples of resonance occurs in the harmonic oscillator with periodic forcing

$$\ddot{x} = -\alpha^2 x - c\dot{x} + \varepsilon \sin(\omega t) \qquad (1)$$

with x, t∈ℝ, where c>0 is the damping and where $\varepsilon \geq 0$ controls the size of the forcing. The forcing has frequency $\omega > 0$ and

Citation: Henk W. Broer, Resonance and Fractal Geometry, DOI: 10.1007/s10440-012-9670-x.

therefore has period $T=2\pi/\omega$. The 'response' solution is periodic

$$x(t) = R\sin(\omega t + \phi)$$

with this same frequency. Its amplitude and phase are given by

$$R = \frac{\varepsilon}{\sqrt{(\omega^2 - \alpha^2)^2 + c^2\omega^2}} \quad \text{and} \quad \tan\phi = \frac{c\omega}{\omega^2 - \alpha^2}.$$

Fixing α,c and ε we consider R as a function of ω, see the amplitude-response diagram in Fig. 1(left) where the response amplitude R is plotted as a function of ω.

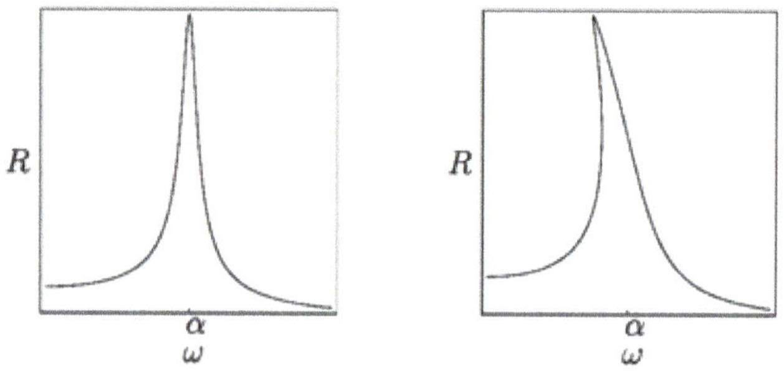

Figure 1: Amplitude-response diagrams in the (ω,R)-plane. Left: the periodically driven harmonic oscillator (1) and right: the driven Duffing oscillator (2)

The Driven Duffing Oscillator

As a non-linear variation on the above we consider the Duffing equation with periodic forcing

$$\ddot{x} = -\alpha^2 x - c\dot{x} - \delta x^3 + \varepsilon\sin(\omega t), \qquad (2)$$

where $\alpha,\omega,c,\varepsilon$ and are positive and where δ with $0<\delta\ll 1$ is considered as a perturbation parameter. Fixing α, c and ε as before, by successive approximation of a solution

$$x(t) = R\sin(\omega t + \phi),$$

we obtain the following relation between ε and R:

$$\varepsilon^2 = \left((\omega^2 - \alpha^2)^2 + c^2\omega^2\right)R^2 + \frac{3}{2}\delta(\omega^2 - \alpha^2)R^4 + \frac{9}{16}\delta^2 R^6 + O(\delta^3),$$

as $\delta\downarrow 0$, with a similar approximation for the phase ϕ, compare with Stoker [84], also see [44]. In Fig. 1 (right) we depict the corresponding curve in the (ω,R)-plane, which now no longer is a graph.

Remarks

- One of the exciting things about resonance concerns the peaks of the amplitude R that can be quite high, even where ε is still moderate. Systems like (1) and (2) form models or metaphors for various resonance phenomena in daily life. In many cases high resonance peaks one needs to 'detune' away from the resonance value corresponding to the peak, think of a marching platoon of soldiers that have to go out of pace when crossing a bridge.
- In other cases, like when 'tuning' the radio receiver to a certain channel, one takes advantage of the peak.
- It should be noted that the nonlinear (2) dynamically is far richer than the linear case (1), e.g., see [58] and references therein.

Geometrical Considerations

In both cases of the driven oscillators witnessed above the state space is $\mathbb{R}^2 \times \mathbb{T}^1 = \{x, y, z\}$., where

$$\dot{x} = y$$

$$\dot{y} = a(x, y, \delta) + \varepsilon\sin z$$

$$\dot{z} = \omega, \qquad (3)$$

with $a(x,y,\delta)=-\alpha^2 x - cy - \delta x^3$, where $\delta=0$ in the harmonic example. The second factor is the circle $\mathbb{T}^1 = \mathbb{R}/(2\pi\mathbb{Z})$ which takes into account the periodicity of the systems in z. The response motion of the form $x(t)=R\sin(\omega t+\phi)$ then corresponds to a closed curve

$$\begin{pmatrix} x \\ y \\ z \end{pmatrix}(t) = \begin{pmatrix} R\sin(\omega t + \phi) \\ \omega R\cos(\omega t + \phi) \\ \omega t + \phi \end{pmatrix}. \qquad (4)$$

This closed curve, when projected onto the (x, y)-plane, exactly forms an ellipse.

We now can describe this motion in terms of a 2-dimensional torus $\mathbb{T}^2 = \mathbb{T}^1 \times \mathbb{T}^1$, parametrized by two variables φ_1 and φ_2 in a system of differential equations of the format

$$\dot{\varphi}_1 = \omega_1$$

$$\dot{\varphi}_2 = \omega_2, \qquad (5)$$

where $\varphi_1=z$ and $\omega_1=\omega$ and where for φ_2 we take the phase of the motion on the ellipse. Thus $\varphi_2$exactly is the time parametrization of this motion, scaled to the period 2π: in this case$\omega_2=\omega_1=\omega$. Therefore in this way, the curve (4) can be seen as a 1:1 torus knot.

To view resonant motion in terms of torus dynamics turns out to be extremely useful and this can also be applied to coupled oscillators. Here a classical example is given by Christiaan Huygens [61], who in 1665 observed the following phenomenon, see Fig. 2. Two nearly identical pendulum clocks mounted on a not completely rigid horizontal beam tend to synchronize. Moreover, when the pendula both move in the vertical plane through the beam, they have a tendency to synchronize in anti-phase motion. A simple model describes this system in the format (5), where the angles φ_1 and φ_2 are the phases of the two oscillators and where again $\omega_1=\omega_2$. Later on we will come back to this and other examples where we will also see other frequency ratios $\omega_1:\omega_2$.

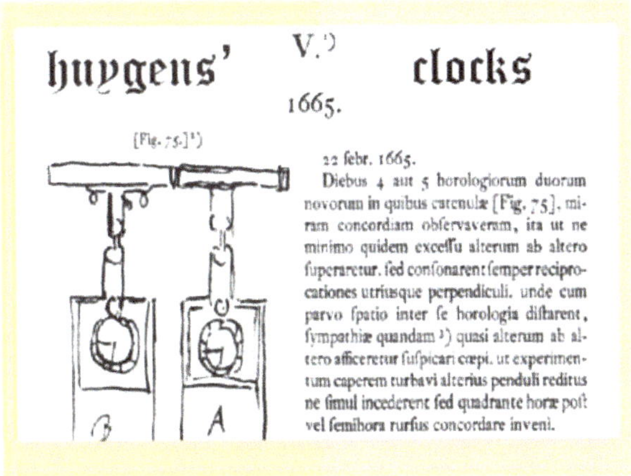

Figure 2: Huygens's synchronizing clocks [61].

Torus Flows and Circle Mappings

In this section we turn to the dynamics on the 2-dimensional torus $\mathbb{T}^2$ from Sect. 1.1.3 for its own sake, introducing the weakly coupled system

$$\dot{\varphi}_1 = \omega_1 + \varepsilon f_1(\varphi_1, \varphi_2)$$
$$\dot{\varphi}_2 = \omega_2 + \varepsilon f_2(\varphi_1, \varphi_2). \tag{6}$$

Here f_1 and f_2 are 2π-periodic functions in both variables. Also we use a parameter ε to control the strength of the coupling. For $\varepsilon=0$ we retrieve the format (5). If T_1 and T_2 are the respective periods of oscillation, then $\omega_1=2\pi/T_1$ and $\omega_2=2\pi/T_2$. We first define the Poincaré mapping from the circle $\mathbb{T}^1$ to itself and then introduce the rotation number.

The Poincaré Mapping

If in (6) the size $|\varepsilon|$ of the coupling is not too large, a first-return Poincaré mapping

$$P : \mathbb{T}^1 \longrightarrow \mathbb{T}^1 \tag{7}$$

is defined, as we shall explain now, also see Fig. 3. Without restricting generality we can take the generating circle $\mathbb{T}^1 \times \{0\}$ for simplicity baptising $\varphi=\varphi_1$. In that case we take the integral curve with initial state $(\varphi_1,\varphi_2)=(\varphi,0)$ and follow this till $(\varphi_1,\varphi_2)=(P(\varphi),0)$, counting mod $2\pi\mathbb{Z}$. It is easy to see that P−Id should be a periodic function in φ, which gives P the general format

$$P : \varphi \mapsto \varphi + 2\pi\alpha + \varepsilon f(\varphi), \tag{8}$$

where $\alpha=\omega_2/\omega_1$ and where f is a 2π-periodic function.

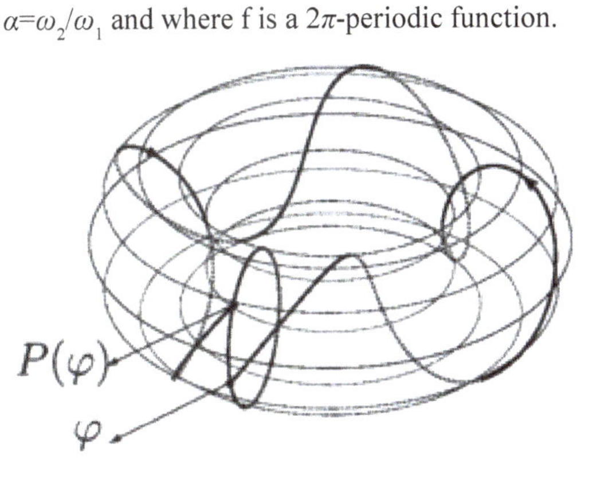

Figure 3: Poincaré mapping of a torus flow [19].

Consideration of the $\mathbb{T}^1$-dynamics generated by iteration of P gives a lot of information about the original $\mathbb{T}^2$-flow, in particular its asymptotic properties as $t\to\infty$. For instance, a fixed point attractor of P corresponds to an attracting periodic orbit of the flow which form a 1:1 torus knot as we saw at the end of Sect. 1.1. Similarly a periodic attractor of P of period q corresponds to an attracting periodic orbit of the flow. In general, periodicity will be related to resonance, but to explain this further we need the notion of rotation number.

Rotation Number

For orientation-preserving homeomorphisms P: $\mathbb{T}^1 \longrightarrow \mathbb{T}^1$ Poincaré has left us the extremely useful concept of rotation number (P), which describes the average amount of rotation as follows:

$$\varrho(P) = \frac{1}{2\pi} \lim_{n\to\infty} \frac{1}{n} (\tilde{P})^n(\varphi) \bmod \mathbb{Z}. \tag{9}$$

Here $\tilde{P} : \mathbb{R} \longrightarrow \mathbb{R}$ is a (non-unique) lift of P which makes the diagram

$$\begin{array}{ccc} \mathbb{R} & \xrightarrow{\tilde{P}} & \mathbb{R} \\ \text{pr} \downarrow & & \downarrow \text{pr} \\ \mathbb{T}^1 & \xrightarrow{P} & \mathbb{T}^1 \end{array}$$

commute, where pr: $\mathbb{R} \longrightarrow \mathbb{T}^1$ is the natural projection $\varphi \mapsto e^{i\varphi}$. This means that in the formula (9) we do not count modulo 2π, but keep counting in $\mathbb{R}$.

From [53, 72] we quote a number of properties of $\varrho(P)$:

- $\varrho(P)$ depends neither on the choice of the lift $\tilde{P}$ nor on the choice of φ;
- $\varrho(P)$ is invariant under topological conjugation. This means that if h: $\mathbb{T}^1 \longrightarrow \mathbb{T}^1$ is another orientation-preserving homeomorphism, then
- $\varrho(hPh^{-1}) = \varrho(P)$;
- If $P:\varphi\mapsto\varphi+2\pi\alpha$ is a rigid rotation then $\varrho(P)=\alpha \bmod \mathbb{Z}$.
- $\varrho(P)\in\mathbb{Q}$ precisely when P has a periodic point. Moreover, $\varrho(P)=p/q$ with p and q relatively prime corresponds to a p:q torus knot.
- If P is of class C^2 and $\varrho(P)=\alpha$ for $\alpha\in\mathbb{R}\setminus\mathbb{Q}$, then, by a result of Denjoy, the mapping P is topologically conjugated to the rigid rotation $\varphi\mapsto\varphi+2\pi\alpha$.
- Recall that in that case any orbit $\{P^n(\varphi)\}_{n\in\mathbb{Z}}$ forms a dense subset of $\mathbb{T}^1$. The corresponding dynamics is called quasi-periodic.
- If P depends continuously on a parameter, then so does $\varrho(P)$.

The Arnold Family of Circle Mappings

A famous example is formed by the Arnold family

$$\mathbb{A}_{\alpha,\varepsilon} : \varphi \mapsto \varphi + 2\pi\alpha + \varepsilon \sin\varphi \tag{10}$$

of circle mappings. So this is the general format (8) where we chose $f(\varphi)=\sin\varphi$.[1]

Periodicity

It is instructive to consider its fixed points, given by the equation

$$\mathbb{A}_{\alpha,\varepsilon}(\varphi) \equiv \varphi,$$

or, equivalently,

$$\sin\varphi = -\frac{2\pi\alpha}{\varepsilon}.$$

A brief graphical inspection reveals that $\bmod 2\pi\mathbb{Z}$ this equation has exactly two solutions for

$$|\varepsilon| > 2\pi\,|\alpha|.$$

In Fig. 4 this region, bounded by the two straight lines $\varepsilon=\pm 2\pi\alpha$, is depicted for $\varepsilon>0$. It is not hard to see that one of the fixed points is attracting and the other repelling. At the boundary $|\varepsilon|=2\pi|\alpha|$ these annihilate one another in a saddle-node bifurcation. For the entire region $|\varepsilon|\geq 2\pi|\alpha|$ one has $\varrho(\mathbb{A}_{\alpha,\varepsilon}) = 0 \bmod \mathbb{Z}$ This region is called the Arnold tongue of rotation number 0.

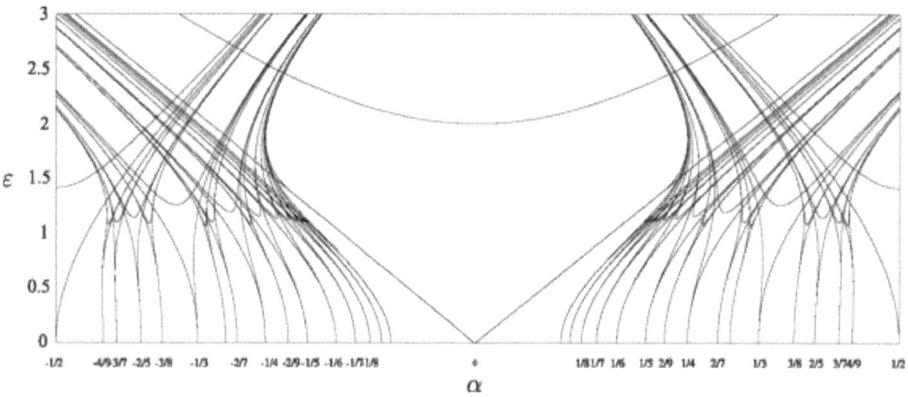

Figure 4: Resonance tongues in the Arnold family (10).

From the properties of Sect. 1.2.2 it follows that, for $(\alpha,\varepsilon)=(p/q,0)$ with p and q relatively prime, one has $\varrho(\mathbb{A}_{\alpha,\varepsilon}) = p/q$. One can show that from each $(\alpha,\varepsilon)=(p/q,0)$ an Arnold tongue emanates, in which for all the parameter points (α,ε) one has $\varrho(\mathbb{A}_{\alpha,\varepsilon}) = p/q$, see Fig. 4. The 'sharpness', i.e., the order of contact of the boundaries of the p/q-tongue at $(\alpha, \varepsilon)=(p/q,0)$ exactly is of order q, see [1, 3, 24].

Fixing $\varepsilon=\varepsilon_0>0$ small, we consider the graph of $\alpha \mapsto \varrho(\mathbb{A}_{\alpha,\varepsilon_0})$. By another general property of Sect. 1.2.2, this function is continuous. Moreover, for every rational value p/q it is constant on some plateau, corresponding to the p/q-tongue, see Fig. 4. The total result is a devil's staircase as depicted in Fig. 5.

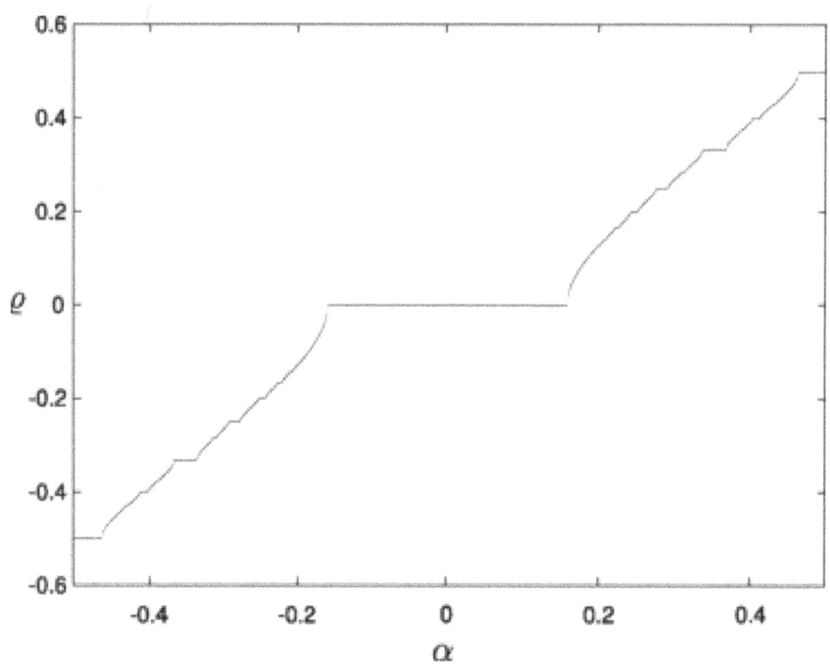

Figure 5: Devil's staircase related to the Arnold family (10). For small $\varepsilon_0>0$ the rotation number $\varrho(\mathbb{A}_{\alpha,\varepsilon_0})$ is depicted as a graph of $\alpha\in[0,1]$.

Quasi-periodicity

In between the tongues the rotation number $\varrho(\mathbb{A}_{\alpha,\varepsilon})$ is irrational and by the properties of Sect. 1.2.2 we know that the corresponding iteration dynamics of $\mathbb{A}$ is quasi-periodic and that each individual orbits densely fills $\mathbb{T}^1$.

Open & Dense Versus Nowhere Dense

In general the (α,ε)-plane of parameters contains a catalogue of the circle dynamics. Again fixing $\varepsilon=\varepsilon_0>0$ small, consider the corresponding horizontal line in the (α,ε)-plane of parameters. We witness the following, also see Fig. 5 and compare with [24] and references therein. The periodic case corresponds to an open & dense subset of the line, and the quasi-periodic case to a nowhere dense subset, which in the 1-dimensional situation is a Cantor set.

Diophantic Rotation Numbers

Quasi-periodicity corresponds to $\varrho = \varrho(P_{\alpha,\varepsilon_0}) \notin \mathbb{Q}$. If we restrict even further to Diophantine ϱby requiring that for constants $\tau>2$ and $\gamma>0$, for all rationals p/q

$$\left|\varrho - \frac{p}{q}\right| \geq \frac{\gamma}{|q|^\tau}, \qquad (11)$$

the conjugations of P_{α,ε_0} with the rigid rotation $\varphi\mapsto\varphi+2\pi\varrho$ can be taken smooth [1, 19]. The rotation numbers ϱ satisfying (11) form a Cantor subset of the former, which has positive Lebesgue measure, which, by choosing $\gamma = \gamma(\varepsilon_0) = \mathcal{O}(\varepsilon_0)$, can be shown to tend to full measure as $\varepsilon_0\to 0$. A fortiori this holds for the original Cantor set given by $\varrho(P_{\alpha,\varepsilon_0}) \notin \mathbb{Q}$.

Fractal Geometry

The Cantor sets under consideration, since they have positive Lebesgue measure, have Hausdorff dimension equal to 1. Moreover Cantor sets have topological dimension 0, since they are totally disconnected: every point has arbitrarily small neighbourhoods with empty boundary. The fact that the Hausdorff dimension strictly exceeds the topological dimension is a characterisation of fractals, see page 15 of [65]. So our Cantor sets are fractals. They also show a lot of self-similarity, a property shared with many other fractals.

Beyond the Arnold Family (10) ...

The organisation of the parameter space in an open & dense set on the one hand, versus a nowhere dense, fractal set of positive measure turns out to be universal, also see [24]. To begin with, any arbitrary smooth (Poincaré) circle mapping of the more general format (8)

$$P_{\alpha,\varepsilon} : \varphi \mapsto \varphi + 2\pi\alpha + \varepsilon f(\varphi)$$

turns out to have an array of resonance tongues similar to the Arnold family (10), forming an open & dense set that corresponds to periodicity, with a fractal complement which is nowhere dense and of positive measure that corresponds to quasi-periodicity.

The only point of difference with (10) is formed by the exact 'sharpness' of the tongues, which depends on the Fourier coefficients of the function f. In particular a tongue at the tip $(\alpha,\varepsilon)=(p/q,0)$ has transverse boundaries if and only if the qth Fourier coefficient does not vanish.

Link with Resonance

Returning to the driven oscillator or the two coupled oscillators we now link periodicity of the Poincaré mapping (8) with resonance. For simplicity we keep $|\varepsilon|$ sufficiently small to ensure this mapping to be a diffeomorphism.

As observed in Sect. 1.2.2 the fact that (α,ε) belongs to the p/q-tongue, i.e., that $\varrho(P_{\alpha,\varepsilon})=p/q$, means that the motion takes place on a p:q torus knot. Generically these periodic orbits come in attracting and repelling pairs and the visible motion takes place on such a periodic attractor. In view of our general 'definition' of resonance in that case we say that the oscillators are in p: qresonance, one also speaks of phase-locking or synchronisation. In the case of 1:1 sometimes the term entrainment is being used.

If (α,ε) is outside the tongues, by the Denjoy theory mentioned before, the torus motion takes place on a dense orbit. We also call this torus motion quasi-periodic. In this case from kamTheory [5, 19, 44] we derive the following. For Diophantine $\varrho=\varrho(P_{\alpha,\varepsilon})$, up to a smooth transformation the Poincaré mapping reads

$$\varphi \mapsto \varphi + 2\pi\varrho$$

and the two oscillators correspondingly have the familiar format

$$\dot{\varphi}_1 = \omega_1$$

$$\dot{\varphi}_2 = \omega_2,$$

such that $\varrho(P_{\alpha,\varepsilon})=\omega_2/\omega_1$ for allε with $|\varepsilon|\ll 1$. In the parameter space this corresponds to a piece of curve through $(\alpha,\varepsilon)=(\varrho,0)$, parametrized by ε. We emphasize that this uncountable union of these curves has positive measure.

Conclusions and Examples

The literature on resonance phenomena is immense, apart from the references already given, for instance see [3, 19, 58, 85, 90] and their bibliographies. For even more references see below. The present point of view models resonant systems in terms of dynamical systems depending on parameters, where resonance takes place in a persistent way.

In the parameter space the resonant set is part of the bifurcation set, which forms a catalogue for transitions to various types of dynamics. What we add to the general discussion on this subject is the overall fractal geometry that usually manifests itself in the complement of all the resonances. We now present a couple of examples.

Huygens's Clocks

Returning to Huygens's synchronizing clocks we first consider the problem from the torus flow point of view. We have two almost identical oscillators that are weakly coupled. This means that in (5) for the frequencies we have $\omega_1\approx\omega_2$ and that $|\varepsilon|$ is small. For the Poincaré mapping (8)

$$P_{\alpha,\epsilon}(\varphi) = \varphi + 2\pi\alpha + \varepsilon f(\varphi)$$

we only make the assumption that the first Fourier coefficient of $f=f(\varphi)$ does not vanish, which ensures that the 1/1-tongue boundaries meet transverse at $(\alpha,\varepsilon)=(0,0)$, where the first 0 has to be taken mod $\mathbb{Z}$. Compare with Fig. 4.

This implies that (α,ε) belongs to the 1/1-tongue, i.e., that the pendulum clocks are in 1:1 resonance, a situation described before as entrainment which is a form of synchronization. This gives a partial explanation of the phenomena discovered by Huygens [61].

Remarks

- Note that the 1:1 resonance of the two clocks could be obtained under quite weak assumptions. If one also wants to understand the phase and anti-phase motions, the coupling between the clocks has to be included into the dynamics, compare with [9, 76] and references therein.
- For another application of these ideas in terms of circadian rhythms and the response to stimuli see [8]. Here it turns out that next to the 1:1 'entrainment' resonance also certain other resonances have biological significance.
- The above ideas can be largely extended to the case of more than two oscillators. For examples in models for the visual neurocortex see [31, 32].

Resonances in the Solar System

From ancient times on resonances have been known to occur in solar system, which are more or less in the spirit of the present section. A well-known example is the orbital 1:2:4 resonance of Jupiter's moons Ganymede, Europa, and Io which was studied by De Sitter [51, 52] using the 'méthodes nouvelles' of Poincaré [77]. The 2:5 orbital resonance between Jupiter and Saturn is described by Moser et al. [67, 68, 81]. These and other resonances by certain authors are being held responsable for gaps in the rings of Saturn and in the asteroid belt.

Another type of resonance is the spin-orbit resonance. As an example thereof, the Moon is 'captured by' the Earth in a 1:1 resonance: the lunar day with respect to the Earth is (approximately) equal to one month. Similarly Pluto and Charon have caught each other in such a 1:1 resonance: as an approximately rigid body the two orbit around the Sun. Interestingly, the planet Mercury is captured in a 3:2 spin-orbit resonance around the Sun [49].

Remarks

- The spin orbit resonances are explained by tidal forces, for instance, the rotation of the Moon has been slowed down to a standstill by tidal friction brought about largely by the reciprocal tidal forces exerted of Earth and Moon. Similarly the rotation of the Earth in the very long run will be put to a stand still by the tidal forces of mainly the Moon. But probably by that time the Sun has already turned into a red giant ...
- This brings us to the subject of adiabatically changing systems as described and summarized by Arnold [2, 3] and which may be used to model such slow changes.[2] One may perhaps expect that the 3:2 spin orbit resonance of Mercury in the very long run, and after quite a number of transitions, will evolve towards another 1:1 resonance. This part of nonlinear dynamical systems still is largely unexplored.

PERIODICALLY DRIVEN OSCILLATORS REVISITED

We now return to periodically driven oscillators, showing that under certain circumstances exactly the set-up of Sect. 1.2 applies.

As a motivating example we discuss the Botafumeiro in the cathedral of Santiago de Compostela, see Fig. 6. Here a large incense container is suspended by a pully in the dome where it can swing in the longitudinal direction of the church. A few men pull up the container when it approaches the ground and let go after, thereby creating a periodic forcing and in this way creating a stable motion of exactly twice the period of the forcing.

Figure 6: Botafumeiro in Santiago de Compostela.

Parametric Resonance

As a another model consider the parametrically driven oscillator

$$\ddot{x} + (a + \varepsilon f(t)) \sin x = 0 \tag{12}$$

with q(t+2π)≡q(t), see [21]. Here a and ε are considered as parameters.[3] For the periodic function f we have studied several examples, namely

$$\begin{aligned} f(t) &= \cos t \quad \text{and} \\ &= \cos t + \frac{3}{2}\cos(2t) \quad \text{and} \\ &= \text{signum}(\cos t), \end{aligned} \tag{13}$$

corresponding to the Mathieu case, the Mathieu case modified by a higher harmonic, and the square case. In this setting the issue is whether the trivial 2π-periodic solution

$$x(t) \equiv 0 \equiv \dot{x}(t)$$

is elliptic, hence stable, or not. For any of such systems (12) from the points

$$(a, \varepsilon) = \left(\frac{1}{4}k^2, 0\right), \quad k = 0, 1, 2, \ldots$$

tongues emanate in the (a,ε)-plane, where inside the trivial 2π-periodic solution is hyperbolic, hence unstable. On the tongue boundaries this solution is parabolic. See Fig. 7 and compare with [14, 17]. We note that this gives rise to a discrete union of tongues, where again the sharpness is governed by the Fourier coefficients of the periodic function f=f(t).

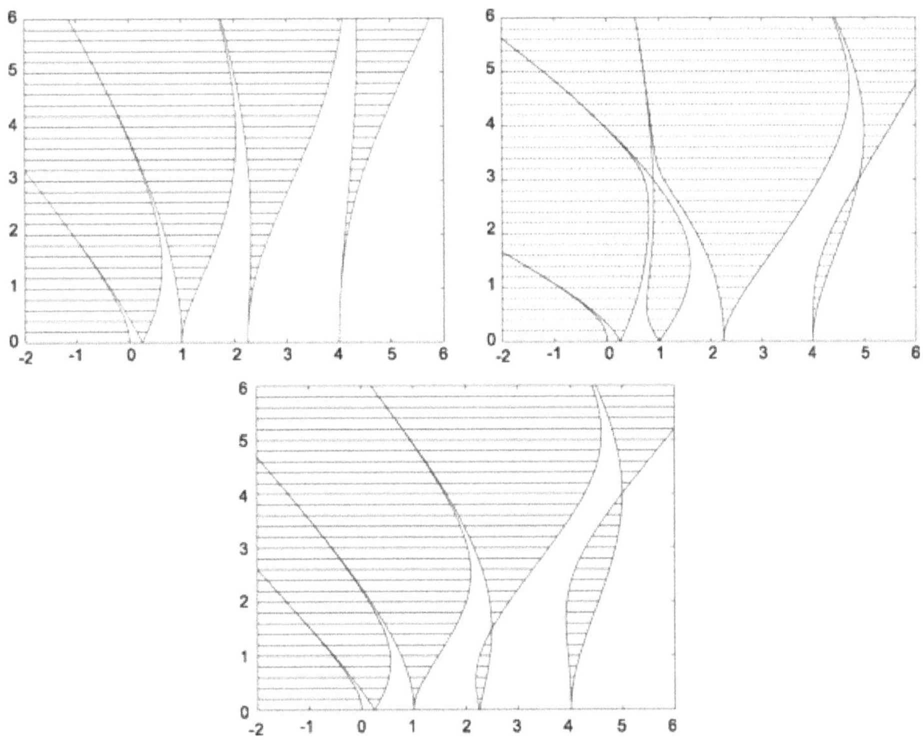

Figure 7: Stability diagrams of Mathieu, modified Mathieu and square case [17].

Subharmonics and Covering Spaces

On the tongue boundaries subharmonic bifurcations occur, see [21, 22, 44] where each bifurcation can be understood in terms of a pitchfork bifurcation on a suitable covering space.

Generally, for the p:q resonance this is constructed as follows. Writing

$$z = \dot{x} + \mathrm{i}\frac{p}{q}x,$$

where we identify $\mathbb{C} \cong \mathbb{R}^2$, the covering map is given by the Van der Pol transformation

$$\Pi : \mathbb{C} \times \mathbb{R}/(2\pi q\mathbb{Z}) \longrightarrow \mathbb{C} \times \mathbb{R}/(2\pi\mathbb{Z})$$

$$(\zeta, t) \mapsto \left(\zeta e^{\mathrm{i}tp/q}, t \,(\mathrm{mod}\ 2\pi\mathbb{Z})\right).$$

On the covering space the group D of decktransformations is generated by

$$(\zeta, t) \mapsto \left(\zeta e^{2\pi p/q}, t - 2\pi\right),$$

which means that Π∘T=Π for any T∈D.[4] Note that D is cyclic of order q, i.e., isomorphic to $\mathbb{Z}_q$, for details see [21, 22]. Instead of the original system in (z, t) we pull the system back along Π, so obtaining a $\mathbb{Z}_q$-equivariant system on the (ζ, t)-covering space. In fact this defines a 1:1 correspondence of systems and usually it is most convenient to work on the covering space. Generally on the covering space equivariant Singularity Theory can be practised, see [26, 56,57] and references therein, as well as equivariant kam Theory [23, 27, 33, 37, 44].

In the present case this construction is only needed for p/q=k/2. The strongest of these resonances occurs inside the tongue labeled by k=1, where a cylinder with $\mathbb{Z}_2$-symmetry is the double cover of a Möbius strip, for a remark in this direction see [58]. This is the setting for the simplest subharmonic bifurcation, namely a period doubling bifurcation: inside the tongue the trivial 2π-periodic solution is unstable but a stable periodic orbit occurs of period 4π. At the boundary a period-doubling bifurcation takes place. The corresponding $\mathbb{Z}_2$-equivariant bifurcation on the covering space exactly is the pitchfork. The ensuing period-doubled periodic motion is exactly the one that occurs in the Botafumeiro example. See Fig. 8 for 'phase portaits' of the Poincaré mapping, left for ε=0.25 and right for ε=0.40

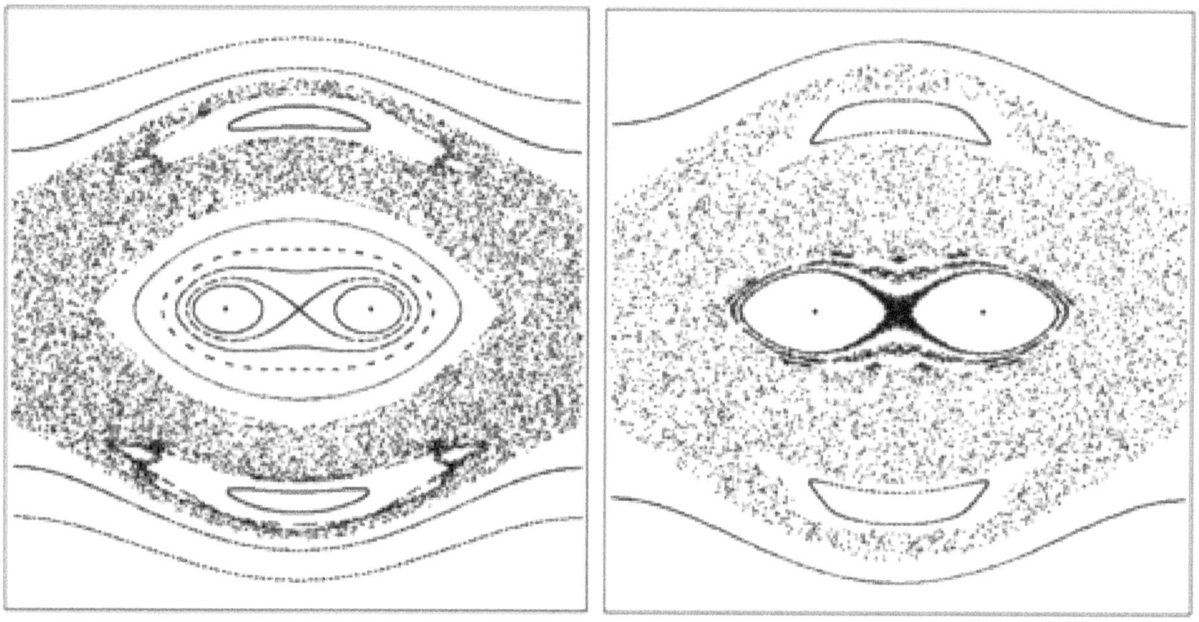

Figure 8: Orbits of the Poincaré mapping of the swing (12).

Remarks

The geometric complexity of the individual tongues in Fig. 7 can be described by Singularity Theory; in fact it turns out that we are dealing with type $\mathbb{A}_{2k-1}$, see [17, 21].

- The parametric 1:2 resonance sometimes also is called the parametric roll. By this mechanism ships have been known to capsize ...
- In Fig. 8 also invariant circles can be witnessed. kam Theory, as discussed before, in particular an application of Moser's Twist Theorem [66], shows that the union of such invariant circles carrying quasi-periodic dynamics has positive measure.
- In both cases the cloud of points[5] is formed by just one or two orbits under the iteration of the Poincaré mapping. These clouds are associated to homoclinic orbits related to the upside down unstable periodic solution, which gives rise to horseshoes. Therefore such an orbit is chaotic since it has positive topological entropy, see [19] and references therein. A classical conjecture is that the cloud densely fills a subset of the plane of positive Lebesgue measure on which the Poincaré mapping is ergodic [4].[6]

The Hill-Schrödinger Equation

Another famous equation is a linearized version of (12) where the forcing term is quasi-periodic in t:

$$\ddot{x} + \big(a + \varepsilon f(t)\big)x = 0, \qquad (14)$$

where now $f(t)=F(\omega_1 t,\omega_2 t,\ldots,\omega_n t)$ for a function $F: \mathbb{T}^n \to \mathbb{R}$, see [28, 54, 69]. As in the case of Sect. 1.2 the countable union of tongues again becomes open & dense and separated by a nowhere dense set of positive measure, determined by Diophantine conditions. The geometry of the individual tongues for small $|\varepsilon|$ is exactly as in the periodic case. For larger values of $|\varepsilon|$ the situation is more complicated also involving non-reducible quasi-periodic tori, compare with [16].

Equation (14) happens to be the eigenvalue equation of the 1-dimensional Schrödinger operator with quasi-periodic potential. We here sketch how our geometric approach fits within the corresponding operator theory. This operator reads

$$(H_{\varepsilon f}x)(t) = -\ddot{x}(t) - \varepsilon f(t)x(t) \qquad (15)$$

with potentialεf; it acts on wave functions $x=x(t)\in L^2(\mathbb{R})$.

We like to note that in the corresponding literature usually the value of $\varepsilon=\varepsilon_0\neq 0$ is fixed and the intersection of the horizontal line $\varepsilon=\varepsilon_0$ with a tongue is referred to as gap: it is a gap in the spectrum of the Schrödinger operator (15). The approach with tongues and the results of [28] regarding the $\mathbb{A}_{2k-1}$-singularity therefore leads to a generic gap closing theory.

Remarks

- In the context of Schrödinger operators the letters are chosen somewhat differently. In particular, instead of x(t) one often considers u(x), which gives this theory a spatial interpretation. Also instead of $\varepsilon f(t)$ one uses V(x), compare with [69].
- For a fixed value $\varepsilon=\varepsilon_0$ the Diophantine Cantor set leads to Cantor spectrum. The total picture is illustrated in the devil's staircase of Fig. 9, where we took n=2, $\omega_1=1$ and $\omega_2 = \frac{1}{2}(\sqrt{5}-1)$. The rotation number ϱ is defined almost as before [28] as a function of a.

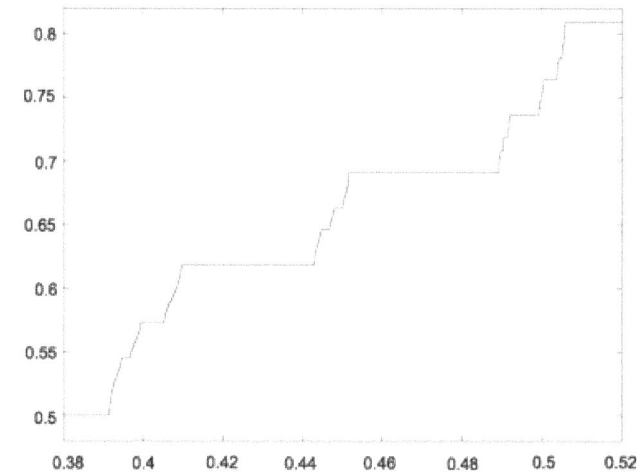

Figure 9: Devil's staircase in the Schrödinger equation with quasi-periodic potential: ω_1=1 and $\omega_2 = \frac{1}{2}(\sqrt{5}-1)$, see [28].

The nonlinear equation

$$\ddot{x} + \big(a + \varepsilon f(t)\big)\sin x = 0,$$

with q quasi-periodic is dealt with in [27]. In comparison with the case of periodic f the averaged, approximating situation, is identical. However, the infinite number of resonances and the Cantorization we saw before leads to an infinite regress of the bifurcation scenarios. For this use was made of equivariant Hamiltonian kam Theory on a suitable covering space [23, 44]. As a consequence the resonant set becomes residual and the quasi-periodic set meagre. Compare this with [7, 45–47] in the dissipative case.

Driven and Coupled Van der Pol-Like Oscillators

The examples of the driven oscillator in Sect. 1.1 were based on approximations of the damped pendulum, the free oscillation of which always tends to the lower equilibrium x=0, $\dot{x} = 0$. Our present interest is formed by Van der Pol-like oscillators that for |x| and $|\dot{x}|$ sufficiently small have negative damping, for this approach compare with [19, 44]. Such oscillators are known to occur in electronics [80, 88, 89].

Therefore our starting point is the periodically driven Van der Pol oscillator in a slightly more general form

$$\ddot{x} = -\alpha^2 x - c\dot{x} - a(x,\dot{x}) + \varepsilon f(x,\dot{x},t;\varepsilon), \tag{16}$$

where the function q is 2π-periodic in the time t. Van der Pol originally considered $a(x,\dot{x}) = bx^2\dot{x}$ and $f(x,\dot{x},t;\varepsilon) = \sin t$. We here assume that a and q are sufficiently smooth, say of class C$^\infty$ or real-analytic. The important restriction on a is that the corresponding free oscillator has a hyperbolic attractor. For a phase portrait in the $(x,\dot{x})$-plane, see Fig. 10. Passing to the system form

$$\begin{aligned}\dot{x} &= y\\ \dot{y} &= -\alpha^2 x - cy - a(x,y) + \varepsilon F(x,y,z;\varepsilon)\\ \dot{z} &= 1,\end{aligned} \tag{17}$$

as before, we get a 3-dimensional state space $\mathbb{R}^2 \times \mathbb{T}^1 = \{(x,y),z\}$, so with angular variable z. Let us denote the corresponding vector field by $X_{\alpha,\varepsilon}$.

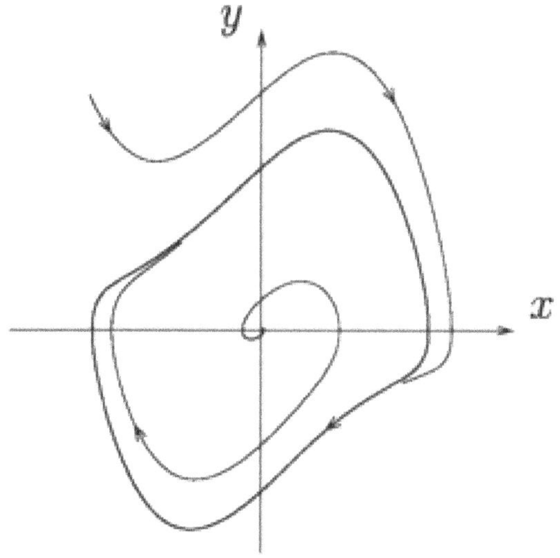

Figure 10: Phase portrait of the free Van der Pol oscillator [19].

This brings us back to the general setting of a 2-torus flow, with two phase angles φ_1,φ_2, e.g., with φ_1 the phase of the free oscillator, i.e., its time parametrization scaled to period 2π, andφ_2=z. For ε=0 we so obtain

$$\dot{\varphi}_1 = \omega_1$$

$$\dot{\varphi}_2 = \omega_2,$$

which is of the familiar format (5). From here the theory of Sect. 1.2 applies in al its complexity, with in the parameter space an open & dense, countable union of resonances and a fractal set of positive measure regarding quasi-periodicity.

Similar results hold for n coupled Van der Pol type oscillators, now with state space $\mathbb{T}^n$, the Cartesian product of n copies of $\mathbb{T}^1$. Next to periodic and quasi-periodic motion, now also chaotic motions occur, see [19] and references therein.

UNIVERSAL STUDIES

Instead of studying classes of driven or coupled oscillators we now turn to a few universal cases of 'generic' bifurcations. The first of these is the Hopf-Neĭmark-Sacker bifurcation for diffeomorphisms, which has occurrence codimension 1. This means that the bifurcation occurs persistently in generic 1-parameter families. However, the open & dense occurrence of countably many resonances and the complementary fractal geometry of positive measure in the bifurcation set are only persistent in generic 2-parameter families. A second bifurcation we study is the Hopf saddle-node bifurcation for diffeomorphisms where we use 3 parameters for describing the persistent complexity of the bifurcation set.

The Hopf-Neĭmark-Sacker Bifurcation

We start with the Hopf-Neĭmark-Sacker bifurcation for diffeomorphisms, but also discuss certain consequences for systems of differential equations. As an example to illustrate our ideas consider the following Duffing–Van der Pol–Liénard type driven oscillator

$$\ddot{x} + (\nu_1 + \nu_3 x^2)\dot{x} + \nu_2 x + \nu_4 x^3 + x^5 = \varepsilon(1 + x^6)\sin t, \quad (18)$$

the coefficients of which can be considered as parameters. Note that for the free oscillator at ν_1=0 the eigenvalues of the linear part at $(x, \dot{x})$=(0,0) cross the imaginary axis at $\pm i\sqrt{\nu_2}$. Excluding the strong resonances where $\sqrt{\nu_2}$ =p/q for p and q relatively prime with q≤5 and assuming that the constants ν_3 and ν_4 are generically chosen, the organization of the (ν_1,ν_2)-plane, for ν_1 consists of an open & dense union of countably many resonance tongues separated by a nowhere dense set of positive measure. This situation is comparable to the Arnold family (10) and Fig. 4, at least for large values of q.

The Nondegerate Case

The general set-up just considers a mapping

$$P : \mathbb{R}^2 \longrightarrow \mathbb{R}^2 \quad (19)$$

around a fixed point, say P(0)=0, where the eigenvalues of the linear part read $e^{\nu_1 \pm 2\pi i \sqrt{\nu_2}}$ with $|\nu_1|$ small. We consider a fixed resonance $(\nu_1,\nu_2)=(0,p^2/q^2)$ with p and q relatively prime, where we will need 2 parameters to versally unfold the linear part [3, 44]. Our main interest is with the periodic points of period q, so in solving the equation

$$P^q(x, y) = (x, y). \quad (20)$$

The zeroes of P^q–Id are studied by a Lyapunov–Schmidt reduction, which leads to a $\mathbb{Z}_q$-equivariant family of functions

$$G_\mu(z) = zB_\mu(u) + C_\mu \bar{z}^{q-1} + \mathcal{O}(|z|^q), \quad (21)$$

where z is an appropriate complex variable and where B_μ is a polynomial in $u=|z|^2$ of degree less than (q−1)/2. We study the corresponding discriminant set given by

$$G_\mu(z) = 0 \quad \text{and} \quad \det D_z G_\mu(z) = 0. \quad (22)$$

Here μ is an unfolding-multiparameter detuning the resonance at hand. The way to study this discriminant set is by $\mathbb{Z}_q$-equivariant contact equivalence [26, 34, 38, 40, 43]. In the present non-degenerate case (22) can be reduced to the polynomial normal form

$$G_\sigma^{\mathrm{NV}}(z) = z(\sigma + |z|^2) + \bar{z}^{q-1}, \quad (23)$$

for a complex parameter σ. In general this set turns out to be a 'tongue' ending in a cusp of sharpness (q−2)/2, which is part of a familiar bifurcation diagram with two periodic orbits of period q inside that annihilate one another at the tongue boundaries in a saddle-node or fold bifurcation [3].

See Fig. 11 which is embedded in the context of (18), of which P is a Poincaré mapping. Here the dynamics of P also has been described in terms of a Poincaré–Takens interpolating normal form approximations, e.g., see [13, 21, 22, 86].

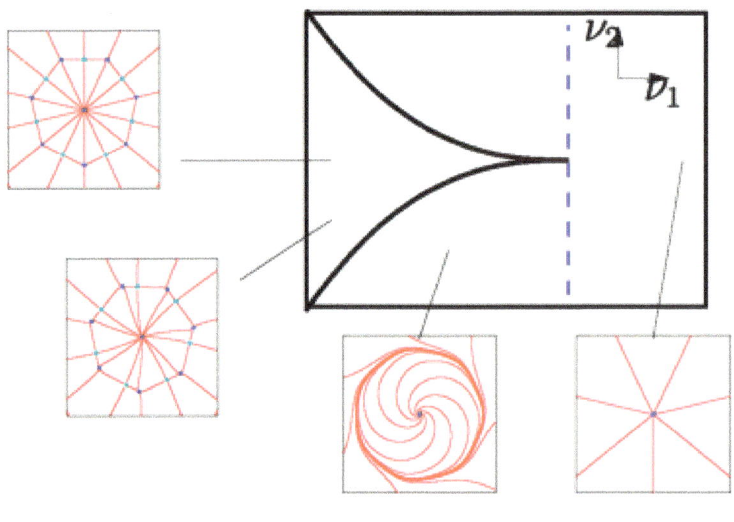

Figure 11: Tongue in nondegenerate resonance q=7 of the Hopf-Neĭmark-Sacker bifurcation [26,34, 38, 43].

Globally a countable union of such cusps is separated by a nowhere dense set of positive measure, corresponding to invariant circles with Diophantine rotation number. As before, see Fig. 4, the latter set contains the fractal geometry.

Remarks

- The above results, summarized from [26, 34, 38, 43], are mainly obtained by $\mathbb{Z}_q$-equivariant Singularity Theory.
- The strong resonances with q=1,2,3 and 4 form a completely different story where the Singularity Theory is far more involved [3, 86]. Still, since the higher order resonances accumulate at the boundaries, there is fractal geometry around, always of positive measure.
- Regarding structural stability of unfoldings of P as in (19) under topological conjugation, all hopes had already disappeared since [71].

A Mildly Degenerate Case

In the mildly degenerate case the 'Hopf' coefficient in the previous example vanishes and is introduced as an extra parameter. This leads to another normal form

$$G_{\sigma,\tau}^{\mathrm{NV}}(z) = z(\sigma + \tau|z|^2 + |z|^4) + \bar{z}^{q-1}, \quad (24)$$

which is now parametrized over $\mathbb{C}^2 \cong \mathbb{R}^4$, hence of codimension 4. As before [26, 40] the normal form (24) is structurally stable under $\mathbb{Z}_q$-equivariant contact equivalence. Here the Singularity Theory is more complex, involving folds, cusps, a swallowtail and a Whitney umbrella. The complete bifurcation diagram is more involved, even at the level of Poincaré–Takens normal form vector field approximations. In Fig. 12, again for the case q=7, a 2-dimensional tomography is shown of the bifurcation set, in which still a tongue-like structure can be discerned, for more details see [38, 43] where also 3-dimensional tomographies are presented.

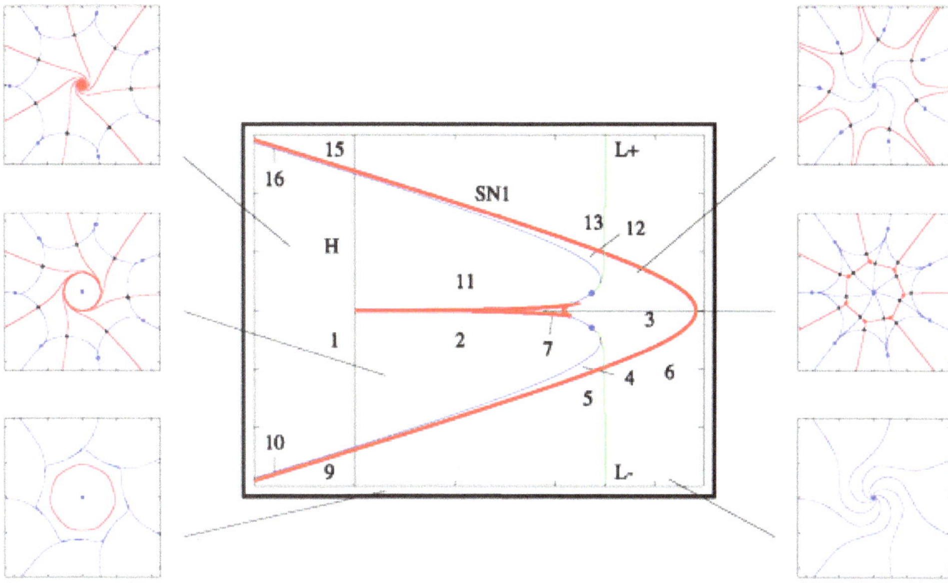

Figure 12: Two-dimensional tomography in a mildly degenerate resonance q=7 of the Hopf–Neĭmark–Sacker bifurcation [26, 34, 38, 43].

To illustrate a mildly degenerate case of the Hopf–Neĭmark–Sacker bifurcation one may well consider the preceding Duffing–Van der Pol–Liénard type driven oscillator (18) where we need all four parameters.

Concluding Remarks

For both cases of the Hopf–Neĭmark–Sacker bifurcation we have a good grip on the part of the bifurcation set that governs the number of periodic points. The full bifurcation set is far more involved and the corresponding dynamics is described only at the level of Poincaré–Takens normal-form vector fields [13, 38, 43]. We note that homo- and heteroclinic phenomena occur at a flat distance in terms of the bifurcation parameters [15, 18, 75].

The Hopf Saddle-Node Bifurcation for Diffeomorphisms

As a continuation of the above programme, we now consider the Hopf saddle-node (or fold Hopf) bifurcation for diffeomorphisms [35, 36, 41], in which the central singularity is a fixed point of a 3-dimensional diffeomorphism, such that the eigenvalues of the linear part at bifurcation are 1 and $e^{2\pi i\alpha}$, where

$$e^{2\pi n i\alpha} \neq 1 \quad \text{for } n = 1, 2, 3 \text{ and } 4, \tag{25}$$

so excluding strong resonances as in the Hopf–Neĭmark–Sacker case of Sect. 3.1. The Hopf saddle-node bifurcation for flows is well-known [20, 58], especially because of the subordinate Hopf–Neĭmark–Sacker and Šilnikov homoclinic bifurcation. Our main interest is how the Hopf–Neĭmark–Sacker bifurcation is being changed into one of the simplest quasi-periodic bifurcations near a 2:5 resonance.

From Vector Fields to Mappings

The linear part of the vector field at bifurcation has eigenvalues 0 and $\pm i\alpha$. This linear part generates an axial symmetry that in a normal form procedure can be pushed stepwise over the entire Taylor series, see [13] and references. This makes it possible to first consider axially symmetric systems, that turn out to be topologically determined by their 3rd-order truncation given by

$$\dot{w} = (-\beta_2 + i\alpha)w - awz - wz^2$$

$$\dot{z} = -\beta_1 - sw\overline{w} - z^2, \tag{26}$$

where $w \in \mathbb{C}$ and $z \in \mathbb{R}$ and where β_1 and β_2 are unfolding parameters [58, 62]. A scaling $\beta_1 = \gamma^2, \beta_2 = \gamma^2\mu$ leads to a vector field

$$Y_{\gamma,\mu,\alpha}(z, w) = \begin{pmatrix} (-\gamma\mu + 2\pi i\alpha/\gamma)w - awz - \gamma wz^2 \\ 1 - z^2 - |w|^2 \end{pmatrix}. \tag{27}$$

From this an axially symmetric map

$$\begin{pmatrix} w \\ z \end{pmatrix} \mapsto \begin{pmatrix} e^{2\pi i\alpha} w[1 - \gamma(\gamma\mu + az + \gamma z^2)] \\ z - \gamma(-1 + |w|^2 + z^2) \end{pmatrix} \tag{28}$$

is obtained in a kind of Euler step. To study a 2:5 resonance we take $\alpha_0 = 2/5$, writing $\alpha = \alpha_0 + \gamma\delta$, and perturb to

$$G_{\mu,\delta}: \begin{pmatrix} w \\ z \end{pmatrix} \mapsto \begin{pmatrix} e^{i\alpha} w[1 - \gamma(\gamma\mu + az + \gamma z^2)] \\ z - \gamma(-1 + |w|^2 + z^2) \end{pmatrix} + \begin{pmatrix} \gamma^3(\varepsilon_1 \overline{w}^4 + \varepsilon_2 z^4) \\ \gamma^4 \varepsilon_3 \operatorname{Re} w^5 \end{pmatrix}, \tag{29}$$

by adding axially non-symmetric order r resonant terms

$$\overline{w}^4 \frac{\partial}{\partial w} \quad \text{and} \quad \operatorname{Re} w^5 \frac{\partial}{\partial z}.$$

The scaling parameter γ and the other constants are fixed suitably.

In the Product of State Space and Parameter Space

In Fig. 13 a Lyapunov diagram is depicted in the parameter plane of the mapping family G. Table 1 contains the corresponding color code. In Fig. 14 we show the dynamics corresponding to two values of $(\mu,\delta/2\pi)$. Let us discuss these numerical data.

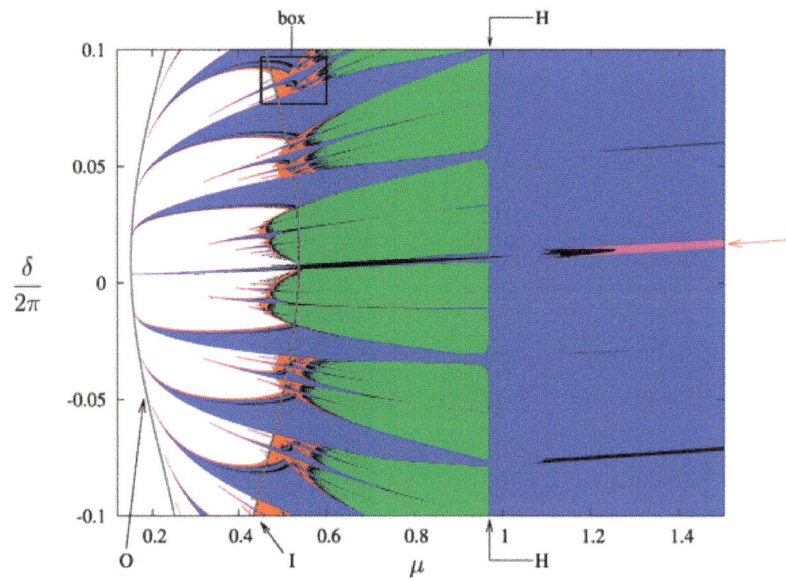

Figure 13: (Color online) Lyapunov diagram of G in the $(\mu,\delta/(2\pi))$-parameter plane. The colors correspond to distinct classes of attractors of G [35, 41]. See Table 1 for the color coding.

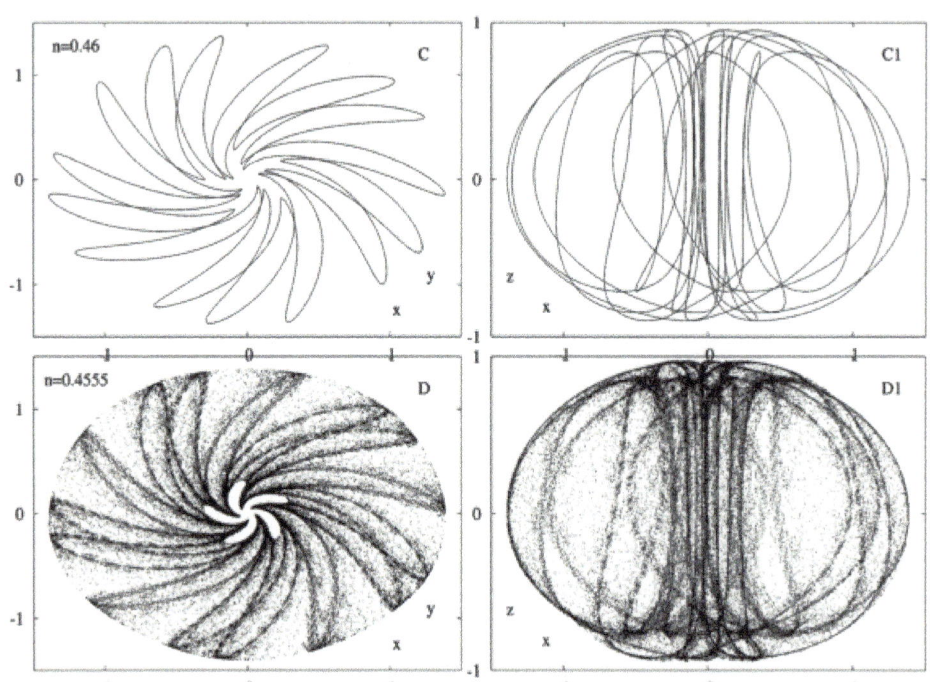

Figure 14: Quasi-periodic (top) and strange attractor (bottom) in the Hopf saddle-node bifurcation for mappings, as seen from two different angles (left and right) [35, 41].

Table 1: Legend of the color coding for Fig. 13, see [35, 41]. The attractors are classified by means of the Lyapunov exponents (ℓ_1,ℓ_2,ℓ_3)

color	Lyapunov exponents	attractor type
red	$\ell_1>0=\ell_2>\ell_3$	strange attractor
yellow	$\ell_1>0>\ell_2>\ell_3$	strange attractor
blue	$\ell_1=0>\ell_2=\ell_3$	invariant circle of focus type
green	$\ell_1=\ell_2=0>\ell_3$	invariant 2-torus
black	$\ell_1=0>\ell_2>\ell_3$	invariant circle of node type
grey	$0>\ell_1>\ell_2=\ell_3$	fixed point of focus type
fuchsia	$0>\ell_1=\ell_2\geq\ell_3$	fixed point of focus type
pale blue	$0>\ell_1>\ell_2>\ell_3$	fixed point of node type
white		no attractor detected

The Parameter Space

On the right-hand-side of the figure this method detects an attracting invariant circle of focus type (blue). In the gaps larger resonances are visible, compare with Fig. 4 for a fixed value of ε. Moving to the left, in the neighbourhood of the line indicated by H a quasi-periodic Hopf bifurcation occurs from a circle attractor to a 2-torus attractor (green). Also here the parameter space is interspersed with a resonance web of which the larger lines are visible. The remaining features, among other things, indicate invariant tori and strange attractors of various types and also more invariant circles.

The State Space

The upper two figures of Fig. 14 show an invariant circle, once seen from the z-direction and once from some w-direction. The lower two figures indicate how this circle has become a strange attractor, from the same two points of view.

Part of these results can be justified mathematically, as seen from the Perturbation Theory point of view. The invariant circles all have one Lyapunov exponent equal to 0 and these are quasi-periodic, perturbations of closed integral curves of a vector field (averaging) approximation, whence their existence can be proven by kam Theory [5, 23]. A similar statement can be made about the 2-tori with two Lyapunov exponents equal to 0. In fact the transition is a quasi-periodic Hopf bifurcation as discussed by Broer et al. [10, 11, 19, 44].

By the same references, this also holds for the quasi-periodic invariant circle in the upper half of Fig. 14. The lower half of this figure is conjectured to show a quasi-periodic Hénon-like attractor, which is the closure of the unstable manifold of an unstable quasi-periodic invariant circle. This is the previous quasi-periodic circle that has become unstable through a quasi-periodic saddle-node bifurcation [11]. For this kind of strange attractor the mathematical background theory largely fails, so the results must remain experimental; for indications in this direction however see [41] and references therein.

For a detailed, computer-assisted bifurcation analysis of the 2:5 resonance 'bubble' we refer to [35]. Compare with earlier work of Chenciner [45–47]. We like to note that the family of mappings G forms a concrete model for the Ruelle–Takens scenario regarding the onset of turbulence. In fact it also illustrates how the earlier scenario of Hopf–Landau–Lifschitz is also included: the present multi-parameter set-up unifies both approaches. For details and background see [19, 44, 59, 60, 63, 64, 70, 78, 79].

Resonance and Fractal Geometry

Interestingly, the blue colors right and left correspond to quasi-periodic circle attractors. The fact that the corresponding regions of the plane look like open sets is misleading. In reality these are meagre sets, dense veined by the residual sets associated to periodicity. These details are just too fine to be detected by the computational precision used.

Particularly in the latter case, in the left half of the diagram, we are dealing with the Arnold resonance web, for a detailed analysis see [36].

CONCLUSIONS

We discuss a number of consequences of the present paper in terms of modelling of increasing complexity.

Next Cases'

The Hopf saddle-node bifurcation for maps, see Sect. 3.2, can be viewed as a 'next case' in the systematic study of bifurcations as compared to, e.g., [58, 62] and many others. The nowhere dense part of parameter space, since it lacks interior points is somewhat problematic to penetrate by numerical continuation methods. Nevertheless, from the 'physical' point of view, this part surely is visible when its measure is positive or, as in the present examples, even close to full measure. Needless to say that this observation already holds for the Hopf–Neĭmark–Sacker bifurcation as described in Sect. 3.1.

Other 'next cases' are formed by the quasi-periodic bifurcations which is a joint application of Kolmogorov–Arnold–Moser Theory [5, 11, 19, 30, 44] and Singularity Theory [55–57, 87]. For overviews see [12, 48, 91]. The quasi-periodic bifurcations are inspired by the classical ones in which equilibria or periodic orbits are replaced by quasi-periodic tori. As an example, in the Hopf saddle node of Sect. 3.2 we met quasi-periodic Hopf bifurcation for mappings from circles to a

2-tori in a subordinate way. Here we witness a global geometry inspired by the classical Hopf bifurcation, which concerns the quasi-periodic dynamics associated to the fractal geometry in the parameter space, compare with Fig. 13. The gaps or tongues in between concern the resonances inside, within which we notice a further 'fractalization' or 'Cantorization'.

A similar 'next case' in complexity is given by the parametrically forced Lagrange top [29, 30], in which a quasi-periodic Hamiltonian Hopf bifurcation occurs. Indeed, we recall from [50] that in the Lagrange top a Hamiltonian Hopf bifurcation occurs, the geometry of which involves a swallowtail catastrophy. By the periodic forcing this geometry is 'Cantorized' yielding countably many tongues with fractal geometry in between.

Remarks

- As said before, in cases with infinite regress the fractal complement is a meagre set which has positive measure. Simon [82] describes a similar situation for 1-dimensional Schrödinger operators. Also see [6].
- It is an interesting property of the real numbers to allow for this kind of dichotomy in measure and topology, compare with Oxtoby [73]. Interestingly, although these properties in the first half of the 20th century were investigated for theoretical reasons, they here naturally show up in the context of resonances and spectra.

Modelling

We like to note that our investigations on the Hopf saddle-node bifurcation for mappings were inspired by climate models [25, 42, 83], where in about 80-dimensional Galerkin projections of PDE models such bifurcations were detected in 3-dimensional center manifolds.

Generally speaking there exists a large-scale programme of modelling in terms of dynamical systems depending on parameters, with applications varying from climate research to mathematical physics and biological cell systems. These models often are high-dimensional and their complexity is partly explained by mechanisms of the present paper, also see [19, 78, 79,91]. In general such models exhibit the coexistence of periodicity (including resonance), quasi-periodicity and chaos, best observed in the product of state- and parameter space.

Footnotes

1. For simplicity we take $|\varepsilon|<1$ which ensures that (10) is a circle diffeomorphism; for $|\varepsilon|\geq 1$ the mapping becomes a circle endomorphism and the current approach breaks down.
2. Mathematical ideas on adiabatic change were used earlier by Rayleigh and Poincaré and by Landau-Lifschitz.
3. For 'historical' reasons we use the letter a instead of α^2 as we did earlier.
4. The co-ordinates (ζ,t) sometimes also are called co-rotating. Also think of the Lagrangean variation of constants.
5. Colloquially often referred to as 'chaotic sea'.
6. Also known as the metric entropy conjecture.

ACKNOWLEDGEMENTS

The author thanks Konstantinos Efstathiou, Aernout van Enter and Ferdinand Verhulst for their help in the preparation of this paper.

REFERENCES

1. Arnol'd, V.I.: Small divisors I: On mappings of the circle onto itself. Am. . Soc. Transl., Ser. 2 **46**, 213–284 (1965)
2. Arnold, V.I.: ematical Methods of Classical Mechanics. GTM, vol. 60. Springer, Berlin (1978)
3. Arnold, V.I.: Geometrical Methods in the Theory of Ordinary Differential Equations. Springer, Berlin (1983)
4. Arnold, V.I., Avez, A.: Probèmes Ergodiques de la Mécanique classique. Gauthier-Villars, Paris (1967). Ergodic Problems of Classical Mechanics. Benjamin (1968)
5. Arnol'd, V.I., Kozlov, V.V., Neishtadt, A.I.: ematical aspects of classical and celestial mechanics. In: Arnol'd, V.I. (ed.) Dynamical Systems III. Springer, Berlin (1988)
6. Avila, A., Damanik, D.: Generic singular spectrum for ergodic Schrödinger operators. Duke . J. **130**(2), 393–400 (2005)
7. Baesens, C., Guckenheimer, J., Kim, S., MacKay, R.S.: Three coupled oscillators: Mode-locking, global bifurcation and toroidal chaos. Physica D **49**(3), 387–475 (1991)
8. Beersma, D.G.M., Broer, H.W., Cargar, K.A., Efstathiou, K., Hoveijn, I.: Pacer cell response to periodic Zeitgebers. Preprint University of Groningen (2011)
9. Bennett, M., Schatz, M.F., Rockwood, H., Wiesenfeld, K.: Huygens clocks. Proc. R. Soc. Lond. A **458**, 563–579 (2002)
10. Braaksma, B.L.J., Broer, H.W.: On a quasi-periodic Hopf bifurcation. Ann. Inst. Henri Poincaré, Anal. Non Linéaire **4**, 115–168 (1987)
11. Braaksma, B.L.J., Broer, H.W., Huitema, G.B.: Towards a quasi-periodic bifurcation theory. Mem. Am. . Soc. **83**(421), 81175 (1990)
12. Broer, H.W.: KAM theory: the legacy of Kolmogorov's 1954 paper. Bull. Am. . Soc. (New Ser.) **41**(4), 507–521 (2004)
13. Broer, H.W.: Normal forms in perturbation theory. In: Meyers, R. (ed.) Encyclopædia of Complexity & System Science, pp. 6310–6329. Springer, Berlin (2009)
14. Broer, H.W., Levi, M.: Geometrical aspects of stability theory for Hill's equations. Arch. Ration. Mech. Anal. **131**, 225–240 (1995)
15. Broer, H.W., Roussarie, R.: Exponential confinement of chaos in the bifurcations sets of real analytic diffeomorphisms. In: Broer, H.W., Krauskopf, B., Vegter, G. (eds.) Global Analysis of Dynamical Systems, pp. 167–210. IoP Publishing, Bristol (2001)
16. Broer, H.W., Simó, C.: Hill's equation with quasi-periodic forcing: Resonance tongues, instability pockets and global phenomena. Bol. Soc. Bras. Mat. **29**, 253–293 (1998)

17. Broer, H.W., Simó, C.: Resonance tongues in Hill's equations: A geometric approach. J. Differ. Equ. **166**, 290–327 (2000)

18. Broer, H.W., Takens, F.: Formally symmetric normal forms and genericity. Dyn. Rep. **2**, 36–60 (1989)

19. Broer, H.W., Takens, F.: Dynamical Systems and Chaos. Appl. . Sc., vol. 172. Springer, Berlin (2011)

20. Broer, H.W., Vegter, G.: Subordinate Sil'nikov bifurcations near some singularities of vector fields having low codimension. Ergod. Theory Dyn. Syst. **4**, 509–525 (1984)

21. Broer, H.W., Vegter, G.: Bifurcational aspects of parametric resonance. Dyn. Rep., New Ser.**1**, 1–51 (1992)

22. Broer, H.W., Vegter, G.: Generic Hopf-Neĭmark-Sacker bifurcations in feed forward systems. Nonlinearity **21**, 1547–1578 (2008)

23. Broer, H.W., Huitema, G.B., Takens, F.: Unfoldings of quasi-periodic tori. Mem. Am. . Soc. **83**(421), 1–82 (1990)

24. Broer, H.W., Simó, C., Tatjer, J.C.: Towards global models near homoclinic tangencies of dissipative diffeomorphisms. Nonlinearity **11**(3), 667–770 (1998)

25. Broer, H.W., Simó, C., Vitolo, R.: Bifurcations and strange attractors in the Lorenz-84 climate model with seasonal forcing. Nonlinearity **15**(4), 1205–1267 (2002)

26. Broer, H.W., Golubitsky, M., Vegter, G.: The geometry of resonance tongues: a singularity theory approach. Nonlinearity **16**, 1511–1538 (2003)

27. Broer, H.W., Hanßmann, H., Jorba, Á., Villanueva, J., Wagener, F.O.O.: Normal-internal resonances in quasi-periodically forces oscillators: a conservative approach. Nonlinearity**16**, 1751–1791 (2003)

28. Broer, H.W., Puig, J., Simó, C.: Resonance tongues and instability pockets in the quasi-periodic Hill-Schrödinger equation. Commun. . Phys. **241**, 467–503 (2003)

29. Broer, H.W., Hanßmann, H., Hoo, J.: The quasi-periodic Hamiltonian Hopf bifurcation. Nonlinearity **20**, 417–460 (2007)

30. Broer, H.W., Hoo, J., Naudot, V.: Normal linear stability of quasi-periodic tori. J. Differ. Equ. **232**, 355–418 (2007)

31. Broer, H.W., Efstathiou, K., Subramanian, E.: Robustness of unstable attractors in arbitrarily sized pulse-coupled systems with delay. Nonlinearity **21**, 13–49 (2008)

32. Broer, H.W., Efstathiou, K., Subramanian, E.: Heteroclinic cycles between unstable attractors. Nonlinearity **21**, 1385–1410 (2008)

33. Broer, H.W., Hanßmann, H., You, J.: On the destruction of resonant Lagrangean tori in Hamiltonian systems. Preprint Universiteit Utrecht (2008)

34. Broer, H.W., Holtman, S.J., Vegter, G.: Recognition of the bifurcation type of resonance in a mildly degenerate Hopf-Neĭmark-Sacker families. Nonlinearity **21**, 2463–2482 (2008)

35. Broer, H.W., Simó, C., Vitolo, R.: The Hopf-saddle-node bifurcation for fixed points of 3D-diffeomorphisms, analysis of a resonance 'bubble'. Physica D **237**, 1773–1799 (2008)

36. Broer, H.W., Simó, C., Vitolo, R.: The Hopf-saddle-node bifurcation for fixed points of 3D-diffeomorphisms, the Arnol'd resonance web. Bull. Belg. . Soc. Simon Stevin **15**, 769–787 (2008)

37. Broer, H.W., Ciocci, M.C., Hanßmann, H., Vanderbauwhede, A.: Quasi-periodic stability of normally resonant tori. Physica D 309–318 (2009)

38. Broer, H.W., Holtman, S.J., Vegter, G., Vitolo, R.: Geometry and dynamics of mildly degenerate Hopf-Neĭmarck-Sacker families near resonance. Nonlinearity **22**, 2161–2200 (2009)

39. Broer, H.W., Hasselblatt, B., Takens, F. (eds.): Handbook of Dynamical Systems, vol. 3. North-Holland, Amsterdam (2010)

40. Broer, H.W., Holtman, S.J., Vegter, G.: Recognition of resonance type in periodically forced oscillators. Physica D **239**(17), 1627–1636 (2010)

41. Broer, H.W., Simó, C., Vitolo, R.: Chaos and quasi-periodicity in diffeomorphisms of the solid torus. Discrete Contin. Dyn. Syst., Ser. B **14**(3), 871–905 (2010)

42. Broer, H.W., Dijkstra, H.A., Simó, C., Sterk, A.E., Vitolo, R.: The dynamics of a low-order model for the Atlantic Multidecadal Oscillation. Discrete Contin. Dyn. Syst., Ser. B **16**(1), 73–102 (2011)

43. Broer, H.W., Holtman, S.J., Vegter, G., Vitolo, R.: Dynamics and geometry near resonant bifurcations. Regul. Chaotic Dyn. **16**(1–2), 39–50 (2011)

44. Broer, H.W., Hanßmann, H., Wagener, F.O.O.: Quasi-periodic bifurcation theory: The geometry of kam. (2012, to appear)

45. Chenciner, A.: Bifurcations de points fixes elliptiques. I. Courbes invariantes. Publ. . IHÉS **61**, 67–127 (1985)

46. Chenciner, A.: Bifurcations de points fixes elliptiques. II. Orbites périodiques et ensembles de Cantor invariants. Invent. . **80**, 81–106 (1985)

47. Chenciner, A.: Bifurcations de points fixes elliptiques. III. Orbites périodiques de "petites" périodes et élimination résonnante des couples de courbes invariantes. Publ. . IHÉS**66**, 5–91 (1988)

48. Ciocci, M.C., Litvak-Hinenzon, A., Broer, H.W.: Survey on dissipative kam theory including quasi-periodic bifurcation theory based on lectures by Henk Broer. In: Montaldi, J., Ratiu, T. (eds.) Geometric Mechanics and Symmetry: The Peyresq Lectures. LMS Lecture Notes Series, vol. 306, pp. 303–355. Cambridge University Press, Cambridge (2005)

49. Correia, A., Laskar, J.: Mercury's capture into the 3/2 spin-orbit resonance as a result of its chaotic dynamics. Nature **429**, 848–850 (2004)

50. Cushman, R.H., van der Meer, J.C.: The Hamiltonian Hopf bifurcation in the Lagrange top. In: Albert, C. (ed.) Géometrie Symplectique et Méchanique, Colloque International La Grande Motte, France, 23–28 Mai 1988. Lecture Notes in ematics, vol. 1416, pp. 26–38 (1990)

51. de Sitter, W.: On the libration of the three inner large satellites of Jupiter. Publ. Astron. Lab. Gron. **17**, 1–119 (1907)

52. de Sitter, W.: New ematical theory of Jupiters satellites. Ann. Sterrewacht Leiden **XII**(1925)

53. Devaney, R.L.: An Introduction to Chaotic Dynamical Systems, 2nd edn. Addison-Wesley, Reading (1989)

54. Eliasson, L.H.: Floquet solutions for the one-dimensional quasi-periodic Schrödinger equation. Commun. . Phys. **146**, 447–482 (1992)

55. Golubitsky, M., Guillemin, V.: Stable Mappings and Their Singularities. Springer, Berlin (1973)

56. Golubitsky, M., Schaeffer, D.G.: Singularities and Groups in Bifurcation Theory, vol. I. Springer, Berlin (1985)

57. Golubitsky, M., Stewart, I., Schaeffer, D.G.: Singularities and Groups in Bifurcation Theory, vol. II. Appl. . Sc., vol. 69. Springer, Berlin (1988)

58. Guckenheimer, J., Holmes, P.: Nonlinear Oscillations, Dynamical Systems, and Bifurcations of Vector Fields, 5th edn. Applied ematical Sciences, vol. 42. Springer, Berlin (1997)

59. Hopf, E.: A ematical example displaying features of turbulence. Commun. Pure Appl. . **1**, 303–322 (1948)

60. Hopf, E.: Repeated branching through loss of stability, an example. In: Diaz, J.B. (ed.) Proceedings of the Conference on Differential Equations, Maryland 1955, pp. 49–56. University Maryland book store, College Park (1956)

61. Huygens, C.: Œvres complètes de Christiaan Huygens (1888–1950), vol. 5, pp. 241–263 and vol. 17, pp. 156–189. Martinus Nijhoff (1888–1950)

62. Kuznetsov, Yu.A.: Elements of Applied Bifurcation Theory, 3rd edn. Applied ematical Sciences, vol. 112. Springer, Berlin (2004)

63. Landau, L.D.: On the problem of turbulence. Dokl. Akad. Nauk SSSR **44**, 339–342 (1944)

64. Landau, L.D., Lifshitz, E.M.: Fluid Mechanics, 2nd edn. Pergamon, Oxford (1987)

65. Mandelbrot, B.B.: The Fractal Geometry of Nature. Freeman, San Francisco (1977)

66. Moser, J.K.: On invariant curves of area-preserving mappings of an annulus. Nachr. Akad. Wiss. Gött., .-Phys. Kl. II. **1**, 1–20 (1962)

67. Moser, J.K.: Lectures on Hamiltonian systems. Mem. Am. . Soc. **81**, 1–60 (1968)

68. Moser, J.K.: Stable and Random Motions in Dynamical Systems, with Special Emphasis to Celestial Mechanics. Annals ematical Studies, vol. 77. Princeton University Press, Princeton (1973)

69. Moser, J.K., Pöschel, J.: An extension of a result by Dinaburg and Sinai on quasi-periodic potentials. Comment. . Helv. **59**, 39–85 (1984)

70. Newhouse, S.E., Ruelle, D., Takens, F.: Occurrence of strange Axiom A attractors near quasi-periodic flows on TmTm, m≤3. Commun. . Phys. **64**, 35–40 (1978)

71. Newhouse, S.E., Palis, J., Takens, F.: Bifurcations and stability of families of diffeomorphisms. Publ. . IHÉS **57**, 5–71 (1983)

72. Nitecki, Z.: Differentiable Dynamics. An Introduction to the Orbit Structure of Diffeomorphisms. Massachusetts Institute of Technology Press, Cambridge (1971)

73. Oxtoby, J.: Measure and Category. Springer, Berlin (1971)

74. Palis, J., de Melo, W.C.: Geometric Theory of Dynamical Systems. Springer, Berlin (1982)

75. Palis, J., Takens, F.: Hyperbolicity & Sensitive Chaotic Dynamics at Homoclinic Bifurcations. Cambridge University Press, Cambridge (1993)

76. Pogromsky, A., Rijlaarsdam, D., Nijmeijer, H.: Experimental Huygens synchronization of oscillators. In: Thiel, M., Kurths, J., Romano, M.C., Moura, A., Károlyi, G. (eds.) Nonlinear Dynamics and Chaos: Advances and Perspectives. Springer Complexity, pp. 195–210 (2010)

77. Poincaré, J.H.: Les Méthodes Nouvelles de la Mécanique Céleste I, II, III, Gauthier-Villars (1892, 1893, 1899). Republished by Blanchard (1987)

78. Ruelle, D., Takens, F.: On the nature of turbulence. Commun. . Phys. **20**, 167–192 (1971)

79. Ruelle, D., Takens, F.: On the nature of turbulence. Commun. . Phys. **23**, 343–344 (1971)

80. Sanders, J.A., Verhulst, F., Murdock, J.: Averaging Methods in Nonlinear Dynamical Systems, Revised 2nd edn. Applied ematical Sciences, vol. 59. Springer, Berlin (2007)

81. Siegel, C.L., Moser, J.K.: Lectures on Celestial Mechanics. Springer, Berlin (1971)

82. Simon, B.: Operators with singular continuous spectrum: I. General operators. Ann. .**141**, 131–145 (1995)

83. Sterk, A.E., Vitolo, R., Broer, H.W., Simó, C., Dijkstra, H.A.: New nonlinear mechanisms of midlatitude atmospheric low-frequency variability. Physica D **239**(10), 702718 (2010)

84. Stoker, J.J.: Nonlinear Vibrations in Mechanical and Electrical Systems, 2nd edn. Wiley, New York (1992)

85. Strogatz, S.H.: Nonlinear Dynamics and Chaos. Addison Wesley, Reading (1994)

86. Takens, F.: Forced oscillations and bifurcations. Applications of Global Analysis I. Comm. of the . Inst. Rijksuniversiteit Utrecht (1974). In: Broer, H.W. Krauskopf, B., Vegter, G. (eds.) Global Analysis of Dynamical Systems, pp. 1–62. IoP Publishing (2001)

87. Thom, R.: Stabilité Structurelle et Morphogénèse. Benjamin, Elmsford (1972)

88. van der Pol, B.: De amplitude van vrije en gedwongen triode-trillingen. Tijdschr. Ned. Radiogenoot. **1**, 3–31 (1920)

89. van der Pol, B.: The nonlinear theory of electric oscillations. Proc. Inst. Radio Engl. **22**, 1051–1086 (1934); Reprinted in: Selected Scientific Papers. North-Holland (1960)

90. Verhulst, F.: Nonlinear Differential Equations and Dynamical Systems, 2nd edn. Universitext. Springer, Berlin (1996)

91. Vitolo, R., Broer, H.W., Simó, C.: Quasi-periodic bifurcations of invariant circles in low-dimensional dissipative dynamical systems. Regul. Chaotic Dyn. **16**(1–2), 154–184 (2011)

CHAPTER 2

APPROXIMATE METHODS FOR NONLINEAR DIFFERENTIAL EQUATIONS WITH OSCILLATIONS

Marwan Alquran and Kamel Al-Khaled

Keywords: Lindstedt-Poincare method, Krylov-Bogoliubov method, Differential transform method, Nonlinear oscillations

INTRODUCTION

When a mathematical model is formulated for a physical problem, it is often represented by equations of a one-mass system with two degrees of freedom that are mostly described using a second-order differential equation, and these equations are not solvable exactly by analytic techniques. If in the differential equation, some small nonlinearities exist, they are introduced in the differential equation of motion as small nonlinear terms, and so the motion of the system is described by a second-order strongly nonlinear differential equation in the form of

$$y''(t) + y(t) + \epsilon f(y(t), y'(t)) = 0, \quad y(0) = 1, \quad y'(0) = 0, \qquad (1.1)$$

where ε is a sufficiently small parameter, so the nonlinear term $\varepsilon f(y(t), y'(t))$ is relatively small. As mentioned in[1], these terms arise because the physical process has small effects. For example, in a fluid flow problem the viscosity may be small, or in the problem of the motion of a projectile, the force due to air resistance may be small. These low order effects are represented by terms in the model equation that, when compared to the other terms, are small and represented by a coefficient parameter ε. Therefore, we must resort to approximation and numerical methods. Foremost, among approximation techniques are the so-called perturbation methods. Perturbation method[2–5] provides the most versatile tools available in nonlinear analysis of physical problems. These methods have their own limitations; for example, the solutions are valid, in most cases, only for small values of the parameter ε. Therefore, we cannot rely fully on the approximations, since there is no criterion on how small the parameters should be. In[6], the authors proved the existence of an approximate solution in the mean for general strongly nonlinear differential equations. They investigate the behavior of the class of solutions. In addition, they implement the homotopy perturbation method and find analytic solutions for strongly nonlinear differential equations of the form

$$x''(t) + ax'^2(t) + bx(t) = F(t) \qquad (1.2)$$

for some constants a, b. As mentioned in[6], numerous methods are developed for analytic solution of strongly nonlinear differential equations describing the vibrations of the oscillator. For example, the harmonic balance method[7–9] which leads to algebraic equations has been used. An analytical approximate approach for determining periodic solutions of nonlinear jerk equations involving third-order time-derivative is presented in[10, 11]. In[12], the authors used the Adomian decomposition method to find the approximate solutions for general form of the second-order ordinary differential equations. In[13], a modified variational approach called global error minimization method is developed for obtaining an approximate closed-form analytical solution for nonlinear oscillator differential equations. In this paper, we propose the use of Lindstedt-Poincare method, the Krylov-Bogoliubov first approximate method, and the differential transform method (DTM) in obtaining the approximate solutions for nonlinear differential equations of the form in Equation 1.1. In order to illustrate the efficiency of these methods, we examine three numerical examples. Comparison is made based on the obtained results. The paper is concluded in the last section. In

Citation: Marwan Alquran, Kamel Al-Khaled, Approximate methods for strongly nonlinear differential equations with oscillations, DOI: 10.1186/2251-7456-6-32.

what follows, we survey briefly the DTM steps while we use directly the other two methods within the text.

METHODS

The goal of this section is to recall notations, definitions, and some theorems of the DTM that will be used in this paper. These are discussed in[12, 14, 15]. Also, we will highlight the main steps of implementing the DTM in solving differential equations. The other two methods (Lindstedt-Poincare and the Krylov-Bogoliubov) are given within the section of Results and discussion.

Differential Transform Method

The differential transform of the kth derivative of a function u(x) is defined to be

$$U(k) = \frac{1}{k!}\left(\frac{d^k}{dx^k}u(x)\right)\Big|_{x=x_0} \tag{2.1}$$

and the inverse transform of U(k) is defined as

$$u(x) = \sum_{k=0}^{\infty} U(k)(x-x_0)^k \tag{2.2}$$

Equation 2.2 is known as the Taylor series expansion of u(x) around $x = x_0$. Also, we need the following theorems. If G(k) is the differential transform of g(x) then the following theorems apply.

Theorem 1. If f(x) = $\frac{d^n g(x)}{dx^n}$, then F(k) = $\frac{(k+n)!}{k!}G(k+n)$.

Theorem 2. If f(x) = $g^2(x)$, then F(k) = $\sum_{i=0}^{k}$G(i)G(k-i).

Theorem 3. If f(x) = $g^3(x)$, then F(k) = $\sum_{i=0}^{k}\sum_{j=0}^{k-i}$G(i)G(j) G(k-i-j).

RESULTS AND DISCUSSION

This section is divided into two parts: weakly and strongly nonlinear oscillations.

Application I: weakly nonlinear oscillation

An example of a weak nonlinear oscillation is the Duffing equation. It is a nonlinear second-order differential equation used to model certain damped and driven oscillators. The equation is given by

$$y''(t) + \delta y'(t) + \alpha y(t) + \beta y^3(t) = 0 \tag{3.1}$$

where the (unknown) function y = y(t) is the displacement at time t; y′is the first derivative of y with respect to time, i.e., velocity; and y″is the second time-derivative of y, i.e., acceleration. The number δ controls the size of the damping (friction), α controls the size of the restoring force, and β controls the amount of nonlinearity in the restoring. The equation describes the motion of a damped oscillator with a more complicated potential than in simple harmonic motion (which corresponds to the case $\beta = \delta = 0$); in physical terms, it models, for example, a spring pendulum whose spring's stiffness does not exactly obey Hooke's law. The Duffing equation is an example of a dynamic system that exhibits chaotic behavior.

In this section, we are concerned about

$$y''(t) + y(t) + \epsilon y^3(t) = 0, \quad y(0) = 1, \quad y'(0) = 0. \tag{3.2}$$

First, by means of Lindstedt-Poincare method[16–18] we use the transformation $\tau = \omega\, t$ which transforms Equation 3.2 into

$$\omega^2 u''(\tau) + u(\tau) + \epsilon u^3(\tau) = 0, \tag{3.2}$$

where u(τ) = y(t). Let

$$\begin{aligned} \omega &= 1 + \epsilon\omega_1, \\ u &= u_0 + \epsilon u_1. \end{aligned} \tag{3.4}$$

If we substitute Equation 3.4 in Equation 3.3 and equate the coefficients of various powers of ε to 0, then we obtain

$$\begin{aligned} &O(1): u_0'' + u_0 = 0; \quad u_0(0) = 1, \quad u_0'(0) = 0, \\ &O(\epsilon): 2\omega_1 u_0'' + u_1'' + u_1 + u_0^3 = 0; \quad u_1(0) = 0, \quad u_1'(0) = 0. \end{aligned} \tag{3.5}$$

Solving the equations in Equation 3.5 and avoiding the occurrence of secular terms in the perturbation solutions yield

$$\begin{aligned} u_0(\tau) &= \cos(\tau), \\ u_1(\tau) &= \frac{1}{32}\left(\cos(\tau) - \cos(3\tau)\right). \end{aligned} \tag{3.6}$$

Therefore, the solution of Equation 3.2 is

$$\begin{aligned} y(t) = \cos\left(\left(1+\frac{3\epsilon}{8}\right)t\right) + \frac{\epsilon}{32}\left(\cos\left(\left(1+\frac{3\epsilon}{8}\right)t\right)\right. \\ \left. - \cos\left(3\left(1+\frac{3\epsilon}{8}\right)t\right)\right). \end{aligned} \tag{3.7}$$

Second, by applying the differential transform to Equation 3.2, we get the recursive formula

$$F(k+2) = -\frac{F(k)}{(k+1)(k+2)} - \frac{\epsilon}{(k+1)(k+2)}\sum_{i=0}^{k}\sum_{j=0}^{k-i} F(i)F(j)F(k-i-j), \tag{3.8}$$

where F(k) is the differential transform of y(t). By the given initial conditions, we have F(0) = 1 and F(1) = 0. Considering N = 8 in the DTM series solution, the approximate solution is

$$\begin{aligned} y_{DTM}(t) &= \sum_{k=0}^{8} F(k)t^k \\ &= 1 - \frac{1}{2}(1+\epsilon)t^2 + \frac{1}{24}(1+\epsilon)(1-3\epsilon)t^4 - \ldots \end{aligned} \tag{3.9}$$

Third, by means of Krylov-Bogoliubov first approximate method[19], the solution of

$$y''(t) + y(t) + \epsilon f(y(t), y'(t)) = 0 \tag{3.10}$$

is sought as

$$y(t) = a(t)\sin(t + \phi(t)), \tag{3.11}$$

where a(t) and ϕ(t) are given by

$$\begin{aligned} a(t) &= \frac{\epsilon}{2\pi}\int_0^{2\pi} f(c\sin(\theta), c\cos(\theta))\cos(\theta)d\theta, \\ \phi(t) &= \frac{\epsilon}{2\pi c}\int_0^{2\pi} f(c\sin(\theta), c\cos(\theta))\sin(\theta)d\theta, \end{aligned} \tag{3.12}$$

and where c is a free constant. Therefore, according to Equation 3.2, the solution is

$$y(t) = A\sin\left(\left(\frac{3\epsilon A^2}{8}+1\right)t+B\right), \tag{3.13}$$

where A and B are constants. Applying the initial conditions given in Equation 3.2 yields A = – 1 and $B=-\frac{\pi}{2}$.

Figure1 shows the plots of the obtained solutions using the proposed methods compared with Mathematica NDSolve tool.

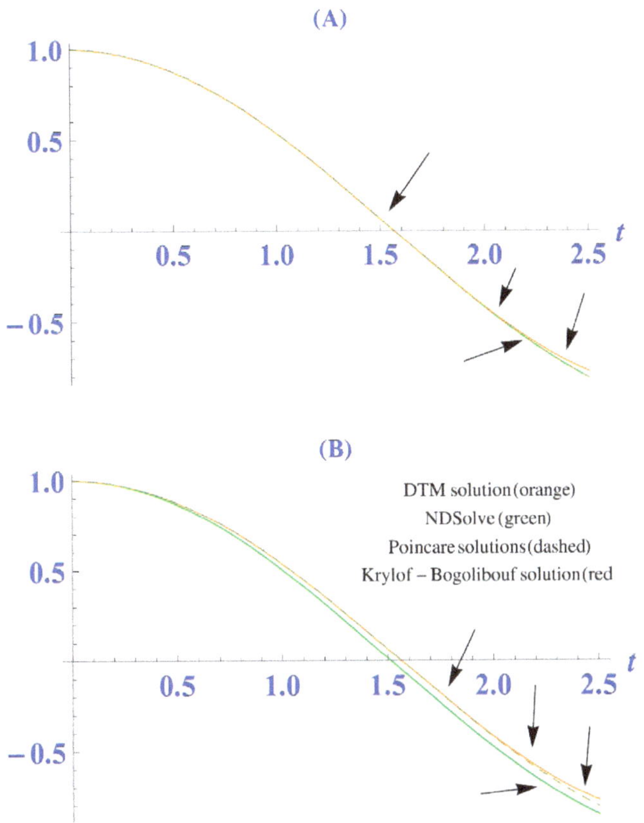

Figure 1: The obtained solutions of the Duffing equation for (A)ε= 0.01 and (B)ε= 0.1.

Application II: strongly nonlinear oscillation

In this section, we apply the Lindstedt-Poincare method and the differential transform method on two strongly nonlinear oscillations.

Example 1.

Consider the nonlinear oscillation

$$y''(t)+y(t)=\epsilon y(t)(1-y'^2(t)),\quad y(0)=1,\quad y'(0)=0. \tag{4.1}$$

First, by means of Lindstedt-Poincare method, the transformation $\tau=\omega t$ converts Equation 4.1 to

$$\omega^2u''(\tau)+u(\tau)-\epsilon u(\tau)+\epsilon\omega^2u(\tau)u'(\tau))^2=0, \tag{4.2}$$

where u(τ) = y(t). Let

$$\omega = 1+\epsilon\omega_1,$$

$$u = u_0+\epsilon u_1. \tag{4.3}$$

Through inserting Equation 4.3 in Equation 4.2 and equating the coefficients of various powers of ε to 0, we yield

$$O(1): u_0''+u_0=0;\quad u_0(0)=1,\quad u_0'(0)=0,$$
$$O(\epsilon): 2\omega_1u_0''+u_1''+u_1-u_0+u_0(u_0')^2=0;\quad u_1(0)=0,\quad u_1'(0)=0. \tag{4.4}$$

Solving the equations in Equation 4.4 and avoiding the occurrence of secular terms in the perturbation solutions yield

$$u_0(\tau)=\cos(\tau),$$
$$u_1(\tau)=-\frac{1}{32}(\cos(\tau)-\cos(3\tau)). \tag{4.5}$$

Therefore, the solution of Equation 4.1 is

$$y(t)=\cos\left(\left(1-\frac{3\epsilon}{8}\right)t\right)-\frac{\epsilon}{32}\left(\cos\left(\left(1+\frac{3\epsilon}{8}\right)t\right)-\cos\left(3\left(1+\frac{3\epsilon}{8}\right)t\right)\right). \tag{4.6}$$

Second, by applying the differential transform to Equation 4.1, we get the recursive formula

$$F(k+2)=\frac{(\epsilon-1)F(k)}{(k+1)(k+2)}-\frac{\epsilon}{(k+1)(k+2)}\times\sum_{i=0}^{k}\sum_{j=0}^{k-i}(i+1)(j+1)F(i)F(j)F(k-i-j), \tag{4.7}$$

where F(k) is the differential transform of y(t). By the given initial conditions, we have F(0) = 1 and F(1) = 0. Considering N = 8 in the DTM series solution, the approximate solution is

$$y_{DTM}(t)=\sum_{k=0}^{8}F(k)t^k = 1-\frac{1}{2}(1-\epsilon)t^2+\frac{1}{24}(1-\epsilon)^2(1-2\epsilon)t^4-\cdots \tag{4.8}$$

The solutions obtained by the proposed methods show an excellent agreement with the one obtained by Mathematica NDSolve tool if we choose ε to be small (see Figure2).

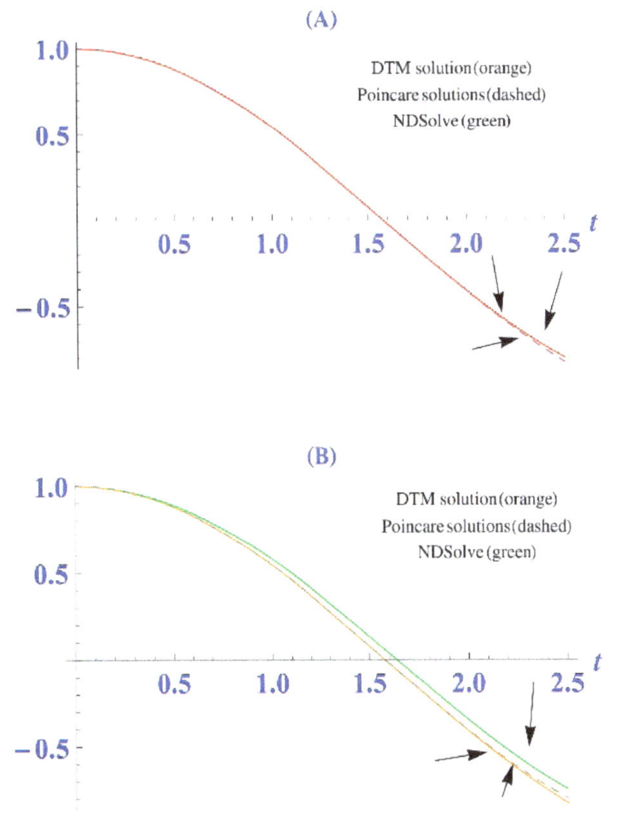

Figure 2: The obtained solutions of Equation4.1for (A)ε= 0.01 and (B)ε= 0.1.

Example 2.

Consider the nonlinear oscillation

$$y''(t) + y(t) = \epsilon y(t) y'^2(t), \quad y(0) = 1, \quad y'(0) = 0. \qquad (4.9)$$

First, by means of Lindstedt-Poincare method, the transformation $\tau = \omega t$ converts Equation 4.9 to

$$\omega^2 u''(\tau) + u(\tau) - \epsilon\omega^2 u(\tau) u'(\tau))^2 = 0, \qquad (4.10)$$

where $u(\tau) = y(t)$. Let

$$\omega = 1 + \epsilon\omega_1,$$

$$u = u_0 + \epsilon u_1. \qquad (4.11)$$

Inserting Equation 4.11 in Equation 4.10 and equating the coefficients of various powers of ε to 0 yield

$$O(1): u_0'' + u_0 = 0; \quad u_0(0) = 1, \quad u_0'(0) = 0,$$

$$O(\epsilon): 2\omega_1 u_0'' + u_1'' + u_1 - u_0(u_0')^2 = 0; \quad u_1(0) = 0, \quad u_1'(0) = 0. \qquad (4.12)$$

Moreover, solving the equations in Equation 4.12 and avoiding the occurrence of secular terms in the perturbation solutions yield

$$u_0(\tau) = \cos(\tau),$$

$$u_1(\tau) = -\frac{1}{32}(\cos(\tau) - \cos(3\tau)). \qquad (4.13)$$

Therefore, the solution of Equation 4.9 is

$$y(t) = \cos\left(\left(1 - \frac{\epsilon}{8}\right) t\right) - \frac{\epsilon}{32}\left(\cos\left(\left(1 - \frac{\epsilon}{8}\right) t\right) - \cos\left(3\left(1 - \frac{\epsilon}{8}\right) t\right)\right). \qquad (4.14)$$

Second, by applying the differential transform to Equation 4.9, we get the recursive formula

$$F(k+2) = -\frac{F(k)}{(k+1)(k+2)} + \frac{\epsilon}{(k+1)(k+2)} \times \sum_{i=0}^{k}\sum_{j=0}^{k-i}(i+1)(j+1)F(i)F(j)F(k-i-j), \qquad (4.15)$$

where F(k) is the differential transform of y(t). By the given initial conditions, we have $F(0) = 1$ and $F(1)=0$. Considering $N = 8$ in the DTM series solution, the approximate solution is

$$y_{DTM}(t) = \sum_{k=0}^{8} F(k) t^k = 1 - \frac{1}{2}t^2 + \frac{1}{24}(1 + 2\epsilon)t^4 - \ldots \qquad (4.16)$$

Figure3 shows the plots of the obtained solutions using the proposed methods compared with Mathematica NDSolve tool.

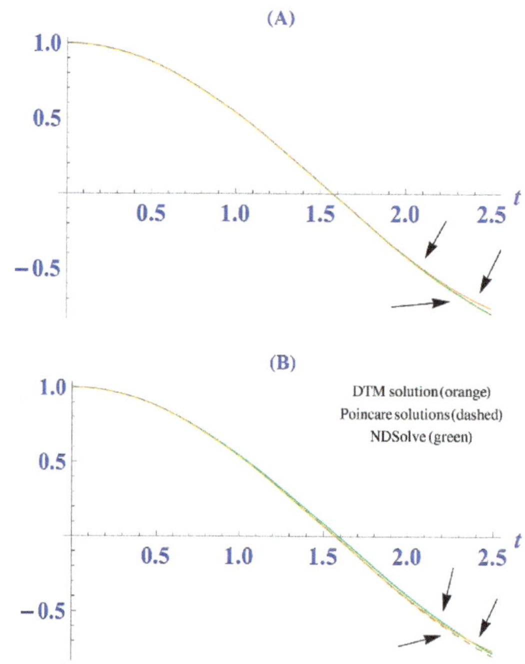

Figure 3: Obtained solutions of Equation4.9for (A)ε= 0.01 and (B)ε= 0.1.

CONCLUSIONS

In this paper, the nonlinear differential equations with oscillator are considered to study the validity of DTM compared with Lindstedt-Poincare method. The most significant features of DTM are its simplicity and its excellent accuracy for different values of the parameter ε. DTM is very effective and convenient for solving truly nonlinear oscillator equations.

ACKNOWLEDGEMENTS

The authors would like to thank the editor and the anonymous referees for their in-depth reading, criticisms, and insightful comments on an earlier version of this paper.

REFERENCES

1. Logan JD: Applied Mathematics. Wiley, New York; 2006.
2. He JH: Some asymptotic methods for strongly nonlinear equations. Int. J. Modern Physics B 2006, 20(10):1141–1199. 10.1142/S0217979206033796
3. Cveticanin L: Homotopy-perturbation method for pure nonlinear differential equation.Chaos, Soliton and Fractals 2006, 30: 1221–1230. 10.1016/j.chaos.2005.08.180

4. 4He JH: A new perturbation technique which is also valid for large parameters. J. Sound Vib2000, 229: 1257–1263. 10.1006/jsvi.1999.2509

5. Hu H: A classical perturbation technique which is valid for large parameters. J. Sound Vib2004, 269: 409–412. 10.1016/S0022-460X(03)00318-3

6. Al-Khaled K, Ali Hajji M: On the existence of solutions for strongly nonlinear differential equations. J. Math. Anal. Appl 2008, 344: 1165–1175. 10.1016/j.jmaa.2008.03.022

7. Guo Z, Leung AY: The iterative homotopy harmonic balance method for conservative Helmholtz-Duffing oscillators. Appl. Math. Comput 2010, 215: 3163–3169.

8. Chen YM, Liu JK: A new method based on the harmonic balance method for nonlinear oscillators. Phys. Lett. A 2007, 368: 371–378.

9. Liu L, Dowell EH, Hall KC: A novel harmonic balance analysis for the Van Der Pol oscillator. Int. J. Non-linear Mech2007,42:2–12. 10.1016/j.ijnonlinmec.2006.09.004

10. Gottlieb HPW: Harmonic balance approach to periodic solutions of nonlinear jerk equations. J. Sound Vib 2004, 271: 671–683. 10.1016/S0022-460X(03)00299-2

11. Hu H: Perturbation method for periodic solutions of nonlinear jerk equations. Phys. Lett. A2008, 271: 4205–4209.

12. Alquran M, Al-Khaled K: Approximate solutions to nonlinear partial integro-differential equations with applications in heat flow. Jordan J. Mathematics Stat 2010, 3(2):93–116.

13. Farzaneh Y, Tootoonchi AA: Global Error Minimization method for solving strongly nonlinear oscillator differential equations. Comp. Math. Appl 2010, 59: 2887–2895. 10.1016/j.camwa.2010.02.006

14. Alquran M, Al-Khaled K: Approximations of Sturm-Liouville Eigenvalues using Sinc-Galerkin and differential transform methods. Appl. Appl. Math 2010, 5(1):128–147.

15. Ayaz F: Solution of the system of differential equations by differential transform method.Appl. Math. Comput 2004, 147: 547–567. 10.1016/S0096-3003(02)00794-4

16. He JH: Modified Lindstedt-Poincare methods for strongly non-linear oscillations: part I: expansion of a constant. Int. J. Non-linear Mech 2002, 37: 309–314. 10.1016/S0020-7462(00)00116-5

17. He JH: Modified Lindstedt-Poincare methods for strongly nonlinear oscillations: part II: a new transformation. Int. J. Non-linear Mech 2002, 37: 315–320. 10.1016/S0020-7462(00)00117-7

18. Cheung YK, Chen SH, Lau SL: A modified Lindstedt-Poincare method for certain strongly non-linear oscillators. Int. J. Non-linear Mech 1991, 26: 367–378. 10.1016/0020-7462(91)90066-3

19. Ross SL: Differential Equations. Blaisdel, New York; 1964.

CHAPTER 3

ANALYSING NONLINEAR MECHANICAL VIBRATIONS

Simon A. Neild, Alan R. Champneys, David J. Wagg, Thomas L. Hill, Andrea Cammarano

INTRODUCTION: NONLINEAR NORMAL MODES AND NORMAL FORMS

The problem addressed here is how to extend the well-established notion of normal modes of linear vibration systems to nonlinear systems in a mathematically consistent way that also allows for practical implementation. In recent years, there has been a lot of research related to the concept of a so-called 'nonlinear normal mode' (NNM). We shall not go into the complete theory here, but refer instead to Kerschen et al. [**1**] and Peeters et al. [**2**]) and Avrimov & Mikhlin [**3**] for surveys of the state of the art. As we shall explain, this concept is of limited use when forcing and damping are present, a restriction that is not shared by the more general concept of thenormal form of a nonlinear system near an equilibrium [**4**–**6**]. At present though, the application of normal form analysis to structural vibration systems [7,**8**], described in more detail in §3 below, is rather technical and not widely adopted. It is also not clear how these structural vibration normal forms relate to the generic theory of normal forms as treated in Murdock [**6**]. Thus, the aim of this paper is to explain the normal form theory for structural vibrations in a more rigorous and historical context, and to show its relevance for a variety of practical problems.

Consider mechanical structures: the dynamics can be modelled as

$$M\ddot{\mathbf{x}} + D\dot{\mathbf{x}} + K\mathbf{x} + N(\mathbf{x}, \dot{\mathbf{x}}, \mathbf{r}) = P\mathbf{r} + \text{c.c.}, \qquad 1.1$$

Citation: Neild SA, Champneys AR, Wagg DJ, Hill TL, Cammarano A. 2015 Analysing nonlinear mechanical vibrations.Phil. Trans. R. Soc. A 373: 20140404. http://dx.doi.org/10.1098/rsta.2014.0404.

where $\mathbf{x} \in \mathbb{R}^n$ represents the displacements of the n degrees of freedom of the structure, M, D and K are, respectively, mass, damping and stiffness matrices and N represents nonlinearity, which is assumed to be sufficiently smooth. The right-hand side represents the forcing on the system $\mathbf{r}(t) \in \mathbb{C}^n$ and can be written as a sum of $e^{i\Omega t}$ terms. The concept of the NNM applies well to the case where there is no forcing or damping, so that P=D=0 in (1.1) and N is a pure function of **x**. In this case, Shaw & Pierre [**9**,**10**] argued that an NNM is just an invariant manifold composed of periodic solutions, whose frequency in the limit of amplitude tending to zero is the same as the linear mode. Here, they can appeal to the Liapunov Centre Theorem that guarantees that such a manifold must exist. Attempts have been made to extend this definition to include damping. Here, things are more troublesome because the Liapunov centre theorem no longer applies and, in general, one should not expect there to be small periodic solutions. Nevertheless, one can still appeal to the theory of invariant manifolds, but unfortunately such manifolds will be non-unique in general and will depend crucially on the relative size of the damping in each mode; see **figure 1** illustrating this point for a general two degrees of freedom system. **Figure 1** a illustrates NNMs the case of zero damping. These manifolds which are shown as projections onto a radial coordinate are composed entirely of periodic orbits, as given by the Liapunov centre theorem. By contrast, **figure 1** b,c shows damped cases where the damping coefficient is greater in either the first or the second mode. In each case, only the most strongly damped mode has a unique invariant manifold tangent to it, with solutions in the manifold now composed of decaying oscillations. For the more weakly damped mode, there are now infinitely many such invariant manifolds and there is no uniquely distinguished one. In general, a tiny perturbation in initial condition will lead to a very different manifold if one computes backwards in time.

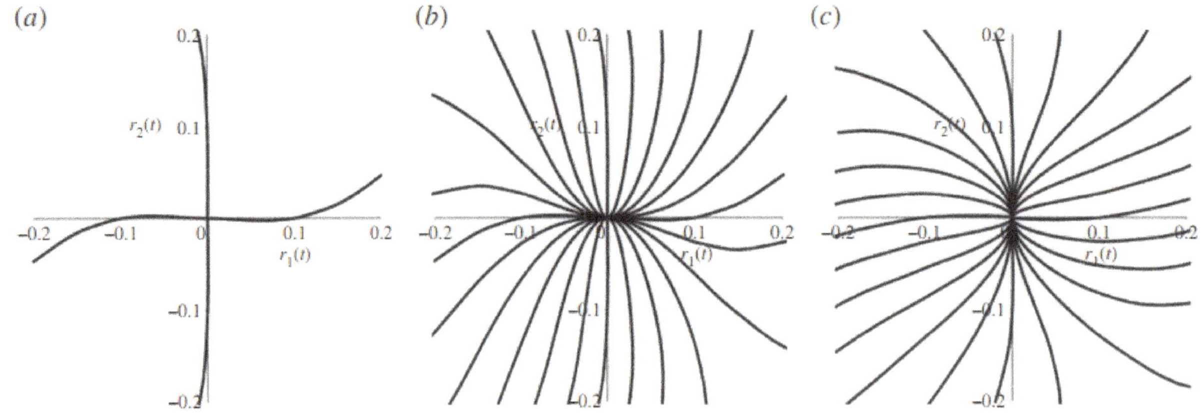

Figure 1. Illustrating the limitation of the NNM concept in the presence of damping. Computations are for the unforced two-degree-of-freedom system whose radial parts are written in polar coordinates as $\dot{r}_1 = -\delta_1 r_1 - r_1^3 + (3 + 0.5\cos\theta_2)r_2^4$, $\dot{r}_2 = -\delta_2 r_2 - r_2^3 + (0.2\sin(\theta_1) - 1)r_1^3$, where (a) $\delta_1=\delta_2=0$, (b) $\delta_1=0.1$, $\delta_2=0.05$ and (c) $\delta_1=0.05$, $\delta_2=0.1$. We show projections, ignoring the angular motion, of the invariant manifolds of the system that are tangent to the linear eigenspaces $r_1=0$ and $r_2=0$.

In the presence of general forcing, there is no guarantee that these invariant manifolds persist. Instead, what one needs is a calculation tool; a way of simplifying the system that respects normal modal analysis in the limit of small amplitude response, but for moderate amplitude also allows for the inclusion of only the necessary nonlinear effects. This is precisely what the theory of normal forms does. A concise description of that theory forms the subject of §2.

The link between normal form theory and NNMs was pointed out by Montaldi et al. [**11**], who showed that NNMs can be constructed directly, as a consequence of the Birkhoff normal form (§2) for undamped systems written in Hamiltonian coordinates. More general normal analysis was carried out on various structural systems in the late 1970s and early 1980s by Holmes and co-workers, see Guckenheimer & Holmes [**4**] and references therein. Jezequel & Lamarque [**7**] showed how forcing and damping can be incorporated in the approximation of normal forms for systems equivalent to (1.1), but written in first-order, i.e. state-space, form. Their method is essentially a hybrid between normal form reduction and the harmonic balance method. Direct harmonic balance and other asymptotic methods, such as multiple scales analysis have been applied by many authors to derive approximations to systems of the form (1.1) [**12**].

Recently, Neild & Wagg [**8**] have extended the work of Jezequel & Lamarque [**7**] to deal directly with second-order systems of the form (1.1); see §1 for the details. One purpose of this paper is to survey their method and to demonstrate its power through applications. Primarily though, we shall show how the second-order method can be recast using the general theory of normal forms, where forcing and damping terms can be treated as specific forms of unfoldings. One does not need to make the additional approximations inherent in the harmonic balance method when deriving the normal form itself. Application of the harmonic balance method to the derived normal form, though, is shown to lead to remarkably powerful predictions for resonant responses, leading to so-called backbone curves in structural analysis, for mode switching in complex multimodal responses and for analytically finding NNMs.

The rest of this paper is outlined as follows. In §2, we describe the historical context of normal form transformations. Section 3 goes on to apply these concepts to engineering vibration problems of the form (1.1). In particular, we show how the Neild & Wagg method can be thought of as the choice of a particular type of normal form. We also explain how to use this method to predict resonant responses, by interpreting the normal form in the context of harmonic balance. Section 4 demonstrates the theory through application to a particular model of a taut cable. The power of the method is illustrated by showing how it can predict secondary bifurcations corresponding to switches in mode shapes. Finally, §5 draws conclusions and points to avenues of future work.

HISTORICAL BACKGROUND TO NORMAL FORM ANALYSIS

Much of our modern-day understanding of dynamical systems can be traced to the work of the French genius Henri Poincaré and the geometric theory he introduced [**13**] to understand systems like the three-body problem in celestial mechanics; see, for example, Barrow-Green [**14**], Diacu & Holmes [**15**], Verhulst [**16**] for historical reviews. His key insight was to work on approximating the system rather than producing series solutions for individual trajectories.

Birkhoff Normal Form

The original context for normal forms was that of conservative systems written in Hamiltonian form. Consider a Hamiltonian system in $\mathbb{R}^{2n}$ close to an elliptic equilibrium point, written in canonical form

$$\begin{pmatrix}\dot{p}\\ \dot{q}\end{pmatrix} = \begin{pmatrix}0 & 1\\ -1 & 0\end{pmatrix}\nabla H(p,q), \quad \text{where } H = \frac{1}{2}\sum_{j=1}^{n}\omega_j(q_j^2 + p_j^2) + H_1(q,p), \tag{2.1}$$

in which $q \in \mathbb{R}^n$ represents the column vector of positions $(q_1,q_2,\ldots,q_n)^T$, p represents the momenta $(p_1,p_2,\ldots,p_n)^T$ and H_1 contains all higher-power terms, starting with cubic. For the purposes of this paper, we shall assume that all vector fields are analytic (infinitely smooth) and we shall suppress any parameter dependence for the time being. Without the nonlinear terms, the system (2.1) is easily solvable and decomposes into quasi-periodic motion with independent frequencies ω_j, j=1,2,…,n. That is, in each degree-of-freedom, the motion comprises periodic orbits of period $2\pi/\omega_j$. We say that the system is completely integrable.

Would not it be nice, at least in a sufficiently small neighbourhood of the zero equilibrium, to find coordinate transformations to successively remove all the nonlinear terms from the equation (cubic and higher terms in H) so that the equation is once again trivially integrable up to any given power? Formally, such reductions can be performed using canonical (i.e. Hamiltonian-structure-preserving) near-identity (i.e. differing only in nonlinear terms) transformations; see, for example, Meyer et al. [17] for details.

But a naive approach can go wrong. Suppose, for example, H_1 contained a term $\gamma_j(p_j^2+q_j^2)^2$ for some j and some given non-zero coefficient γ_j, leading to terms $4\gamma_j p_j(p_j+q_j)^2$ in the $\dot{q}_j$ equation and $4\gamma_j q_j(p_j+q_j)^2$ in the $\dot{p}_j$ equation. Such a term is called resonant because, no matter what near-identity transformation is chosen, a nonlinear component of this term remains. The resonance occurs in terms that contain powers $p_j^{\alpha_j}$ and $q_j^{\beta_j}$, where certain integer relationships existing between the corresponding eigenvalues $\pm i\omega_j$, j=1,…,n. The specific fourth-order expression $\gamma_j(p_j^2+q_j^2)^2$ contains resonant terms where $\alpha_j(i\omega_j)^2+\beta_j(-i\omega_j)^2=0$ with $\alpha_j=\beta_j=2$. Other resonances can arise due to internal resonances where two or more of the frequencies ω_j are rationally related (which leads to the so-called small divisors problem in Hamiltonian systems that had greatly exercised Poincaré and his contemporaries). A good choice of transformed variables are the so-called action-angle coordinates, and the resulting system is often called the Birkhoff normal form [18].

General First-Order Normal Forms

The problem is that many vibration problems (1.1) cannot be written in Hamiltonian form owing to the presence of non-conservative forces due to excitation, damping, friction or gyroscopic effects. Even if a Hamiltonian formation does exist, the mechanical interpretation in terms of velocities and displacements is likely to be lost. Instead, there is a more general theory of normal forms that can be applied to arbitrary-dimensional dynamical systems of the form

$$\dot{x} = Ax + N_x(x), \quad x \in \mathbb{R}^p \qquad 2.2$$

in the neighbourhood of an equilibrium point, where Nx represents nonlinear terms (with the subscript reminding us that we are in the original unscaled coordinates $\boldsymbol{x}$). Here, we shall be interested in the case that the Jacobian is non-hyperbolic (that is, there are eigenvalues on the imaginary axis). Often this theory is implemented in conjunction with centre manifold theory so that only the nonlinear terms associated with the non-hyperbolic degrees of freedom are retained [4,19].

Note that the unforced version of (1.1) can be written in this form by setting the number of states p=2n and

$$A = \begin{bmatrix} 0 & I \\ -M^{-1}K & -M^{-1}D \end{bmatrix} \quad \text{and} \quad N_x = \begin{bmatrix} 0 \\ -M^{-1}N(x) \end{bmatrix}, \qquad 2.3$$

in which $x = (\mathbf{x}, \dot{\mathbf{x}})^T$, as pointed out by Jezequel & Lamarque [7].

Now, let us suppose that a linear transformation $\boldsymbol{x}=\Phi\mathbf{q}$ is applied to (2.2) so that it is written in the form

$$\dot{\mathbf{q}} = \Lambda\mathbf{q} + N_q(\mathbf{q}), \quad q \in \mathbb{C}^p, \qquad 2.4$$

where $Nq(\mathbf{q})=\Phi^{-1}Nx(\Phi\mathbf{q})$ and Λ contains the linearization in the simplest (Jordan canonical) form.

We seek to systematically change variables by a sequence of near-identity transformations to remove non-resonant terms of successively higher powers in the Taylor series expansion of the nonlinear term Nq. That is, we seek a new variable $\mathbf{u} \in \mathbb{C}^p$ with

$$\dot{\mathbf{q}} = \Lambda\mathbf{q} + N_q(\mathbf{q}) \xrightarrow{\mathbf{q}=\mathbf{u}+\mathbf{h}(\mathbf{u})} \dot{\mathbf{u}} = \Lambda\mathbf{u} + N_u(\mathbf{u}), \qquad 2.5$$

where $\mathbf{h}=o(\mathbf{u})$, and Nu contains only resonant terms (in a sense to be made precise shortly). To find the unknown function $\mathbf{h}(\mathbf{u})$, we differentiate the transform, $\dot{\mathbf{q}} = (I + D\mathbf{h}(\mathbf{u}))\dot{\mathbf{u}}$, and substitute for $\dot{\mathbf{q}}$ and $\dot{\mathbf{u}}$ to give

$$\Lambda\mathbf{u} + \Lambda\mathbf{h}(\mathbf{u}) + N_q(\mathbf{u} + \mathbf{h}(\mathbf{u})) = (I + D\mathbf{h}(\mathbf{u}))(\Lambda\mathbf{u} + N_u(\mathbf{u})). \qquad 2.6$$

On its own, (2.6) is a complicated, nonlinear functional equation for $\mathbf{h}(\mathbf{u})$. But we simplify it by expanding $\mathbf{h}(\mathbf{u})$ as a Taylor series and solve, where possible, for the unknown coefficients of $\mathbf{h}$, term by term. Thus, let $\mathbf{h}(\mathbf{u})=\mathbf{h}^{(2)}(\mathbf{u})+\mathbf{h}^{(3)}(\mathbf{u})+\mathbf{h}^{(4)}(\mathbf{u})+\cdots$, where each $\mathbf{h}^{(k)}$ is a sum of homogeneous monomial terms of degree k. That is, the ith component of $\mathbf{h}^{(k)}$ takes the form

$$h_i^{(k)} = \sum_{\mathbf{m}_k} h_{i,\mathbf{m}_k}^{(k)} \prod_{j=1}^{p} u_j^{m_j}, \qquad 2.7$$

where we use a vector notation for multi-indices such that $\mathbf{m}k=(m_1,m_2,m_3,\ldots m_p)$, where m_j is a whole number in the range $0\leq m_j\leq k$ with the additional condition that $\sum_{j=1}^{p} m_j = k$ and the sum in (2.7) is over the set of all such indices. So, for example, for p=2 and k=2, we have $\mathbf{m}k$=(2,0), (1,1) or (0,2) such that

$$h_i^{(2)} = h_{i,2,0}^{(2)}u_1^2 + h_{i,1,1}^{(2)}u_1u_2 + h_{i,0,2}^{(2)}u_2^2.$$

Now we are in a position to simplify (2.6). Suppose that coordinate transformations have taken place to remove all non-resonant terms up to k−1. Then, at $\mathcal{O}(k)$, (2.6) reads

$$\Lambda\mathbf{u} + \Lambda\mathbf{h}^{(k)}(\mathbf{u}) + N_q^{(k)}(\mathbf{u}) = \Lambda\mathbf{u} + D\mathbf{h}^{(k)}(\mathbf{u})\Lambda\mathbf{u} + N_u^{(k)}(\mathbf{u}) + \text{h.o.t.},$$

where again a superscript (k) on N_q and N_u indicates terms that are homogeneous polynomials of degree k and the terms involve (k+1)-powers and higher. This expression can be rearranged to read

$$N_u^{(k)}(\mathbf{u}) - N_q^{(k)}(\mathbf{u}) = \Lambda\mathbf{h}^{(k)}(\mathbf{u}) - D\mathbf{h}^{(k)}(\mathbf{u})\Lambda\mathbf{u} := [\Lambda, \mathbf{h}^{(k)}](\mathbf{u}), \qquad 2.8$$

which is known as the homological equation associated with the linear operator Λ. The operator $[\cdot,\cdot]$ on the right-hand side is known as the Lie bracket between Λ and $\mathbf{h}$ (which is closely related to the so-called Poisson bracket in the case that the system is Hamiltonian).

It might be worthwhile to remind ourselves what we are trying to do. We want to choose the coefficients of all the terms in $\mathbf{h}^{(k)}$ to make $N_u^{(k)}$ as simple as possible. Note that the Lie bracket can be thought of as a linear operator acting on the set of homogeneous polynomials of degree k. If this linear operator is invertible, then the homological equation will have a unique solution for any choice of $N_u^{(k)}$. In particular, we are free to choose $N_u^{(k)} \equiv 0$, and then we will have found a unique solution for the coefficients of $\mathbf{h}^{(k)}$ to make all the kth power terms disappear in the simplified system. Terms that cannot be removed in $N_u^{(k)}$, the resonant terms, arise precisely when the Lie bracket is non-invertible.

Note that the Lie bracket as an operator depends on the matrix Λ, hence this matrix must contain all information necessary to define which terms, in the space of homogeneous polynomials of degree k, are resonant. To spell out which these terms are, it is most convenient to look at the homological equation (2.8) term by term to end up with an indicial form of the equation in terms of the indices m_j.

The indicial equation can be derived in complete generality for any Jordan canonical form of the matrix Λ [6]. But for ease of explanation, we shall treat the simplified case where the eigenvalues of A are semi-simple. That is, the geometric multiplicity of each eigenvalue is equal to its algebraic multiplicity, and hence Λ is diagonalizable so that

$$\Lambda = \mathrm{diag}\{\lambda_1, \lambda_2, \ldots \lambda_p\}, \qquad 2.9$$

where $\lambda_i \in \mathbb{C}$ are the eigenvalues of Λ allowing for multiplicity. Note how the unforced general vibration system (2.3) can be transformed using eigenvectors into (2.4) with Λ of the form (2.9) with $(\lambda_{2i-1},\lambda_{2i})=(+\omega_i,-\omega_i)$, for i=1,…n.

Under the assumption (2.9), the indicial form of the homological equation reads

$$N_{u,i,\mathbf{m}_k}^{(k)}\prod_{j=1}^{p}u_j^{m_j} - N_{q,i,\mathbf{m}_k}^{(k)}\prod_{j=1}^{p}u_j^{m_j} = \lambda_i h_{i,\mathbf{m}_k}^{(k)}\prod_{j=1}^{p}u_j^{m_j} - h_{i,\mathbf{m}_k}^{(k)}\sum_{l=1}^{p}\frac{\partial}{\partial u_l}\left(\prod_{j=1}^{p}u_j^{m_j}\right)\lambda_l u_l$$
$$= h_{i,\mathbf{m}_k}^{(k)}\left(\lambda_i - \sum_{l=1}^{p}m_l\lambda_l\right)\prod_{j=1}^{p}u_j^{m_j}, \qquad 2.10$$

where a subscript $(i,\mathbf{m}_k)$ on N_q and N_u represents the corresponding coefficient of the kth power term with vector index **mk**.

Looking at the form of equation (2.10), we find that we are free to choose $N_{u,i,\mathbf{m}_k}^{(k)}=0$ and

$$h_{i,\mathbf{m}_k}^{(k)} = \frac{N_{q,i,\mathbf{m}_k}^{(k)}}{(\lambda_i - \sum_{l=1}^{p}m_l\lambda_l)}, \qquad 2.11$$

thus removing the $\Pi_j u_j^{m_j}$ term from the simplified equation (the normal form), unless

$$\lambda_i - \sum_{l=1}^{p}m_l\lambda_l = 0. \qquad 2.12$$

The condition (2.12) is thus precisely the condition for $\prod_{j=1}^{p}u_j^{m_j}$ to be a resonant term of the ith equation. For resonant terms, we instead choose $h_{i,\mathbf{m}_k}^{(k)}=0$ and then $N_{u,i,\mathbf{m}_k}^{(k)} = N_{q,i,\mathbf{m}_k}^{(k)}$ and this term remains in the simplified normal form.

The normal form method then carries out the above procedure successively for each value of k, starting with k=2. For each k-value, we step over all possible indices $\{i,\mathbf{m}_k\}$ and choose the coefficient $h_{i,\mathbf{m}_k}^{(k)}$ to satisfy (2.11), unless $\{i,\mathbf{m}_k\}$ satisfies the resonance condition (2.12) in which case we choose $h_{i,\mathbf{m}_k}^{(k)}=0$. Note that at each level k, the near-identity transformations used to remove the non-resonant terms introduces new terms with powers greater than k, so before proceeding to level k+1, the new system must be fully calculated. The value of k>2 at which we stop this process is called the degree of the normal form. Such normal forms are also sometimes called truncated normal forms.

Unfolding, Truncation and Dynamics of the Normal Forms

The notion of a normal form is often applied to situations where there is a local bifurcation of the equilibrium point **q=0** (e.g. [**19**] or [**4**]). In this case, we think of the system (2.4) as depending on parameters $\mu \in \mathbb{C}^r$ (typically the parameters are real, but they need not be) and r corresponds to the codimension of the bifurcation which occurs at μ=0. We then define an extended set of unknowns $\tilde{\mathbf{q}} = (\mathbf{x}, \mu) \in \mathbb{C}^{p+r}$ with the additional r trivial equations $\dot{\mu}=0$. Thus, we get

$$\dot{\tilde{\mathbf{q}}} = \tilde{\Lambda}\tilde{\mathbf{q}} + \tilde{N}_{\tilde{q}}(\tilde{\mathbf{q}}), \quad \text{where } \tilde{\Lambda} \in \mathbb{C}^{(p+r)\times(p+r)} = \begin{pmatrix}\Lambda & 0\\ 0 & 0\end{pmatrix} \qquad 2.13$$

and all μ-dependence appears inside the first p components of the new nonlinear term $\tilde{N}_{\tilde{q}}$. The final r components of $\tilde{N}_{\tilde{q}}$ are precisely zero. The normal form method now proceeds exactly as defined above with p everywhere replaced by p+r.

We can then apply the normal form method as outlined above to the system (2.13). The resulting normal form is called an unfolded normal form. Sometimes judicious choices can be made for certain coefficients of a normal form, or by scaling arguments certain coefficients can be set to unity or zero. Such systems are sometimes called hyper-normal forms [**6**].

The normal form provides a way of simplifying the dynamics of a system near a non-hyperbolic equilibrium, and to classify the dynamics as belonging to one of a few pre-analysed cases according to the sign of certain critical coefficients in the normal form. One of the problems with using normal forms to provide more precise details is that, in general, they represent an asymptotic expansion only. Hence, regardless to which degree k they are truncated, they ignore the beyond-all-orders terms which can lead, for example, to transverse homoclinic and heteroclinic tangles. These will never be captured by the normal form, see, for example, Champneys & Kirk [**20**] and references therein for the case of the normal form of the codimension-two saddle-node/Hopf bifurcation (also known as the Gavrilov–Guckenheimer bifurcation). Occasionally (roughly speaking, when the dynamics of the normal form does not include any structurally unstable homo-/heteroclinic connections), an unfolded, truncated (hyper)-normal form can be shown to contain dynamics that are topologically equivalent to all possible structurally stable dynamics that can occur in a neighbourhood of the bifurcation point. Such a normal form is called a versal unfolding [**21**] or a topological normal form [**19**], an example of which is the normal form of a Hopf bifurcation, truncated after third-degree terms.

Consider the damped Duffing equation

$$\ddot{x} + c\dot{x} + \omega^2 x + \alpha x^3 = 0. \qquad 2.14$$

We can include this in the normal form via an additional equation $\dot{c}=0$. Carrying out the above steps, we first diagonalize the linear part by writing $q_1=\frac{1}{2\omega}(\omega x-\mathrm{i}\dot{x})$, $q_2=\frac{1}{2\omega}(\omega x+\mathrm{i}\dot{x})$ and $q_3=c$. Then, we end up with a system of the form (2.13) for $\mathbf{q}=[q_1,q_2,q_3]^T$ with

$$\Lambda=\begin{bmatrix}\mathrm{i}\omega & 0 & 0\\ 0 & -\mathrm{i}\omega & 0\\ 0 & 0 & 0\end{bmatrix} \quad \text{and} \quad N_q=\begin{bmatrix}\frac{\mathrm{i}\alpha}{2\omega}(q_1+q_2)^3-\frac{q_3}{2}(q_1-q_2)\\ -\frac{\mathrm{i}\alpha}{2\omega}(q_1+q_2)^3+\frac{q_3}{2}(q_1-q_2)\\ 0\end{bmatrix}.$$

Now, from the form of the linearization of the extended system up to $\mathcal{O}(3)$, we see from (2.12) that quadratic terms of the form $q_3 q_i$ are resonant, as are cubic terms of the form $\alpha q_1^2 q_2$ in the $\dot{q}_1$ equation or of the form $\alpha q_1 q_2^2$ in the $\dot{q}_2$ one. We can remove all non-resonant cubic terms with a transformation of the form

$$q_1=\frac{\alpha}{8\omega}(-2u_1^3+3u_1u_2^2+u_2^3) \quad \text{and} \quad q_2=\frac{\alpha}{8\omega}(u_1^3+3u_2u_1^2-2u_2^3).$$

Hence, the third-degree normal form of the damped, unforced Duffing equation can be written as

$$\dot{u}_1=\mathrm{i}\omega u_1+\frac{3\alpha\mathrm{i}}{2\omega}u_1^2u_2-\frac{c}{2}(u_1-u_2) \quad \text{and} \quad \dot{u}_2=-\mathrm{i}\omega u_2-\frac{3\alpha\mathrm{i}}{2\omega}u_1u_2^2+\frac{c}{2}(u_1-u_2), \quad \dot{c}=0, \qquad 2.15$$

cf. Jezequel & Lamarque [7].

Jezequel & Lamarque [7] extended this normal form method in combination with the harmonic balance method to consider forced vibration problems of the form (1.1). This enabled them to predict resonant responses in the case that the forcing frequency Ω is close to one of the natural frequencies of the system. There remains a philosophical weakness that their method does not produce an approximation to the system itself, as Poincaré's envisaged, it simply uses the normal form method to predict resonant (periodic) responses within a harmonic balance framework. This does not allow the complete dynamics close to certain multiple resonances to be explored, which in general will include quasi-periodic and possibly chaotic motion.

NORMAL FORMS FOR SECOND-ORDER MECHANICAL SYSTEMS

One of the weaknesses of using the general normal form method for vibration problems, as in Jezequel & Lamarque [7], is that this does not use the specific structure of the matrix A in (2.3). In view of this, Neild & Wagg [8] produced an extension to the method that can be applied directly to second-order differential equations of the form (1.1), preserving the decomposition into velocity and position variables. We term this the second-order normal forms, referring to the order of the differential equations rather than the level of accuracy achieved. As shown in §3c, this can lead to more accurate predictions of resonant amplitudes. However, their method still relies on the harmonic balance method and so one cannot appeal directly to the mathematical theory of normal forms. The key contribution of this paper then is to show this second-order normal form method can be re-derived in a similar spirit to the analysis of the previous section.

Theoretical Development

Consider a system of the form (1.1). We shall specifically treat the case where $r(t)=e^{\mathrm{i}\Omega t}$ for some fixed forcing frequency Ω, although the method is easily generalizable to quasi-periodic forcing. The first step is to diagonalize the system as much as possible. To do this, we set $\mathbf{x}=\Phi\mathbf{q}$, where Φ is a matrix of eigenvectors of $M^{-1}K$, giving $\ddot{\mathbf{q}}+\Lambda\mathbf{q}+N_q(\mathbf{q},\dot{\mathbf{q}})=-\delta_1\hat{D}\dot{\mathbf{q}}+\delta_2(\hat{P}r(t)+\text{c.c.})$, with: $\Lambda=\operatorname{diag}\{\omega_1^2,\omega_2^2,\dots\omega_n^2\}$, and where $\hat{D}$, $\hat{P}$ and Nq are the transformed damping matrix, forcing vector and nonlinearity, respectively, and ωi are the linear frequencies of the system. Here, we have introduced two small parameters δ_1 and δ_2 which will play the role of unfolding parameters. Furthermore, we shall be interested in the critical situation where there is a resonance between the forcing frequency and one of the natural, undamped frequencies of the system, without loss of generality, ω_1. That is, we assume that $\Omega=\omega_1+\delta_3$, where δ_3 is a third unfolding parameter. Finally, we suppose that there are no other linear resonances when $\boldsymbol{\delta}=(\delta_1,\delta_2,\delta_3)=\mathbf{0}$; that is $\Omega\neq\omega i$ for i=2,…n. Then, we can perform a further linear change of variables to remove the non-resonant forcing terms by writing

$$\ddot{\mathbf{q}}+\Lambda\mathbf{q}+\sum_k N_q^{(k)}(\mathbf{q},\dot{\mathbf{q}})=P_q r(t)\xrightarrow{\mathbf{q}=\mathbf{v}+\mathbf{e}r}\ddot{\mathbf{v}}+\Lambda\mathbf{v}+\sum_k N_v^{(k)}(\mathbf{v},\dot{\mathbf{v}},r)=P_v r(t), \qquad 3.1$$

where $e_1=0$ and $e_k=P_k/(\omega_k^2-\Omega^2)$ so that the forcing term is in the first, resonant equation only. Here, nonlinear terms have been expressed in monomial form.

We shall now use a tilde to represent an extended variable, so that $\tilde{\mathbf{v}}\in\mathbb{C}^{n+4}$ is the vector $(\mathbf{v},r,\boldsymbol{\delta})$ and write the complete system in the form

$$\ddot{\tilde{\mathbf{v}}}+\tilde{\Lambda}\tilde{\mathbf{u}}+\sum_k \tilde{N}_v^{(k)}(\tilde{\mathbf{v}},\dot{\tilde{\mathbf{v}}})=\tilde{P}_v r(t), \qquad 3.2$$

where the first n components of $\ddot{\tilde{\mathbf{v}}}$ are given by the transformed equation in (3.1) and the last four components by $\ddot{r}=-\Omega^2 r$ and $\ddot{\delta}=0$. Here, the first n components of $\tilde{N}_v\in\mathbb{C}^{n+4}$ are equal to the nonlinearity Nv and the last four components are zero. Also,

$$\tilde{\Lambda}\in\mathbb{C}^{n+4}\times\mathbb{C}^{n+4}=\operatorname{diag}\{\omega_1^2,\omega_2^2,\dots\omega_n^2,\Omega^2,0,0,0\}. \qquad 3.3$$

Note that forcing, damping and frequency mismatch are all now thought of as nonlinear terms, because they contain terms that involve an unfolding parameter δi times a state variable u or r or their derivatives, and we think of the δs as state variables too. Observe how the system (3.2) looks remarkably similar to the unfolded first-order normal form (2.13) apart from the second derivatives on the left-hand side. This difference will be important when we try to compute the normal form for such a system.

Now we focus on transforming the equations of motion to the form

$$\ddot{\tilde{\mathbf{v}}}+\tilde{\Lambda}\tilde{\mathbf{u}}+\sum_k \tilde{N}_v^{(k)}(\tilde{\mathbf{v}},\dot{\tilde{\mathbf{v}}})=\tilde{P}_v r\xrightarrow{\tilde{\mathbf{v}}=\tilde{\mathbf{u}}+\sum_k\tilde{\mathbf{h}}^{(k)}(\tilde{\mathbf{u}},\dot{\tilde{\mathbf{u}}},r)}\ddot{\tilde{\mathbf{u}}}+\tilde{\Lambda}\tilde{\mathbf{u}}+\sum_k \tilde{N}_u^{(k)}(\tilde{\mathbf{u}},\dot{\tilde{\mathbf{u}}},r)=\tilde{P}_u r, \qquad 3.4$$

where, as with $\tilde{\mathbf{v}}$, $\tilde{\mathbf{u}}=(\mathbf{u},r,\delta)$. As before, we suppose that all non-resonant nonlinear terms have been removed up to degree k−1 using the transformation to remove terms at $\mathcal{O}(k)$. Eliminating $\tilde{\mathbf{v}}$ leads to the $\mathcal{O}(k)$ equation

$$\tilde{N}_u^{(k)}(\tilde{\mathbf{u}}, \dot{\tilde{\mathbf{u}}}) - \tilde{N}_v^{(k)}(\tilde{\mathbf{u}}, \dot{\tilde{\mathbf{u}}}) = \tilde{\Lambda}\tilde{\mathbf{h}}^{(k)}(\tilde{\mathbf{u}}, \dot{\tilde{\mathbf{u}}}) + \frac{d^2}{dt^2}\tilde{\mathbf{h}}^{(k)}(\tilde{\mathbf{u}}, \dot{\tilde{\mathbf{u}}}). \quad 3.5$$

To proceed, we next complexify the system by writing

$$u_i = \mathrm{u}_i + \bar{\mathrm{u}}_i, \quad \dot{u}_i = \mathrm{i}\omega_i(\mathrm{u}_i - \bar{\mathrm{u}}_i) \quad \text{and} \quad r = \mathrm{r} + \bar{\mathrm{r}}, \quad 3.6$$

where ui and u_i are the ith elements in **u** and $\mathbf{u}$, respectively, and the bar indicates the complex conjugate. This results in 2n+5 arguments for $\tilde{\mathbf{h}}^{(k)}$, $\tilde{N}_u$ and $\tilde{N}_v$ which we temporarily call **z** such that $\mathbf{z} = (\mathbf{u}, \bar{\mathbf{u}}, \mathrm{r}, \bar{\mathrm{r}}, \delta_1, \delta_2, \delta_3)$. As with the first-order case, we wish to write (3.5) in indicial form and solve the corresponding homological equation. To do this, we again assume a particular component for the pre- and post-transformed nonlinear terms and the transform term of

$$\tilde{N}_{v,i,\tilde{\mathbf{m}}_k}^{(k)} \prod_{j=1}^{2n+5} z_j^{\tilde{m}_j}, \quad \tilde{N}_{u,i,\tilde{\mathbf{m}}_k}^{(k)} \prod_{j=1}^{2n+5} z_j^{\tilde{m}_j} \quad \text{and} \quad \tilde{h}_{i,\tilde{\mathbf{m}}_k}^{(k)} \prod_{j=1}^{2n+5} z_j^{\tilde{m}_j}, \quad 3.7$$

respectively. Here, we define the augmented vector $\tilde{\mathbf{m}}_k$ of length 2n+5, corresponding to the 2n+5 arguments, **z**, and may be written using the notation $\tilde{\mathbf{m}} = (m_1, \ldots m_n, m_{-1}, \ldots m_{-n}, m_r, m_{-r}, m_{\delta 1}, m_{\delta 2}, m_{\delta 3})$, with the jth element referred to as $\tilde{m}_j$. Using this indicial representation of the n+4 element vectors $\tilde{N}_v^{(k)}$, $\tilde{N}_u^{(k)}$ and $\tilde{h}^{(k)}$, (3.5) may be written in homological form as

$$\tilde{N}_{u,i,\tilde{\mathbf{m}}_k}^{(k)} \prod_{j=1}^{2n+5} z_j^{\tilde{m}_j} - \tilde{N}_{v,i,\tilde{\mathbf{m}}_k}^{(k)} \prod_{j=1}^{2n+5} z_j^{\tilde{m}_j} = \tilde{\lambda}_i \tilde{h}_{i,\tilde{\mathbf{m}}_k}^{(k)} \prod_{j=1}^{2n+5} z_j^{\tilde{m}_j} + \tilde{h}_{i,\tilde{\mathbf{m}}_k}^{(k)} \sum_{l=1}^{2n+5} \frac{\partial}{\partial z_l}\left(\sum_{l=1}^{2n+5} \frac{\partial}{\partial z_l}\left(\prod_{j=1}^{2n+5} z_j^{\tilde{m}_j}\right)\gamma_l z_l\right)\gamma_l z_l$$
$$= \tilde{h}_{i,\tilde{\mathbf{m}}_k}^{(k)}\left(\lambda_i + \left[\sum_{j=1}^{2n+5} \tilde{m}_j \gamma_j\right]^2\right)\prod_{j=1}^{2n+5} z_j^{\tilde{m}_j}. \quad 3.8$$

Here, $\tilde{\lambda}_i$ is the ith diagonal element in $\tilde{\Lambda}$, see (3.3), and $\dot{z}_i = \gamma_i z_i$ such that γi is given by

$$\gamma_i = \begin{cases} \mathrm{i}\omega_i & \text{for: } 1 \le i \le n, \\ -\mathrm{i}\omega_{i-n} & \text{for: } n+1 \le i \le 2n, \\ \mathrm{i}\Omega & \text{for: } i = 2n+1, \\ -\mathrm{i}\Omega & \text{for: } i = 2n+2, \\ 0 & \text{for: } 2n+3 \le i \le 2n+5. \end{cases} \quad 3.9$$

By considering the form of $\tilde{\Lambda}$ and γi, equation (3.8) can be simplified to give for 1≤i≤n

$$N_{u,i,\tilde{\mathbf{m}}_k}^{(k)} \prod_{j=1}^{2n+5} z_j^{\tilde{m}_j} - N_{v,i,\tilde{\mathbf{m}}_k}^{(k)} \prod_{j=1}^{2n+5} z_j^{\tilde{m}_j} = h_{i,\tilde{\mathbf{m}}_k}^{(k)}\left(\omega_i^2 - \left[(m_r - m_{-r})\Omega + \sum_{j=1}^{n}(m_j - m_{-j})\omega_j\right]^2\right)\prod_{j=1}^{2n+5} z_j^{\tilde{m}_j}, \quad 3.10$$

where, for these values of i, $N_{v,i,\tilde{\mathbf{m}}_k}^{(k)} = \tilde{N}_{v,i,\tilde{\mathbf{m}}_k}^{(k)}$ and we define $N_{u,i,\tilde{\mathbf{m}}_k}^{(k)} = \tilde{N}_{u,i,\tilde{\mathbf{m}}_k}^{(k)}$ and $h_{u,i,\tilde{\mathbf{m}}_k}^{(k)} = \tilde{h}_{u,i,\tilde{\mathbf{m}}_k}^{(k)}$. For n+1≤i≤n+4, $\tilde{N}_{v,i,\tilde{\mathbf{m}}_k}^{(k)} = 0$, so we set $\tilde{N}_{u,i,\tilde{\mathbf{m}}_k}^{(k)} = \tilde{h}_{i,\tilde{\mathbf{m}}_k}^{(k)} = 0$.

The homological equation can be inverted and solved for $h_{i,\tilde{\mathbf{m}}_k}^{(k)}$ provided we avoid the resonant terms for which

$$\omega_i^2 = \left[(m_r - m_{-r})\Omega + \sum_{j=1}^{n}(m_j - m_{-j})\omega_j\right]^2. \quad 3.11$$

Note that m_r or m_{-r}=1 with $m_{\delta,1}$=1 and i=1 automatically leads to resonance because of the fundamental resonance we have assumed between the forcing and natural frequency ω_1; that is, $\omega_1=\Omega$. But there will in general be other resonances; for example, whenever $m_i=m_{-i}+1$ as with the normal form of the Duffing equation. As we shall see in later examples, there can be further internal resonances if $k\omega_i=l\omega_j$ for some integers i, j, k and l.

Practical Implementation

Having shown in principle that a second-order version of the normal form can be derived for structural vibration systems, we now turn to a practical implementation of it, as discussed in detail in Neild & Wagg **[8]** and Neild **[22]**. This implementation avoids the need to explicitly form the extended 2n+5 dimensional state vectors. The key difference in this implementation is that, rather than seeking a specific normal form, we seek to project the system onto a nonlinear equivalent of normal modes, in a harmonic-balance-like approach. As we shall see this approach relies on computation of exactly the same normal-form coefficients as in the previous subsection. So, we specifically suppose that in the complexification step (3.6), we seek a solution of the form $u_i = \mathrm{u}_i + \bar{\mathrm{u}}_i$, with $\mathrm{u}_i = (U_i \mathrm{e}^{-\mathrm{i}\phi_i}/2)\mathrm{e}^{\mathrm{i}\omega_{ri}}$, where ω_{ri} is the response frequency in the ith modal coordinate direction. Here, ωri is, in general, a nonlinear function of the variables u that must be determined as part of the calculation. It corresponds, for an unforced system to the nonlinear natural frequency for moderate amplitude periodic orbits. This concept can be made precise using the Liapunov centre theorem $\omega_{ri}=2\pi/T$, where T is the period for a given amplitude periodic orbit in the NNM invariant manifold (in the sense of Shaw & Pierre **[9]**) tangent to the ωi eigenspaces. In the presence of forcing, the response frequency of a resonant mode is taken to be $\omega_{ri}=\Omega$.

Now the analysis is the same as before using (3.2) and then the equivalent to (3.1), but without using the extended states. The equivalent of the expression (3.5) for the k=1 case reads

$$N_v^{(1)}(\mathbf{u},\dot{\mathbf{u}},r)+\ddot{\mathbf{h}}^{(1)}(\mathbf{u},\dot{\mathbf{u}},r)+\Gamma\mathbf{h}^{(1)}(\mathbf{u},\dot{\mathbf{u}},r)=N_u^{(1)}(\mathbf{u},\dot{\mathbf{u}},r). \quad 3.12$$

Here, we have introduced diagonal matrix Γ with the ith diagonal element being ω_{ri}^2. To first approximation, k=1, this equals Λ as $\omega_{ri}^2\approx\omega_i^2$. Making this substitution has no algebraic effect on the lowest degree approximation to the normal form, but results in more accurate prediction of the harmonics of the responses [23]. As a consequence, the k>1 equations contain $\Gamma-\Lambda$ correction terms (although an acceptable approximation is normally achieved without evaluating these equations).

As before, after complexification of the equations, we now seek to solve (3.12). This can be for each of the terms in the $o(k)$ nonlinearity $N_u^{(k)}$. Now, instead of introducing multi-indices $\{i,\mathbf{m}_k\}$, we simply place all the combinations of states that exist in $N^{(k)}$ into vector $\mathbf{u}^*$ (of length ℓ) and introduce coefficient matrices $\mathbf{h}^*$, $\mathbf{n}^*$ and $\mathbf{n}_u^*$ in $\mathbb{C}^{n\times\ell}$, so that

$$\left.\begin{aligned}&N_v^{(1)}(\mathbf{u},\dot{\mathbf{u}},r)=\mathbf{n}^*\mathbf{u}^*(\mathfrak{u},\bar{\mathfrak{u}},\mathfrak{r},\bar{\mathfrak{r}}),\quad \mathbf{h}^{(1)}(\mathbf{u},\dot{\mathbf{u}},r)=\mathbf{h}^*\mathbf{u}^*(\mathfrak{u},\bar{\mathfrak{u}},\mathfrak{r},\bar{\mathfrak{r}})\\ \text{and}\quad &N_u^{(1)}(\mathbf{u},\dot{\mathbf{u}},r)=\mathbf{n}_u^*\mathbf{u}^*(\mathfrak{u},\bar{\mathfrak{u}},\mathfrak{r},\bar{\mathfrak{r}}).\end{aligned}\right\} \quad 3.13$$

This is an alternative to the arbitrary product notation used earlier, see (3.7) for the indicial form of this. Using the new representation, (3.12) may be rewritten as

$$\mathbf{n}_u^*-\mathbf{n}^*=\tilde{\mathbf{h}}^* \quad \text{where, in indicial form: } \tilde{h}_{i,\ell}^*=-\beta_{i,\ell}^*h_{i,\ell}^*. \quad 3.14$$

Considering the indicial form, it can be shown that

$$\beta_{i,\ell}^*=\left([m_r-m_{-r}]\Omega+\sum_{j=1}^{n}[(m_j-m_{-j})\omega_{rj}]\right)^2-\omega_{ri}^2, \quad 3.15$$

where the form of the ℓth term in $\mathbf{u}^*$ is given by

$$u_\ell^*=\mathfrak{r}^{m_r}\bar{\mathfrak{r}}^{m_{-r}}\prod_{j=1}^{n}(\mathfrak{u}_k^{m_k}\bar{\mathfrak{u}}_k^{m_{-k}}). \quad 3.16$$

Noting the minus sign in (3.14), introduced to maintain consistency of the definition of β^* with previous publications, (3.15) is exactly equivalent to (3.10). The appearance of ω_{ri} rather than ω_i is due to the introduction of Γ.

Considering the indicial version of (3.14), $n_{u,i,\ell}^*+\beta_{i,\ell}^*h_{i,\ell}^*=n_{i,\ell}^*$, if $\beta_{i,\ell}^*$ is zero, it is resonant and must be retained in the post-transformed nonlinear vector, otherwise the nonlinear term can be removed using the transform such that

$$\left.\begin{aligned}&\text{if } \beta_{i,\ell}^*=0:\quad n_{u,i,\ell}^*=n_{i,\ell}^*,\quad h_{i,\ell}^*=0 \qquad \text{(a resonant term)}\\ \text{and}\quad &\text{otherwise:}\quad n_{u,i,\ell}^*=0,\quad h_{i,\ell}^*=\frac{n_{i,\ell}^*}{\beta_{i,\ell}^*} \qquad \text{(a non-resonant term).}\end{aligned}\right\} \quad 3.17$$

Having identified $\mathbf{h}^*$ and $\mathbf{n}_u^*$, and hence $\mathbf{h}^{(1)}$ and $N_u^{(1)}$ using (3.13), we can approximate the transform and transformed dynamic equations to

$$\mathbf{v}=\mathbf{u}+\mathbf{h}^{(1)}(\mathfrak{r},\bar{\mathfrak{r}},r),\quad \ddot{\mathbf{u}}+\Lambda\tilde{\mathbf{u}}+N_u^{(1)}(\mathfrak{r},\bar{\mathfrak{r}},r)=\tilde{P}_u r. \quad 3.18$$

The approximation here is that we have used only the k=1 terms, $\mathbf{h}^{(1)}$ and $N_u^{(1)}$ rather than a summation over all k as was used in (3.4). To refine these expressions, the higher k terms can be included using a similar approach (e.g. [23]), however this is not normally necessary.

In the case of forced systems, when the forcing is close to resonance for all the modes, the forcing transform (3.1) reduces to $\mathbf{q}=\mathbf{v}$ and the nonlinear transform $\mathbf{v}\rightarrow\mathbf{u}$ remains the same as the unforced case, this is discussed further in the example that follows. When non-resonant forcing occurs in any mode then additional terms appear in $\mathbf{u}^*$ due to the forcing transform—for examples which include non-resonant forcing, see Wagg & Neild [24] and Neild & Wagg [8], in which a two-mode forced system is assessed. Wagg & Neild [24] also contains a discussion and comparison of the second-order normal forms alongside the harmonic balance, averaging and multiple scales techniques for the vibration of a forced single mode system. The relationship between the forced response and the NNMs of the unforced system is discussed in Hill et al.[25].

Two examples are now considered, the first is the Duffing oscillator which will be used to compare results from the first-order, state-space, variant of the normal forms with the second-order, oscillator equation, variant. Both responses will be considered using just the k=1 terms. This example will also be used to demonstrate the forced and damped case. The second example considers the NNMs for a two-mode model of a cable under free vibration. Both the resonant and the harmonic responses are calculated to yield the response frequencies and NNM mode-shapes.

Example: the Duffing equation

Let us consider the unforced, undamped Duffing equation

$$\ddot{x}+\omega_n x+\alpha x^3=0. \quad 3.19$$

As this system has one degree-of-freedom $x=\mathbf{x}=\mathbf{q}=\mathbf{v}$. Using the matrix formulation, we keep the cubic term αx^3, so that $N_v^{(3)}(\mathbf{u})=\alpha u_1^3$. Using (3.13), $\mathbf{n}^*$ and $\mathbf{u}^*$ may be defined, then $\boldsymbol{\beta}^*$ can be calculated using (3.15) and from this the transformed nonlinear terms and the transform can be found using (3.17).

This gives

$$\mathbf{n}^*=\begin{bmatrix}\alpha\\3\alpha\\3\alpha\\\alpha\end{bmatrix}^T,\quad \mathbf{u}^*=\begin{bmatrix}u_1^3\\u_1^2\bar{u}_1\\u_1\bar{u}_1^2\\\bar{u}_1^3\end{bmatrix}\to\boldsymbol{\beta}^*=\omega_{r1}^2\begin{bmatrix}8\\0\\0\\8\end{bmatrix}^T\to\mathbf{n}_u^*=\begin{bmatrix}0\\3\alpha\\3\alpha\\0\end{bmatrix}^T,\quad \mathbf{h}^*=\frac{\alpha}{8\omega_{r1}^2}\begin{bmatrix}1\\0\\0\\1\end{bmatrix}^T,$$

Using (3.1), (3.4) and (3.13) allows the transformed equation of motion along with the transform to be written as

$$\ddot{u}_1+\omega_{n1}^2u_1+3\alpha(u_1^2\bar{u}_1+u_1\bar{u}_1^2)=0,\quad x=q_1=u_1+\frac{\alpha}{8\omega_{r1}^2}(u_1^3+\bar{u}_1^3),$$

respectively. As the response for the ith mode will just be at the resonant frequency ω_{ri}, we can write the steady-state solution

$$u_i=u_i+\bar{u}_i=\left(\frac{U_i}{2}e^{-i\phi_i}\right)e^{i\omega_{ri}t}+\left(\frac{U_i}{2}e^{i\phi_i}\right)e^{-i\omega_{ri}t}. \qquad 3.20$$

Hence,

$$x=U_1\cos(\omega_{r1}t)+\frac{\alpha}{32\omega_{r1}^2}U_1^3\cos(3\omega_{r1}t),\quad \text{with: }\omega_{r1}^2=\omega_1^2+\frac{3\alpha}{4}U_1^2, \qquad 3.21$$

where we have defined t such that ϕ_1=0. Note that as the harmonic terms were removed using the near-identify transform, U_1 conveniently represents the amplitude of the resonant response.

Considering the case where the system is damped and forced at a frequency close to resonance, such that $\ddot{x}+2\zeta\omega_n\dot{x}+\omega_n x+\alpha x^3=R\cos(\Omega t)$, the forcing transform is **q**=**u**. As the forcing is near-resonant, the response frequency is set as $\omega r_1=\Omega$ and the near-identity transform is the same as that derived above, it is unaffected by the introduction of the forcing or the damping (which can be thought of as a resonant term). The resulting resonant dynamics and transform are

$$\ddot{u}_1+2\zeta\omega_{n1}u_1+\omega_{n1}^2u_1+3\alpha(u_1^2\bar{u}_1+u_1\bar{u}_1^2)=R\cos(\Omega t)\quad\text{and}\quad x=q_1=u_1+\frac{\alpha}{8\omega_{r1}^2}(u_1^3+\bar{u}_1^3),$$

respectively. Using the steady-state solution (3.20) and balancing the complex exponential terms allows the amplitude relationship $[(\omega_n^2-\Omega^2)U_1+(3\alpha/4)U_1^3]^2+[2\zeta\omega_n\Omega U_1]^2=R^2$ and phase relationship $\tan(\phi_1)=2\zeta\omega_n\Omega/(\omega_n^2-\Omega^2+(3\alpha/4)U_1^2)$ to be found. Once U_1 and ϕ_1 have been found for a given forcing, then the response, including harmonics, can be calculated using the transform equation and (3.20).

The unforced solution can be compared with the solution derived using the first-order version of the normal form technique in which the state-space, rather than oscillator, form of the equations are used. Here, the first step is to rewrite the equation of motion in state-space form using $x=(x,\dot{x})$ and then to apply a linear transform giving

$$\dot{x}=\begin{bmatrix}0&1\\-\omega^2&0\end{bmatrix}x+\begin{bmatrix}0\\-\alpha x_1^3\end{bmatrix}\to\dot{\mathbf{q}}=\begin{bmatrix}i\omega&0\\0&-i\omega\end{bmatrix}\mathbf{q}+\frac{i\alpha}{2\omega}\begin{bmatrix}(q_1+q_2)^3\\-(q_1+q_2)^3\end{bmatrix},$$

where, as the system is written in first-order form, we have $x=\begin{bmatrix}1&1\\i\omega_n&-i\omega_n\end{bmatrix}\mathbf{q}.$

Applying the normal form transformation in a similar way to the second-order matrix formulation, now using $\mathbf{u}=[u\ \bar{u}]^T$ gives

$$x=U_1\left(1-\frac{3\alpha U_1^2}{16\omega_{n1}^2}\right)\cos(\omega_{r1}t)+\frac{\alpha}{32\omega_{n1}^2}U_1^3\cos(3\omega_{r1}t),\quad\text{with: }\omega_{r1}=\omega_{n1}+\frac{3\alpha}{8\omega_{n1}}U_1^2 \qquad 3.22$$

[**7,12**]. Note that here U_1 does not fully represent the amplitude of the resonant response. This is because, when identifying the transform, only the $e^{i\omega r_1 t}$ terms are kept in the first equation of motion in **U** with $e^{-i\omega r_1 t}$ terms being represented in the transform and vice versa for the second equation of motion [**8**].

The response (or backbone) curves for both the second- and the first-order variants of the normal forms, (3.21) and (3.22), respectively, are shown in **figure 2**a. For comparison, the solution derived using numerical continuation, using AUTO [**26**], in which no assumptions are made regarding the smallness of terms, is also shown. It can be seen that at low amplitudes, corresponding to weaker nonlinear terms, there is good agreement in the prediction of the response frequency, however as the amplitude increases the higher power terms become significant, firstly for the first-order normal form approximation at about 0.6 and then also for the second-order version at about 1.8. **Figure 2**b compares the second-order normal form and the AUTO predictions of a forced response, again it can be seen that good agreement is achieved.

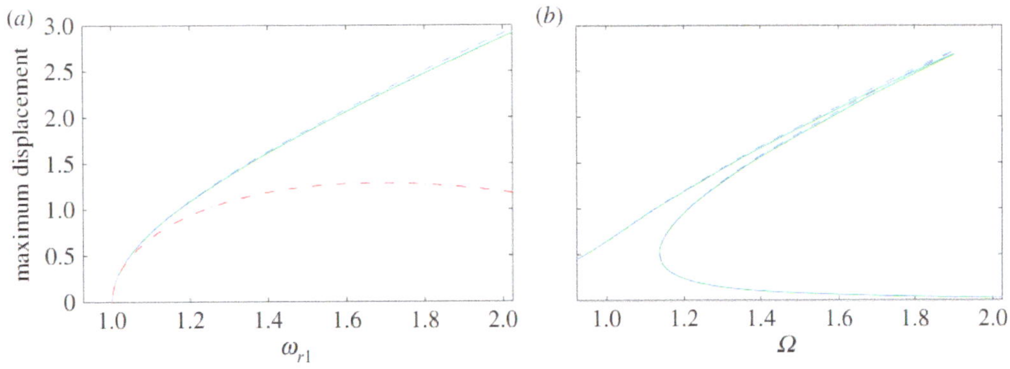

Figure 2. Duffing oscillator example with ω=1 and α=0.5: (a) backbone curves and (b) forced response curves when ζ=0.01 and R=0.1. The solid line shows the exact computation using AUTO; the dashed line shows the result of the second-order normal form calculation; and the dashed-dotted line the result of the first-order normal form.

APPLICATION TO NONLINEAR NORMAL MODES

We now consider how the application of normal forms to unforced, undamped second-order differential equations can be used to calculate the NNMs of a system. Second-order normal forms can be used to derive the backbone curves of a system via the resonant responses (e.g. Wagg & Neild [24] or Hill et al. [27]). Here, we extend this to finding the NNM mode shapes, which requires both the resonant response and also the harmonics captured in **h**. While the technique is general, to facilitate the discussion we apply it to a well-known nonlinear system—the dynamics of a taut cable. Specifically, we will consider the interactions between the first out-of- and in-plane modes (we use mode to indicate the modes of the linearized system). In doing this, to calculate the associated NNMs, we will derive the harmonics excited in all modes due to resonant responses in the first out-of- and in-plane modes.

To examine the dynamics of a taut cable, we use the modal equations of motion derived by Warnitchai et al. [28] and discussed in detail in Wagg & Neild [24], which accounts for gravitational sag and tension effects. The motion of the cable out-of- and in-plane (where in-plane relates to the plane in which gravitational sag occurs) are represented as

$$v(x,t)=\sum_n \phi_n(x)y_n(t) \quad \text{and} \quad w(x,t)=w_s(x)+\sum_n \psi_n(x)z_n(t), \qquad 4.1$$

where x is the distance from one of the supports and the other support is positioned at x=ℓ. Here, $\phi_n(x)$ and $\psi_n(x)$ are the nth out-of- and in-plane linear mode shapes, respectively, $y_n(t)$ and $z_n(t)$ represent the modal contributions for the nth modes and $w_s(x)$ captures the static sag in the cable (noting that v and w are defined as zero on the chord line between the supports). The dynamics for the nth out-of- and in-plane modes may be expressed as

$$\ddot{y}_n+\omega_{yn}^2 y_n+\sum_k \frac{\nu_{nk}}{m} y_n(y_k^2+z_k^2)+\sum_k 2\frac{\beta_{nk}}{m} y_n z_k=0$$

and

$$\ddot{z}_n+\omega_{zn}^2 z_n+\sum_k \frac{\nu_{nk}}{m} z_n(y_k^2+z_k^2)+\sum_k 2\frac{\beta_{nk}}{m} z_n z_k+\sum_k \frac{\beta_{kn}}{m}(y_k^2+z_k^2)=0, \qquad 4.2$$

respectively. Here, modal damping and external forcing terms have been removed from the original derivation in line with investigating the backbone curves of the system. Parameters m, β_{ij} and ν_{ij} are given in Warnitchai et al. [28] and Gonzalez-Buelga et al. [29], but importantly β_{ij} is zero for even j. The natural frequencies of the out-of-plane modes are proportional to the mode number n, and for even n, natural frequencies for the in-plane modes match those of the out-of-plane ones. For odd n, the in-plane natural frequencies are slightly higher than the out-of-plane ones, again see [28,29]. The in-plane axis is defined as positive down, and so gravitational sag is positive.

Gonzalez-Buelga et al. [29] considered these equations in terms of internal resonance showing that 1 : 1 resonance occurs between the second out-of- and in-plane modes and Macdonald et al. [30] generalized this for the nth out-of- and in-plane modes. In addition, they both show that 2 : 1 resonance can occur, but only for the case where the cable is inclined. Hill et al. [27] identified the backbone curves when the system is reduced to the first out-of- and in-plane modes. Here, we build on this work by deriving algebraic expressions for the NNMs associated with these backbone curves, which requires not only the resonant responses of the two modes but also the harmonic response in all modes. To do this, we first calculate the backbone curves along with expressions that capture the harmonics contained in the response. These are then used to find the NNMs in terms of just the two first modes and finally the additional modal contribution due to harmonics in other modes is added to give the full NNM expressions.

Resonant Equation of Motion and Harmonic Response

Letting $\mathbf{q}=(y_1\ z_1)^T$ the modal equation of motion for the reduced two-mode model may be written in the form $\ddot{\mathbf{q}}+\Lambda\mathbf{q}+N_q(\mathbf{q})=0$, an unforced version of the left-hand expression in (3.1), where

$$\mathbf{q}=\begin{pmatrix} q_1 \\ q_2 \end{pmatrix}, \quad \Lambda=\begin{bmatrix} cc\omega_{y1}^2 & 0 \\ 0 & \omega_{z1}^2 \end{bmatrix} \quad \text{and} \quad \mathbf{N}_q^{(1)}=\begin{pmatrix} c\left(\frac{\nu_{11}}{m}\right)(q_1^3+q_1q_2^2)+2\left(\frac{\beta_{11}}{m}\right)(q_1q_2) \\ \left(\frac{\nu_{11}}{m}\right)(q_2^3+q_1^2q_2)+\left(\frac{\beta_{11}}{m}\right)(q_1^2+3q_2^2) \end{pmatrix}.$$

Here, taking $N_q(\mathbf{q})=\sum_k N_q^{(k)}(\mathbf{q})$, all the nonlinear terms have been placed in $N_q^{(1)}$. As there is no forcing, **v**=**q** and so $N_v^{(1)}(\mathbf{v})=N_q^{(1)}(\mathbf{q})$. To calculate the nonlinear transform, see (3.4), $N_v^{(1)}(\mathbf{u})$ is considered. Rewriting this nonlinear vector in terms of $\mathbf{u}$, where, for the kth coordinate $u_k=\mathrm{u}_k+\bar{\mathrm{u}}_k$, gives

$$N_v^{(1)}(\mathbf{u}) = \frac{\nu_{11}}{m}\begin{pmatrix}(u_1+\bar{u}_1)^3+(u_1+\bar{u}_1)(u_2+\bar{u}_2)^2\\ (u_2+\bar{u}_2)^3+(u_1+\bar{u}_1)^2(u_2+\bar{u}_2)\end{pmatrix}+\frac{\beta_{11}}{m}\begin{pmatrix}2(u_1+\bar{u}_1)(u_2+\bar{u}_2)\\ (u_1+\bar{u}_1)^2+3(u_2+\bar{u}_2)^2\end{pmatrix}.$$

This can now be expressed in matrix form $\mathbf{n}_v^*\mathbf{u}^*(\mathbf{u},\bar{\mathbf{u}},\mathfrak{r},\bar{\mathfrak{r}})$, see (3.13) along with (3.6), and the matrix $\boldsymbol{\beta}^*$ can be derived using (3.15), where it can be assumed that, due to the closeness of the natural frequencies, we can write the response frequencies as $\omega r_1=\omega r_2$. Using (3.17), the nonlinear vector in the transformed equation of motion and the transformation vector can be written as

$$N_u^{(1)}=\frac{\nu_{11}}{m}\begin{pmatrix}3u_1\bar{u}_1u_1+2u_2\bar{u}_2u_1+u_1\bar{u}_2^2+\bar{u}_1u_2^2\\ 3u_2\bar{u}_2u_2+2u_1\bar{u}_1u_2+u_1^2\bar{u}_2+\bar{u}_1^2u_2\end{pmatrix} \qquad 4.3$$

And

$$\mathbf{h}^{(1)}=\frac{\nu_{11}}{8m\omega_{r1}^2}\begin{pmatrix}u_1^3+\bar{u}_1^3+u_1u_2^2+\bar{u}_1\bar{u}_2^2\\ u_2^3+\bar{u}_2^3+u_1^2u_2+\bar{u}_1^2\bar{u}_2\end{pmatrix}+\frac{\beta_{11}}{3m\omega_{r1}^2}\begin{pmatrix}2(u_1u_2+\bar{u}_1\bar{u}_2)-6(u_1\bar{u}_2+\bar{u}_1u_2)\\ u_1^2+\bar{u}_1^2+3(u_2^2+\bar{u}_2^2)-6u_1\bar{u}_1-18u_2\bar{u}_2\end{pmatrix}, \qquad 4.4$$

respectively. The modal response of the system can now be written as $\mathbf{q}=\mathbf{v}=\mathbf{u}+\mathbf{h}^{(1)}$, where $\mathbf{h}^{(1)}$captures the harmonic content of the response.

Using the steady-state solution (3.20), the transformed equations of motion, $\ddot{\mathbf{u}}+\Lambda\mathbf{u}+N_u^{(1)}=0$, can be written as

$$\left\{\omega_{y1}^2-\omega_{r1}^2+\frac{\nu_{11}}{4m}[3U_1^2+(2+p)U_2^2]\right\}U_1=0 \qquad 4.5$$

And

$$\left\{\omega_{z1}^2-\omega_{r1}^2+\frac{\nu_{11}}{4m}[(2+p)U_1^2+3U_2^2]\right\}U_2=0, \qquad 4.6$$

where $p=\cos(2(\phi_1-\phi_2))$ with the condition $\sin(2(\phi_1-\phi_2))=0$ to ensure (4.3) is real. This allows possible solutions p=±1. It can be shown that only p=−1 gives physically meaningful solutions (as $\omega z_1\neq\omega y_1$) which means that when both linear modes are present, they are ±90° out-of-phase, see Hill et al. [27] for a more detailed discussion of this.

Backbone Curves and Modal Response

Taking the physically meaningful p=−1 case, there are two semi-trivial solutions for (4.5) and (4.6) in which only one of the two linear modes is resonant

$$S1:\quad \omega_{r1}^2=\omega_{y1}^2+\frac{3\nu_{11}}{4m}U_1^2,\quad U_2=0 \qquad 4.7$$

And

$$S2:\quad \omega_{r1}^2=\omega_{z1}^2+\frac{3\nu_{11}}{4m}U_2^2,\quad U_1=0, \qquad 4.8$$

and two further solutions exist in which both linear modes are resonant

$$S3^{\pm}:\quad \omega_{r1}^2=\omega_{y1}^2+\frac{\nu_{11}}{4m}(3U_1^2+U_2^2),\quad U_2^2=U_1^2-\frac{2m}{\nu_{11}}(\omega_{z1}^2-\omega_{y1}^2). \qquad 4.9$$

As $\omega z_1>\omega y_1$, it can be seen that valid solutions for S3 only exist if $U_1\geq 2m(\omega_{z1}^2-\omega_{y1}^2)/\nu_{11}$. The point at which $U_1=2m(\omega_{z1}^2-\omega_{y1}^2)/\nu_{11}$ results in $U_2=0$ and lies on backbone curve S1, see (4.8). Hence these S3 solutions (there are two solutions relating to the relative phase of the two linear modes being ±90°, solutions S3±) are branches from S1 following a bifurcation at $U_2=0$. This bifurcation from the S1 solution is shown in **figure 3** using the parameters defined in Gonzalez-Buelga et al. [29]. Note that the S3± backbone curves lie on top of each other in this projection.

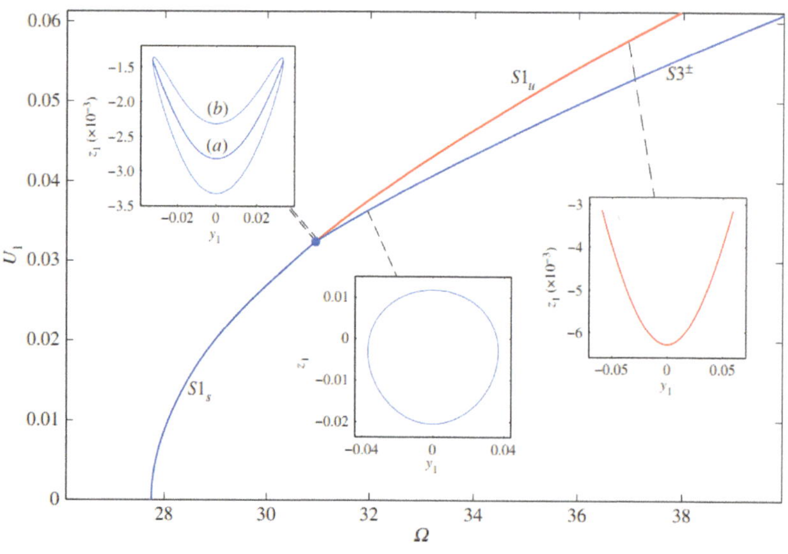

Figure 3. Backbone curves S1 and S3± for a cable with inset panels showing the motion in projection $q_1=y_1$versus $q_2=z_1$. The inset panel for the response near the bifurcation contains two responses; the response line (a) corresponding to a point on S1s just below the bifurcation and the response loop (b) corresponding to a point on S3+ just above the bifurcation. Subscripts s and u indicate that backbone curve S1 is stable below and unstable above the bifurcation.

Using the transform equation, $\mathbf{q}=\mathbf{v}=\mathbf{u}+\mathbf{h}^{(1)}$, and the steady-state solution (3.20) gives the response of the modal coordinates

$$\left.\begin{aligned} y_1 &= U_1\cos(\omega_{r1}t)+\frac{\nu_{11}}{32m\omega_{r1}^2}U_1(U_1^2-U_2^2)\cos(3\omega_{r1}t)\\ &\quad+\frac{\beta_{11}}{3m\omega_{r1}^2}U_1U_2\sin(2\omega_{r1}t)\\ \text{and}\qquad z_1 &= U_2\sin(\omega_{r1}t)+\frac{\nu_{11}}{32m\omega_{r1}^2}U_2(U_1^2-U_2^2)\sin(3\omega_{r1}t)\\ &\quad+\frac{\beta_{11}}{6m\omega_{r1}^2}(U_1^2-3U_2^2)\cos(2\omega_{r1}t)-\frac{\beta_{11}}{2m\omega_{r1}^2}(U_1^2+3U_2^2), \end{aligned}\right\} \tag{4.10}$$

when the phase difference is $\phi_2-\phi_1=90°$ and ϕ_1 has been set to zero.

The S2 solution consists of purely in-plane resonant motion as $U_1=0$. In this case, the response given by (4.10) reduces to purely in-plane motion. Harmonics exist in this motion including a static displacement which, as it is negative, reduces the apparent sag of the cable during resonance. For the S1 solution, which is resonant purely in the out-of-plane direction as $U_2=0$, there are also harmonic components in the out-of-plane response. In addition, there is an amplitude-dependent in-plane non-resonant response. For the S1 solution example of the motion of the mid-span of the cable (where the mode shapes are unity) in the y_1 versus z_1 plane are shown as in **figure 3** as inset plots. Note that in contrast to Hill et al. [**27**], these inset panels include the harmonic components as well as the resonant ones. Examples of the more complex motion of the S3 solution, in which there is 90° out-of-phase resonant motion in both planes, are also shown in **figure 3**. Here, it can be seen that away from the bifurcation (indicated with a dot), the cable is whirling in a near elliptical shape—it would be exactly circular if $\omega y_1=\omega z_1$, albeit offset from the origin in the vertical direction.

These inset panels show the motion at the mid-span assuming that the harmonics of the cable response are limited to the first out-of- and in-plane modes. Once additional harmonics from other modes are included in the response, the full NNM response can be derived. This is done in §4c for the NNMs that contain resonant responses solely in either or in both the first out-of- or in-plane modes.

Additional Modal Contributions

In the previous analysis, only the first out-of- and in-plane modes have been considered. It has been shown that in-plane motion is present throughout (**figure 3**), even for the S1 solution, where there is no resonant in-plane response, but there are twice- and zero-frequency components. By inspection of (4.2), it can be seen that this is due to the term $\beta_{11}y_1^2/m$ that arises from the $\sum_k(\beta_{kn}/m)(y_k^2+z_k^2)$ term which is due to the variation in tension in the cable during oscillations. For the S1 solution, where the first in-plane mode is non-resonant, this term may loosely be thought of as a non-resonant 'forcing' of the in-plane mode giving rise to twice- and zero-frequency response components—being non-resonant components they appear in $\mathbf{h}^{(1)}$, (4.4), as $\beta_{11}(u_1^2+\bar{u}_1^2-6u_1\bar{u}_1)/(3m\omega_{r1}^2)$.

We now assess whether non-resonant response terms can occur in higher modes of the cable due to the resonant response of the first out-of- and in-plane modes. For the nth out-of-plane mode, all the terms in the equation of motion (4.2) are multiples of yn, hence there can only be a response if the mode is resonant. For the in-plane modes, a response is possible due to the $\beta_{n1}(y_1^2+z_1^2)/m$ terms from the summation $\sum_k(\beta_{kn}/m)(y_k^2+z_k^2)$. These terms can result in a non-zero $\mathbf{h}^{(1)}$, even with zero resonant response in the nth in-plane mode, that is captured by

$$\left.\begin{aligned}\text{non-resonant } n\text{th}\\ \text{in-plane mode}\end{aligned}\right\}\quad \mathbf{h}^{(1)}_{\text{non-res}}=\frac{\beta_{n1}}{3m\omega_{r1}^2}[u_1^2+\bar{u}_1^2+u_2^2+\bar{u}_2^2-6(u_1\bar{u}_1+u_2\bar{u}_2)], \tag{4.11}$$

following the same approach as was used to calculate (4.4) and then setting $u_n=0$. Applying the resonant response solution for the first in- and out-of-plane modes using (3.20), noting that $\beta_1 n=\beta_{11}/n$ for odd n and zero for even n, and applying the phase condition that $\phi_2-\phi_1=90°$, and $\phi_1=0$, gives the response for the nth in-plane mode

$$z_n=\begin{cases}-\dfrac{\beta_{11}}{2n(4-n^2)m\omega_{r1}^2}U_1^2\cos(2\omega_{r1}t)-\dfrac{\beta_{11}}{2n^3m\omega_{r1}^2}U_1^2 & \text{odd } n \text{ with } n>1\\ 0 & \text{even } n\end{cases}. \tag{4.12}$$

Here, we have noted that, when n>1, $z_n=\mathbf{h}^{(1)}_{\text{non-res}}$ if only the resonant response of the first-in- and out-of-plane modes are considered.

Nonlinear Normal Modes

These modal contributions can be added to those from the first-in- and out-of-plane modes, (4.10). Using (4.1), along with the mode shapes, the NNMs can be identified in terms of the lateral displacement v(x,y) and the vertical displacement w(x,y). By approximating the in-plane modes as sinusoidal, an acceptable approximation if the cable is taut [**24**], the response may be written in terms of amplitude of the resonant responses U_1 and U_2 as

$$v(x,t)=\sin\left(\frac{\pi x}{\ell}\right)\left[U_1\cos(\omega_{r1}t)+\frac{\nu_{11}}{32m\omega_{r1}^2}U_1(U_1^2-U_2^2)\cos(3\omega_{r1}t)+\frac{\beta_{11}}{3m\omega_{r1}^2}U_1U_2\sin(2\omega_{r1}t)\right], \quad 4.13$$

in the out-of-plane direction and

$$w(x,t)=w_s(x)+\sin\left(\frac{\pi x}{\ell}\right)$$
$$\times\left[U_2\sin(\omega_{r1}t)+\frac{\nu_{11}}{32m\omega_{r1}^2}U_2(U_1^2-U_2^2)\sin(3\omega_{r1}t)-\frac{\beta_{11}}{3m\omega_{r1}^2}U_2^2\cos(2\omega_{r1}t)-\frac{\beta_{11}}{m\omega_{r1}^2}U_2^2\right]$$
$$-\frac{\beta_{11}}{2m\omega_{r1}^2}(U_1^2-U_2^2)\cos(2\omega_{r1}t)\sum_{\bar{n}=1}^{\infty}\frac{1}{(2\bar{n}-1)(4-(2\bar{n}-1)^2)}\sin\left((2\bar{n}-1)\frac{\pi x}{\ell}\right)$$
$$-\frac{\beta_{11}}{2m\omega_{r1}^2}(U_1^2+U_2^2)\sum_{\bar{n}=1}^{\infty}\frac{1}{(2\bar{n}-1)^3}\sin\left((2\bar{n}-1)\frac{\pi x}{\ell}\right), \quad 4.14$$

in the in-plane direction. These expressions along with the response frequency and resonant amplitude relationships, (4.7), (4.8) and (4.9), define the NNMs.

In the out-of-plane direction, the NNM mode shape is straightforward, a half sine wave matching the linear mode shape; however, the response contains not only the response frequency but also harmonics. In the in-plane direction, the NNM is more complex with modal components from all the odd in-plane modes. This expression can be significantly simplified using the Fourier series expansions

$$\sum_{\bar{n}=1}^{\infty}\frac{1}{(2\bar{n}-1)(4-(2\bar{n}-1)^2)}\sin\left((2\bar{n}-1)\frac{\pi x}{\ell}\right)=\frac{\pi}{8}\sin^2\left(\frac{\pi x}{\ell}\right),$$

$$\sum_{\bar{n}=1}^{\infty}\frac{1}{(2\bar{n}-1)^3}\sin\left((2\bar{n}-1)\frac{\pi x}{\ell}\right)=\frac{\pi^3}{8\ell^2}(x\ell-x^2),$$

such that

$$w(x,t)=w_s(x)+\sin\left(\frac{\pi x}{\ell}\right)$$
$$\times\left[U_2\sin(\omega_{r1}t)+\frac{\nu_{11}}{32m\omega_{r1}^2}U_2(U_1^2-U_2^2)\sin(3\omega_{r1}t)-\frac{\beta_{11}}{3m\omega_{r1}^2}U_2^2\cos(2\omega_{r1}t)-\frac{\beta_{11}}{m\omega_{r1}^2}U_2^2\right]$$
$$-\frac{\beta_{11}\pi}{16m\omega_{r1}^2}(U_1^2-U_2^2)\cos(2\omega_{r1}t)\sin^2\left(\frac{\pi x}{\ell}\right)-\frac{\beta_{11}\pi^3}{16m\omega_{r1}^2}(U_1^2+U_2^2)\left[\frac{x}{\ell}-\left(\frac{x}{\ell}\right)^2\right]. \quad 4.15$$

It is, perhaps, interesting to note that the last term, which is not a function of time, takes the same shape along the cable as the static sag ($w_s(x)\propto(\ell x-x^2)$ from **[24]**). It can therefore be seen as reducing the sag effect due to the increased average dynamic tension in the cable as it oscillates. The in-plane NNM response, at the level of accuracy of the normal form transformation, therefore consists of the linear mode shape oscillating at three frequencies along with a dynamically corrected static sag and a further term in the shape of the square of the linear mode shape. Due to this final term, the shape of the NNM in the in-plane direction is dependent on the amplitudes of the resonant responses in the first-in- and out-of-plane modes and varies over each oscillation.

To examine these NNMs, first consider the S1 solution, in which $U_2=0$. **Figure 4** shows the trajectory of the mid-span deflections $v(\ell/2,t)$ and $w(\ell/2,t)$ for a range of values of U_1 up to the bifurcation amplitude in the y versus −z (to account for the fact that z is defined positive downwards) plane using (4.13) and (4.15). For each trajectory, the resonant frequency is different, as shown in **figure 3**. The dashed line represents the maxima of the trajectories and corresponds to the initial deflection points from which, provided the initial shape is imposed, the system can be released into the pure S1 NNM. This initial shape is a half sine wave out-of-plane regardless of the amplitude of U_1 but is a more complex weakly amplitude-dependent shape in-plane. Note that the in-plane deflection plotted here includes static sag—the value of which corresponds to the minimum of the dashed line where the initial out-of-plane deflection is zero. Any deflection along the NNM from this point results in the cable oscillating above the static sag position. For the S2 solution, as $U_1=0$, the motion is confined to the vertical plane such that both the trajectories and the initial deflection points would lie on the vertical axis in the figure.

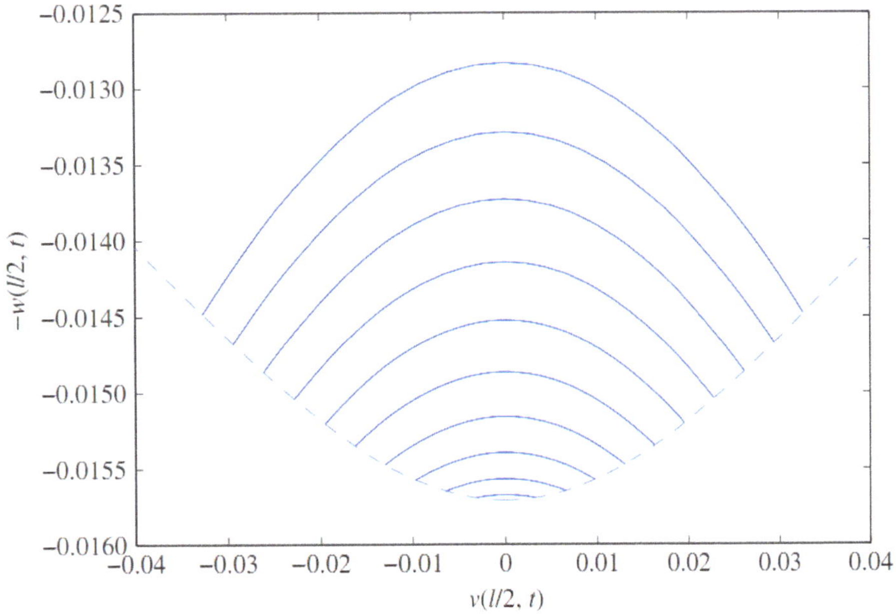

Figure 4. Trajectory of mid-span deflections for a range of U_1 values on the S1 solution up to the bifurcation amplitude (solid lines) and the locus of points from which a stationary cable can be released onto a pure S1 solution provided the deflected shape along the length of the cable is appropriate (dashed line).

Considering the $S3^{\pm}$ solution, **figure 5** shows the full cable response in space when U_1 is (a) at the bifurcation point and (b) at 0.1% greater than and (c) 1% greater than the amplitude at which the bifurcation occurs. Here, the vertical displacements are plotted relative to the static sag and hence the sag equilibrium is the straight (red) line between the supports. The cable location is shown at quarter time-period spacing as thick (green) curves and the trace of the cable over all time at eighth-span locations are shown as thin (magenta) loops. The initial transition into whirling, panel (b), is a solution that closely envelops the pre-whirling response, panel (a), as is suggested by the responses (a) and (b) in **figure 3**. **Figure 5** a,b also shows that the motion over all time occurs above the sag location. This is perhaps contrary to the expectation that whirling is initiated when the vertical momentum becomes sufficient for the cable to wrap around the zero deflection point rather like a pendulum going 'over the top' as it undergoes a transition from swinging to rotating motion. Instead, the initial whirling motion, panel (b), envelops the pre-whirling solution, and then rapidly becomes a near-elliptical orbit that encircles the sag equilibria as can be seen in panel (c).

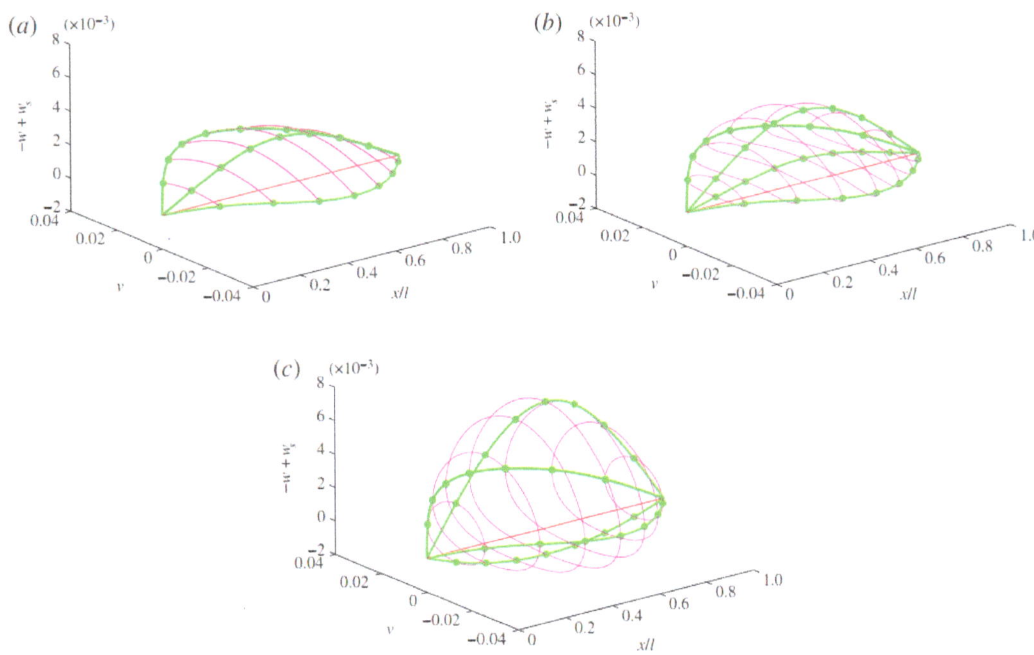

Figure 5. Trajectory of the cable for the S3 solution (a) at the bifurcation point and for a U_1 amplitude (b) 0.1% and (c) 1% greater than that at which the bifurcation occurs. Vertical deflections are plotted relative to the static sag shape. Green lines show cable locations at quarter time period points, magenta lines shown the eighth-span traces of the cable for all time, the red line shown the sag shape and the blue line the chord between the two supports.

CONCLUDING REMARKS

The purpose of this paper has been to review the method of normal forms in the context of second-order mechanical vibration problems, to demonstrate its power, and to show how recent developments fit with the historical account of the subject. To this end, the main novel contribution is the re-derivation of the second-order normal form technique using the general theory (§3). This shows that, provided forcing and damping are considered to be weak, then forced-damped systems can be treated in a mathematically consistent way. Moreover, this treatment provides insight into the region of validity of the method. Frequency detuning, forcing amplitude and damping coefficients can be treated as unfolding terms of the same order of magnitude as the nonlinear terms. While this assumption of weak forcing and damping may appear restrictive, for smooth nonlinear systems close to resonance, one is usually dealing with internal forces generated from structural stiffnesses that are much larger than excitation and damping forces.

Other novelties of this work include a re-derivation of the homological equation (3.11) without reference to harmonic balance or an artificial book-keeping parameter ε as was used in Neild & Wagg [**22**]. Also, we have applied the method to the computation of the NNMs of a multiple-degree-of-freedom model of a cable close to the resonance between in-phase and out-of-phase fundamental modes. This has enabled a general expression for the responses, (4.13) and (4.15), including contributions from arbitrarily many higher order modes to be derived. These fully parametrized expressions have enabled us to draw novel insight into the mechanism of transition from planar to whirling solutions in the problem.

As discussed elsewhere, the second-order normal form method keeps the transform physically relevant. It could be argued that the concept of the normal form is the natural generalization of linear modal analysis for finite degree of freedom systems. Linear modal analysis essentially identifies eigenvectors and their associated natural frequencies and damping constants. In addition, the normal form identifies the key nonlinear terms that must be included for a complete description of the dynamics. Normal form analysis can be compared to other methods for providing approximate solutions to weakly nonlinear vibration problems. There are a wide class of perturbation methods; averaging, harmonic balance, multiple scales, etc. (e.g. [**31**] or [**32**]). These have been applied successfully by many authors to produce approximate solutions close to many different types of resonance. While the methods often give the same algebraic solutions, for example for the prediction of backbone curves in the absence of forcing and damping, it is often hard to predict the region of validity of the analysis, and they tend not to explain the full dynamics including harmonics. The normal form is philosophically different; the method works not by looking for specific kinds of solution, but by providing sequences of approximations to the equations of motion themselves. Thus, one still has access to the complete dynamics. Moreover, the method is algorithmic and can readily be carried out computationally.

It is also useful to compare normal form analysis with direct calculation of NNMs. Our application to cable dynamics in §4 has shown that the normal form provides a natural way to compute local (i.e. weak amplitude) approximations to NNMs in undamped systems. However, as argued in the caption to **figure 1**, NNMs are not uniquely defined in the presence of damping, which provides both conceptual and computational problems. For example, in a recent paper by Renson et al.[**33**], a finite-element method was proposed for computing NNMs directly. In example 5.2 of that paper, a large-amplitude approximation to a NNM is computed under two different assumptions on the magnitude of damping for a two-degree-of-freedom nonlinear oscillator. Not only is it found that the NNM shape depends strongly on the damping coefficient, but the NNM is only computed for the most strongly damped mode; it is not uniquely defined for the other mode. Unfortunately, in reality, the least damped mode is the one that is most likely to be excited in practice.

By contrast, the normal form method is not limited by restriction to the most heavily damped mode. Indeed, normal form methods are routinely used in conjunction with Melnikov's method where the forcing and damping terms play the role of unfolding parameters. This enables the complete dynamics (not just that restricted to putative NNMs) to be studied in the neighbourhood of a resonance (e.g. [**34**]). Furthermore, in this paper, it has been shown that damping and forcing can naturally be included as unfolding parameters in the second-order normal form. Examples of the second-order normal form method applied to forced and damped vibration problems were presented in Wagg & Neild [**24**].

Finally, we note that the second-order normal form method is naturally generalizable to higher degrees of freedom, and to cases of multi-frequency excitation, parametric resonance and multiple or internal resonances. These will simply result in different detuning parameters δ in the notation of §3a. An exploration of such cases is left for future work, as is application of the method to systems of higher degrees of freedom using computational methods.

REFERENCES

1. Kerschen G, Peeters M, Golinval JC, Vakakis AF. 2009 Nonlinear normal modes, Part I: a useful framework for the structural dynamicist. Mech. Syst. Signal. Process. 23, 170–194. (doi:10.1016/j.ymssp.2008.04.002)
2. Peeters M, Viguié R, Sérandour G, Kerschen G, Golinval JC. 2009 Nonlinear normal modes, Part II: toward a practical computation using numerical continuation techniques. Mech. Syst. Signal. Process. 23, 195–216. (doi:10.1016/j.ymssp.2008.04.003)
3. Avramov KV, Mikhlin YV. 2013 Review of applications of nonlinear normal modes for vibrating mechanical systems. Appl. Mech. Rev. 65, 020801. (doi:10.1115/1.4023533)
4. Guckenheimer J, Holmes PJ. 1983 Nonlinear oscillations, dynamical systems, and bifurcations of vector fields. New York, NY: Springer.
5. Chow SN, Li C, Wang D. 1994 Normal forms and bifurcations of planar vector fields. Cambridge, UK: Cambridge University Press.
6. Murdock J. 2002 Normal forms and unfoldings for local dynamical systems. New York, NY: Springer.

7. Jezequel L, Lamarque CH. 1991 Analysis of non-linear dynamic systems by the normal form theory. J. Sound Vib. 149, 429–459. (doi:10.1016/0022-460X(91)90446-Q)
8. Neild SA, Wagg DJ. 2011 Applying the method of normal forms to second order nonlinear vibration problems. Proc. R. Soc. A 467, 1141–1163. (doi:10.1098/rspa.2010.0270)
9. Shaw S, Pierre C. 1991 Non-linear normal modes and invariant manifolds. J. Sound Vib. 150,
10. 170–173. (doi:10.1016/0022-460X(91)90412-D)
11. Shaw SW, Pierre C. 1994 Normal modes of vibration for non-linear continuous systems. J. Sound Vib. 169, 319 347. (doi:10.1006/jsvi.1994.1021)
12. Montaldi J, Roberts M, Stewart I. 1990 Existence of nonlinear normal modes of symmetric Hamiltonian systems. Nonlinearity 3, 695–730. (doi:10.1088/0951-7715/3/3/009)
13. Nayfeh AH. 1993 Method of normal forms. New York, NY: Wiley.
14. Poincaré H. 1892–9 New methods of celestial mechanics. New York, NY: American Institute of Physics. [English transl. edited by Goroff DL 1993.]
15. Barrow-Green J. 1996 Poincaré and the three body problem. History of Mathematics, vol. 11. Providence, RI: American Mathematical Society.
16. Diacu F, Holmes PJ. 1999 Celestial encounters. Princeton, NJ: Princeton University Press.
17. Verhulst F. 2012 Henri Poincaré: impatient genius. New York, NY: Springer.
18. Meyer KR, Hall GR, Offin D. 2009 Introduction to Hamiltonian dynamical systems and the N-body problem, 2nd edn. New York, NY: Springer.
19. Birkhoff GD. 1927 Dynamical systems, vol. IX. Providence, RI: American Mathematical Society Colloquium Publication.
20. Kuznetsov YuA. 2004 Elements of applied bifurcation theory. New York, NY: Springer.
21. Champneys AR, Kirk V. 2004 The entwined wiggling of homoclinic curves emerging from
22. saddle-node/Hopf instabilities. Phys. D 195, 77–105. (doi:10.1016/j.physd.2004.03.004)
23. Arnol'd VI. 1988 Geometrical methods in the theory of ordinary differential equations. New York,
24. NY: Springer.
25. Neild SA. 2012 Approximate methods for analysing nonlinear structures. In Exploiting nonlinear behaviour in structural dynamics (eds LN Virgin, DJ Wagg), pp. 77–84. New York, NY: Springer.
26. Neild SA, Wagg DJ. 2013 A generalized frequency detuning method for multidegreeof-freedom oscillators with nonlinear stiffness. Nonlinear Dyn. 73, 649–663. (doi:10.1007/s11071-013-0818-7)
27. Wagg DJ, Neild SA. 2014 Nonlinear vibration with control: for flexible and adaptive structures, 2nd edn. New York, NY: Springer.
28. Hill TL, Cammarano A, Neild SA, Wagg DJ. 2015 Interpreting the forced responses of a twodegree-of-freedom nonlinear oscillator using backbone curves. J. Sound Vib. 349, 276–288. (doi:10.1016/j.jsv.2015.03.030)
29. Doedel E, Oldeman BE, Champneys AR, Dercole F, Fairgrieve T, Kuznetsov Yu, Wang X, Zhang C 2007 AUTO-07P: Continuation and bifurcation software for ordinary differential equations. See http://sourceforge.net/projects/auto-07p/.
30. Hill TL, Cammarano A, Neild SA, Wagg DJ. 2015 Out-of-unison resonance in weakly nonlinear coupled oscillators. Proc. R. Soc. A 471, 20140659. (doi:10.1098/rspa.2014.0659)
31. Warnitchai Y, Fujino T, Susumpov A. 1995 A nonlinear dynamic model for cables and its application to a cable structure-system. J. Sound Vib. 187, 695–712. (doi:10.1006/jsvi.1995.0553)
32. Gonzalez-Buelga A, Neild SA, Wagg DJ, Macdonald JHG. 2008 Modal stability of inclined cables subjected to vertical support excitation. J. Sound Vib. 318, 565–579. (doi:10.1016/ j.jsv.2008.04.031)
33. Macdonald JHG, Dietz MS, Neild SA, Gonzalez-Buelga A, Crewe AJ, Wagg DJ. 2010 Generalised modal stability of inclined cables subjected to support excitations. J. Sound Vib. 329, 4515–4533. (doi:10.1016/j.jsv.2010.05.002)
34. Nayfeh AH. 2000 Perturbation methods. New York, NY: Wiley.
35. Holmes HM 2012 Introduction to perturbation methods, 2nd edn. New York, NY: Springer.
36. Renson L, Deliege G, Kerschen G. 2014 An effective finite-element-based method for the computation of nonlinear normal modes of nonconservative systems. Meccanica 49, 1901–1916. (doi:10.1007/s11012-014-9875-3)
37. Haller G 1999 Chaos near resonance. New York, NY: Springer.

CHAPTER 4 SOME INTERESTING BIFURCATIONS OF NONLINEAR WAVES FOR THE GENERALIZED DRINFELD-SOKOLOV SYSTEM

Huixian Cai, Chaohong Pan and Zhengrong Liu

INTRODUCTION

The D(m, n) system [1, 2] is read as

$$u_t + (v^m)_x = 0,$$

$$v_t + a(v^n)_{xxx} + bu_x v + cuv_x = 0, \tag{1}$$

where a, b, c, m, and n are constants. Through some transformations, Xie and Yan [3] got some compacton and solitary pattern solutions of system (1) with m=n-1, including

$$v_a(\xi) = \left[\frac{2n(\lambda - cd_0)}{(n+1)a\alpha_m}\cos^2\left(\frac{(n-1)\sqrt{\alpha_m}}{2n}\xi\right)\right]^{1/(n-1)}, \tag{2}$$

$$u_a(\xi) = \frac{v_a^{n-1}(\xi)}{\lambda} + d_0,$$

$$v_b(\xi) = \left[\frac{2n(cd_0 - \lambda)}{(n+1)a\alpha_m}\sinh^2\left(\frac{(n-1)\sqrt{\alpha_m}}{2n}\xi\right)\right]^{1/(n-1)},$$

$$u_b(\xi) = \frac{v_b^{n-1}(\xi)}{\lambda} + d_0, \tag{3}$$

where $\lambda \neq 0, d_0$ are constants,

$$\xi = x - \lambda t, \qquad \alpha_m = \frac{bm + c}{a\lambda(m+1)}. \tag{4}$$

Deng et al. [4], by using the Weierstrass elliptic function method, presented many solutions of system (1) with n=m+1. It also includes the above solutions (2) and (3). Zhang et al. [5] showed some solutions of system (1) under the special parameters via employing the bifurcation method. By means of the complete discrimination system for polynomial method, many solutions of system (1) were acquired in [6].

In [2], the system

$$u_t + (v^2)_x = 0,$$

$$v_t + av_{xxx} + bu_x v + cuv_x = 0 \tag{5}$$

was introduced. Wang [7] gave recursion, Hamiltonian, symplectic and cosymplectic operator, roots of symmetries, and scaling symmetry for system (5). Wazwaz [8], by using the tanh method and the sine-cosine method, obtained many solutions with compact and noncompact structures of system (5), including

$$v_c(\xi) = \lambda\sqrt{-\frac{6}{2b+c}}\operatorname{sech}\left(\sqrt{\frac{\lambda}{a}}\xi\right), \qquad u_c(\xi) = \frac{1}{\lambda}v_c^2(\xi), \qquad \left(\frac{\lambda}{a} > 0\right),$$

$$v_d(\xi) = \lambda\sqrt{-\frac{3}{2b+c}}\tanh\left(\sqrt{-\frac{\lambda}{2a}}\xi\right), \qquad u_d(\xi) = \frac{1}{\lambda}v_d^2(\xi), \qquad \left(\frac{\lambda}{a} < 0\right),$$

$$v_e(\xi) = \lambda\sqrt{-\frac{3}{2b+c}}\coth\left(\sqrt{-\frac{\lambda}{2a}}\xi\right), \qquad u_e(\xi) = \frac{1}{\lambda}v_e^2(\xi), \qquad \left(\frac{\lambda}{a} < 0\right). \tag{6}$$

Citation: Huixian Cai, Chaohong Pan, and Zhengrong Liu, "Some Interesting Bifurcations of Nonlinear Waves for the Generalized Drinfeld-Sokolov System," Abstract and Applied Analysis, vol. 2014, Article ID 189486, 20 pages, 2014. doi:10.1155/2014/189486.

Biazar and Ayati [9] obtained some solutions of system (5) through Exp-function method and modification of Exp-function method. Zhang et al. [10], by employing the complex method, gained all meromorphic exact solutions of system (5). Applying the auxiliary equation method, some exact solutions of system (5) were given in [11]. El-Wakil and Abdou [12] got some new exact solutions of system (5) by means of modified extended tanh-function method.

The other generalized Drinfel'd-Sokolov system [13]

$$u_t + a_1 uu_x + b_1 u_{xxx} + c_0 (v^\sigma)_x = 0,$$
$$v_t + a_2 uv_x + b_2 v_{xxx} = 0 \quad (7)$$

is considered in [14–22]. In [23–31], many exact solutions of system (7) were obtained. Clearly, system (1) and system (7) are two different systems.

In this paper, we are interested in system (1). We study the bifurcations of nonlinear waves for system (1).

Under the transformations

$$\xi = x - \lambda t, \qquad u(x,t) = \psi(\xi), \qquad v(x,t) = \varphi(\xi), \quad (8)$$

system (1) is reduced to

$$\lambda\psi' - (\varphi^m)' = 0,$$
$$-\lambda\varphi' + a(\varphi^n)''' + b\psi'\varphi + c\psi\varphi' = 0. \quad (9)$$

Integrating the first equation of system (9) once, we have

$$\psi = \frac{1}{\lambda}\varphi^m + d, \quad (10)$$

where d is an integral constant. Substituting (10) into the second equation of system (9) and integrating it once yield the following equation:

$$g + (cd - \lambda)\varphi + \frac{bm + c}{\lambda(m+1)}\varphi^{m+1}$$
$$+ a\left[n(n-1)\varphi^{n-2}(\varphi')^2 + n\varphi^{n-1}\varphi''\right] = 0, \quad (11)$$

where g is another integral constant. Letting $\varphi'(\xi) = y$, we obtain the planar system

$$\frac{d\varphi}{d\xi} = y,$$
$$\frac{dy}{d\xi} = -\frac{n(n-1)\varphi^{n-2}y^2 + \alpha_m\varphi^{m+1} + \beta\varphi + \gamma}{n\varphi^{n-1}}, \quad (12)$$

where α_m is given in (4) and

$$\beta = \frac{cd - \lambda}{a},$$
$$\gamma = \frac{g}{a}. \quad (13)$$

Letting $n\varphi^{n-1}d\tau = d\xi$, system (12) becomes

$$\frac{d\varphi}{d\tau} = n\varphi^{n-1}y,$$
$$\frac{dy}{d\tau} = -n(n-1)\varphi^{n-2}y^2 - \alpha_m\varphi^{m+1} - \beta\varphi - \gamma, \quad (14)$$

which is called P(m,n) system. Clearly, systems (12) and (14) possess the same first integral

$$\varphi^n\left(\frac{n}{2}\varphi^{n-2}y^2 + \frac{\alpha_m}{m+n+1}\varphi^{m+1} + \frac{\beta}{n+1}\varphi + \frac{\gamma}{n}\right) = h. \quad (15)$$

Employing (14) and (15), we reveal some interesting bifurcation phenomena listed in the above abstract.

In Section 2, we will consider D(2, 1) system. Firstly, we will show that the fractional solitary waves can be bifurcated from the trigonometric periodic waves and the elliptic periodic waves. Secondly, we will demonstrate that the kink waves can be bifurcated from the smooth solitary waves and the singular waves. In Section 3, we will consider D(1, 2) system. Firstly, we will confirm that the compactons can be bifurcated from the smooth solitary waves. Secondly, we will clarify that the peakons can be bifurcated from the solitary waves and the singular cusp waves. In Section 4, D(2, 2) system will be considered. We will verify that the solitary waves can be bifurcated from the smooth periodic waves and the periodic singular waves. A short conclusion will be given in Section 5.

THE BIFURCATIONS OF SOLITARY WAVES AND KINK WAVES FOR D(2, 1) SYSTEM

When m=2 and n=1, system (1) becomes D(2,1) system:

$$u_t + (v^2)_x = 0,$$
$$v_t + av_{xxx} + bu_x v + cuv_x = 0. \quad (16)$$

We will reveal two kinds of interesting bifurcation phenomena to system (16). The first phenomenon is that fractional solitary waves can be bifurcated from two types of smooth periodic waves: trigonometric periodic waves and elliptic periodic waves. The second phenomenon is that the kink waves can be bifurcated from the smooth solitary waves and the singular waves. We state these results and give proof as follows.

Note that P(2,1) system is read as

$$\frac{d\varphi}{d\tau} = y,$$
$$\frac{dy}{d\tau} = -\alpha_2\varphi^3 - \beta\varphi - \gamma, \quad (17)$$

Where

$$\alpha_2 = \frac{2b + c}{3a\lambda}, \quad (18)$$

and β, γ are given in (13). When $27\alpha_2\gamma^2 > -4\beta^3$, let

$$\varphi_{01} = \frac{1}{\alpha_2}\left(\frac{\Omega}{18}\right)^{1/3} - \beta\left(\frac{2}{3\Omega}\right)^{1/3},$$

$$\varphi_{02} = \beta\left(1-\sqrt{3}i\right)\left(\frac{1}{12\Omega}\right)^{1/3} - \frac{1+\sqrt{3}i}{2\alpha_2}\left(\frac{\Omega}{18}\right)^{1/3},$$

$$\varphi_{03} = \beta\left(1+\sqrt{3}i\right)\left(\frac{1}{12\Omega}\right)^{1/3} - \frac{1-\sqrt{3}i}{2\alpha_2}\left(\frac{\Omega}{18}\right)^{1/3},$$

$$\Omega = \sqrt{12\alpha_2^3\beta^3 + 81\alpha_2^4\gamma^2} - 9\alpha_2^2\gamma, \tag{19}$$

$$\gamma_1(\beta) = \frac{2}{3\sqrt{3}}\left(-\frac{\beta^3}{\alpha_2}\right)^{1/2}. \tag{20}$$

Proposition 1. For given $\alpha_2 > 0 \; and \; \beta < 0, \; if \; -\gamma_1(\beta) < \gamma < 0$, if, then $\varphi_{01}, \varphi_{02}, and \; \varphi_{03}$ are real and system (16) has two types of special periodic wave solutions which become the fractional solitary wave solution

$$v_k(\xi) = \sqrt{-\frac{3\beta}{\alpha_2}\frac{9+2\beta\xi^2}{9-6\beta\xi^2}}, \qquad u_k(\xi) = \frac{1}{\lambda}v_k^2(\xi) + d, \tag{21}$$

when $\gamma \to -\gamma_1(\beta)+0$. These two types of special periodic wave solutions are as follows.

(1)Trigonometric periodic wave solution:

$$v_1(\xi) = \varphi_{03} + \frac{2A_1}{\sqrt{\Delta_1}\cos(\eta_1\xi) - B_1},$$

$$u_1(\xi) = \frac{1}{\lambda}v_1^2(\xi) + d, \tag{22}$$

Where

$$\varphi_1 = -\varphi_{03} + \sqrt{-\frac{2\beta}{\alpha_2} - 2\varphi_{03}^2},$$

$$\varphi_2 = -\varphi_{03} - \sqrt{-\frac{2\beta}{\alpha_2} - 2\varphi_{03}^2}. \tag{23}$$

(2)Elliptic periodic wave solution:

$$v_2(\xi) = \frac{\varphi_3(\varphi_4-\varphi_6) + \varphi_6(\varphi_3-\varphi_4)sn^2(\eta_2\xi, k_1)}{(\varphi_4-\varphi_6) + (\varphi_3-\varphi_4)sn^2(\eta_2\xi, k_1)},$$

$$u_2(\xi) = \frac{1}{\lambda}v_2^2(\xi) + d, \tag{24}$$

Where

$$\eta_2 = \sqrt{\frac{\alpha_2}{8}(\varphi_3-\varphi_5)(\varphi_4-\varphi_6)},$$

$$k_1^2 = \frac{(\varphi_3-\varphi_4)(\varphi_5-\varphi_6)}{(\varphi_3-\varphi_5)(\varphi_4-\varphi_6)},$$

$$\varphi_3 = \frac{1}{2}\left(-\varphi_5 - \varphi_6 + \sqrt{-8\frac{\beta}{\alpha_2} - 3\varphi_5^2 - 2\varphi_5\varphi_6 - 3\varphi_6^2}\right),$$

$$\varphi_4 = \frac{1}{2}\left(-\varphi_5 - \varphi_6 - \sqrt{-8\frac{\beta}{\alpha_2} - 3\varphi_5^2 - 2\varphi_5\varphi_6 - 3\varphi_6^2}\right),$$

$$\varphi_5 = -\frac{1}{3}\varphi_6 + \left(2\beta + \frac{2}{3}\alpha_2\varphi_6^2\right)\left(\frac{1}{4\theta_2}\right)^{1/3}\left(1+i\sqrt{3}\right) - \frac{1}{6\alpha_2}\left(\frac{\theta_2}{2}\right)^{1/3}\left(1-i\sqrt{3}\right),$$

$$\varphi_{03}^* < \varphi_6 < \varphi_{03},$$

$$\varphi_{03}^* = -\varphi_{02} - \sqrt{-\frac{2\beta}{\alpha_2} - 2\varphi_{02}^2},$$

$$\theta_1 = -108\alpha_2^2\gamma - 36\alpha_2^2\beta\varphi_6 - 20\alpha_2^3\varphi_6^3,$$

$$\theta_2 = \theta_1 + \sqrt{4(6\alpha_2\beta + 2\alpha_2^2\varphi_6^2)^3 + \theta_1^2}. \tag{25}$$

For the varying process, see Figure 1.

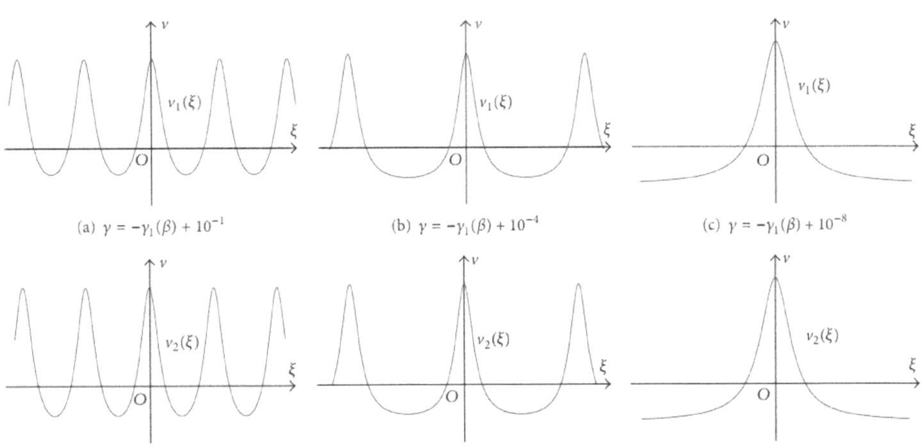

(a) $\gamma = -\gamma_1(\beta) + 10^{-1}$ (b) $\gamma = -\gamma_1(\beta) + 10^{-4}$ (c) $\gamma = -\gamma_1(\beta) + 10^{-8}$

(d) $\gamma = -\gamma_1(\beta) + 10^{-1}$ (e) $\gamma = -\gamma_1(\beta) + 10^{-4}$ (f) $\gamma = -\gamma_1(\beta) + 10^{-8}$

Figure 1: The solitary wave is bifurcated from two types of periodic waves. The varying process for the figures of $v_1(\xi)$ and $v_2(\xi)$ when $\alpha_2 > 0$, $\beta < 0$, and $\gamma \to -\gamma_1(\beta) + 0$ where $\alpha_2 = 4$ and $\beta = -9$.

Proposition 2. For given $\alpha_2 > 0$ and $\beta < 0$, if $0 < \gamma < \gamma_1(\beta)$, then φ_{01}, φ_{02}, and φ_{03} are real and system (16) has two types of special periodic wave solutions which become a fractional solitary wave solution:

$$v_y(\xi) = \sqrt{-\frac{3\beta}{\alpha_2}\frac{2\beta\xi^2+9}{6\beta\xi^2-9}}, \qquad u_y(\xi) = \frac{1}{\lambda}v_y^2(\xi) + d, \tag{26}$$

when $\gamma \to \gamma_1(\beta) - 0$. These two types of special periodic wave solutions are as follows.

(1)Trigonometric periodic wave solution:

$$v_3(\xi) = \varphi_{01} + \frac{2A_2}{B_2 - \sqrt{\Delta_2}\cos(\eta_3\xi)},$$

$$u_3(\xi) = \frac{1}{\lambda}v_3^2(\xi) + d, \tag{27}$$

Where

$$A_2 = (\varphi_7 - \varphi_{01})(\varphi_{01} - \varphi_8),$$

$$B_2 = 2\varphi_{01} - \varphi_7 - \varphi_8,$$

$$\Delta_2 = B_2^2 + 4A_2,$$

$$\eta_3 = \sqrt{-\frac{A_2\alpha_2}{2}},$$

$$\varphi_7 = -\varphi_{01} + \sqrt{-\frac{2\beta}{\alpha_2} - 2\varphi_{01}^2},$$

$$\varphi_8 = -\varphi_{01} - \sqrt{-\frac{2\beta}{\alpha_2} - 2\varphi_{01}^2}. \tag{28}$$

(2) Elliptic periodic wave solution:

$$v_4(\xi) = \frac{\varphi_{12}(\varphi_9 - \varphi_{11}) + \varphi_9(\varphi_{11} - \varphi_{12})sn^2(\eta_4\xi, k_2)}{(\varphi_9 - \varphi_{11}) + (\varphi_{11} - \varphi_{12})sn^2(\eta_4\xi, k_2)},$$

$$u_4(\xi) = \frac{1}{\lambda}v_4^2(\xi) + d, \tag{29}$$

Where

$$\eta_4 = \sqrt{\frac{\alpha_2}{8}(\varphi_9 - \varphi_{11})(\varphi_{10} - \varphi_{12})},$$

$$k_2^2 = \frac{(\varphi_9 - \varphi_{10})(\varphi_{11} - \varphi_{12})}{(\varphi_9 - \varphi_{11})(\varphi_{10} - \varphi_{12})},$$

$$\varphi_{01}^* = -\varphi_{02} + \sqrt{-\frac{2\beta}{\alpha_2} - 2\varphi_{02}^2},$$

$$\varphi_{01} < \varphi_9 < \varphi_{01}^*,$$

$$\theta_3 = -20v_9^3\alpha_2^3 - 36v_9\alpha_2^2\beta - 108\alpha_2^2\gamma,$$

$$\theta_4 = \theta_3 + \sqrt{4(2v_9^2\alpha_2^2 + 6\alpha_2\beta)^3 + \theta_3^2},$$

$$\varphi_{10} = -\frac{\varphi_9}{3} - \left(\frac{2}{3}\varphi_9^2\alpha_2 + 2\beta\right)\left(\frac{2}{\theta_4}\right)^{1/3} + \frac{1}{3\alpha_2}\left(\frac{\theta_4}{2}\right)^{1/3},$$

$$\varphi_{11} = \frac{1}{2}\left(-\varphi_9 - \varphi_{10} + \sqrt{-3\varphi_9^2 - 2\varphi_9\varphi_{10} - 3\varphi_{10}^2 - 8\frac{\beta}{\alpha_2}}\right),$$

$$\varphi_{12} = \frac{1}{2}\left(-\varphi_9 - \varphi_{10} - \sqrt{-3\varphi_9^2 - 2\varphi_9\varphi_{10} - 3\varphi_{10}^2 - 8\frac{\beta}{\alpha_2}}\right).$$

For the varying process, see Figure 2.

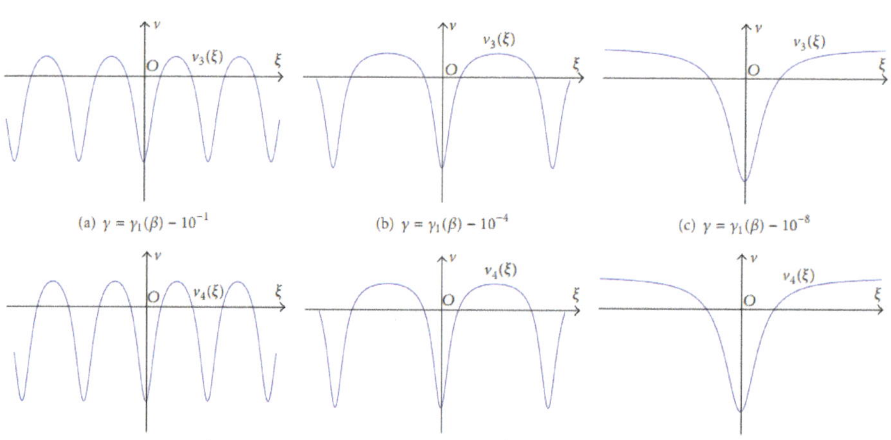

Figure 2: The solitary wave is bifurcated from two types of periodic waves. The varying process for the figures of $v_3(\xi)$ and $v_4(\xi)$ when $\alpha_2 > 0$, $\beta < 0$, and $\gamma \to \gamma_1(\beta) - 0$ where $\alpha_2 = 4$ and $\beta = -9$.

Proposition 3. For given $\alpha_2 < 0$ *and* $\beta > 0$, *if* $-\gamma_1(\beta) < \gamma < \gamma_1(\beta)$, then system (16) has four nonlinear wave solutions which become two kink wave solutions:

$$v_\delta^{\pm}(\xi) = \pm\sqrt{-\frac{\beta}{\alpha_2}}\tanh\left(\sqrt{\frac{\beta}{2}}\xi\right), \qquad u_\delta(\xi) = \frac{1}{\lambda}v_\delta^2(\xi) + d, \tag{31}$$

When $\gamma \to 0$. These four nonlinear wave solutions are as follows:

$$v_5(\xi) = \frac{4A_3\left(\varphi_{01} - \delta e^{-\eta_5\xi}\right) - \varphi_{01}\left(\delta e^{-\eta_5\xi} - B_3\right)^2}{4A_3 - \left(\delta e^{-\eta_5\xi} - B_3\right)^2},$$

$$u_5(\xi) = \frac{1}{\lambda}v_5^2(\xi) + d,$$

$$v_6(\xi) = \frac{4A_3\left(\varphi_{01} - \delta e^{\eta_5\xi}\right) - \varphi_{01}\left(\delta e^{\eta_5\xi} - B_3\right)^2}{4A_3 - \left(\delta e^{\eta_5\xi} - B_3\right)^2},$$

$$u_6(\xi) = \frac{1}{\lambda}v_6^2(\xi) + d,$$

$$v_7(\xi) = \frac{4A_4\left(\varphi_{03} - \mu e^{\eta_6\xi}\right) - \varphi_{03}\left(\mu e^{\eta_6\xi} - B_4\right)^2}{4A_4 - \left(\mu e^{\eta_6\xi} - B_4\right)^2},$$

$$u_7(\xi) = \frac{1}{\lambda}v_7^2(\xi) + d,$$

$$v_8(\xi) = \frac{4A_4\left(\varphi_{03} - \mu e^{-\eta_6\xi}\right) - \varphi_{03}\left(\mu e^{-\eta_6\xi} - B_4\right)^2}{4A_4 - \left(\mu e^{-\eta_6\xi} - B_4\right)^2},$$

$$u_8(\xi) = \frac{1}{\lambda}v_8^2(\xi) + d, \tag{32}$$

Where

$$\delta = 8\sqrt{-\frac{\beta}{\alpha_2}},$$

$$A_3 = 6\varphi_{01}^2 + \frac{2\beta}{\alpha_2},$$

$$\eta_5 = \sqrt{-\frac{A_3\alpha_2}{2}},$$

$$B_3 = 4\varphi_{01},$$

$$\mu = -8\sqrt{-\frac{\beta}{\alpha_2}},$$

$$A_4 = 6\varphi_{03}^2 + \frac{2\beta}{\alpha_2},$$

$$\eta_6 = \sqrt{-\frac{A_4\alpha_2}{2}},$$

$$B_4 = 4\varphi_{03}. \tag{33}$$

These four nonlinear wave solutions possess the following properties.

(a) If $0 < \gamma < \gamma_1(\beta)$, then $v_5(\xi)$ *and* $v_6(\xi)$ represent two solitary waves which tend to two kink waves $v_\delta^{\pm}(\xi)$ (see Figure 3) when $\gamma \to 0+0$.

(b) If $-\gamma_1(\beta) < \gamma < 0$, then $v_5(\xi)$ *and* $v_6(\xi)$ represent two singular waves which tend to two kink waves $v_\delta^{\pm}(\xi)$ (see Figure 4) when $\gamma \to 0-0$.

(c) If $0 < \gamma < \gamma_1(\beta)$, then $v_7(\xi)$ *and* $v_8(\xi)$ represent two singular waves which tend to two kink waves $v_\delta^{\pm}(\xi)$ (see Figure 5) when .

(d) If $-\gamma_1(\beta) < \gamma < 0$, then $v_7(\xi)$ *and* $v_8(\xi)$ represent two solitary waves which tend to two kink waves $v_\delta^{\pm}(\xi)$ (see Figure 6) when .

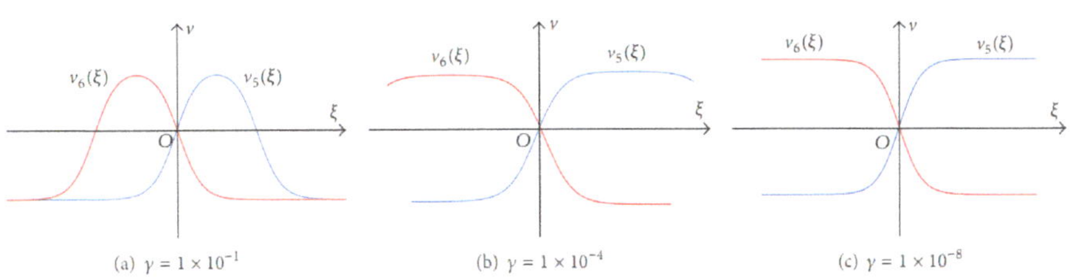

Figure 3: Two kink waves are bifurcated from two peak-solitary waves. The varying process for the figures of $v_5(\xi)$ and $v_6(\xi)$ when $\alpha_2 < 0$, $\beta > 0$, and $\gamma \to 0+0$ where $\alpha_2 = -2$ and $\beta = 9$.

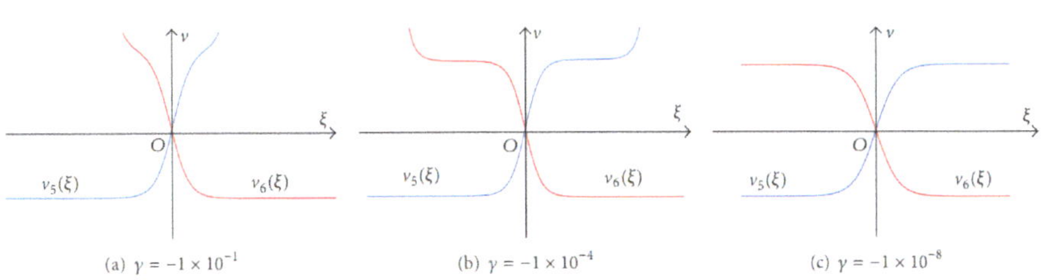

Figure 4: Two kink waves are bifurcated from two singular waves. The varying process for the figures of $v_5(\xi)$ and $v_6(\xi)$when $\alpha_2 < 0, \beta > 0$, and and $\gamma \to 0-0$ where $\alpha_2 = -2$ and $\beta = 9$.

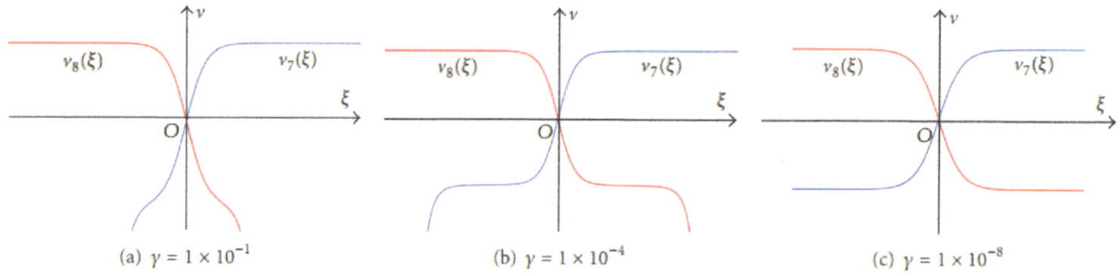

Figure 5: Two kink waves are bifurcated from two singular waves. The varying process for the figures of $v_7(\xi)$ and $v_8(\xi)$ when $\alpha_2 < 0, \beta > 0$, and $\gamma \to 0+0$ where $\alpha_2 = -2$ and $\beta = 9$.

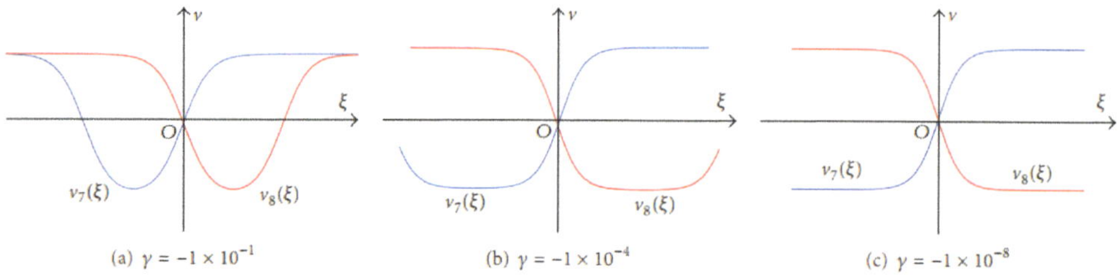

Figure 6: Two kink waves are bifurcated from two valley-solitary waves. The varying process for the figures of $v_7(\xi)$ and $v_8(\xi)$ when $\alpha_2 < 0$, , and $\beta > 0$, and $\gamma \to 0-0$ where $\alpha_2 = -2$ and $\beta = 9$.

The Derivations of Propositions 1–3. According to the qualitative theory, we obtain the bifurcation phase portraits of system (17) as in Figure 7. Employing some orbits in Figure 7, we derive the results of Propositions1–3 as follows.

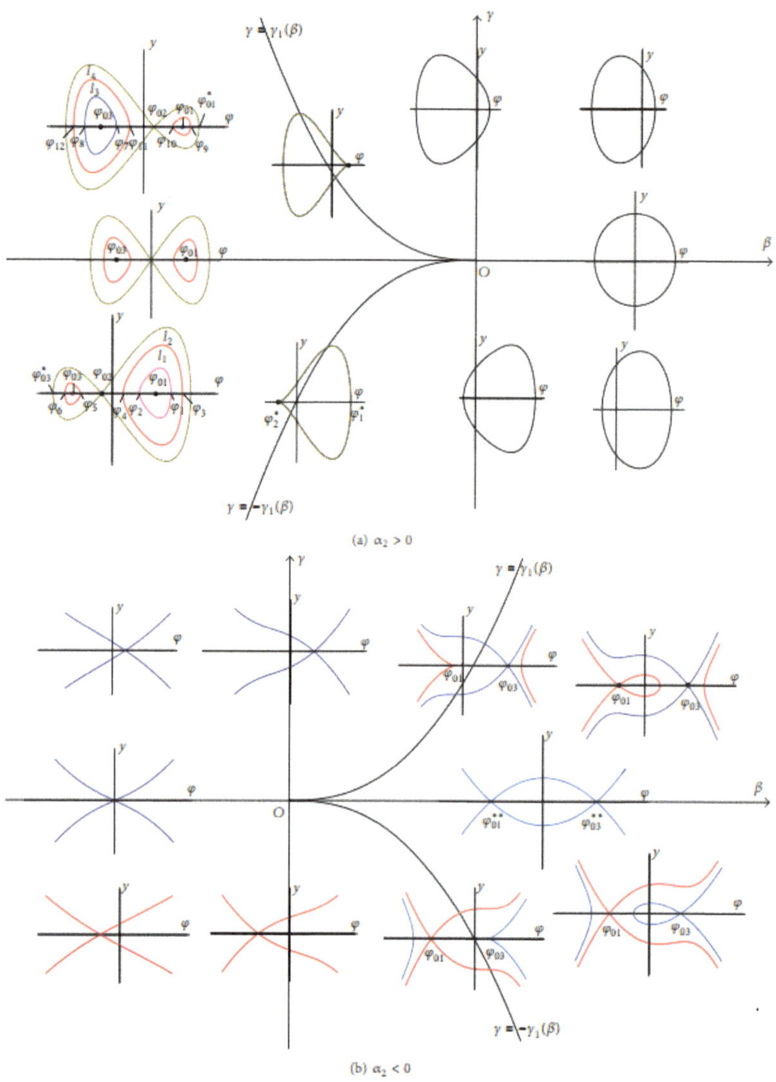

Figure 7: The bifurcation phase portraits of system (17).

(1) When $\alpha_2 > 0$, $\beta < 0$, and $-\gamma_1(\beta) < \gamma < 0$ (as in Figure 7(a)), the closed curves l_1 and l_2 possess the following expressions:

$$l_1: y^2 = \frac{\alpha_2}{2}(\varphi - \varphi_{03})^2 (\varphi_1 - \varphi)(\varphi - \varphi_2)$$
$$(\varphi_{03} < \varphi_2 \le \varphi \le \varphi_1),$$
$$l_2: y^2 = \frac{\alpha_2}{2}(\varphi_3 - \varphi)(\varphi - \varphi_4)(\varphi - \varphi_5)(\varphi - \varphi_6)$$
$$(\varphi_6 < \varphi_5 < \varphi_4 \le \varphi \le \varphi_3). \tag{34}$$

Substituting (34) into $d\varphi/d\xi = y$ and integrating them along l_1 and l_2, respectively, it follows that

$$\int_{\varphi}^{\varphi_1} \frac{ds}{(s - \varphi_{03})\sqrt{(\varphi_1 - s)(s - \varphi_2)}} = \sqrt{\frac{\alpha}{2}}|\xi| \quad (\text{along } l_1),$$
$$\int_{\varphi}^{\varphi_3} \frac{ds}{\sqrt{(\varphi_3 - s)(s - \varphi_4)(s - \varphi_5)(s - \varphi_6)}} = \sqrt{\frac{\alpha}{2}}|\xi|$$
$$(\text{along } l_2). \tag{35}$$

Completing the integrals above and solving the equations for φ, we obtain the solutions $v_1(\xi)$, $u_1(\xi)$ (see (22)) and $v_2(\xi)$ $u_2(\xi)$ (see (24)).

(2) When $\alpha_2 > 0$, $\beta < 0$, and $0 < \gamma < \gamma_1(\beta)$ (as in Figure 7(a)), the closed curves l_3 and l_4 possess the following expressions:

$$l_3: y^2 = \frac{\alpha}{2}(\varphi_{01} - \varphi)^2 (\varphi_7 - \varphi)(\varphi - \varphi_8)$$
$$(\varphi_8 \le \varphi \le \varphi_7 < \varphi_{01}),$$
$$l_4: y^2 = \frac{\alpha}{2}(\varphi_9 - \varphi)(\varphi_{10} - \varphi)(\varphi_{11} - \varphi)(\varphi - \varphi_{12})$$
$$(\varphi_{12} \le \varphi \le \varphi_{11} < \varphi_{10} < \varphi_9). \tag{36}$$

Substituting (36) into $d\varphi/d\xi = y$ and integrating them along l_3 and l_4, respectively, it follows that

$$\int_{\varphi_8}^{\varphi} \frac{ds}{(\varphi_{01} - s)\sqrt{(\varphi_7 - s)(s - \varphi_8)}} = \sqrt{\frac{\alpha}{2}}|\xi| \quad (\text{along } l_3),$$
$$\int_{\varphi_{12}}^{\varphi} \frac{ds}{\sqrt{(\varphi_9 - s)(\varphi_{10} - s)(\varphi_{11} - s)(s - \varphi_{12})}} = \sqrt{\frac{\alpha}{2}}|\xi|$$
$$(\text{along } l_4). \tag{37}$$

Completing the integrals above and solving the equations for φ, we obtain the solutions $v_3(\xi)$, $u_3(\xi)$ (see (27)) and $v_4(\xi)$ $u_4(\xi)$ (see (29)).

When $\gamma \to -\gamma_1(\beta) + 0$, it follows that φ_2, φ_{03}, and φ_{03}^* tend to $\varphi_2^* = -\sqrt{-\beta/3\alpha_2}$ and φ_1, φ_3 tend to $\varphi_1^* = \sqrt{-3\beta/\alpha_2}$. Further, we have

$$\varphi_4 \longrightarrow \varphi_2^*, \qquad \varphi_5 \longrightarrow \varphi_2^*, \qquad \varphi_6 \longrightarrow \varphi_2^*,$$
$$A_1 = (\varphi_1 - \varphi_{03})(\varphi_{03} - \varphi_2) \longrightarrow 0,$$
$$B_1 = \varphi_1 + \varphi_2 - 2\varphi_{03} \longrightarrow \varphi_1^* - \varphi_2^*,$$
$$\eta_1 = \sqrt{-\frac{A_1\alpha_2}{2}} \longrightarrow 0,$$
$$\eta_2 = \sqrt{\frac{\alpha_2}{8}(\varphi_3 - \varphi_5)(\varphi_4 - \varphi_6)} \longrightarrow 0,$$
$$\sqrt{\Delta_1} = \sqrt{B_1^2 + 4A_1} = B_1 + \frac{2A_1}{B_1} + \cdots,$$
$$\cos(\eta_1\xi) = 1 - \frac{(\eta_1\xi)^2}{2!} + \frac{(\eta_1\xi)^4}{4!} + \cdots,$$
$$\operatorname{sn}(\eta_2\xi, k_1) = \eta_2\xi - (1 + k_1^2)\frac{(\eta_2\xi)^3}{3!} + \cdots. \tag{38}$$

Thus, we have

$$\begin{aligned}
&\lim_{\gamma\to-\gamma_1+0} v_1(\xi) \\
&= \lim_{\gamma\to-\gamma_1+0} \varphi_{03} + \frac{2A_1}{\sqrt{\Delta_1}\cos(\eta_1\xi) - B_1} \\
&= \lim_{\gamma\to-\gamma_1+0} \varphi_{03} \\
&\quad + 2\left(A_1 + o\left(A_1^2\right)\right) \\
&\quad \times \left(\left(B_1 + \frac{2A_1}{B_1} + o\left(A_1^2\right)\right)\right. \\
&\quad \left.\times \left(1 + \frac{\alpha_2 A_1 \xi^2}{4} + o\left(A_1^2\right)\right) - B_1\right)^{-1} \\
&= \lim_{\gamma\to-\gamma_1+0} \varphi_{03} + \frac{2A_1 + o\left(A_1^2\right)}{(2A_1/B_1) + (\alpha_2 A_1 B_1 \xi^2/4) + o\left(A_1^2\right)} \\
&= \varphi_2^* + \frac{8(\varphi_1^* - \varphi_2^*)}{\alpha_2(\varphi_1^* - \varphi_2^*)^2\xi^2 + 8} \\
&= \sqrt{-\frac{3\beta}{\alpha_2}\frac{9 + 2\beta\xi^2}{9 - 6\beta\xi^2}} \\
&= v_k(\xi) \quad (\text{see } (21)), \\
&\lim_{\gamma\to-\gamma_1+0} v_2(\xi) \\
&= \lim_{\gamma\to-\gamma_1+0} \frac{\varphi_3(\varphi_4 - \varphi_6) + \varphi_6(\varphi_3 - \varphi_4)\operatorname{sn}^2(\eta_2\xi, k_1)}{(\varphi_4 - \varphi_6) + (\varphi_3 - \varphi_4)\operatorname{sn}^2(\eta_2\xi, k_1)} \\
&= \lim_{\gamma\to-\gamma_1+0} \left(\left(\varphi_3(\varphi_4 - \varphi_6) + \varphi_6(\varphi_3 - \varphi_4)\right.\right. \\
&\quad \times \left[\sqrt{\frac{\alpha_2}{8}(\varphi_3 - \varphi_5)(\varphi_4 - \varphi_6)}\xi\right. \\
&\quad \left.\left. + o(\varphi_4 - \varphi_6)\right]^2\right) \\
&\quad \times \left((\varphi_4 - \varphi_6) + (\varphi_3 - \varphi_4)\right. \\
&\quad \times \left[\sqrt{\frac{\alpha_2}{8}(\varphi_3 - \varphi_5)(\varphi_4 - \varphi_6)}\xi\right. \\
&\quad \left.\left.\left. + o(\varphi_4 - \varphi_6)\right]^2\right)^{-1}\right) \\
&= \lim_{\gamma\to-\gamma_1+0} \left(\left(8\varphi_3(\varphi_4 - \varphi_6) + \varphi_6(\varphi_3 - \varphi_4)(\varphi_3 - \varphi_5)\right.\right. \\
&\quad \left.\times (\varphi_4 - \varphi_6)\alpha_2\xi^2 + o\left((\varphi_4 - \varphi_6)^{3/2}\right)\right) \\
&\quad \times \left(8(\varphi_4 - \varphi_6) + (\varphi_3 - \varphi_4)(\varphi_3 - \varphi_5)\right. \\
&\quad \left.\left.\times (\varphi_4 - \varphi_6)\alpha_2\xi^2 + o\left((\varphi_4 - \varphi_6)^{3/2}\right)\right)^{-1}\right) \\
&= \frac{8\varphi_1^* + \varphi_2^*(\varphi_1^* - \varphi_2^*)^2\alpha_2\xi^2}{8 + (\varphi_1^* - \varphi_2^*)^2\alpha_2\xi^2} \\
&= \sqrt{-\frac{3\beta}{\alpha_2}\frac{9 + 2\beta\xi^2}{9 - 6\beta\xi^2}} \\
&= v_k(\xi) \quad (\text{see } (21)).
\end{aligned} \tag{39}$$

Similarly, we can prove the limit property of $v_3(\xi)$ and $v_4(\xi)$ when $\gamma \to \gamma_1(\beta) - 0$.

(3) When $\alpha_2 < 0$, $\beta > 0$, and $-\gamma_1(\beta) < \gamma < \gamma_1(\beta)$ (as in Figure 7(b)), the curve connecting with $(\varphi_{01}, 0)$ embraces the expression

$$y^2 = -\frac{\alpha_2}{2}(\varphi - \varphi_{01})^2\left(\varphi^2 + 2\varphi_{01}\varphi + 3\varphi_{01}^2 + \frac{2\beta}{\alpha_2}\right). \tag{40}$$

Substituting (40) into $d\varphi/d\xi = y$ and integrating it, we have

$$\int_p^\varphi \frac{ds}{\sqrt{(s - \varphi_{01})^2(s^2 + 2\varphi_{01}s + 3\varphi_{01}^2 + (2\beta/\alpha_2))}} = \sqrt{-\frac{\alpha_2}{2}}\xi, \tag{41}$$

Where

$$p = \varphi_{01} - \sqrt{-\frac{\beta}{\alpha_2}\frac{24\alpha_2\varphi_{01}^2 + 8\beta}{\alpha_2\varphi_{01}^2 - 8\varphi_{01}\sqrt{-\alpha_2\beta} + 9\beta}}. \tag{42}$$

Completing the integral above and solving the equation for φ, we obtain

$$v_5(\xi) = \frac{4A_3\left(\varphi_{01} - \delta e^{-\eta_5\xi}\right) - \varphi_{01}\left(\delta e^{-\eta_5\xi} - B_3\right)^2}{4A_3 - \left(\delta e^{-\eta_5\xi} - B_3\right)^2}. \tag{43}$$

From $v_5(\xi)$, we get

$$v_6(\xi) = v_5(-\xi) = \frac{4A_3\left(\varphi_{01} - \delta e^{\eta_5\xi}\right) - \varphi_{01}\left(\delta e^{\eta_5\xi} - B_3\right)^2}{4A_3 - \left(\delta e^{\eta_5\xi} - B_3\right)^2}. \tag{44}$$

(4) When $\alpha_2 < 0, \beta > 0$, and $-\gamma_1(\beta) < \gamma < \gamma_1(\beta)$ (as in Figure 7(b)), the curve connecting with $(\varphi_{03}, 0)$ embraces the expression

$$y^2 = -\frac{\alpha_2}{2}(\varphi_{03} - \varphi)^2\left(\varphi^2 + 2\varphi_{03}\varphi + 3\varphi_{03}^2 + \frac{2\beta}{\alpha_2}\right). \tag{45}$$

Substituting (45) into $d\varphi/d\xi = y$ and integrating it, we have

$$\int_q^{\varphi} \frac{ds}{\sqrt{(\varphi_{03} - s)^2\left(s^2 + 2\varphi_{03}s + 3\varphi_{03}^2 + (2\beta/\alpha_2)\right)}} = \sqrt{-\frac{\alpha_2}{2}}\xi, \tag{46}$$

where

$$q = \varphi_{03} + \sqrt{-\frac{\beta}{\alpha_2}}\frac{24\alpha_2\varphi_{03}^2 + 8\beta}{\alpha_2\varphi_{03}^2 + 8\varphi_{03}\sqrt{-\alpha_2\beta} + 9\beta}. \tag{47}$$

Completing the integral above and solving the equation for φ, we obtain

$$v_7(\xi) = \frac{4A_4\left(\varphi_{03} - \mu e^{\eta_6\xi}\right) - \varphi_{03}\left(\mu e^{\eta_6\xi} - B_4\right)^2}{4A_4 - \left(\mu e^{\eta_6\xi} - B_4\right)^2}. \tag{48}$$

From $v_7(\xi)$, we get

$$\begin{aligned} v_8(\xi) &= v_7(-\xi) \\ &= \frac{4A_4\left(\varphi_{03} - \mu e^{-\eta_6\xi}\right) - \varphi_{03}\left(\mu e^{-\eta_6\xi} - B_4\right)^2}{4A_4 - \left(\mu e^{-\eta_6\xi} - B_4\right)^2}. \end{aligned} \tag{49}$$

Note that when $\gamma \to 0$, it follows that $\varphi_{01} \to \varphi_{01}^{**} = -\sqrt{-\beta/\alpha_2}$, $A_3 \to -4\beta/\alpha_2$, $B_3 \to -4\sqrt{-\beta/\alpha_2}$, and $\eta_5 \to \sqrt{-2\beta}$. Thus, we have

$$\begin{aligned} \lim_{\gamma\to 0} v_5(\xi) &= \lim_{\gamma\to 0}\frac{4A_3\left(\varphi_{01} - \delta e^{-\eta_5\xi}\right) - \varphi_{01}\left(\delta e^{-\eta_5\xi} - B_3\right)^2}{4A_3 - \left(\delta e^{-\eta_5\xi} - B_3\right)^2} \\ &= \lim_{\gamma\to 0}\frac{\delta\varphi_{01}e^{-2\eta_5\xi} + (4A_3 - 2\varphi_{01}B_3)e^{-\eta_5\xi}}{\delta e^{-2\eta_5\xi} - 2B_3e^{-\eta_5\xi}} \\ &= \frac{\delta\sqrt{-\alpha_2\beta}e^{-\sqrt{2\beta}\xi} - 8\beta}{\delta\alpha_2 e^{-\sqrt{2\beta}\xi} - 8\sqrt{-\alpha_2\beta}} \\ &= \sqrt{-\frac{\beta}{\alpha_2}}\frac{e^{\sqrt{\beta/2}\xi} - e^{-\sqrt{\beta/2}\xi}}{e^{\sqrt{\beta/2}\xi} + e^{-\sqrt{\beta/2}\xi}} \end{aligned}$$

$$= \sqrt{-\frac{\beta}{\alpha_2}} \tanh\left(\sqrt{\frac{\beta}{2}}\xi\right)$$

$$= v_\delta^+(\xi) \quad (\text{see (31)}). \tag{50}$$

Similarly, we can also get $v_6(\xi) \to v_\delta^-(\xi), v_7(\xi) \to v_\delta^+(\xi)$, and $v_8(\xi) \to v_\delta^-(\xi)$ when $\gamma \to 0$. These complete the derivations of Propositions 1–3.

THE BIFURCATIONS OF COMPACTONS AND PEAKONS FOR D(1, 2) SYSTEM

When m=1 and n=2, system (1) becomes

$$u_t + v_x = 0,$$

$$v_t + a\left(v^2\right)_{xxx} + bu_x v + cuv_x = 0. \tag{51}$$

We will reveal two kinds of interesting bifurcation phenomena to system (51). The first phenomenon is that the smooth solitary waves can turn into the compactons. The second phenomenon is that the peakons can be bifurcated from the singular cusp waves and the solitary waves. The concrete results are stated as follows.

Note that P(1, 2) system is read as

$$\frac{d\varphi}{d\tau} = 2\varphi y,$$

$$\frac{dy}{d\tau} = -2y^2 - \alpha_1\varphi^2 - \beta\varphi - \gamma, \tag{52}$$

Where

$$\alpha_1 = \frac{b+c}{2a\lambda}, \tag{53}$$

and β, γ are given in (13).

For fixed α_1, put

$$\gamma_2(\beta) = \frac{2\beta^2}{9\alpha_1},$$

$$\gamma_3(\beta) = \frac{\beta^2}{4\alpha_1}. \tag{54}$$

Proposition 4. For given $\alpha_1 > 0$ and $\beta \neq 0$, *if* $0 < \gamma < \gamma_3(\beta)$, then system (51) has a family of solitary wave solutions which become the compacton solution

$$v_o(\xi) = \begin{cases} -\frac{4\beta}{3\alpha_1}\cos^2\left(\frac{\sqrt{\alpha_1}}{4}\xi\right), & |\xi| < \frac{2\pi}{\sqrt{\alpha_1}}, \\ 0, & |\xi| \geq \frac{2\pi}{\sqrt{\alpha_1}}, \end{cases}$$

$$u_o(\xi) = \frac{1}{\lambda}v_o(\xi) + d, \tag{55}$$

when $\gamma \to 0+0$. The solitary wave solutions are as follows.

(1) When $\beta < 0$, the solitary wave solutions possess the expressions $v_9(\xi) = \varphi(\xi)$ and $u_9(\xi) = (1/\lambda)v_9(\xi) + d$, *and* $\varphi(\xi)$ is given by the implicit function

$$\frac{\sqrt{\alpha_1}|\xi| - \pi}{2} = \arcsin\left(\frac{r_1 + r_3 - 2\varphi}{r_1 - r_3}\right) - \frac{r_2}{\sqrt{A_5}}\ln|\Phi_1|, \tag{56}$$

Where

$$r_1 = -\frac{1}{3\alpha_1}\left(3\alpha_1 r_2 + 2\beta - \sqrt{6\alpha_1\beta r_2 + 4\beta^2}\right),$$

$$r_2 = -\frac{1}{2\alpha_1}\left(\beta + \sqrt{\beta^2 - 4\alpha_1\gamma}\right),$$

$$r_3 = -\frac{1}{3\alpha_1}\left(3\alpha_1 r_2 + 2\beta + \sqrt{6\alpha_1\beta r_2 + 4\beta^2}\right),$$

$$A_5 = (r_1 - r_2)(r_2 - r_3),$$

$$B_5 = r_1 + r_3 - 2r_2,$$

$$\Phi_1 = \frac{(r_1 - r_3)(\varphi - r_2)}{2\sqrt{A_5(r_1 - \varphi)(\varphi - r_3)} + B_5(\varphi - r_2) + 2A_5}. \tag{57}$$

(2) When $\beta > 0$, the solitary wave solutions possess the expressions $v_{10}(\xi) = \varphi(\xi)$ and $u_{10}(\xi) = (1/\lambda)v_{10}(\xi) + d$ is given by the implicit function

$$\frac{\pi - \sqrt{\alpha_1}|\xi|}{2} = \arcsin\left(\frac{r_6 + r_4 - 2\varphi}{r_6 - r_4}\right) + \frac{r_5}{\sqrt{A_6}}\ln|\Phi_2|, \tag{58}$$

Where

$$r_4 = -\frac{1}{3\alpha_1}\left(3\alpha_1 r_5 + 2\beta + \sqrt{6\alpha_1\beta r_5 + 4\beta^2}\right),$$

$$r_5 = -\frac{1}{2\alpha_1}\left(\beta - \sqrt{\beta^2 - 4\alpha_1\gamma}\right),$$

$$r_6 = -\frac{1}{3\alpha_1}\left(3\alpha_1 r_5 + 2\beta - \sqrt{6\alpha_1\beta r_5 + 4\beta^2}\right),$$

$$A_6 = (r_6 - r_5)(r_5 - r_4),$$

$$B_6 = 2r_5 - r_6 - r_4,$$

$$\Phi_2 = \frac{2\sqrt{A_6(r_6 - \varphi)(\varphi - r_4)} + B_6(r_5 - \varphi) + 2A_6}{(r_6 - r_4)(r_5 - \varphi)}. \tag{59}$$

For the varying process, see Figure 8.

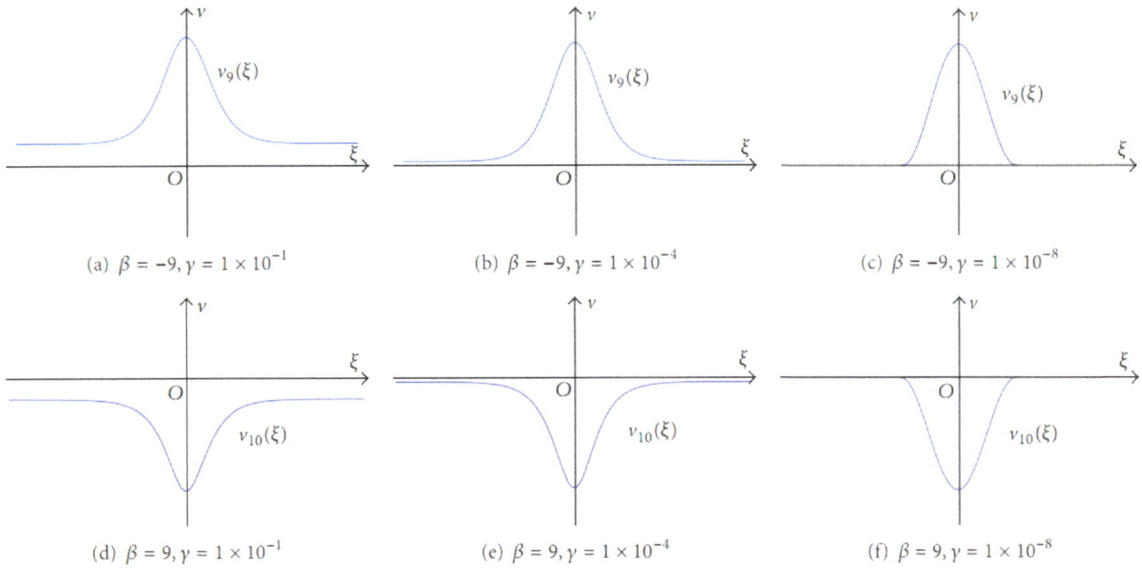

Figure 8: The solitary waves turn into the compactons. The varying process for the figures of $v_9(\xi)$ and $v_{10}(\xi)$ when $\alpha_1 > 0, \beta \neq 0$, and $\gamma \to 0+0$ where $\alpha_1 = 2$.

Proposition 5. For given $\alpha_1 < 0$ *and* $\beta \neq 0$, *if* $\gamma_3(\beta) < \gamma < \gamma_2(\beta)$, then system (51) has two types of nonlinear wave solutions which tend to the peakon solution

$$v_m(\xi) = \frac{2\beta}{3\alpha_1}\left(e^{-\sqrt{|\alpha_1|/4}|\xi|} - 1\right),$$

$$u_m(\xi) = \frac{1}{\lambda} v_m(\xi) + d, \tag{60}$$

when $\gamma \to \gamma_2(\beta) - 0$. For the varying process, see Figures 9 and 10. These two types of nonlinear wave solutions are singular cusp wave solutions and solitary wave solutions of the following expressions.

(1) When , the singular cusp wave solutions possess the expression

$$v_{11}(\xi) = \left[\left(\frac{\beta}{3\alpha_1} - \sqrt{\frac{\gamma}{2\alpha_1}}\right) - \left(\frac{\beta}{3\alpha_1} + \sqrt{\frac{\gamma}{2\alpha_1}}\right)e^{-\sqrt{|\alpha_1|/4}|\xi|}\right] \times \left(e^{\sqrt{|\alpha_1|/4}|\xi|} - 1\right),$$

$$u_{11}(\xi) = \frac{1}{\lambda} v_{11}(\xi) + d, \tag{61}$$

and the solitary wave solutions possess the expressions $v_{12}(\xi) = \varphi(\xi)$ *and* $u_{12}(\xi) = (1/\lambda)v_{12}(\xi) + d$, *and* $\varphi(\xi)$ is given by the implicit function

$$|\xi| = -\sqrt{\frac{4}{|\alpha_1|}}\left(\ln\left|\frac{2\sqrt{\Phi_3} - M_1}{r_7 - r_8}\right| + \frac{r_9}{\sqrt{A_7}} \ln\left|\frac{2\sqrt{A_7\Phi_3} + B_7(r_9 - \varphi) + 2A_7}{(r_7 - r_8)(r_9 - \varphi)}\right|\right), \tag{62}$$

Where

$$\Phi_3 = (r_7 - \varphi)(r_8 - \varphi),$$

$$M_1 = 2\varphi - r_7 - r_8,$$

$$r_7 = -\frac{1}{3\alpha_1}\left(3\alpha_1 r_9 + 2\beta + \sqrt{6\alpha_1\beta r_9 + 4\beta^2}\right),$$

$$r_8 = -\frac{1}{3\alpha_1}\left(3\alpha_1 r_9 + 2\beta - \sqrt{6\alpha_1\beta r_9 + 4\beta^2}\right),$$

$$r_9 = -\frac{1}{2\alpha_1}\left(\beta - \sqrt{\beta^2 - 4\alpha_1\gamma}\right),$$

$$A_7 = (r_8 - r_9)(r_7 - r_9),$$

$$B_7 = r_7 + r_8 - 2r_9. \tag{63}$$

(2) When $\beta > 0$, the singular cusp wave solutions possess the expression

$$v_{13}(\xi) = \left[\left(\frac{\beta}{3\alpha_1} - \sqrt{\frac{\gamma}{2\alpha_1}}\right) - \left(\frac{\beta}{3\alpha_1} + \sqrt{\frac{\gamma}{2\alpha_1}}\right)e^{\sqrt{|\alpha_1|/4}|\xi|}\right] \times \left(e^{-\sqrt{|\alpha_1|/4}|\xi|} - 1\right),$$

$$u_{13}(\xi) = \frac{1}{\lambda}v_{13}(\xi) + d, \tag{64}$$

and the solitary wave solutions possess the expressions $v_{14}(\xi) = \varphi(\xi)$ *and* $u_{14}(\xi) = (1/\lambda)v_{14}(\xi) + d$, *and* $\varphi(\xi)$ is given by the implicit function

$$|\xi| = -\sqrt{\frac{4}{|\alpha_1|}}\left(\ln\left|\frac{2\sqrt{\Phi_4} + M_2}{r_{11} - r_{12}}\right| - \frac{r_{10}}{\sqrt{A_8}}\ln\left|\frac{2\sqrt{A_8\Phi_4} + B_8(\varphi - r_{10}) + 2A_8}{(r_{11} - r_{12})(\varphi - r_{10})}\right|\right), \tag{65}$$

where

$$\Phi_4 = (\varphi - r_{11})(\varphi - r_{12}),$$

$$M_2 = 2\varphi - r_{11} - r_{12},$$

$$r_{10} = -\frac{1}{2\alpha_1}\left(\beta + \sqrt{\beta^2 - 4\alpha_1\gamma}\right),$$

$$r_{11} = -\frac{1}{3\alpha_1}\left(3\alpha_1 r_{10} + 2\beta + \sqrt{6\alpha_1\beta r_{10} + 4\beta^2}\right),$$

$$r_{12} = -\frac{1}{3\alpha_1}\left(3\alpha_1 r_{10} + 2\beta - \sqrt{6\alpha_1\beta r_{10} + 4\beta^2}\right),$$

$$A_8 = (r_{10} - r_{11})(r_{10} - r_{12}),$$

$$B_8 = 2r_{10} - r_{11} - r_{12}. \tag{66}$$

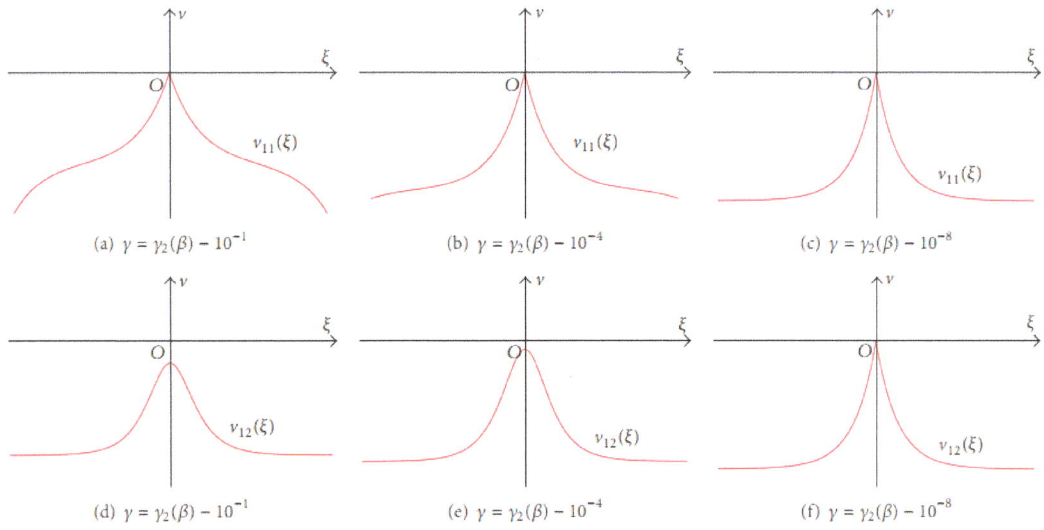

Figure 9: The peakon is bifurcated from the singular cusp wave and the solitary wave. The varying process for the figures of $v_{11}(\xi)$ and $v_{12}(\xi)$ when $\alpha_1 < 0, \beta < 0$, and $\gamma \to \gamma_2(\beta) - 0$ where $\alpha_1 = -2$ and $\beta = -9$.

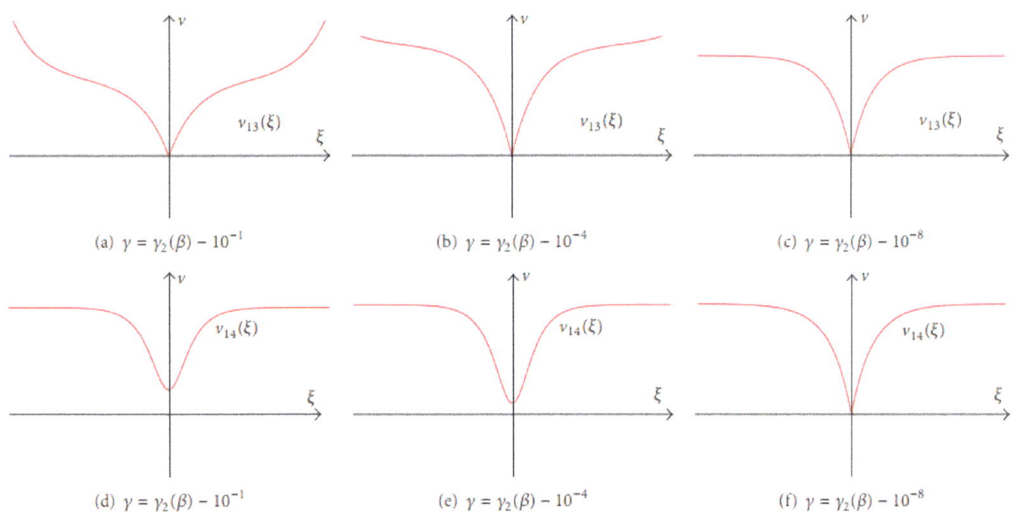

Figure 10: The peakon is bifurcated from the singular cusp wave and the solitary wave. The varying process for the figures of $v_{13}(\xi)$ and $v_{14}(\xi)$ when $\alpha_1 < 0, \beta > 0$, and $\gamma \to \gamma_2(\beta) - 0$ where $\alpha_1 = -2$ and $\beta = 9$.

The Derivations of Propositions 4 and 5. According to the qualitative theory, we obtain the bifurcation phase portraits of system (52) as in Figure 11. Through some orbits in Figure 11, we derive the results of Propositions4 and 5 as follows.

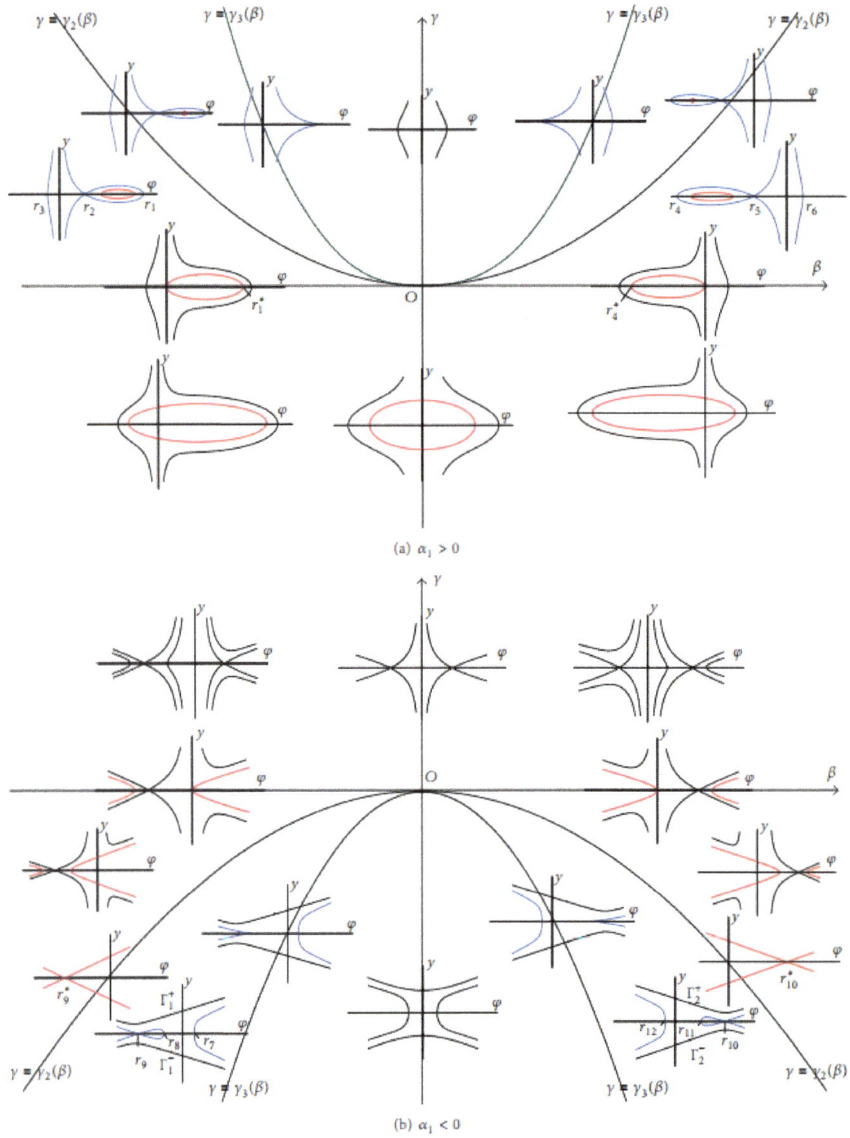

Figure 11: The bifurcation phase portraits of system (52).

(1) When $\alpha_1 > 0, \beta < 0$, and $0 < \gamma < \gamma_3(\beta)$ (as in Figure 11(a)), the homoclinic orbit owns the expression

$$y^2 = \frac{\alpha_1}{4\varphi^2}(\varphi - r_2)^2 (r_1 - \varphi)(\varphi - r_3) \quad (r_3 < r_2 < \varphi \le r_1). \tag{67}$$

Substituting (67) into $d\varphi/d\xi = y$ and integrating it, we have

$$|\xi| = \int_{\varphi}^{r_1} \sqrt{\frac{4s^2}{\alpha_1 (s - r_2)^2 (r_1 - s)(s - r_3)}} ds. \tag{68}$$

Completing the integral above, we get

$$|\xi| = \sqrt{\frac{4}{\alpha_1}} \left[\arcsin\left(\frac{r_1 + r_3 - 2\varphi}{r_1 - r_3} \right) + \frac{\pi}{2} - \frac{r_2}{\sqrt{A_5}} \ln|\Phi_1| \right]. \tag{69}$$

(2) When $\alpha_1 > 0, \beta > 0$, and $0 < \gamma < \gamma_3(\beta)$ (as in Figure 11(a)), the homoclinic orbit is expressed by

$$y^2 = \frac{\alpha_1}{4\varphi^2}(r_5 - \varphi)^2 (\varphi - r_4)(r_6 - \varphi) \quad (r_4 \le \varphi < r_5 < r_6). \tag{70}$$

Substituting (70) into $d\varphi/d\xi = y$ and integrating it, we have

$$|\xi| = \int_{r_4}^{\varphi} \sqrt{\frac{4s^2}{\alpha_1 (r_5 - s)^2 (s - r_4)(r_6 - s)}}. \tag{71}$$

Completing the integral above, we get

$$|\xi| = -\sqrt{\frac{4}{\alpha_1}}\left[\arcsin\left(\frac{r_6+r_4-2\varphi}{r_6-r_4}\right)-\frac{\pi}{2}+\frac{r_5}{\sqrt{A_6}}\ln|\Phi_2|\right]. \tag{72}$$

Note that when $\beta < 0$ and $\gamma \to 0+0$, it follows that $r_1 \to r_1^* = -4\beta/3\alpha_1$, $r_2 \to 0$, and $r_3 \to 0$. Further, we have

$$B_5 = r_1 + r_3 - 2r_2 \longrightarrow r_1^*,$$
$$A_5 = (r_1 - r_2)(r_2 - r_3) \longrightarrow 0,$$

$$\begin{aligned}\lim_{\gamma\to 0+0}\frac{r_3}{r_2} &= \lim_{\gamma\to 0+0} -\frac{3\alpha_1 r_2 + 2\beta + \sqrt{6\alpha_1\beta r_2 + 4\beta^2}}{3\alpha_1 r_2} \\ &= \lim_{\gamma\to 0+0} -\frac{1}{3\alpha_1 r_2} \\ &\quad\times\left[3\alpha_1 r_2 + 2\beta - 2\beta\left(1+\frac{3\alpha_1 r_2}{4\beta}+o\left(r_2^2\right)\right)\right] \\ &= \lim_{\gamma\to 0+0} -\frac{1}{3\alpha_1 r_2}\left(3\alpha_1 r_2 - \frac{3\alpha_1 r_2}{2}+o\left(r_2^2\right)\right) = -\frac{1}{2},\end{aligned}$$

$$\begin{aligned}\lim_{\gamma\to 0+0}\frac{r_2}{\sqrt{A_5}} &= \lim_{\gamma\to 0+0}\frac{r_2}{\sqrt{(r_1-r_2)(r_2-r_3)}} \\ &= \lim_{\gamma\to 0+0}\sqrt{\frac{r_2}{r_1+r_3-r_2-r_1(r_3/r_2)}} \\ &= \sqrt{\frac{0}{(3/2)r_1^*}} = 0,\end{aligned}$$

$$\begin{aligned}\lim_{\gamma\to 0+0}\Phi_1 &= \lim_{\gamma\to 0+0}\frac{(r_1-r_3)(\varphi-r_2)}{2\sqrt{A_5(r_1-\varphi)(\varphi-r_3)}+B_5(\varphi-r_2)+2A_5} \\ &= \frac{r_1^*\varphi}{r_1^*\varphi} = 1.\end{aligned} \tag{73}$$

Thus, from

$$|\xi| = \lim_{\gamma\to 0+0}\sqrt{\frac{4}{\alpha_1}}\left[\arcsin\left(\frac{r_1+r_3-2\varphi}{r_1-r_3}\right)+\frac{\pi}{2}-\frac{r_2}{\sqrt{A_5}}\ln|\Phi_1|\right], \tag{74}$$

we have

$$|\xi| = \sqrt{\frac{4}{\alpha_1}}\left[\arcsin\left(\frac{r_1^*-2\varphi}{r_1^*}\right)+\frac{\pi}{2}\right]. \tag{75}$$

Solving (75) for φ, we can get $\varphi = v_o(\xi)$ (see (55)), which is the same as (2).

Similarly, from (72), we can also get solution $\varphi \to v_o(\xi)$ when $\beta > 0$ and $\gamma \to 0+0$.

(3) When $\alpha_1 < 0$, $\beta \neq 0$, and $\gamma_3(\beta) < \gamma < \gamma_2(\beta)$ (as in Figure 11(b)), the curves $\Gamma_1^{\pm}$ and $\Gamma_2^{\pm}$ own the expression

$$y^2 = -\frac{\alpha_1}{4}\left(\varphi^2 + \frac{4\beta}{3\alpha_1}\varphi + \frac{2\gamma}{\alpha_1}\right). \tag{76}$$

Substituting (76) into $d\varphi/d\xi = y$ and integrating it, we have

$$\int_\varphi^0 \frac{ds}{\sqrt{s^2+(4\beta/3\alpha_1)s+(2\gamma/\alpha_1)}} = \sqrt{-\frac{\alpha_1}{4}}|\xi| \quad (\beta<0),$$
$$\int_0^\varphi \frac{ds}{\sqrt{s^2+(4\beta/3\alpha_1)s+(2\gamma/\alpha_1)}} = \sqrt{-\frac{\alpha_1}{4}}|\xi| \quad (\beta>0). \tag{77}$$

Completing the integrals above and solving the equations for φ, we get $\varphi = v_{11}(\xi)$ (see (61)) and $\varphi = v_{13}(\xi)$ (see (64)).

(4) When $\alpha_1 < 0$, $\beta \neq 0$, and $\gamma_3(\beta) < \gamma < \gamma_2(\beta)$ (as in Figure 11(b)), two homoclinic orbits can be expressed as

$$y^2 = -\frac{\alpha_1}{4\varphi^2}(\varphi-r_9)^2(r_7-\varphi)(r_8-\varphi) \quad (\beta<0), \tag{78}$$

$$y^2 = -\frac{\alpha_1}{4\varphi^2}(r_{10}-\varphi)^2(\varphi-r_{11})(\varphi-r_{12}) \quad (\beta>0) \tag{79}$$

Substituting (78) and (79) into $d\varphi/d\xi = y$ and integrating it, we have

$$|\xi| = \int_\varphi^{r_8}\sqrt{\frac{4s^2}{|\alpha_1|(s-r_9)^2(r_7-s)(r_8-s)}}ds,$$
$$|\xi| = \int_{r_{11}}^{\varphi}\sqrt{\frac{4s^2}{|\alpha_1|(r_{10}-s)^2(s-r_{11})(s-r_{12})}}ds. \tag{80}$$

Completing the integrals above, we get (62) and (65).

Note that when $\beta < 0$ and $\gamma \to \gamma_2(\beta) - 0$, it follows that $\sqrt{\gamma/2\alpha_1} \to \beta/3\alpha_1$, $r_7 \to 0$, $r_8 \to 0$, $r_9 \to r_9^* = -2\beta/3\alpha_1$ and

$$M_1 = 2\varphi - r_7 - r_8 \longrightarrow 2\varphi,$$
$$B_7 = r_8 + r_7 - 2r_9 \longrightarrow -2r_9^*,$$
$$\sqrt{A_7} = \sqrt{(r_8-r_9)(r_7-r_9)} \longrightarrow -r_9^*,$$
$$\sqrt{\Phi_3} = \sqrt{(r_7-\varphi)(r_8-\varphi)} \longrightarrow -\varphi. \tag{81}$$

Thus, we have

$$\lim_{\gamma \to \gamma_2 - 0} v_{13}(\xi)$$
$$= \lim_{\gamma \to \gamma_2 - 0} \left[\left(\frac{\beta}{3\alpha_1} - \sqrt{\frac{\gamma}{2\alpha_1}} \right) - \left(\frac{\beta}{3\alpha_1} + \sqrt{\frac{\gamma}{2\alpha_1}} \right) e^{-\sqrt{|\alpha_1|/4}|\xi|} \right]$$
$$\times \left(e^{\sqrt{|\alpha_1|/4}|\xi|} - 1 \right)$$
$$= \frac{2\beta}{3\alpha_1} \left(e^{-\sqrt{|\alpha_1|/4}|\xi|} - 1 \right)$$
$$= v_m(\xi) \quad (\text{see } (60)),$$

$$|\xi| = \lim_{\gamma \to \gamma_2 - 0} - \sqrt{\frac{4}{|\alpha_1|}} \times \left(\ln \left| \frac{2\sqrt{\Phi_3} - M_1}{r_7 - r_8} \right| + \frac{r_9}{\sqrt{A_7}} \ln \left| \frac{2\sqrt{A_7 \Phi_3} + B_7 (r_9 - \varphi) + 2A_7}{(r_7 - r_8)(r_9 - \varphi)} \right| \right)$$
$$= \lim_{\gamma \to \gamma_2 - 0} - \sqrt{\frac{4}{|\alpha_1|}} \times \ln \left| \left(\frac{2\sqrt{\Phi_3} - M_1}{r_7 - r_8} \right) \right. \tag{82}$$

$$|\xi| = \lim_{\gamma \to \gamma_2 - 0} - \sqrt{\frac{4}{|\alpha_1|}} \times \left(\ln \left| \frac{2\sqrt{\Phi_3} - M_1}{r_7 - r_8} \right| + \frac{r_9}{\sqrt{A_7}} \ln \left| \frac{2\sqrt{A_7 \Phi_3} + B_7 (r_9 - \varphi) + 2A_7}{(r_7 - r_8)(r_9 - \varphi)} \right| \right)$$
$$= \lim_{\gamma \to \gamma_2 - 0} - \sqrt{\frac{4}{|\alpha_1|}} \times \ln \left| \left(\frac{2\sqrt{\Phi_3} - M_1}{r_7 - r_8} \right) \times \left(\frac{2\sqrt{A_7 \Phi_3} + B_7 (r_9 - \varphi) + 2A_7}{(r_7 - r_8)(r_9 - \varphi)} \right)^{r_9/\sqrt{A_7}} \right|$$
$$= -\sqrt{\frac{4}{|\alpha_1|}} \ln \left| \frac{-4\varphi (r_9^* - \varphi)}{4 r_9^* \varphi} \right|$$
$$= -\sqrt{\frac{4}{|\alpha_1|}} \ln \left| \frac{r_9^* - \varphi}{r_9^*} \right|. \tag{83}$$

Solving (78) for φ, we obtain $\varphi = v_m(\xi)$ (see (60)).

Similarly, from (65), we can also get $\varphi = v_m(\xi)$ (see (60)) when $\beta > 0$ and $\gamma \to \gamma_2(\beta) - 0$. These complete the derivations of Propositions 4 and 5.

The Bifurcations of Solitary Waves for D(2, 2) System

When m=2 and n=2, system (1) becomes

$$u_t + \left(v^2\right)_x = 0,$$
$$v_t + a\left(v^2\right)_{xxx} + b u_x v + c u v_x = 0. \tag{84}$$

We will reveal the interesting bifurcation phenomenon to system (84). That is, the solitary waves can be bifurcated from the smooth periodic waves and the singular periodic waves. The concrete results are stated as follows.

Note that P(2, 2) system is read as

$$\frac{d\varphi}{d\tau} = 2\varphi y,$$
$$\frac{dy}{d\tau} = -2y^2 - \alpha_2 \varphi^3 - \beta\varphi - \gamma, \tag{85}$$

where α_2 is as (18) and β, γ are given in (13). Let

$$r_a = \frac{1}{3}\left[\frac{1}{\alpha_2}\left(\frac{5Q}{4}\right)^{1/3} - \beta\left(\frac{100}{Q}\right)^{1/3}\right],$$

$$r_b = \frac{\beta}{3}\left(\frac{25}{2Q}\right)^{1/3}\left(1 - i\sqrt{3}\right) - \frac{1}{6\alpha_2}\left(\frac{5Q}{4}\right)^{1/3}\left(1 + i\sqrt{3}\right),$$

$$r_c = \frac{\beta}{3}\left(\frac{25}{2Q}\right)^{1/3}\left(1 + i\sqrt{3}\right) - \frac{1}{6\alpha_2}\left(\frac{5Q}{4}\right)^{1/3}\left(1 - i\sqrt{3}\right),$$

$$Q = -27\alpha_2^2\gamma + \sqrt{80\alpha_2^3\beta^3 + 729\alpha_2^4\gamma^2},$$

$$\gamma_4(\beta) = \frac{4\sqrt{5}}{27}\left(-\frac{\beta^3}{\alpha_2}\right)^{1/2}, \tag{86}$$

and let $\gamma_1(\beta)$ be as (20).

Proposition 6. For fixed α_2 *and* β, system (84) has solitary wave solutions

$$v_i^{\pm}(\xi) = \pm\sqrt{-\frac{5\beta}{\alpha_2}\left[\frac{2}{3} - \tanh^2\left(\sqrt[4]{-\frac{\alpha_2\beta}{80}}\xi\right)\right]},$$

$$u_i(\xi) = \frac{1}{\lambda}v_i^2(\xi) + d, \tag{87}$$

which can be bifurcated from the following two types of nonlinear wave solutions.

(1) Smooth periodic wave solution:

$$v_{15}(\xi) = r_a - (r_a - r_b)\, sn^2(\eta_7\xi, k_3),$$

$$u_{15}(\xi) = \frac{1}{\lambda}v_{15}^2(\xi) + d, \tag{88}$$

Where

$$-\gamma_4(\beta) < \gamma < \gamma_4(\beta)$$

$$\eta_7 = \sqrt{\frac{\alpha_2}{20}}(r_a - r_c),$$

$$k_3 = \sqrt{\frac{r_a - r_b}{r_a - r_c}}. \tag{89}$$

(2)Singular periodic wave solutions:

$$v_{16}(\xi) = \frac{(A_9 + r_a)\, cn(\eta_8\xi, k_4) - A_9 + r_a}{cn(\eta_8\xi, k_4) + 1}, \tag{91}$$

$$u_{16}(\xi) = \frac{1}{\lambda}v_{16}^2(\xi) + d, \quad (\alpha_2 > 0, \beta < 0),$$

$$v_{17}(\xi) = \frac{A_9 + r_a - (A_9 - r_a)\, cn(\eta_8\xi, k_5)}{1 + cn(\eta_8\xi, k_5)},$$

$$u_{17}(\xi) = \frac{1}{\lambda}v_{17}^2(\xi) + d, \quad (\alpha_2 < 0, \beta > 0), \tag{92}$$

Where

$$
\begin{aligned}
&-\gamma_1(\beta) < \gamma < -\gamma_4(\beta),\\
&A_9^2 = (b_0 - r_a)^2 + a_0^2,\\
&a_0^2 = -\frac{1}{4}(r_b - r_c)^2,\\
&b_0 = \frac{1}{2}(r_b + r_c),\\
&\eta_8 = \sqrt{\frac{1}{5}A_9|\alpha_2|},\\
&k_4^2 = \frac{A_9 - b_0 + r_a}{2A_9},\\
&k_5^2 = \frac{A_9 + b_0 - r_a}{2A_9}.
\end{aligned} \tag{93}
$$

For the varying process, see Figures 12 and 13.

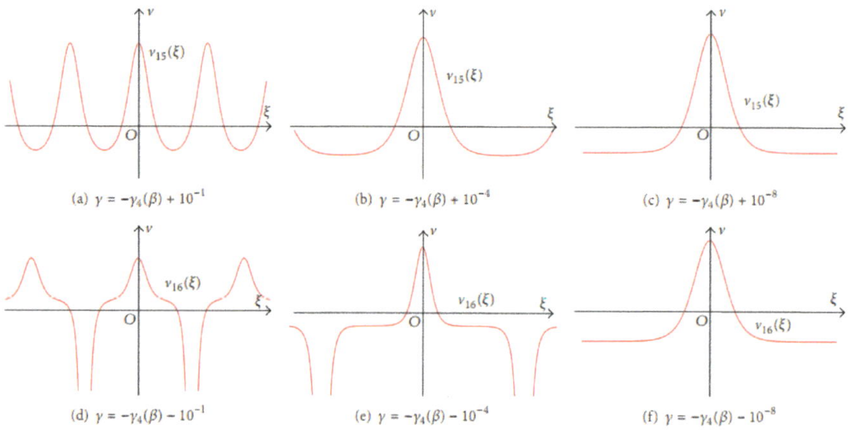

Figure 12: The solitary wave is bifurcated from the smooth periodic wave and the singular periodic wave. The varying process for the figures of $v_{15}(\xi)$ and $v_{16}(\xi)$ when $\alpha_2 = 5, \beta = -9, \gamma \rightarrow \gamma_4(\beta) + 0$, or $\gamma \rightarrow \gamma_4(\beta) - 0$.

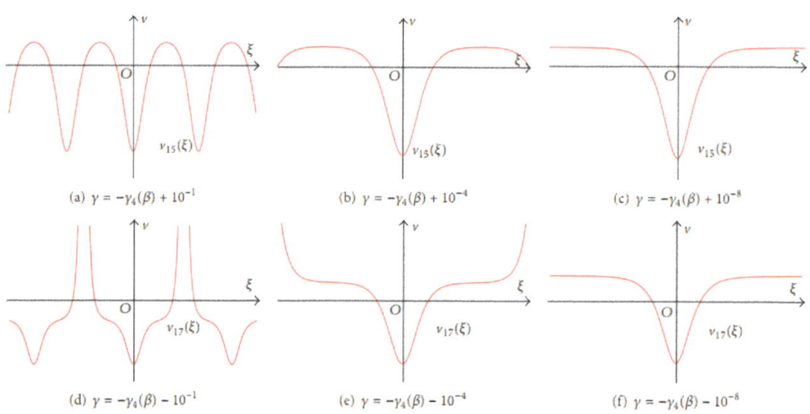

Figure 13: The solitary wave is bifurcated from the smooth periodic wave and the singular periodic wave. The varying process for the figures of $v_{15}(\xi)$ and $v_{17}(\xi)$ when $\alpha_2 = -5, \beta = 9, \gamma \rightarrow \gamma_4(\beta) + 0$, or $\gamma \rightarrow \gamma_4(\beta) - 0$.

The Derivations of Proposition 6. According to the qualitative theory, we also obtain the bifurcation phase portraits of system (85) as in Figure 14. Employing some orbits in Figure 14, we derive the results of Proposition 6 as follows.

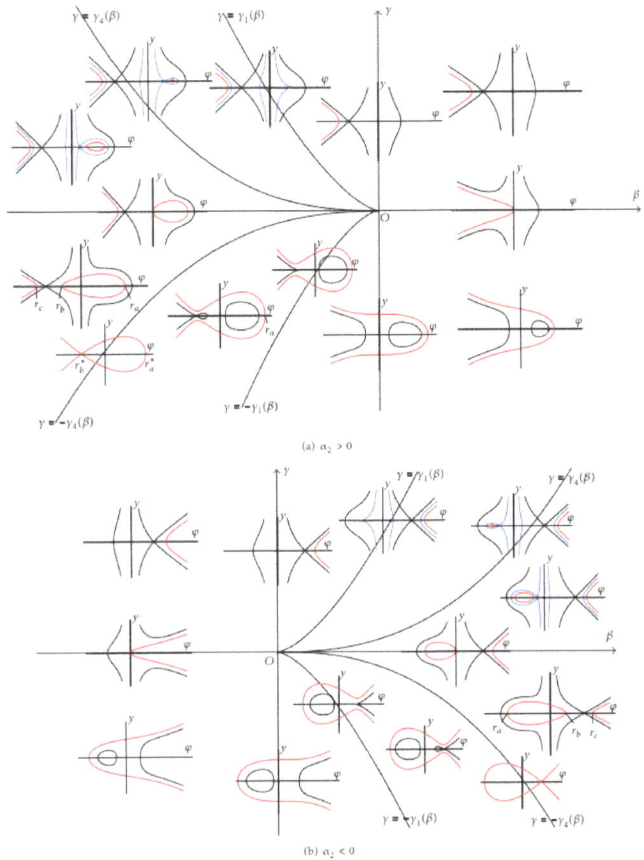

Figure 14: The bifurcation phase portraits of system (85).

(1) When $\alpha_2 > 0, \beta < 0$, and $-\gamma_4(\beta) < \gamma < 0$ (as in Figure 14(a)), the closed orbit owns the expression

$$y^2 = \frac{\alpha_2}{5}(r_a - \varphi)(\varphi - r_b)(\varphi - r_c) \quad (r_c < r_b \le \varphi \le r_a).$$

(93)

Substituting (93) into $d\varphi/d\xi = y$ and integrating it, we have

$$\int_{\varphi}^{r_a} \frac{ds}{\sqrt{(r_a - s)(s - r_b)(s - r_c)}} = \sqrt{\frac{\alpha_2}{5}}|\xi|. \tag{94}$$

Completing the integral above and solving the equation for φ, we get $v_{15}(\xi)$ (see (88)).

When $\alpha_2 > 0$ and $-\gamma_1(\beta) < \gamma < -\gamma_4(\beta)$, r_a is real and r_b and r_c become a pair of conjugate complex in (94). Completing the integral (94) and solving the equation for φ, we get $v_{16}(\xi)$ (see (90)).

(2) When $\alpha_2 < 0, \beta > 0$, similarly, we get $v_{15}(\xi)$ (see (88)) and $v_{17}(\xi)$ (see (91)).

Note that when $\alpha_2 > 0$ and $\gamma \to -\gamma_4(\beta) + 0$, it follows that $r_a \to r_a^* = (2/3)\sqrt{-5\beta/\alpha_2}$, $r_b \to r_b^* = -(1/3)\sqrt{-5\beta/\alpha_2}$ and $r_c \to r_b^* = -(1/3)\sqrt{-5\beta/\alpha_2}$. Further,

$$k_3 = \sqrt{\frac{r_a - r_b}{r_a - r_c}} \longrightarrow 1,$$

$$\eta_7 = \sqrt{\frac{\alpha_2}{20}(r_a - r_c)} \longrightarrow \sqrt[4]{-\frac{\alpha_2\beta}{80}}. \tag{95}$$

Thus, we have

$$\lim_{\gamma\to-\gamma_4+0} v_{15}(\xi) = \lim_{\gamma\to-\gamma_4+0} r_a - (r_a - r_b)\,\mathrm{sn}^2(\eta_7\xi, k_3)$$

$$= r_a^* - (r_a^* - r_b^*)\tanh^2\left(\sqrt[4]{-\frac{\alpha_2\beta}{80}}\xi\right)$$

$$= \sqrt{-\frac{5\beta}{\alpha_2}}\left[\frac{2}{3} - \tanh^2\left(\sqrt[4]{-\frac{\alpha_2\beta}{80}}\xi\right)\right]$$

$$= v_i^+(\xi) \quad (\text{see } (87)).$$

If $\alpha_2 > 0$ and $\gamma \to -\gamma_4(\beta) - 0$, then it follows that $r_a \to r_a^* = (2/3)\sqrt{-5\beta/\alpha_2}$, $r_b \to r_b^* = -(1/3)\sqrt{-5\beta/\alpha_2}$ and $r_c \to r_b^* = -(1/3)\sqrt{-5\beta/\alpha_2}$. Further, it follows that

$$a_0^2 = -\frac{1}{4}(r_b - r_c)^2 \longrightarrow 0,$$

$$b_0 = \frac{1}{2}(r_b + r_c) \longrightarrow r_b^*,$$

$$A_9 = \sqrt{(b_0 - r_a)^2 + a_0^2} \longrightarrow r_a^* - r_b^*,$$

$$k_4^2 = \frac{A_9 - b_0 + r_a}{2A_9} \longrightarrow 1,$$

$$\eta_8 = \sqrt{\frac{1}{5}A_9\alpha_2} \longrightarrow 2\sqrt[4]{-\frac{\alpha_2\beta}{80}}. \tag{97}$$

Thus, we get

$$\lim_{\gamma\to-\gamma_4-0} v_{16}(\xi) = \lim_{\gamma\to-\gamma_4-0} \frac{(A_9 + r_a)\,\mathrm{cn}(\eta_8\xi, k_4) - A_9 + r_a}{\mathrm{cn}(\eta_8\xi, k_4) + 1}$$

$$= \frac{(2r_a^* - r_b^*)\,\mathrm{sech}\left(2\sqrt[4]{-\alpha_2\beta/80}\xi\right) + r_b^*}{1 + \mathrm{sech}\left(2\sqrt[4]{-\alpha_2\beta/80}\xi\right)}$$

$$= \frac{(2r_a^* - r_b^*) + r_b^*\left[2\cosh^2\left(\sqrt[4]{-\alpha_2\beta/80}\xi\right) - 1\right]}{2\cosh^2\left(\sqrt[4]{-\alpha_2\beta/80}\xi\right)}$$

$$= (r_a^* - r_b^*)\,\mathrm{sech}^2\left(\sqrt[4]{-\frac{\alpha_2\beta}{80}}\xi\right) + r_b^*$$

$$= \sqrt{-\frac{5\beta}{\alpha_2}}\left[\frac{2}{3} - \tanh^2\left(\sqrt[4]{-\frac{\alpha_2\beta}{80}}\xi\right)\right]$$

$$= v_i^+(\xi) \quad (\text{see } (87)). \tag{98}$$

Similarly, we can also get

$v_{15}(\xi) \to v_i^-(\xi)$ and $v_{17}(\xi) \to v_i^-(\xi)$ when $\alpha_2 < 0$, $\beta > 0$, and $\gamma \to -\gamma_4(\beta) \pm 0$, and . Hereto, we have completed all of the derivations.

CONCLUSIONS

In this paper, by employing the bifurcation method and qualitative theory of dynamical systems, we have revealed some interesting bifurcation phenomena of nonlinear waves for the D(m, n) system (1). Firstly, for D(2, 1) system, we have pointed out that the fractional solitary waves can be bifurcated from the trigonometric periodic waves and the elliptic periodic waves (see Figures 1 and 2). In the meantime, the kink waves can be bifurcated from the solitary waves and the singular waves (see Figures 3–6). Secondly, for D(1, 2) system, we have showed that the solitary waves can turn into the compactons (see Figure 8) and the peakons can be bifurcated from the singular cusp waves and the solitary waves (see Figures 9 and 10). Thirdly, for D(2, 2) system, we have confirmed that the solitary waves can be bifurcated from the smooth periodic waves and the singular periodic waves (see Figures 12 and 13).

Conflict of Interests

The authors declare that there is no conflict of interests regarding the publication of this paper.

ACKNOWLEDGMENT

This research is supported by the National Natural Science Foundation of China (no. 11171115).

REFERENCES

1. P. J. Olver, Applications of Lie Groups to Differential Equations, vol. 107 of Graduate Texts in Mathematics, Springer, New York, NY, USA, 2nd edition, 1993.
2. Ü. Göktaş and W. Hereman, "Symbolic computation of conserved densities for systems of nonlinear evolution equations," Journal of Symbolic Computation, vol. 24, no. 5, pp. 591–622, 1997.
3. F. D. Xie and Z. Y. Yan, "New exact solution profiles of the nonlinear dispersion Drinfel'd-Sokolov (D(m,n)) system," Chaos, Solitons and Fractals, vol. 39, no. 2, pp. 866–872, 2009.
4. X. J. Deng, J. L. Cao, and X. Li, "Travelling wave solutions for the nonlinear dispersion Drinfel'd-Sokolov (D(m,n)) system," Communications in Nonlinear Science and Numerical Simulation, vol. 15, no. 2, pp. 281–290, 2010.
5. K. L. Zhang, S. Q. Tang, and Z. J. Wang, "Bifurcations of travelling wave solutions for the nonlinear dispersion Drinfel'd-Sokolov (D(m,n)) system," Applied Mathematics and Computation, vol. 217, no. 4, pp. 1620–1631, 2010.
6. J. Y. Hu, "Classification of single traveling wave solutions to the nonlinear dispersion Drinfel'd-Sokolov system," Applied Mathematics and Computation, vol. 219, no. 4, pp. 2017–2025, 2012.
7. J. P. Wang, "A list of 1 + 1 dimensional intergrable equations and their propertise," Journal of Mathematical Physics, vol. 9, pp. 213–233, 2002.
8. A. M. Wazwaz, "Exact and explicit travelling wave solutions for the nonlinear Drinfeld-Sokolov system,"Communications in Nonlinear Science and Numerical Simulation, vol. 11, no. 3, pp. 311–325, 2006.
9. J. Biazar and Z. Ayati, "Exp and modified Exp function methods for nonlinear Drinfeld-Sokolov system,"Journal of King Saud University—Science, vol. 24, no. 4, pp. 315–318, 2012.
10. F. Zhang, J. M. Qi, and W. J. Yuan, "Further results about traveling wave exact solutions of the Drinfeld-Sokolov Equations," Journal of Applied Mathematics, vol. 2013, Article ID 523732, 6 pages, 2013.
11. A. Bekir, "New exact travelling wave solutions for regularized long-wave, phi-four and Drinfeld-Sokolov equations," International Journal of Nonlinear Science, vol. 6, pp. 46–52, 2008.
12. S. A. El-Wakil and M. A. Abdou, "Modified extended tanh-function method for solving nonlinear partial differential equations," Chaos, Solitons and Fractals, vol. 31, no. 5, pp. 1256–1264, 2007. ·
13. M. Gürses and A. Karasu, "Integrable KdV systems: recursion operators of degree four," Physics Letters A: General, Atomic and Solid State Physics, vol. 251, no. 4, pp. 247–249, 1999.
14. P. Casati and M. Pedroni, "Drinfeld-Sokolov reduction on a simple lie algebra from the bihamiltonian point of view," Letters in Mathematical Physics, vol. 25, no. 2, pp. 89–101, 1992.
15. F. Delduc, L. Fehér, and L. Gallot, "Nonstandard Drinfeld-Sokolov reduction," Journal of Physics A: Mathematical and General, vol. 31, no. 25, pp. 5545–5563, 1998.
16. L. Fehér, J. Harnad, and I. Marshall, "Generalized Drinfeld-Sokolov reductions and KdV type hierarchies," Communications in Mathematical Physics, vol. 154, no. 1, pp. 181–214, 1993.
17. L. Fehér and I. Marshall, "Extensions of the matrix Gelfand-Dickey hierarchy from generalized Drinfeld-Sokolov reduction," Communications in Mathematical Physics, vol. 183, no. 2, pp. 423–461, 1997.
18. B. Feigin, "Quantization of the Drinfeld-Sokolov reduction," Physics Letters B: Nuclear, Elementary Particle and High-Energy Physics, vol. 246, no. 1-2, pp. 75–81, 1990.
19. E. Frenkel, V. Kac, and M. Wakimoto, "Characters and fusion rules for W-algebras via quantized Drinfeld-Sokolov reduction," Communications in Mathematical Physics, vol. 147, no. 2, pp. 295–328, 1992.
20. E. Frenkel, N. Reshetikhin, and M. A. Semenov-Tian-Shansky, "Drinfeld-Sokolov reduction for difference operators and deformations of W-algebras I. The case of Virasoro algebra," Communications in Mathematical Physics, vol. 192, no. 3, pp. 605–629, 1998. ·
21. M. A. Semenov-Tian-Shansky and A. V. Sevostyanov, "Drinfeld-Sokolov reduction for difference operators and deformations of W-algebras II. The general semisimple case," Communications in Mathematical Physics, vol. 192, no. 3, pp. 631–647, 1998. ·
22. A. D. Sole, V. G. Kac, and D. Valeri, "Classical W-algebras and generalized Drinfeld-Sokolov Bi-

Hamiltonian systems within the theory of poisson vertex algebras," Communications in Mathematical Physics, vol. 323, pp. 663–711, 2013.

23. Y. Long, W.-G. Rui, B. He, and C. Chen, "Bifurcations of travelling wave solutions for generalized Drinfeld-Sokolov equations," Applied Mathematics and Mechanics, vol. 27, no. 11, pp. 1549–1555, 2006. ·
24. Y. Ugurlu and D. Kaya, "Exact and numerical solutions of generalized Drinfeld-Sokolov equations,"Physics Letters A: General, Atomic and Solid State Physics, vol. 372, no. 16, pp. 2867–2873, 2008.
25. Z. S. Wen, Z. R. Liu, and M. Song, "New exact solutions for the classical Drinfel'd-Sokolov-Wilson equation," Applied Mathematics and Computation, vol. 215, no. 6, pp. 2349–2358, 2009. ·
26. E. Sweet and R. A. van Gorder, "Analytical solutions to a generalized Drinfel'd-Sokolov equation related to DSSH and KdV6," Applied Mathematics and Computation, vol. 216, no. 10, pp. 2783–2791, 2010.
27. E. Sweet and R. A. van Gorder, "Trigonometric and hyperbolic type solutions to a generalized Drinfel'd-Sokolov equation," Applied Mathematics and Computation, vol. 217, no. 8, pp. 4147–4166, 2010.
28. E. Sweet and R. A. van Gorder, "Exponential-type solutions to a generalized Drinfel'd-Sokolov equation,"Physica Scripta, vol. 82, no. 3, Article ID 035006, 2010.
29. E. Sweet and R. A. van Gorder, "Traveling wave solutions (μ,v) to a generalized Drinfel'd-Sokolov system which satisfy u=a1vm+a0," Applied Mathematics and Computation, vol. 218, no. 19, pp. 9911–9921, 2012. ·
30. L. P. Wu, S. F. Chen, and C. P. Pang, "Traveling wave solutions for generalized Drinfeld-Sokolov equations," Applied Mathematical Modelling, vol. 33, no. 11, pp. 4126–4130, 2009. ·
31. A. Karasu and S. Y. Sakovich, "Bäcklund transformation and special solutions for the Drinfeld-Sokolov-Satsuma-Hirota system of coupled equations," Journal of Physics A: Mathematical and General, vol. 34, no. 36, pp. 7355–7358, 2001. ·

CHAPTER 5

BEYOND IN-PHASE AND ANTI-PHASE COORDINATION IN A MODEL OF JOINT ACTION

Daniele Avitabile, Piotr Słowi´nski, Benoit Bardy, Krasimira Tsaneva-Atanasova

Keywords: Coupled oscillators, Dynamical system, Bifurcation analysis, Coordination regimes, Numerical continuation, Parameter dependence

INTRODUCTION

Many body movements are periodic in their nature [18]. For example, postural sway [2], walking [8, 20], running [37], swimming [54], galloping [8, 20] and juggling [58] have a cyclic pattern in the position of the end effectors or joint angles. Synchronisation is a fundamental aspect of oscillatory coordination dynamics in human and animal body movements [33] and has been found in many different situations [48]. Coordination is characterised by a bounded temporal relationship created by a convergent dynamical process [26, 44]. Coordination regimes depend on symmetries and couplings between oscillators. Frequency entrainment, where two oscillators adopt a central frequency, occurs even with a very weak coupling. With a relatively strong coupling or if the system is symmetrical, phase entrainment can also take place. These processes may be continuous or intermittent; that is, the phases of the two oscillators may also align periodically [18, 35, 47].

In the case of two coupled oscillators, the regular patterns of coordination are well captured by the properties of the relative phase between the periodic movements of the two subsystems [32, 33]. The simplest pattern is observed when the phase of the two oscillators coincide to give in-phase (0°) monostable coordination pattern. An example of this behaviour is given by iso-lateral limb movements [6]. Monostable anti-phase (180°) coordination can also occur, and an example of such behaviour is observed in team sports (competitive games) [5, 10, 12, 15]. In many real systems, anti-phase stability coexists with in-phase stability [22, 33, 44,56]. Previous studies address the modelling of the two coupled oscillators as a nonlinear dynamical system, the fitting of its periodic orbits to human movements [28] and the systematic analysis of the effects of linear and nonlinear terms to the observed limit cycles [4]. The observed relations between frequency and amplitude [22] as well as peak velocities [44] in many but not all [3] oscillatory movements turned out to be well represented by a hybrid oscillator [22] formed by a combination of Van der Pol and Rayleigh nonlinear damping terms.

A classical example of model exhibiting bistability is the so-called HKB model proposed in the seminal work by Haken, Kelso and Bunz [18, 22]. The model, which was originally developed for bimanual finger coordination [30], was found to be representative of a wide range of applications in human movement [7,18], suggesting that the dynamics that it produces are somehow fundamental and make formal construct for the study of coordination dynamics [32, 33]. Although the model was originally developed in order to account for intra-personal phenomena, the same patterns have been shown to be representative of both sensorimotor and interpersonal behaviours [31, 49, 50]. The model successfully reproduces not only the patterns of stability observed in bimanual coordination experiments but also their dependence upon frequency [22]. The HKB model admits a potential function that yields the experimentally observed change in attractors' landscapes. Furthermore, the HKB model and its stochastic extension reproduced the characteristic fluctuation increase and slowing down observed experimentally near instabilities [51].

Citation: Daniele Avitabile, Piotr Słowiński, Benoit Bardy, and Krasimira Tsaneva-Atanasova, Beyond in-phase and anti-phase coordination in a model of joint action, doi: 10.1007/s00422-016-0691-9.

The development of the HKB model has been inspired by the in-phase and anti-phase coordination dynamics observed in bimanual coordination in the context of the finger movements experiment [30, 52, 53]. Therefore, most previous research has focused on a fixed set of model parameters that guarantees the stability of these particular dynamics. Furthermore, significant contributions to understanding these coordination patterns (albeit in a narrow parameter range and with limiting assumptions on the parameters controlling the coupling strength) have been made for different oscillator frequencies and inputs [1, 3, 7, 17, 19, 25] as well as noise in the system [9, 49, 50]. All previous mathematical analyses of the HKB model have focused on the relative phase dynamics, under the assumption that the amplitude of the coupled oscillators is constant [1, 3,9, 17, 19, 22]. Several recent articles have studied the phase-approximation dynamics in the HKB model by considering the multiple stable states of the system and the ability to switch between them by changing the frequency and the coupling parameters [38]. The bifurcations leading to transitions between anti-phase and in-phase dynamics in a reduced phase approximation of the HKB model [16] have been also studied. To our knowledge, however, a bifurcation analysis of the full four-dimensional HKB model, considering all model parameters as well as general (i.e. weak and strong) coupling strengths, has not been performed. Such analysis could provide an insight into other possible qualitative behaviours that the solutions of the model might exhibit, as well as characterise the possible changes in the dynamics of the solutions corresponding to any changes in the parameter values of the model. We also note recent further developments of dynamical systems' approaches for studying sensorimotor dynamics, involving dynamical repertoires, hierarchies of timescales and structured flows on manifolds [23].

Given that the HKB model is a widely accepted tool in this field, it is imperative to examine systematically all the possible coordination regimes supported by this system. In addition, classifying changes of dynamical regimes in terms of positions and velocities of the two coupled oscillators would undoubtedly shed light on the HKB model's applicability to explain movement coordination in joint actions and human interactions with an adaptive virtual partner (VP)[13, 34, 40, 59–61]. In the present paper, we take a different approach in analysing the HKB model, as we study the full four-dimensional system of first order differential equations describing the evolution of the positions and velocities of the two coupled oscillators. We begin by characterising the local and global dynamics of the single HKB oscillator and reveal a global transition in the model that governs the existence of periodic solutions in a range of the oscillator's parameter values. We proceed by systematically characterising the full HKB model dynamics not only by varying the coupling strength parameters but also the rest of the model parameters, i.e. the parameters governing the single oscillator's properties. In addition to the very well-studied coordination patterns, we find a stable phase-locked solution that spans a wide range of relative phases and persists for a wide range of model parameters' values. We also show that relaxing the constant amplitude assumption allows for much richer coordination dynamics and coexistence of various stable coordination attractors (multi-stability regimes).

RESULTS

Intrinsic Properties of the Oscillator in the HKB Model

Recently, a significant scientific effort has been put towards the development of VP interaction systems. In particular, the single HKB oscillator is being used to drive the movement dynamics exhibited by the VP [13,34, 59–61]. The dynamics of the model is an important consideration in designing such systems and in particular for parametrising the ordinary differential equation that governs the behaviour of the VP. For example depending on the constraints of the experimental set-up, a certain range of amplitude and/or frequency for the VP periodic behaviour might be desirable. Although some properties of the HKB oscillator have been measured and studied both experimentally and analytically [28, 29], the dynamics of the single HKB oscillator has not been systematically investigated theoretically. To address this gap, we begin by examining a single HKB oscillator:

$$\ddot{x} = -\dot{x}\left(\alpha x^2 + \beta \dot{x}^2 - \gamma\right) - \omega^2 x,$$

which could be written as a planar autonomous dynamical system of the form:

$$\dot{x} = y,$$
$$\dot{y} = -y\left(\alpha x^2 + \beta y^2 - \gamma\right) - \omega^2 x, \qquad (1)$$

where x represents the position, y the velocity, $\omega \in \mathbb{R}^+$ is related to the natural frequency of the oscillator andα, β, $\gamma \in \mathbb{R}$ are parameters governing the intrinsic dynamics of Eq. (1).

The single HKB oscillator is a hybrid Rayleigh–Van der Pol [22] planar system, and although the analysis of planar systems of ordinary differential equations is very well established [21, 27, 43], it has not been applied to the single HKB oscillator model. Furthermore, whenever planar systems are coupled, they are often studied in the weak coupling limit, which we don't require for the numerical continuation analysis presented here. In our analysis, we focus on the global dynamics of the system and aim to characterise all possible dynamic states that the single HKB oscillator model supports, as well as their dependence on all model parameters. System (1) admits the origin (0, 0) as a trivial steady state for any parameter value $\omega \in \mathbb{R}^+$. Given $\omega > 0$, the Jacobian matrix at the trivial equilibrium $(x, y) = (0, 0)$ is

$$J = \begin{bmatrix} 0 & 1 \\ -\omega^2 & \gamma \end{bmatrix}$$

For $|\gamma| \geq 2\omega$, the Jacobian has a pair of nonzero real eigenvalues:

$$\lambda = \frac{\gamma \pm \sqrt{\gamma^2 - 4\omega^2}}{2}$$

Thus, the equilibrium is a stable node (sink) for $\gamma < 0$ and unstable node (source) for $\gamma > 0$. For $|\gamma| < 2\omega$, the Jacobian has a pair of complex conjugate eigenvalues of the form:

$$\lambda = \frac{\gamma}{2} \pm i\frac{\sqrt{4\omega^2 - \gamma^2}}{2}$$

Hence, the equilibrium is a stable focus (spiral sink) for $-2\omega < \gamma < 0$ and unstable focus (spiral source) for $0 < \gamma < 2\omega$.

Changing the value of the parameter γ near $\gamma = 0$ leads to a change in the sign of the eigenvalues' real part, which is associated with loss or gain of stability. The system undergoes a Hopf bifurcation at $\gamma = 0$, which gives rise to oscillations. We could analytically verify further the sufficient conditions for the existence of Hopf bifurcation by showing that:

$$\left.\frac{\partial \lambda_r(\gamma)}{\partial \gamma}\right|_{\gamma=0} = \frac{1}{2} \neq 0,$$

$$l_1(\gamma)|_{\gamma=0} = \frac{-(\alpha + 3\beta\omega^2)}{2\omega(\omega^2+1)} \neq 0$$

$$\iff \alpha + 3\beta\omega^2 \neq 0,$$

where λ_r and l_1 are the real part of the eigenvalues and the first Lyapunov coefficient [36], respectively. The sign of the first Lyapunov coefficient [36] determines whether the Hopf bifurcation is subcritical or supercritical; hence, we are in the supercritical (subcritical) case if $\alpha + 3\beta\omega^2 > 0$ (< 0). The system (1) has a degenerate Hopf bifurcation when $\alpha + 3\beta\omega^2 = 0$.

Next we carry out bifurcation analysis using numerical continuation in AUTO [11]. We set NTST = 50,NCOL = 4 for the mesh, and EPSL = 10^{-9}, EPSU = 10^{-9} for the tolerances of the Newton solver. In Fig. 1a, we continue the trivial steady state $(x, y) = (0, 0)$ in γ: oscillations arise at a supercritical Hopf bifurcation at $\gamma = 0$ since for $\alpha = 1$ and $\beta = 1$, $\alpha + 3\beta\omega^2 > 0$. Stable periodic solutions exist for various values of the intrinsic frequency ω: the lower ω, the larger the oscillations amplitude and the longer the period (see inset in Fig. 1b). Similar scenarios are found for various combinations of α and β (Fig. 1c, d). The amplitude of the periodic solutions increase as either α or β are decreased. For $\alpha = -1$, the Hopf bifurcation is supercritical and the oscillatory branch is stable, whereas for $\beta = -0.1$ the Hopf bifurcation is subcritical and oscillatory branches, which are originally unstable, restabilise at a saddle node. We note that system (1) exhibits bistability between a stable equilibrium and a stable periodic states in the case of $\beta = -0.1$ (Fig. 1d).

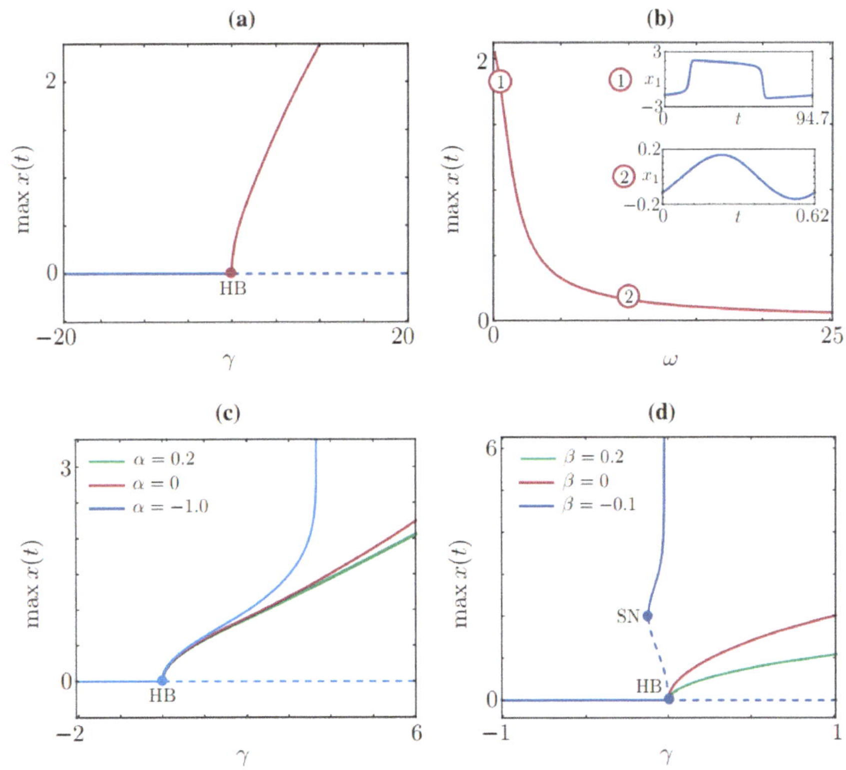

Figure 1: Bifurcation diagrams for a single HKB oscillator. a The trivial equilibrium becomes unstable at a supercritical Hopf bifurcation (HB) in the continuation parameter γ for $\omega = 2$, $\alpha = 1$, $\beta = 1$. b The periodic orbit for $\gamma = 2$ is continued in the parameter ω. The lower ω, the larger the oscillations amplitude and the longer the period. c–d Continuations in γ are repeated for various values of α and β. In panel c, for $\alpha = -1$, the periodic branch undergoes a global bifurcation (vertical asymptote), whereas in panel d, for $\beta = -0.1$, the Hopf bifurcation is subcritical, and the emanating branch restabilises at a saddle-node bifurcation, before disappearing in a global bifurcation. Solid (dashed) lines represent stable (unstable) states of (1).

The above analysis reveals that when parameters α and β have opposite signs (Fig. 1c, d), there is a critical value for γ at which the amplitude (and period) of the stable limit cycle solutions in the model rapidly increase to infinity. As a result, all periodic solutions vanish for values of γ above this critical value. Furthermore, such transitions occur robustly for a large range of α and β parameter values. We believe it is important to understand where and why this singularity occurs, as it corresponds to a non-physical behaviour. Since this feature has not previously been reported in the literature on the HKB model, we present a thorough investigation of this phenomenon. In order to analyse the behaviour of the system at infinity, we employ methods presented in Chapter 3.10 of reference [46]. We start by projecting system (1) on the Poincaré sphere using the following transformation:

$$X = \frac{x}{\sqrt{1+x^2+y^2}}, \quad Y = \frac{y}{\sqrt{1+x^2+y^2}},$$
$$Z = \frac{1}{\sqrt{1+x^2+y^2}},$$

which defines one-to-one correspondence between points (X, Y, Z) on the upper hemisphere S^2 with $Z > 0$and points (x, y) in the plane defined by:

$$x = \frac{X}{Z}, \quad y = \frac{Y}{Z},$$

The points on the equator of S^2 correspond to points at infinity of $\mathbb{R}^2$. Under the transformation above, the HKB oscillator on S^2 with $Z > 0$ is given by:

$$\begin{aligned}
\dot{X} &= \frac{Y}{Z^2}\left[\alpha X^3Y+\beta XY^3+Z^2(-\gamma XY+(\omega^2-1)X^2+1)\right] \\
\dot{Y} &= \frac{1}{Z^2}\Big[(Y^2-1)Y(\alpha X^2+\beta Y^2) \\
&\quad + Z^2\big(Y(-\gamma(Y^2-1)-XY)+\omega^2X(Y^2-1)\big)\Big] \\
\dot{Z} &= \frac{Y}{Z}\left[\alpha X^2Y+\beta Y^3+Z^2((\omega^2-1)X-\gamma Y)\right]
\end{aligned} \qquad 2$$

System (1) has eight equilibria on the equator $X^2 + Y^2 = 1$ of S^2 (see Theorem 1 from Chapter 3.10 of [46]) that represents the limit $x, y \to \infty$. In general, the equilibria are given by the solutions of the following equation:

$$XQ_m(X, Y) - YP_m(X, Y) = 0 \qquad 3$$

where P_m and Q_m are homogeneous m-th degree polynomials in x and y according to the following representation of the system (1):

$$\begin{aligned}
\dot{x} &= P(x, y) = P_1(x, y) + \cdots + P_m(x, y) \\
\dot{y} &= Q(x, y) = Q_1(x, y) + \cdots + Q_m(x, y)
\end{aligned} \qquad 4$$

In our case, the highest degree homogeneous polynomials are:

$$\begin{aligned}
P_3(x, y) &= 0 \\
Q_3(x, y) &= -\alpha x^2 y - \beta y^3
\end{aligned} \qquad 5$$

Hence, all equilibria at the equator are the solutions of the following system of equations:

$$\begin{aligned}
&X^2 + Y^2 = 1 \\
&XQ_3(X, Y) - YP_3(X, Y) = -\alpha X^3 Y - \beta Y^4 = 0
\end{aligned} \qquad 6$$

and are given by:

$$\begin{aligned}
X &= 0, & Y &= \pm 1, \\
X &= \pm 1, & Y &= 0, \\
X &= \pm\frac{\sqrt{\alpha}}{\sqrt{\alpha-\beta}}, & Y &= \pm\frac{\sqrt{\beta}}{\sqrt{\beta-\alpha}}.
\end{aligned} \qquad 7$$

The flow between the nodes is determined using the following equation (see Theorem 1 from Chapter 3.10 of [46]):

$$G_{m+1} = \cos\theta\, Q_m(\cos\theta, \sin\theta) - \sin\theta\, P_m(\cos\theta, \sin\theta) = 0, \qquad 8$$

where θ is an angle along the equator.

The flow between the equilibria on the equator of the Poincaré sphere is counterclockwise if $G_{m+1} > 0$ and clockwise where $G_{m+1} < 0$. We find that only the equilibria $X = 0$, $Y = \pm 1$ are hyperbolic. They are stable nodes for $\beta < 0$ and are unstable nodes for $\beta > 0$. The other six equilibria as given in (7) are non-hyperbolic. We established their types by combining information gathered from the flow on the equator and from numerical integration of the transformed system (2). We summarise our findings in two representative cases in which, as the parameter γ increases, the period and amplitude of the stable periodic orbit grows to infinity exponentially fast and the periodic orbit disappears. More specifically, at the critical value γ^*, the stable periodic orbit becomes a heteroclinic cycle connecting four equilibria of saddle type at the equator on the Poincaré sphere.

In Fig. 2, we illustrate how the structure of the global phase portrait of the system (2), projected on the (X, Y)-plane, changes with increasing γ. Fig. 2a–c, for $\alpha = -1$, $\beta = 1$, shows the transition occurring as γ is varied in the bifurcation diagram of Fig. 1c; and Fig. 2d–f, for $\alpha = 1$, $\beta = -0.1$ shows the transition occurring as γ is varied in the bifurcation diagram of Fig. 1d. In both cases, the disappearance of the stable limit cycle solution in the model is due to the same mechanism. However, depending on the signs of the parameters α, β, different invariant objects are involved in the transition. Panels (a–c) in Fig. 2 show that there are two types of connecting orbits in the phase space of the HKB oscillator. The first type connects the unstable equilibria $(0, \pm 1)$ (red dots) with the saddle points $(\pm 1, 0)$ (black dots) and the second connects the saddle points $(\mp\sqrt{\alpha}/\sqrt{\alpha-\beta}, \pm\sqrt{\beta}/\sqrt{\beta-\alpha})$ (black dots) with the stable periodic orbit surrounding the unstable equilibrium at the origin $(0, 0)$. As the parameter γ increases, the two types of connections become tangent and the periodic orbit stretches along the X axis as depicted in panel (b) for value of $\gamma = 3.65860608978$ (just before the transition). At the critical value, $\gamma = \gamma^*$, the periodic orbit becomes a heteroclinic cycle connecting four saddle equilibria. After the transition, the heteroclinic cycle disappears and the global phase portrait changes. In panel (c), we show that after the transition there are connections between the saddle points $(\mp\sqrt{\alpha}/\sqrt{\alpha-\beta}, \pm\sqrt{\beta}/\sqrt{\beta-\alpha})$ and the stable equilibria $(\pm\sqrt{\alpha}/\sqrt{\alpha-\beta}, \pm\sqrt{\beta}/\sqrt{\beta-\alpha})$, and between the saddle points $(\pm 1, 0)$ and the unstable equilibrium at the origin $(0, 0)$. In this case, the single HKB oscillator has stable periodic solutions only for $\gamma \in (0, \gamma^*)$. Panels (d, e) in Fig. Fig.22 demonstrate that, for $\alpha = 1$, $\beta = -0.1$, in addition to the stable periodic orbit there is also an unstable periodic orbit (red loop) surrounding the stable equilibrium at the origin (0, 0) (green dot). Although unstable periodic orbits could not be observed experimentally, such objects are important from dynamical systems point of view. For example, in this case the branch of unstable periodic orbits forms the boundary between the basins of attraction of the coexisting stable equilibrium and stable periodic orbit for $\gamma \in (\gamma_{SN}, \gamma^*)$ (see Fig. Fig.1d).1d). Irrespective of the presence of unstable limit cycle, we find again two types of connecting orbits in the phase space for $\gamma < \gamma^*$ as shown in panel (d). The first type connects the unstable equilibria $(\mp\sqrt{\alpha}/\sqrt{\alpha-\beta}, \pm\sqrt{\beta}/\sqrt{\beta-\alpha})$ (red dots) with the saddle points $(\pm\sqrt{\alpha}/\sqrt{\alpha-\beta}, \pm\sqrt{\beta}/\sqrt{\beta-\alpha})$ (black dots). The second type connects the saddle points $(\pm 1, 0)$(black dots) with the stable periodic orbit. Here we observe again that the connections become tangent to the periodic orbit, as it stretches along the X-axis growing into a heteroclinic cycle between four saddle equilibria (black dots) for $\gamma = \gamma^*$, as depicted in panel (e) where $\gamma = -0.022618$ (just before the transition). After the transition $\gamma > \gamma^*$, the heteroclinic cycle disappears and the invariant objects of the system reconnect. This, however, occurs in a different manner compared to the case presented in panels (a–c). The saddle equilibria$(\pm 1, 0)$ are now connected with stable nodes $(0, \pm 1)$ and the unstable periodic orbit is connected to the saddle points $(\pm\sqrt{\alpha}/\sqrt{\alpha-\beta}, \pm\sqrt{\beta}/\sqrt{\beta-\alpha})$. In panel (f), we show the phase portrait for $\gamma = 0.5$, which illustrates the connections after the unstable periodic orbit disappeared in a subcritical Hopf bifurcation (at $\gamma = 0$, compare with Fig. 1). In this case, the single HKB oscillator has stable periodic solutions only for $\gamma \in (\gamma_{SN}, \gamma^*)$.

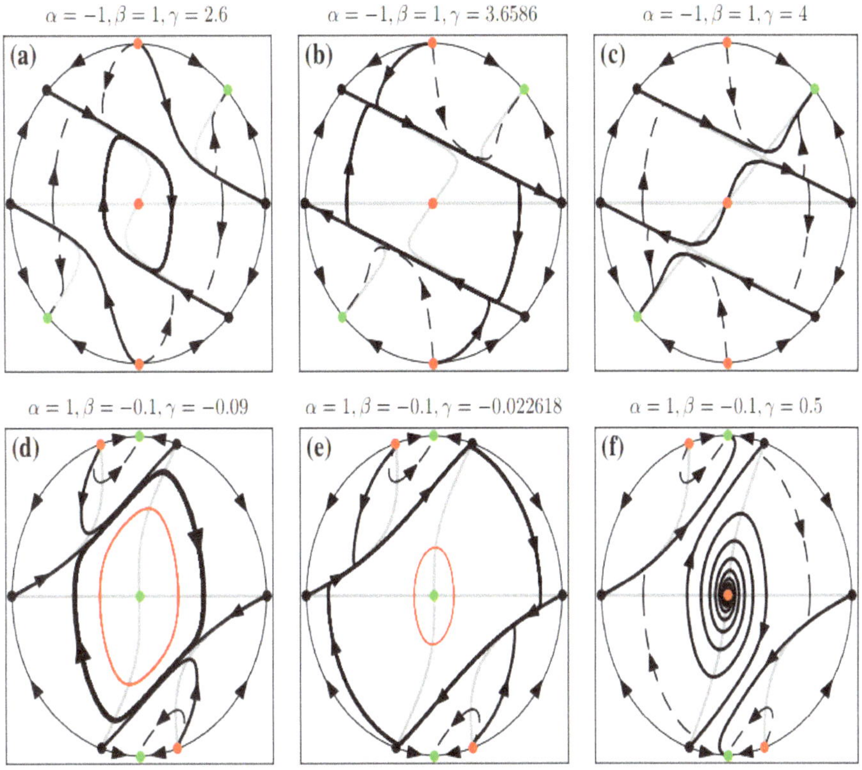

Figure 2: Global phase portraits of the system Eq. (2), projected on the (X, Y)-plane, for different parameters values. Green dots indicate stable equilibria; red dots indicate unstable equilibria; black dots indicate equilibria of a saddle type; red line indicates unstable periodic orbit; thick black lines indicate heteroclinic connections between different equilibria or between equilibria and stable periodic orbits; grey lines indicate nullclines; dashed lines examples of trajectories; arrows indicate direction of the flow (colour figure online).

Bifurcation Analysis of the Full HKB Model

Full System Model Equations

Previous analysis of the HKB model has focussed on the dynamics of the relative phase that is given by the difference of the two oscillators' phases. However, in applications involving VP interaction environments [13, 60, 61], other properties of the HKB model dynamics become crucial. Such properties include the amplitude and phase of the oscillatory solutions, as well as their existence, parameter dependence and stability. In order to address these questions, we focus below on the full HKB system. The original HKB model evolves in time (measured in seconds) according to a set of nonlinear differential Equations [22]:

$$\ddot{x}_1 + \dot{x}_1\left(\alpha x_1^2 + \beta \dot{x}_1^2 - \gamma\right) + \omega^2 x_1 = I_{12}(\dot{x}_1, \dot{x}_2, x_1, x_2)$$
$$\ddot{x}_2 + \dot{x}_2\left(\alpha x_2^2 + \beta \dot{x}_2^2 - \gamma\right) + \omega^2 x_2 = I_{21}(\dot{x}_1, \dot{x}_2, x_1, x_2), \quad 9$$

where x_1 and x_2 represent the position of the two agents' end effectors and

$$I_{12}(\dot{x}_1, \dot{x}_2, x_1, x_2) = (a + b(x_1 - x_2)^2)(\dot{x}_1 - \dot{x}_2)$$
$$I_{21}(\dot{x}_1, \dot{x}_2, x_1, x_2) = (a + b(x_2 - x_1)^2)(\dot{x}_2 - \dot{x}_1), \quad 10$$

are coupling functions with coefficients $a, b \in \mathbb{R}$. The above system of two coupled second order ordinary differential equations (ODEs) (9) can be written as a four-dimensional autonomous system of first order ODEs:

$$\dot{x}_1 = y_1$$
$$\dot{x}_2 = y_2$$
$$\dot{y}_1 = (a + b(x_1 - x_2)^2)(y_1 - y_2) - (y_1\left(\alpha x_1^2 + \beta y_1^2 - \gamma\right) + \omega^2 x_1)$$
$$\dot{y}_2 = (a + b(x_2 - x_1)^2)(y_2 - y_1) - (y_2\left(\alpha x_2^2 + \beta y_2^2 - \gamma\right) + \omega^2 x_2), \quad 11$$

where x_i and y_i represent position and velocity of the ith agent's end effector, respectively. The resulting dynamical system has a four-dimensional state space [7]. The parameter ω (commonly referred to as *eigen frequency*) defines, in conjunction with α, β and γ, the intrinsic dynamics of the two coupled oscillators. The oscillators' positions and velocities are coupled via the parameters a and b, commonly referred to as *coupling strengths*. The HKB model behaviour then depends on the intrinsic dynamics parameters as well as the coupling strengths. Although coordination/synchronisation in system (11) emerges as a consequence of coupling, its dynamics (i.e. number, type and stability of coordination patterns) depends not only on the nature of the coupling but also on the intrinsic properties of each coupled oscillator. In the HKB system (11), both the intrinsic dynamics and the couplings are highly nonlinear, opening up the possibility of obtaining multi-stability and hence multi-functionality.

Coordination Regimes in the HKB Model

In this section, we study the existence and stability of the possible coordination regimes in the full HKB model (11) by conducting a systematic analysis in all model (control) parameters. The numerical bifurcation analysis is carried out using numerical continuation in AUTO [11]. We set NTST = 50, NCOL = 4 for the mesh, and EPSL = 10^{-9}, EPSU = 10^{-9} for the tolerances of the Newton solver. We perform time-stepping simulations of the model (11) in matlab [39], using the ode45 solver with default numerical settings. In the simulations presented below, we use the following typical intrinsic dynamics parameter values as default, $\alpha = 1, \beta = 1, \gamma = 1$ and $\omega = 0.2$, unless otherwise stated in the figure legends.

In agreement with previously performed analysis on the HKB relative phase dynamics [1, 3, 9, 17, 19, 22], we confirm existence and study the stability of the well-characterised in-phase and anti-phase oscillatory solutions. Moreover, we find a new family of stable periodic phase-locked solutions characterised by relative phase in the interval (0°, 180°). These solutions are found to be stable in a wide range of parameter values. We note that this family of solutions is unstable for the commonly used set of model parameters based on [22]. Examples of the three solution types described above are plotted in Fig. 3. We show how such solutions are born when we vary γ for various combinations of the parameters a and b in the bifurcation diagrams of Fig. 4, whose branches are colour-coded as in Fig. 3. Here and henceforth, we use subscripts I, A, L (or combinations thereof) to indicate bifurcations occurring on solution branches of in-phase, anti-phase and phase-locked solutions, respectively. We also keep the corresponding colour-code convention for branches of solutions and solutions profiles of in-phase, anti-phase and phase-locked type.

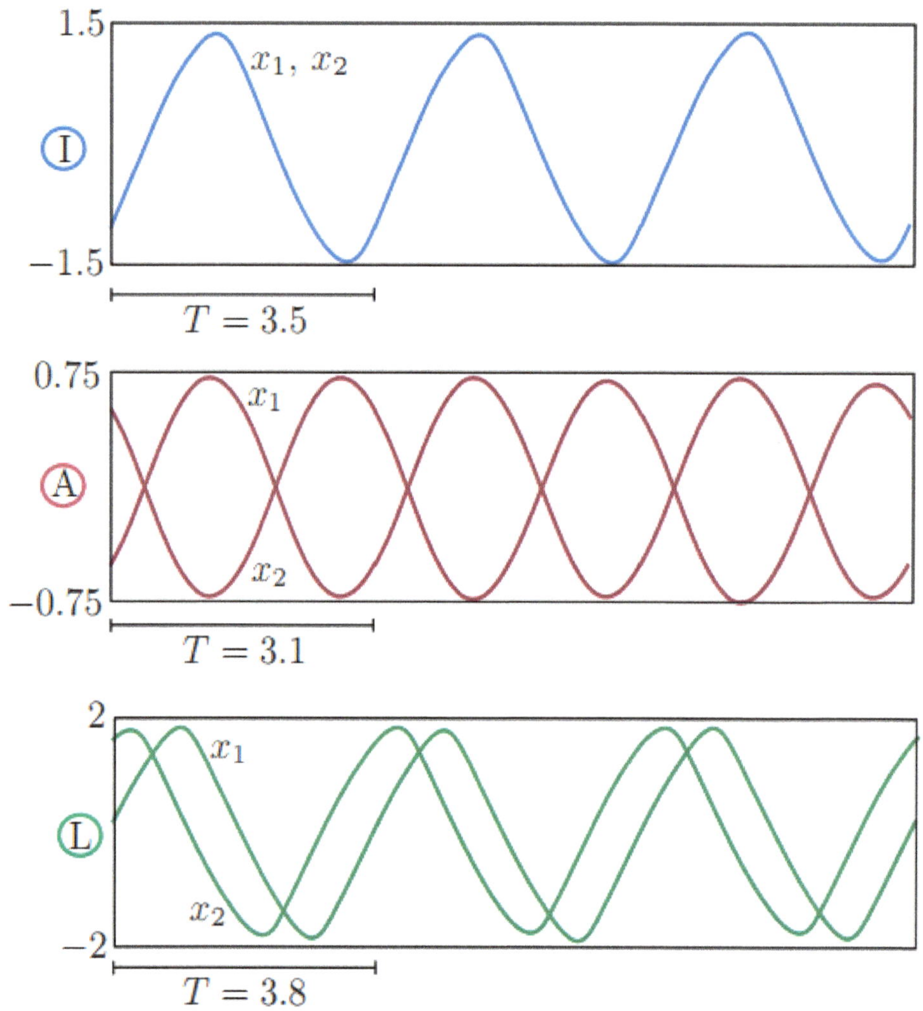

Figure 3: Examples of stable in-phase (I), anti-phase (A) and phase-locked (L) solutions. Solutions and parameter values are also indicated in the bifurcation diagrams of Fig. 4.

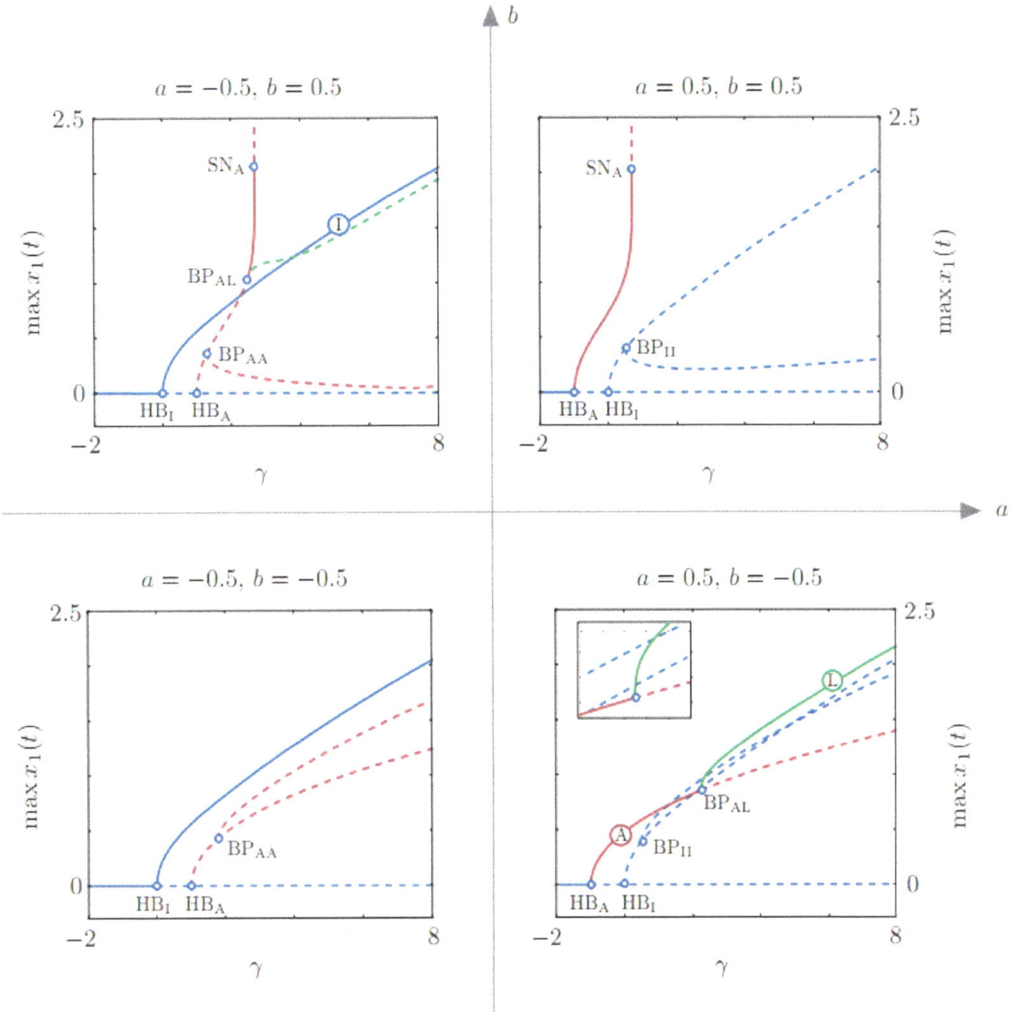

Figure 4: Representative bifurcation diagrams in the parameter γ for all possible combinations of coupling strengths, a and b. *Solid* (*dashed*) lines represent stable (unstable) states of (11).

In-phase and anti-phase coordination regimes are born via Hopf bifurcations (HB_I, HB_A) of the trivial steady state. In the first quadrant (where the coupling strength parameters are both positive), anti-phase coordination is the only stable state: a branch of unstable in-phase solutions is born at HB_I and bifurcates at a symmetry-breaking bifurcation, BP_{II}, giving rise to a secondary branch of unstable in-phase solutions where the oscillation amplitudes for agent 1 and 2 differ. In the third quadrant (where the coupling strength parameters are both negative), the scenario is specular: in-phase oscillations are now stable, while anti-phase solutions are unstable and bifurcate at BP_{AA}. In the first and third quadrants of the (a, b)-plane ($a = 0.5$, $b = 0.5$ and $a = -0.5$, $b = -0.5$, respectively), there are no branches of phase-locked solutions.

Phase-locked coordination regimes arise in the second and fourth quadrants of the (a, b)-plane, at symmetry-breaking bifurcations of anti-phase solutions (BP_{AL}). Phase-locked solutions are found to be always stable (unstable) in the fourth (second) quadrant. We note that, in these quadrants, the coupling non-linearities I_{12} and I_{21}, as functions of $x_1 - x_2$, attain both negative and positive values as opposed to what happens in the first and third quadrants, where such functions are strictly positive and strictly negative, respectively. Stable phase-locked solutions, spanning relative phases in the range of (0°, 180°), exist for $a > 0$ and $b < 0$. Such parameter settings could be used to model experiments displaying coordination regimes different from the canonical in-phase and anti-phase ones (see Discussion section for more details).

In Fig. 5, we summarise the behaviour of the representative examples reported above, for selected values of a and b, by continuing in (γ, a) and (γ, b) all bifurcation points found in Fig. 4. The two-parameter continuations are performed so as to show how the solution landscape changes as we pass from the first to the second quadrant (continuation in (γ, a)-plane) and from the first to the fourth quadrant (continuation in(γ, b)-plane). In these two-parameter bifurcation diagrams, we highlight areas where stationary and oscillatory solutions are stable. In the (γ, a)-plane, the organising centre is at $\gamma = 0$, $a = 0$: at this point, the eigenvalues of the linearised Jacobian at the trivial state (x_1, x_2, x_3, x_4) = (0, 0, 0, 0) are purely imaginary, equal to $\pm 2i$, each with multiplicity 2, corresponding to eigenvalues (0, $\mp i/2$, 0, 1) and ($\mp i/2$, 0, 1, 0). For low positive values of the damping γ, the system supports stable in-phase solutions (for negative values of a) and stable anti-phase solutions (in a wedge delimited by the locus of SN_A and BP_{AA}). In a sizeable region of parameter space, stable in-phase and anti-phase solutions coexist (see intersection between magenta- and blue-shaded areas). We note that the original set of parameter values based on [22] could be found in this region.

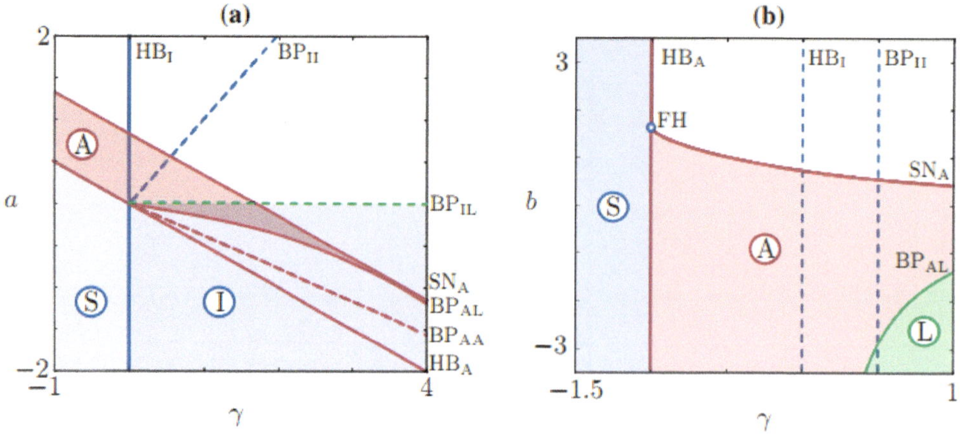

Figure 5: Two-parameter continuations of bifurcations occurring in Fig. 4. Panel a We fix b = 0.5 and continue in the (γ , a)-plane the bifurcations of the top panels of Fig. 4; shaded areas represent regions of stability for steady states (S), anti-phase (A) and in-phase (I) periodic solutions. Panel b We fix a = 0.5 and continue in the (γ, b)-plane the bifurcations in the right panels of Fig. 4; stable phase-locked solutions are indicated by (L).

In the (γ, b)-plane, the organising centre is a fold–Hopf bifurcation around γ = - 1, $b \approx 1.625$ (FH in Fig. 5) where the locus of saddle nodes of the anti-phase solutions, SN_A, collides with the locus of Hopf bifurcations HB_A. In this region of parameter space, phase-locked solutions are found for sufficiently high damping and sufficiently negative values of b. It should be noted, however, that phase-locked and anti-phase oscillations do not coexist for the default choice of intrinsic dynamics parameter values. As it can also be verified analytically, the locus of bifurcations HB_A and HB_B of the stationary steady state do not depend on b. It should be also noted that, for suitable combination of the parameters, it is possible to visit stable in-phase, anti-phase and phase-locked oscillations by varying a. An example of continuation in a for b = - 0.5 and γ= 1, ω = 2, a = b = 1 is depicted in Fig. 6. It can be clearly seen in the inset of Fig. 6 that, as a increases, the stable in-phase coordination regime (characterised by relative phase 0°) loses stability; then, a phase-locked coordination regime emerges (ranging over relative phases in (0°, 180°) in a continuous fashion) and eventually a stable anti-phase coordination regime (characterised by relative phase 180°) is established. The bifurcation diagram implies that, in an experimental set-up where a were to be assigned randomly, we would observe trajectories with relative phases distributed in the interval (0°, 180°), and with peaks at 0° and 180°. To verify this prediction, we performed an uncertainty quantification study, in which a is assigned randomly, near the inset of Fig. 6, and histograms of relative phases between $x_1(t)$ and $x_2(t)$ are computed a posteriori. In Fig. 7a, we perform 2000 independent simulations, where a is sampled from the uniform distribution between 0.05 and 0.2 and plot the resulting phase lag histogram: the distribution for the phase difference φ_1- φ_2 (right) is bimodal and sharply peaked around 0° and 180°, as expected, with a small but nonzero probability of finding an intermediate phase lag. The likelihood that experiments display such intermediate relative phases is deeply affected by the distribution of a: if we pass from a uniform to a normal distribution for a, (Fig. 7b), the resulting phase lag distribution develops also a peak around 90°, which is inherited from the parameter distribution.

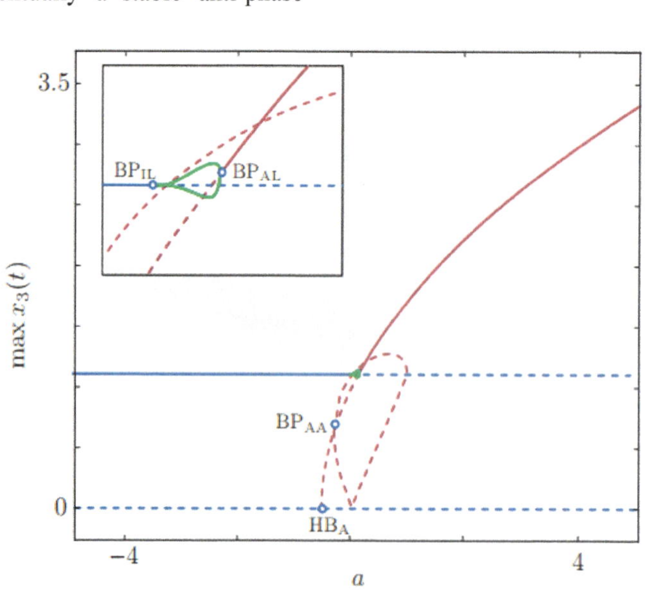

Figure 6: Continuation in the parameter a, for b = −0.5 and γ = 1, ω = 2, α = β = 1. The branches show that, with suitable combination of the parameters, it is possible to have stable in-phase, anti-phase and phase-locked oscillations by varying a. Solid lines represent stable and dashed lines unstable states of (11).

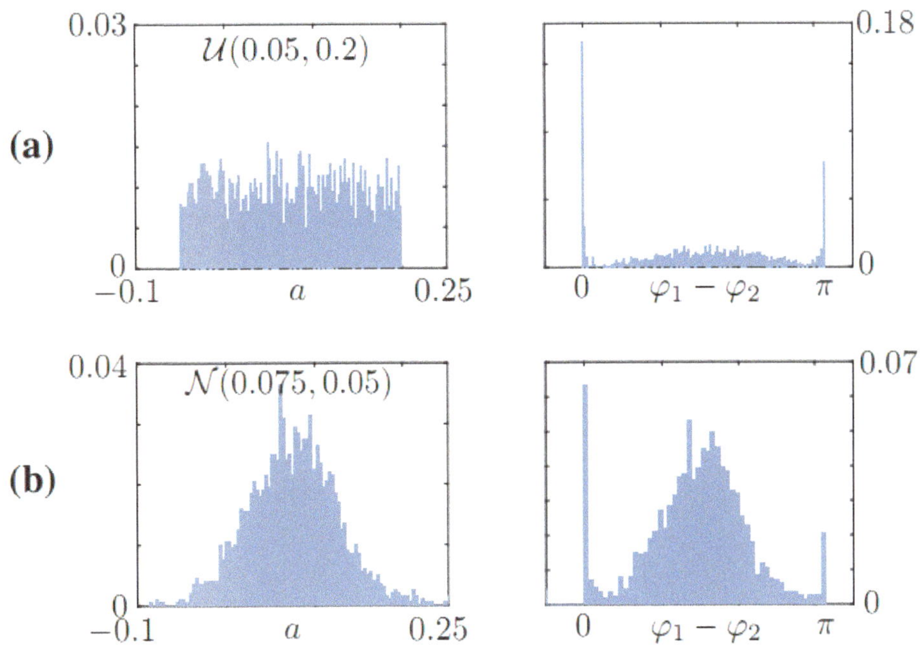

Figure 7: Distribution of the phase difference between $x_1(t)$ and $x_2(t)$ (left panels) obtained when the control parameter a is randomly distributed (right panels) with values close to the inset of Fig. 6. a The parameter a is sampled from a uniform distribution (left) and 2000 independent simulations are performed; the histogram for the relative phase $\phi_1 - \phi_2$ is bimodal and sharply peaked around 0° and 180°, with a small but nonzero probability of finding an intermediate phase lag. b The experiment is repeated with a normal distribution, which causes a third peak to develop around 90° in the distribution for the phase lag; the latter peak is inherited from the distribution of the control parameter a.

It is interesting to study the behaviour of various periodic solutions as the common eigenfrequency of the oscillators, ω, varies. We selected stable in-phase, anti-phase and phase-locked solutions and continued them in ω (Fig. 8). We found that such solutions behave essentially as in the single oscillator case (Fig. 1b): low frequencies elicit large amplitude oscillations with abrupt time transitions, whereas large frequencies induce smoother small amplitude oscillations. In this case the changes in the oscillation patterns occur to both agents, with various phase differences. The branch of phase-locked solutions undergoes a series of saddle-node bifurcations, giving rise to stable solutions in which the phase difference is reversed (see solutions profiles$L_{1,2}$ in Fig. 8).

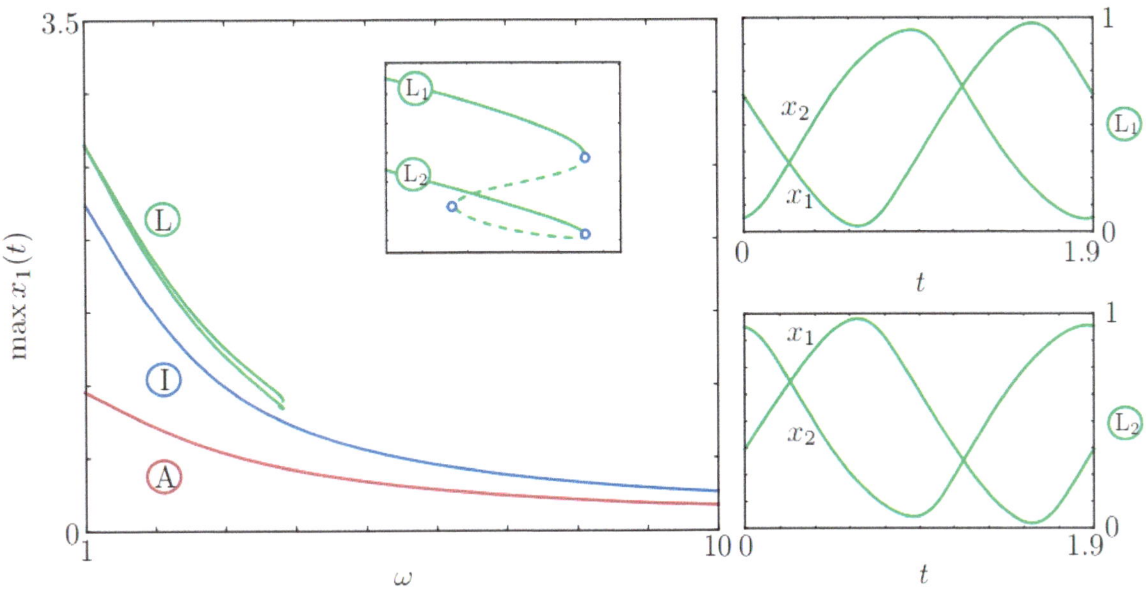

Figure 8: Continuation of in-phase, anti-phase and phase-locked solutions in the frequency ω. Solutions behave similarly to the single HKB oscillator case (Fig. 1b), except they have various phase behaviours. The branch of phase-locked solutions undergoes a series of saddle-node bifurcation, giving rise to stable solutions in which the phase difference is reversed (see solutions profiles L1,2). We note that the branches in this figure do not coexist, as they are found in different regions of parameter space: $\alpha = \beta = 1$ and $a = -0.5$, $b = -0.5$, $\gamma = 5$ (in-phase), $a = 0.5$, $b = -0.5$, $\gamma = 1.2$ (anti-phase) and $a = 0.5$, $b = -0.5$, $\gamma = 6.2$ (phase-locked).

Finally we investigate the impact of intrinsic oscillator dynamics on the collective behaviour of the HKB model by performing bifurcation analysis in the intrinsic dynamics parameters α and β. Instead of presenting two-parameter bifurcation diagrams for different cases, we report here only notable examples of our computations (see Figs. 9 and and10a).10a). The bifurcation structures found in these cases have common traits with the ones discussed above for the coupling strengths parameters a and b, that is, the trivial steady state undergoes Hopf bifurcations to anti-phase and in-phase periodic states, and various symmetry-breaking bifurcations give rise to phase-locked solutions. Interestingly, when varying α and β we could find period-doubling cascades, which are found robustly when α and β have opposite signs, as evidenced in Fig. 9a, where α = - 0.5, β = 0.5, and Fig. 10a, where α = 0.5, β = - 0.05. Representative stable solutions on the period-doubling cascade are also shown in Fig. 10a.

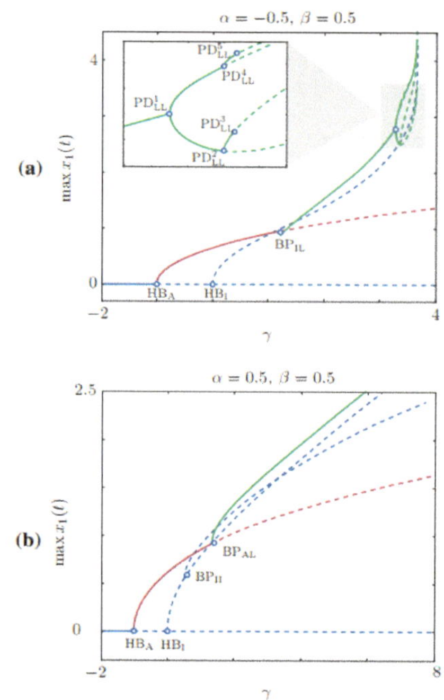

Figure 9: Bifurcation diagram in γ for ω = 2, a = 0.5, b = - 0.5 and various values of α and β. *Solid lines* represent stable and *dashed lines* unstable states of (11)

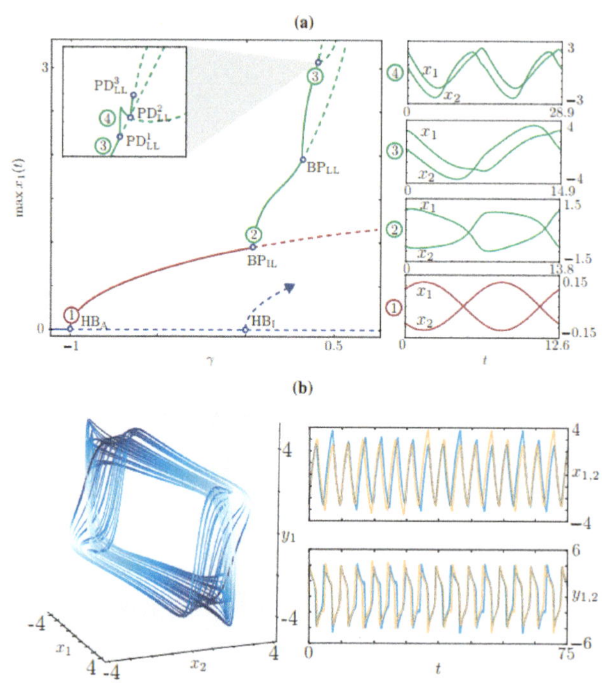

Figure 10: Period-doubling cascade. a Branch with period-doubling cascade and stable non-trivial periodic solutions. We plot one period of several stable solutions along the branch, whereas we omit the unstable branch emanating from HBI. Solutions feature increasing solution periods, $T_0 \approx 12.6$, $T_2 \approx 13.8$, $T_3 \approx 14.9$, $T_4 \approx 28.9$, corresponding to $\Omega_1 \approx 0.500$, $\Omega_2 \approx 0.455$, $\Omega_3 \approx 0.422$, $\Omega_4 \approx 0.217$, respectively. Parameters ω = 0.5, a = 0.5, b = −0.5 α = 0.5 and β = −0.05. b Attractor found for γ = 3.42; the coordination regime shows erratic phase changes, during which x_1 and x_2 alternate in the leading position. This regime involves fast velocity switches, as evidenced by the time traces of y_1 and y_2. Solid lines represent stable and dashed lines unstable states of (11).

Using direct numerical simulations we explored the system behaviour close to the period-doubling cascade, finding chaotic regimes (see Fig. 10b) in which the solution remains bounded and features sudden erratic phase transitions, during which the agents alternate as leaders and followers. In this regime, the velocities y_1 and y_2 undergo fast switches. The existence of such complex solutions is perhaps not surprising from a dynamical systems viewpoint; however, the behaviour described above has not been reported nor investigated previously, and can be used to model experiments where the movement coordination is irregular in nature. Last but not least, knowledge about the existence of such solutions is critical when designing virtual player interaction environments [34, 59–61] and/or planning human dynamic clamp experiments based on the HKB model [13].

Bistability and Hysteresis

In this section, we explore further the dependence of the HKB model dynamics on the intrinsic properties of the coupled oscillators. In suitable regions of parameter space we find coexisting stable periodic states characterised by different relative phases or phase lags. In Fig. 11a, we run a continuation similar to the ones presented above, but we set $\alpha = -1.7$. The branches of this bifurcation diagram are qualitatively similar to the ones of the previous sections; however, the in-phase periodic branch originating at the Hopf bifurcation HB_I undergoes a symmetry-breaking bifurcation (BP_{IL}). Such branch is initially unstable, undergoes 2 other symmetry-breaking bifurcations, restabilises at a saddle-node bifurcation and then features a period-doubling cascade. The stable portion of this branch (solid green branch between SN_L and PD_{LL}) coexists with a branch of stable anti-phase solutions originating from the trivial state at HB_A (red branch).

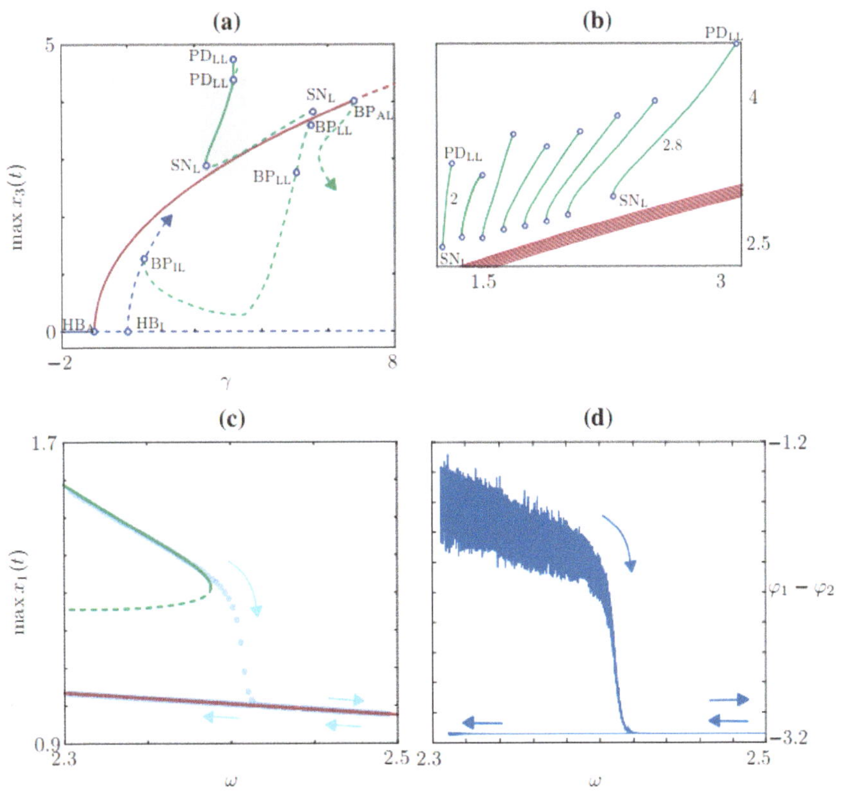

Figure 11: Bistability and hysteresis between anti-phase and phaselocked solutions. a The in-phase branch (blue) undergoes a symmetry breaking bifurcation (BP_{IL}) and the resulting unstable phase-locked branch, featuring two further symmetry-breaking bifurcations (BP_{IL}), restabilises at a saddle-node bifurcation, before a period-doubling cascade takes place. A stable portion of the phase-locked branch (solid green line between SN_L and PD_{LL}) coexists with the anti-phase branch originating at HBI (solid red branch). Parameters: a = 0.5, b = 0.5, ω = 3, α = −1.7, β = 0.5. b We repeat the experiment for ω ∈ [2, 2.8] and plot stable branches to highlight the bistability region. c ω is varied by continuation and by quasi-static sweeps in direct numerical simulations (blue dots), for γ = 1.7; the time simulation follows the phase-locked branch up to the saddle node at ω ≈ 2.4, where an abrupt and hysteretic transition to an anti-phase solution is observed. d Phase lag during numerical simulation in c (colour figure online).

This bifurcation structure opens up the possibility of observing abrupt relative phase transitions between phase-locked (at any relative phase between 0° and 180°) and anti-phase (at relative phase equal to 180°) coordination regimes as a function of the eigenfrequency ω. We find that bistability is observed in a significant region of parameter space: in the inset of Fig. 11b we report overlapping stable portions of phase-locked and anti-phase branches as we vary the eigenfrequencies ω. As ω is varied, the anti-phase branch (red) changes only slightly, while the stable phase-locked branch moves to the left and expands. Our analysis predicts coexistence in the region $(\gamma, \omega) \in [1.2, 3.2] \times [2, 2.8]$ (which was found robustly in other parameter regions, not shown). It is important to note that this phase transition is qualitatively different from the transition addressed by the original HKB model [22], where an increase in frequency leads to transition from anti-phase to in-phase coordination. In the parameter regime described above, an increase in frequency leads to transition from phase-locked to anti-phase coordination behaviour.

To illustrate the dynamical switch between solution types, we perform time-stepping simulations in which the eigenfrequency ω is varied quasi-statically and compare with the bifurcation analysis. In Fig. 11c, we continued an anti-phase (red) and phase-locked (green) solution for $\gamma = 1.7$, $\omega = 2.3$ in the parameter ω; the phase-locked branch destabilises at a saddle-node bifurcation, whereas the anti-phase branch remains stable for $\omega \in [2.3, 2.5]$. We then initialised a time simulation on the phase-locked branch (blue dots in Fig. 11c) and changed ω in slow increments (from $\omega = 2.3$ up to $\omega = 2.5$) followed by small decrements (from $\omega = 2.5$down to $\omega = 2.3$). The time simulation shows an abrupt and hysteretic change in the solution type. This could be further appreciated in Fig. 11d where we plot the time simulation using the phase lag $\varphi_1 - \varphi_2$. The x_2 is delayed with respect to x_1, with an initial phase lag $\varphi_1 - \varphi_2 \approx 90°$; when $\omega \approx 2.4$, we observe a transition to an orbit with $\varphi_1 - \varphi_2 \approx 180°$ (anti-phase solution).

Effect of Heterogeneity in Eigenfrequencies on the Coordination Regimes

In the computations shown so far, the two oscillators possess a common eigenfrequency ω. In order to study the effect of heterogeneities on coordination, we introduce two parameters ω_1, ω_2, then fix ω_1 to the nominal value $\omega_1 = 2$ and use the ratio ω_1/ω_2 as a continuation parameter. The difference in eigenfrequencies introduces a heterogeneity in the system and has the potential to turn in-phase solutions into phase-locked solutions and vice-versa. In order to illustrate this idea, we performed bifurcation analysis in the parameter ω_1/ω_2 investigating the in-phase solutions which exist for parameter values within the range of those used in the original HKB model [22], as well the stable phase-locked solutions which we reported above for $a > 0$ and $b < 0$ (see Fig. 4). In Fig. 12, we initialise the continuation with an in-phase and a phase-locked periodic solution. We plot the bifurcation diagram in terms of the phase lag (measured in radians), by computing the approximate phases times $\varphi_i = t_i/T$, for $i = 1, 2$, where t_i is the time at which the orbit $x_i(t)$ attains its maximum and T is the solution period. Fig. 12a depicts a stable, initially in-phase, solution at $\omega_1/\omega_2 = 1$ that turns into a phase-locked solution asω_1/ω_2 is increased/decreased, losing stability at saddle-node bifurcation. In Fig. 12b, we show how the phase lag is reduced when the frequency ratio is varied and an in-phase (albeit unstable) solution is eventually attained, before a new phase-locked solution arises.

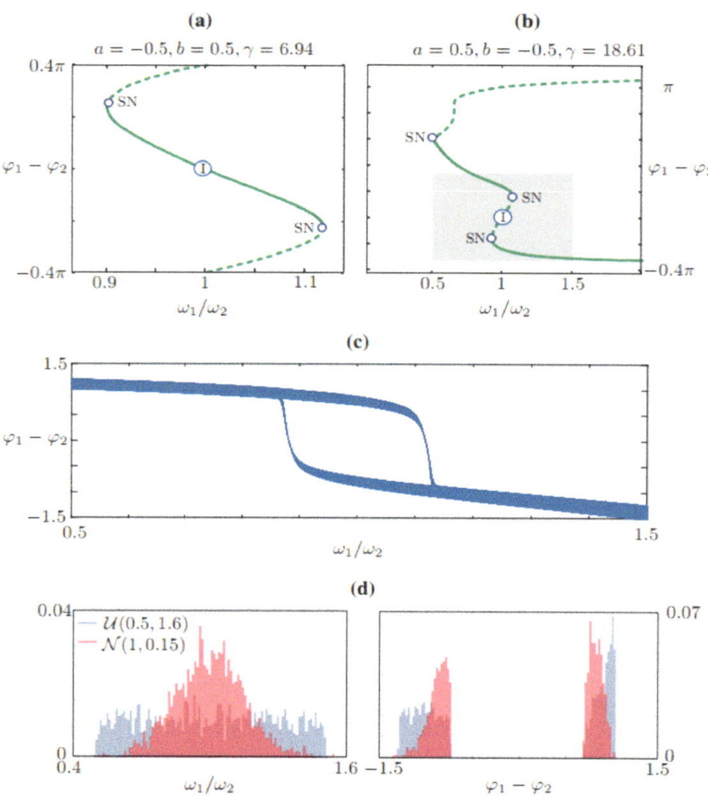

Figure 12: Phase difference between periodic solutions x1(t) and x2(t) as a function of the ratio ω1/ω2. a A stable solution on the in-phase branch in the fourth quadrant of Fig. 4 is continued in ω_1/ω_2. b The continuation is repeated starting from a solution on the phase-locked branch in the second quadrant of Fig. 4. In both cases, heterogeneity in the eigenfrequencies impacts the phase lag of the solution. c When the ratio ω_1/ω_2 is modulated with a slowly varying sinusoidal function, the actors alternate in the leading position with hysteretic cycles, which follow the branches in the inset of b and jump at the corresponding saddle-node bifurcations. d We perform an experiment similar to the one in Fig. 7; when the parameter ω_1/ω_2 is drawn randomly near the shaded area in b from a uniform (blue) or a normal (red) distribution, the resulting phase lag distribution is bimodal, with peaks at ±57.30°, as predicted by the bifurcation diagram in b and by the parameter sweep in c (colour figure online).

The bifurcation structure in Fig. 12b implies that hysteresis between phase-locked solutions with opposite phase lags (relative phases) is possible in the model. To illustrate this, we perform time-stepping simulations in which the ratio is varied quasi-statically as $\omega_1(t) = \omega_2(t)[1 + \sin(0.005t)]$ and plot the results in Fig. 12c. The two oscillators swap in the leader and follower role, following the branches of Fig. 12b and switching roles at the corresponding saddle-node bifurcations. This numerical experiment could be interpreted in the light of the joint improvisation scenario in the "mirror game", a recently proposed paradigm for studying the dynamics of two people improvising motion together [45]. In particular, as the participants are asked to imitate each other and create synchronised and interesting motions, they would be naturally trying to adjust their movement velocities and thus eigenfrequencies to each other. This would lead to variation in the ratio of their eigenfrequencies and, respectively, exchange of leader and follower roles while playing the game. Indeed, observations based on our data collected in a "mirror game" setting [55] indicate that the distribution of relative phase during a typical joint improvisation sessions are bimodal pointing to possible hysteretic dynamics. As we see in Fig. 12d, the bimodal distribution emerges also in the case of randomly assigned frequency ratio ω_1/ω_2: when the value of this parameter is drawn randomly (close to the hysteretic region) from a uniform or a normal distribution, the resulting phase lag distribution is bimodal, with peaks at $\pm 57.30°$, as predicted by the bifurcation diagram in Fig. 12b and by the parameter sweep in Fig. 12c.

DISCUSSION

In this paper, we have systematically investigated the dynamics of the HKB model in the state space spanned by the position and velocities of the coupled oscillators. Furthermore, we go beyond the weakly coupled regime and consider the coupling strength parameters as generic. We show that stable periodic solutions in the single HKB oscillator model are born via a Hopf bifurcation as the damping parameter γ becomes positive. Furthermore, we reveal that under certain intrinsic oscillator properties the periodic solutions of the single HKB model could disappear via a heteroclinic cycle, associated with rapid increase in the magnitude of the state variables. Although such behaviour cannot be observed in a physical system, it can have significant consequences for the design and development of the virtual players. Bifurcation analysis of the full four-dimensional HKB model reveals a variety of different coordination dynamics. Attractors at a constant relative phase of 0° (in-phase) and/or 180° (anti-phase) are born via Hopf bifurcations detected in the damping parameter γ. We find symmetrical attractors of phase-locked solutions at intermediate values of relative phase (between 0° and 180°), which increase or decrease gradually as γ is increased. We demonstrate that the phase-locked solutions are born in a symmetry-breaking bifurcation of periodic orbit in which the anti-phase periodic attractor loses stability as the damping parameter γ is varied. Changing the sign of the coupling strengths has the effect of shifting the attractors by 180°, thus changing the phase that remains stable at high frequencies, from 0° to 180° or vice-versa. We also show that change in the intrinsic oscillators' properties (i.e. varying the parameter α) can lead to complex dynamics mediated via a period-doubling cascade. Furthermore, different intrinsic dynamics can also bring about a variety of bistability modes, which are different that the type of bistability described in the original HKB model study [22]. Finally we consider a case of a heterogeneity in the system by introducing difference in the eigenfrequency of the coupled oscillators. We demonstrate how this results in bistability and hysteresis. Our uncertainty quantification simulations presented in Fig. 12d confirm that in the case of heterogeneous oscillators hysteresis loops and phase-locked coordination modes should be expected in experiments, as suggested in [2]. What is more, existence of such hysteresis loop provides an excellent opportunity for a quantitative experimental validation of the HKB model using two heterogeneous coupled oscillators, e.g. by putting weights on body parts as suggested in [2] or by using heterogeneous pendula as in [50].

In a large number of multi-stable examples observed experimentally, the patterns of stability change under different conditions. Bimanual finger coordination is bistable at low frequencies, but above a critical frequency the anti-phase pattern is no longer sustainable [22]. Similarly postural sway is bistable at low frequencies (20° and 180°), but the phase-locked (20°) mode looses stability at high frequencies or when other behaviours, such as reaching, are incorporated in the task [2]. These transitions between stable states, and particularly the loss of stability of the anti-phase mode at high frequencies, appear to be a fundamental feature of human coordination [33]. The hypothesis that these real-world patterns and transitions between them are emergent phenomena due to a self-organised dynamical system are substantiated by experimental results such as critical fluctuations, critical slowing down and hysteresis between modes [2, 18, 51]. In this paper, we make the first step towards identifying parameter regimes and dynamics that would allow to model a variety of different experimental observations using the same modelling framework.

Many recent experimental studies of human movement coordination [5, 10, 12, 15, 24, 57] have reported persistent movement coordination dynamics other than the well-known in-phase and anti-phase synchronisation behaviour that have inspired the development of the HKB model [22]. Despite the large number of behaviours whose dynamics are well represented by the theoretically predicted in-phase and anti-phase stability, there are several counter examples where human body movements show evidence of stability at different or additional intermediate phases. Examples of real-world systems with stabilities at other relative phases include: the human postural system (stability at 20°) [2], amble to walk gait in quadrupeds (stability at90°) [8], the bipedal skipping gait [42], coordination tendencies of successful defences (30°) and unsuccessful defences (90°) in soccer [12], squash (135°) [41] and butterfly stroke swimming (90°) [14] as well as variety of relative phase distributions in other team sports [10]. There is evidence that other phases can be stable simultaneously with 0° and 180°. These multi-stable dynamics can exist naturally or be learnt [57]. Our results about the existence of stable phase-locked periodic solutions in the HKB model that span all possible relative phases between 0° and 180° could be related to some of the above mentioned experimental observations. In particular, analysis of the data collected from interactions between player dyads allowed for a description of the space-time dynamics of basketball match play. In the longitudinal direction, a strong attraction to in-phase was reported for all possible dyads but not so for the lateral direction. Instead, attractions to in-phase or anti-phase were observed among most dyads with the player vs. player dyads tending on balance to demonstrate less pronounced attractions or repulsions to certain relative phases than the player–opponent dyads [5, 15]. Interpersonal coordination tendencies of 1-vs-1 subphases were investigated in [12]. The experimental results presented in Fig. 2 in [12] could be, for example, qualitatively accounted for by the type of coordination stability dependence on the coupling parameter a found in the HKB model. Specifically (see Fig. Fig.66 for $b = -0.5$ and $\gamma = 1$, $\omega = 2$, $\alpha = \beta = 1$), as the coupling strength between the velocity components of the two oscillators increases, the stable coordination regime exhibited by the HKB model undergoes transitions from stable in-phase coordination for $a < 0$ through stable phase-locked coordination spanning all possible relative phases in (0°, 180°); as a increases, to a stable anti-phase coordination regime with increasing amplitudes.

Last but not least, very recent experiments involving the use of virtual partner interaction [13, 34, 59–61] have employed to various degree the HKB model in order to study social interactions and interpersonal coordination. These studies have used adaptation in the HKB parameter values in their implementations. Knowledge about how the type and stability of the possible HKB model solutions depend on the model parameters could greatly facilitate the design and ensure robustness of such hybrid systems where a human interacts with a virtual partner whose movements are driven by the HKB model. Furthermore, comparison of the theoretical predictions and dynamics observed in experiments with a virtual partner could allow for quantitative, rather than qualitative, validation of different models of motor coordination. Although deficits of the HKB model are well known, see, for example, discussion in [3], our analysis demonstrates that this model has much richer dynamics than previously considered and showcases mathematical tools that could be very useful in future studies of human movement coordination.

ACKNOWLEDGMENTS

The authors would like to thank Ed Rooke (University of Bristol, UK) for initial discussions on the HKB model analysis and Pablo Aguirre (Universidad Tecnica Federico Santa Maria, Chile) for helpful discussions on the global transition in the single HKB oscillator. This work was funded by the European Project AlterEgo FP7 ICT 2.9 - Cognitive Sciences and Robotics, Grant Number 600610. The research of KT-A was supported by grants EP/L000296/1 and EP/N014391/1 of the Engineering and Physical Sciences Research Council (EPSRC).

REFERENCES

1. Assisi CG, Jirsa VK, Kelso JAS. Dynamics of multifrequency coordination using parametric driving: theory and experiment. Biol Cybern. 2005;93(1):6–21. doi: 10.1007/s00422-005-0558-y.
2. Bardy B, Oullier O, Bootsma RJ, Stoffregen TA. Dynamics of human postural transitions. J Exp Psychol Hum Percept Perform. 2002;28(3):499. doi: 10.1037/0096-1523.28.3.499.
3. Beek P, Peper C, Daffertshofer A. Modeling rhythmic interlimb coordination: beyond the Haken–Kelso–Bunz model. Brain Cognit. 2002;48(1):149–165. doi: 10.1006/brcg.2001.1310.
4. Beek PJ, Schmidt RC, Morris AW, Sim MY, Turvey MT. Linear and nonlinear stiffness and friction in biological rhythmic movements. Biol Cybern. 1995;73(6):499–507. doi: 10.1007/BF00199542.
5. BourboussonJ,SeveC,McGarryT.Spacetimecoordination dynamics in basketball: part 1. Intra-and inter-couplings among player dyads. J Sports Sci. 2010;28(3):339–347. doi: 10.1080/02640410903503632.
6. Buchanan JJ, Ryu YU. One-to-one and polyrhythmic temporal coordination in bimanual circle tracing. J Mot Behav. 2006;38(3):163–184. doi: 10.3200/JMBR.38.3.163-184.
7. Calvin S, Jirsa VK. Perspectives on the dynamic nature of coupling in human coordination. Berlin: Springer; 2011.
8. Collins JJ, Stewart IN. Coupled nonlinear oscillators and the symmetries of animal gaits. J Nonlinear Sci.1993;3(1):349–392. doi: 10.1007/BF02429870.
9. Daffertshofer A. A dynamical model for mirror movements. Phys D. 1999;132(1–2):243–266. doi: 10.1016/S0167-2789(99)00044-5.
10. Davids K. Complex systems in sport. Abingdon: Routledge; 2013.
11. Doedel EJ. Auto: a program for the automatic bifurcation analysis of autonomous systems. Congr Numer.1981;30:265–284.
12. Duarte R, Arajo D, Davids K, Travassos B, Gazimba V, Sampaio J. Interpersonal coordination tendencies shape 1-vs-1 sub-phase performance outcomes in youth soccer. J Sports sci. 2012;30(9):871–877. doi: 10.1080/02640414.2012.675081.
13. Dumas G, de Guzman GC, Tognoli E, Kelso JS. The human dynamic clamp as a paradigm for social interaction. Proc Natl Acad Sci. 2014;111(35):E3726–E3734. doi: 10.1073/pnas.1407486111.
14. Ehrlacher C, Bardy BG, Faugloire E, Stoffregen TA (2003) Sports expertise influences learning of postural coordination. In: Studies in perception and action VII: twelfth international conference on perception and Action, Gold Coast, Queensland, Australia, Lawrence Erlbaum, p 104, 13–18 July 2003
15. Esteves PT, Arajo D, Davids K, Vilar L, Travassos B, Esteves C. Interpersonal dynamics and relative positioning to scoring target of performers in 1 vs. 1 sub-phases of team sports. J sports sci.2012;30(12):1285–1293. doi: 10.1080/02640414.2012.707327.
16. Frank TD, Silva PL, Turvey MT. Symmetry axiom of Haken–Kelso–Bunz coordination dynamics revisited in the context of cognitive activity. J Math Psychol. 2012;56(3):149–165. doi: 10.1016/j.jmp.2012.03.001.
17. Fuchs A, Jirsa VK. The HKB model revisited: how varying the degree of symmetry controls dynamics.Hum

Mov Sci. 2000;19(4):425–449. doi: 10.1016/S0167-9457(00)00025-7.

18. Fuchs A, Jirsa VK. Coordination: neural, behavioral and social dynamics. Berlin: Springer; 2008.
19. Fuchs A, Jirsa VK, Haken H, Kelso JA. Extending the HKB model of coordinated movement to oscillators with different eigenfrequencies. Biol Cybern. 1996;74(1):21–30. doi: 10.1007/BF00199134.
20. Golubitsky M, Stewart I, Buono PL, Collins J. Symmetry in locomotor central pattern generators and animal gaits. Nature. 1999;401(6754):693–695. doi: 10.1038/44416.
21. Guckenheimer J, Holmes P. Nonlinear oscillations, dynamical systems, and bifurcations of vector fields. New York: Springer; 2013.
22. Haken H, Kelso JS, Bunz H. A theoretical model of phase transitions in human hand movements. Biol Cybern. 1985;51(5):347–356. doi: 10.1007/BF00336922.
23. Huys R, Perdikis D, Jirsa VK. Functional architectures and structured flows on manifolds: a dynamical framework for motor behavior. Psychol Rev. 2014;121(3):302. doi: 10.1037/a0037014.
24. Issartel J, Marin L, Cadopi M. Unintended interpersonal co-ordination:can we march to the beat of our own drum? Neurosci Lett. 2007;411(3):174–179. doi: 10.1016/j.neulet.2006.09.086.
25. Jirsa V, Fink P, Foo P, Kelso J. Parametric stabilization of biological coordination: a theoretical model. J Biol Phys. 2000;26(2):85–112. doi: 10.1023/A:1005208122449.
26. Jirsa VK, Scott Kelso J. The excitator as a minimal model for the coordination dynamics of discrete and rhythmic movement generation. J Mot Behav. 2005;37(1):35–51. doi: 10.3200/JMBR.37.1.35-51.
27. Jordan DW, Smith P. Nonlinear ordinary differential equations. Oxford: Oxford University Press; 2007.
28. Kay BA, Kelso JA, Saltzman EL, Schöner G. Space-time behavior of single and bimanual rhythmical movements: data and limit cycle model. J Exp Psychol Hum Percept Perform. 1987;13(2):178–192. doi: 10.1037/0096-1523.13.2.178.
29. Kay BA, Saltzman EL, Kelso J. Steady-state and perturbed rhythmical movements: a dynamical analysis.J Exp Psychol Hum Percept Perform. 1991;17(1):183. doi: 10.1037/0096-1523.17.1.183.
30. Kelso J. Phase transitions and critical behavior in human bimanual coordination. Am J Physiol Regul Integr Comp Physiol. 1984;246(6):R1000–R1004.
31. Kelso J, DelColle J, Schöner G. Action-perception as a pattern formation process. Atten Perform XIII.1990;5:139–169.
32. Kelso J, Schöner G, Scholz J, Haken H. Phase-locked modes, phase transitions and component oscillators in biological motion. Phys Scr. 1987;35(1):79. doi: 10.1088/0031-8949/35/1/020.
33. Kelso JS. Dynamic patterns: the self-organization of brain and behavior. Cambridge: MIT press; 1997.
34. Kelso JS, de Guzman GC, Reveley C, Tognoli E. Virtual partner interaction (VPI): exploring novel behaviors via coordination dynamics. PLoS One. 2009;4(6):e5749. doi: 10.1371/journal.pone.0005749.
35. Kurths J. Synchronization: a universal concept in nonlinear sciences. Cambridge: Cambridge University Press; 2003.
36. Kuznetsov YA. Elements of applied bifurcation theory. New York: Springer; 2013.
37. Laffaye G, Bardy B, Durey A. Principal component structure and sport-specific differences in the running one-leg vertical jump. Int J Sports Med. 2007;28(05):420–425. doi: 10.1055/s-2006-924507.
38. Leise T, Cohen A. Nonlinear oscillators at our fingertips. Am Math Mon. 2007;114(1):14–28.
39. MATLAB (2012) Version 8.0.0.783 (R2012b). The MathWorks Inc., Natick
40. Mattout J. Brain-computer interfaces: a neuroscience paradigm of social interaction? A matter of perspective. Front Hum Neurosci. 2012;6:114. doi: 10.3389/fnhum.2012.00114.
41. McGarry T. Identifying patterns in squash contests using dynamical analysis and human perception. Int J Perform Anal Sport. 2006;6(2):134–147.
42. Minetti AE. The biomechanics of skipping gaits: a third locomotion paradigm? Proc R Soc Lond Ser B Biol Sci. 1998;265(1402):1227–1233. doi: 10.1098/rspb.1998.0424.
43. Mook D, Nayfeh A. Nonlinear Oscil. New York: Wiley; 1979.
44. Mortl A, Lorenz T, Vlaskamp BN, Gusrialdi A, Schuba A, Hirche S. Modeling inter-human movement coordination: synchronization governs joint task dynamics. Biol Cybern. 2012;106(4–5):241–259. doi: 10.1007/s00422-012-0492-8.
45. Noy L, Dekel E, Alon U. The mirror game as a paradigm for studying the dynamics of two people improvising motion together. Proc Natl Acad Sci. 2011;108(52):20947–20952. doi: 10.1073/pnas.1108155108.
46. Perko L. Differential equations and dynamical systems. New York: Springer; 2001.
47. Pikovsky A, Rosenblum M, Kurths J, Hilborn RC. Synchronization: a universal concept in nonlinear science. Am J Phys. 2002;70(6):655–655. doi: 10.1119/1.1475332.
48. Saltzman E, Kelso J. Skilled actions: a task-dynamic approach. Psychol Rev. 1987;94(1):84. doi: 10.1037/0033-295X.94.1.84.
49. Schmidt RC, Carello C, Turvey MT. Phase transitions and critical fluctuations in the visual coordination of rhythmic movements between people. J Exp Psychol Hum Percept Perform. 1990;16(2):227–247. doi: 10.1037/0096-1523.16.2.227.
50. Schmidt RC, Richardson MJ. Dynamics of interpersonal coordination. Coord Neural Behav Soc Dyna Soc. 2008;4:281–308. doi: 10.1007/978-3-540-74479-5_14.

51. Schoener G, Haken H, Kelso J. A stochastic theory of phase transitions in human hand movement. Biol Cybern. 1986;53(4):247–257. doi: 10.1007/BF00336995.

52. Scott Kelso J, Holt KG, Rubin P, Kugler PN. Patterns of human interlimb coordination emerge from the properties of non-linear, limit cycle oscillatory processes: theory and data. J Mot Behav. 1981;13(4):226–261. doi: 10.1080/00222895.1981.10735251.

53. Scott Kelso J, Putnam CA, Goodman D. On the space-time structure of human interlimb co-ordination.Q J Exp Psychol. 1983;35(2):347–375. doi: 10.1080/14640748308402139.

54. Seifert L, Chollet D, Bardy B. Effect of swimming velocity on arm coordination in the front crawl: a dynamic analysis. J Sports Sci. 2004;22(7):651–660. doi: 10.1080/02640410310001655787.

55. Słowiński P, Zhai C, Alderisio F, Salesse R, Gueugnon M, Marin L, Bardy BG, Di Bernardo M, Tsaneva-Atanasova K. Dynamic similarity promotes interpersonal coordination in joint action. J R Soc Interface. 2016;13(116):20151,093. doi: 10.1098/rsif.2015.1093.

56. Warren WH. The dynamics of perception and action. Psychol Rev. 2006;113(2):358–389. doi: 10.1037/0033-295X.113.2.358.

57. Zanone PG, Kelso JAS. Coordination dynamics of learning and transfer: collective and component levels. J Exp Psychol Hum Percept Perform. 1997;23(5):1454–1480. doi: 10.1037/0096-1523.23.5.1454.

58. Zelic G, Mottet D, Lagarde J. Behavioral impact of unisensory and multisensory audio-tactile events: pros and cons for interlimb coordination in juggling. PloS One. 2012;7(2):e32,308. doi: 10.1371/journal.pone.0032308.

59. Zhai C, Alderisio F, Słowiński P, Tsaneva-Atanasova K, di Bernardo M. Design of a virtual player for joint improvisation with humans in the mirror game. PloS One. 2016;11(4):e0154,361. doi: 10.1371/journal.pone.0154361.

60. Zhai C, Alderisio F, Tsaneva-Atanasova, K, di Bernardo M (2014) Adaptive tracking control of a virtual player in the mirror game: decision and Control (CDC). In: 2014 IEEE 53rd Annual Conference on, pp 7005–7010

61. Zhai C, Alderisio F, Tsaneva-Atanasova K, di Bernardo M (2014) A novel cognitive architecture for a human-like virtual player in the mirror game. In: Systems, Man and Cybernetics (SMC), 2014 IEEE International Conference, pp 754–759

SECTION II - PERTURBATION THEORY AND HOMOTOPY PERTURBATION METHODS

CHAPTER 6

APPLICATION OF THE PERTURBATION ITERATION METHOD TO BOUNDARY LAYER TYPE PROBLEMS

Mehmet Pakdemirli

Keywords: Perturbation methods, Perturbation–iteration algorithm, Boundary layer problems, Ordinary differential equations, Singular perturbation problems

BACKGROUND

The small parameter assumption is one of the most characteristic limitations of the perturbation methods. For nonlinear problems, this corresponds to a weakly nonlinear system. In recent decades, efforts have been intensified to overcome the limitation of the perturbation methods and to produce uniform solutions for large perturbation parameters.

One such attempt is to incorporate perturbation solutions with the iteration procedures to advance to the real solutions for large parameters. For algebraic equations, the systematic way of constructing single point iteration algorithms for root finding was outlined in the pioneering work due to Pakdemirli and Boyacı (2007). It was shown that many different root finding algorithms, some of which are original can be constructed by the perturbation–iteration algorithms. The algorithms are classified as PIA(n, m), n representing the number of correction terms in the perturbation expansion and m representing the highest order of derivatives in the Taylor series expansions. Although PIA(1,1) corresponds to the well-known Newton–Raphson algorithm, many different algorithms non-existent in the literature can be constructed with different selections of the numbers n and m with $n \leq m$ (Pakdemirli and Boyacı 2007; Pakdemirli et al. 2007, 2008).

The classification of such algorithms for algebraic equations turns out to be quite original. Motivated by the successful results, the systematic and algorithmic way of incorporating the methods were applied to first order ordinary differential equations in the pioneering work of Pakdemirli et al. (2011). The extension of the method to differential equations is non-trivial as the dependent variable and its derivatives are considered as separate variables inspired from the Lie Group theory applied to differential equations. The method is called as "the perturbation iteration method" to distinguish it from the past literature on the so-called "iteration-perturbation methods" (He 2001; Mickens 1987, 2005, 2006) which are not systematic approaches and does not produce general algorithms valid for various types of differential equations. The new perturbation–iteration method is directly applicable in a systematic algorithmic way, does not require special transformations or *ad hoc* assumptions. Second order differential equations were treated by the same method by Aksoy and Pakdemirli (2010) for Bratu type equations. Aksoy et al. (2012) further solved some nonlinear heat transfer equations. Dolapci et al. (2013) applied the method to Fredholm and Volterra integral equations. Şenol et al. (2013) treated the first order differential equation systems using the method. Pakdemirli (2013) reviewed the mentioned work. The basic idea behind the new method is to construct an approximate solution as in the classical perturbation methods, but then to iterate over this approximate solution to converge to the real solution of the problem. The iterations indeed enable one to converge to the real solutions for large perturbation parameters also. The mentioned work contains only applications to the regular perturbation problems which do not have secular terms in the analysis. In a very recent work, Pakdemirli (2015) successfully applied the method to problems with secularities also. The work considers the modifications of the method for non-regular problems containing secular terms which possess

Citation: Mehmet Pakdemirli, Application of the perturbation iteration method to boundary layer type problems, DOI 10.1186/s40064-016-1859-4.

slow convergence problems and truncation leads to unphysical unbounded solutions.

So far, the method has not been applied to boundary layer type singular problems which remain an open area of research. The work presented here is an initial attempt to address the implementation of the method to such singular problems. It is well known that for such problems, the method of matched asymptotic expansions is the most reliable method. In the method of matched asymptotic expansions, a solution valid outside the boundary layer is constructed first, which is the regular solution of the original problem, named the outer solution. Then another solution inside the boundary layer region where abrupt changes in the solutions is observed is constructed, called the inner solution. The outer and inner solutions are matched over the overlapping region and a composite expansion valid throughout the whole domain is constructed as a final step.

In the previous work, the systematic way of constructing perturbation–iteration algorithms are named as PIA(n,m), n representing the number of correction terms in the perturbation expansion and m representing the order of highest derivatives in the Taylor series expansion. The simplest algorithm is the PIA(1,1) with one correction term in the perturbation expansion and first order derivatives in the Taylor series expansion. For consistency and solvable iteration equations, n should be equal or less than m ($n \leq m$). As n or m increases, the convergence became faster but the algebraic complexity of the equations may end up with analytically unsolvable equations. In the previous work on differential equations, solutions are presented for PIA(1,1), PIA(1,2), PIA(1,3) and PIA(2,2) algorithms. In this pioneering and preliminary work on boundary layers, only the PIA(1,1) algorithms are applied to boundary layer type problems. Linear and nonlinear problems are treated with the PIA(1,1) method and solutions are contrasted with the available exact or numerical solutions of the original problem. For PIA(1,1), since the order of equations to be solved is less than the number of boundary conditions, the outer and inner solutions must be found by separate iterations and matched together similar to the method of matched asymptotic expansions.

Finally, as a last comment, the Lie Group methods of constructing exact solutions to differential equations cannot be applied to the problems considered here. The main reason is that the problems are defined over finite domains. For such domains, the symmetries of the equations do not remain stable after the boundary conditions are applied. To obtain group invariant solutions (similarity solutions) the equations as well as the boundary conditions and boundaries should remain invariant under the transformations especially for nonlinear problems (Bluman and Kumei 1989). For infinite or semi-infinite domains, this is more likely to occur but for finite domains, the symmetries are lost. For the singular problems with finite domains, the perturbation solutions, especially the method of matched asymptotic expansions is one of the most reliable analytical method. Here, an alternative approach which is a modification of the perturbation–iteration method to handle such problems is developed.

THEORY OF THE PERTURBATION ITERATION ALGORITHM FOR SECOND ORDER DIFFERENTIAL EQUATIONS

In this section, PIA(1,1) algorithm is derived for second order differential equations. In accordance, only one correction term in the perturbation expansion and Taylor expansions up to first order derivatives are taken in constructing the iteration procedure. Consider a general second order differential equation,

$$F(y'', y', y, \varepsilon) = 0 \tag{1}$$

with $y = y(x)$ and ε is the perturbation parameter. Note that a parameter such as ε may not exist at all in the original equation. In that case, the parameter should carefully be inserted into the appropriate part of the equation. The role of the perturbation parameter is to simplify the resulting iteration equations which cannot be solved otherwise. If the parameter is artificially introduced, it can be taken as 1 at the end. Only one correction term is taken in the perturbation expansion

$$y_{n+1} = y_n + \varepsilon (y_c)_n \tag{2}$$

Upon substitution of (2) into (1) and expanding in a Taylor series with first order derivatives only yields

$$F(y_n'', y_n', y_n, 0) + F_y(y_n'', y_n', y_n, 0)\, \varepsilon\, (y_c)_n + F_{y'}(y_n'', y_n', y_n, 0)\varepsilon\, (y_c')_n + F_{y''}(y_n'', y_n', y_n, 0)\varepsilon\, (y_c'')_n + F_\varepsilon(y_n'', y_n', y_n, 0)\, \varepsilon = 0 \tag{3}$$

where $F_y = \frac{\partial F}{\partial y}$, $F_{y'} = \frac{\partial F}{\partial y'}$, $F_{y''} = \frac{\partial F}{\partial y''}$, $F_\varepsilon = \frac{\partial F}{\partial \varepsilon}$ and all derivatives are evaluated at $\varepsilon = 0$. It is readily observed that the above equation is a variable coefficient non-homogenous linear second order differential equation with respect to the unknown $(y_c)_n$ in its most general form. Starting with an initial guess y_0, first $(y_c)_0$ is calculated from (3) and then substituted into (2) to calculate y_1. The iteration procedure is repeated using (3) and (2) until a satisfactory result is obtained. In solving boundary layer problems, for outer solution, the original equation is iterated. However for constructing the inner solution, first the equation is expressed in terms of the boundary layer variable and then the iteration is performed over the transformed equation. The last step is to match the solutions and construct a composite expansion valid within all the domain. This last step is similar to the calculations of the method of matched asymptotic expansions. If the inner and outer solutions cannot be matched, possible reasons may be the wrong location of the boundary layer or a bad initial guess for the iterations.

APPLICATIONS TO LINEAR DIFFERENTIAL EQUATIONS

Two linear second order constant coefficient problems, one homogenous and the other non-homogenous are treated using the PIA(1,1) algorithm.

Example Problem 1

Consider the singular perturbation problem

$$\varepsilon y'' + 2y' - y = 0 \quad y(0) = 0, \quad y(1) = 1 \tag{4}$$

For the problem, an outer solution and an inner solution will be sought and both solutions will be matched to construct a composite solution.

The Outer Solution

To find the outer solution, the original equation is taken in the analysis

$$F(y, y', y'', \varepsilon) = \varepsilon y'' + 2y' - y = 0 \tag{5}$$

Before starting, one of the important issues is the location of the boundary layer. This requires knowledge of the physics of the problem. For variable coefficient linear second order singular equations, a systematic way of locating the boundary condition indeed exists (Nayfeh 1981). For the problem considered, the boundary layer is located near $x \approx 0$. Therefore, the outer solution is not expected to satisfy the condition at $x = 0$. Substituting (5) into (3) and re-arranging, the iteration equation is

$$2y'_n - y_n - \varepsilon(y_c)_n + 2\varepsilon(y'_c)_n + \varepsilon y''_n = 0, \quad n = 0, 1, 2, \ldots \tag{6}$$

Note that due to the small term multiplying the highest derivative, the original equation yields a first order iteration equation which leads to inconsistencies if both conditions are forced to satisfy the solution. This justifies an outer and inner solutions to be constructed separately.

Inspired by the condition at the right, a simple initial guess is suggested

$$y_0 = 1 \tag{7}$$

to start the iteration. Substituting this initial guess into (6) and solving for $(y_c)_0$ yields

$$(y_c)_0 = -\frac{1}{\varepsilon} + \frac{c_1}{2} e^{x/2} \tag{8}$$

Hence, the first iteration result is

$$y_1 = y_0 + \varepsilon(y_c)_0 = \frac{c_1 \varepsilon}{2} e^{x/2} \tag{9}$$

and imposing the right hand side boundary condition yields

$$y_1 = e^{(x-1)/2} \tag{10}$$

Progressing in a similar way, the second iteration result is

$$y_2 = e^{(x-1)/2} + \frac{\varepsilon}{8}(1 - x)e^{(x-1)/2} \tag{11}$$

This result can also be found with the method of matched asymptotic expansions. However, with PIA(1,1), the same solution is retrieved from a very simple initial guess. This solution does not satisfy the boundary condition at the left hand side i.e. $y_2(0) \neq 0$.

The Inner Solution

The inner solution is valid within a small region called the boundary layer where the solution has a sharp turn. The first step is to determine the boundary layer variable by stretching the coordinate inside this layer

$$\xi = \frac{x}{\varepsilon^{\nu}} \tag{12}$$

Substituting the transformed variable into the original equation yields

$$\varepsilon^{1-2\nu} \frac{d^2 Y}{d\xi^2} + 2\varepsilon^{-\nu} \frac{dY}{d\xi} - Y = 0 \tag{13}$$

where $Y = Y(\xi)$. Balancing should be performed for the terms. There are two options: Balancing the first and second term or the first and third term. Balancing the second and third term is not an option because $\nu = 0$ and the original equation is retrieved with no stretching.

1. First and second term

For this case $1-2\nu = -\nu \Rightarrow \nu = 1$. Equation (13) after multiplication with ε is

$$\frac{d^2 Y}{d\xi^2} + 2\frac{dY}{d\xi} - \varepsilon Y = 0 \tag{14}$$

The goal is to retain as much terms as one can at the leading order. In this case, two terms remain in the first order and the choice is admissible.

2. First and third term

Balancing the first and third term requires $1-2\nu = 0 \Rightarrow \nu = 1/2$. Equation (13) is

$$\frac{d^2 Y}{d\xi^2} + 2\varepsilon^{-1/2} \frac{dY}{d\xi} - Y = 0 \tag{15}$$

Since the middle term is the only leading dominant term, this choice is not an admissible choice and hence discarded.

To summarize, the distinguished limit is $\nu = 1$ and the boundary layer variable is

$$\xi = \frac{x}{\varepsilon} \tag{16}$$

and the transformed equation to be solved is

$$F(Y, Y', Y'', \varepsilon) = Y'' + 2Y' - \varepsilon Y = 0 \tag{17}$$

Equation (3) takes the special form then

$$Y''_n + 2Y'_n + 2\varepsilon(Y'_c)_n + \varepsilon(Y''_c)_n - \varepsilon Y_n = 0, \quad n = 0, 1, 2, \ldots \tag{18}$$

Using (18) and (2) and applying the boundary condition at the left, i.e. $Y(0) = 0$, an iteration process can be constructed. Starting from a very simple guess

$$Y_0 = 0 \tag{19}$$

The successive two iteration are

$$Y_1 = \varepsilon c_3 (1 - e^{-2\xi}) \tag{20}$$

$$Y_2 = \varepsilon c_3 (1 - e^{-2\xi}) + \varepsilon \left\{ c_7 (1 - e^{-2\xi}) + \frac{\varepsilon c_3}{2} \xi \left(1 + e^{-2\xi}\right) \right\} \tag{21}$$

Note that this solution is valid in the neighborhood of $x = 0$ and therefore is not expected to satisfy the boundary condition at the right hand side. The equations solved are second order while the conditions are less than the required value of 2. The undetermined coefficients will be determined by the matching conditions.

Matching

The inner and outer solutions should smoothly combine in the overlapping region which requires solutions to be matched. Employing the Van Dyke's (1975) matching principle

$$(y_2)^i = (Y_2)^0 \tag{22}$$

in the overlapping region. In accordance, the outer expansion is written in terms of the inner variable, approximated and equated to the inner expansion written in terms of the outer variable and approximated. Hence

$$(y_2)^i = e^{(\varepsilon\xi-1)/2} + \frac{\varepsilon}{8}(1-\varepsilon\xi)e^{(\varepsilon\xi-1)/2} \tag{23}$$

and then approximation is taken up to two terms for fixed ξ

$$(y_2)^i \cong e^{-1/2}\left(1+\varepsilon\frac{\xi}{2}\right) + \frac{\varepsilon}{8}e^{-1/2} + \dots \tag{24}$$

Returning back to the original variable x now

$$(y_2)^i \cong e^{-1/2}\left(1+\frac{x}{2}\right) + \frac{\varepsilon}{8}e^{-1/2} + \dots \tag{25}$$

For the right hand side of the equation in the matching, the inner expansion is represented in terms of the outer variable x

$$(Y_2)^0 = \varepsilon c_3(1-e^{-2x/\varepsilon}) + \varepsilon\left(c_7(1-e^{-2x/\varepsilon}) + \frac{\varepsilon c_3}{2}\frac{x}{\varepsilon}\left(1+e^{-2x/\varepsilon}\right)\right) \tag{26}$$

and approximated up to two terms for fixed x

$$(Y_2)^0 \cong \varepsilon c_3(1+\frac{x}{2}) + \varepsilon c_7 \tag{27}$$

Note that for x kept fixed and not negligibly small, $e^{-2x/\varepsilon}$ is an exponentially small term which can be neglected. Equating (27) and (25), the constants

$$\varepsilon c_3 = e^{-1/2}, \quad c_7 = \frac{1}{8}e^{-1/2} \tag{28}$$

are determined from the matching conditions. Hence the final inner and outer solutions in terms of the original spatial variable are

$$Y_2 = e^{-1/2}(1-e^{-2x/\varepsilon}) + \frac{1}{2}e^{-1/2}x(1+e^{-2x/\varepsilon}) + \frac{\varepsilon}{8}e^{-1/2}\left(1-e^{-2x/\varepsilon}\right) \tag{29}$$

$$y_2 = e^{(x-1)/2} + \frac{\varepsilon}{8}(1-x)e^{(x-1)/2} \tag{30}$$

The inner solution is valid in the neighborhood of $x = 0$ (inside the boundary layer) and the outer solution is valid outside the boundary layer. A solution which is valid throughout the domain is desirable and one can construct a composite solution valid within all the domain of interest

$$y = Y_2 + y_2 - (y_2)^i \tag{31}$$

To construct the composite expansion, one simply adds the inner and outer solutions and subtracts the common overlapping part which is counted twice from the solution. Substituting (29), (30) and (25) into (31), the composite expansion is

$$y = e^{(x-1)/2} - \left(1-\frac{x}{2}\right)e^{-1/2-2x/\varepsilon} + \frac{\varepsilon}{8}\left\{(1-x)e^{(x-1)/2} - e^{-1/2-2x/\varepsilon}\right\} \tag{32}$$

which is a valid approximation throughout the whole domain. This solution satisfies both of the boundary conditions and can also be retrieved by the method of matched asymptotic expansions. To compare with the exact solution of the problem

$$y = \frac{e^{(-1+\sqrt{1+\varepsilon})x/\varepsilon} - e^{(-1-\sqrt{1+\varepsilon})x/\varepsilon}}{e^{(-1+\sqrt{1+\varepsilon})/\varepsilon} - e^{(-1-\sqrt{1+\varepsilon})/\varepsilon}} \tag{33}$$

first, the Taylor series expansions are written

$$\sqrt{1+\varepsilon} \cong 1 + \frac{1}{2}\varepsilon - \frac{1}{8}\varepsilon^2 \tag{34}$$

$$e^{(-1+\sqrt{1+\varepsilon})/\varepsilon} \cong e^{1/2-\varepsilon/8} \tag{35}$$

$$e^{(-1-\sqrt{1+\varepsilon})/\varepsilon} \cong e^{-2/\varepsilon-1/2+\varepsilon/8} \cong 0 \tag{36}$$

The second term is an exponentially small term which can be neglected. Under the approximations, the exact solution is

$$y^e \cong e^{-1/2+\varepsilon/8}e^{x/2-\varepsilon x/8} - e^{-1/2+\varepsilon/8}e^{-2x/\varepsilon-x/2+\varepsilon x/8} \tag{37}$$

Approximating further the exponential terms with the perturbation parameter

$$e^{\varepsilon/8} \cong 1 + \frac{\varepsilon}{8} \tag{38}$$

$$e^{-\varepsilon x/8} \cong 1 - \frac{\varepsilon}{8}x \tag{39}$$

the result is

$$y \cong e^{-1/2+x/2} - e^{-1/2-2x/\varepsilon-x/2} + \frac{\varepsilon}{8}\left\{(1-x)e^{-1/2+x/2} - (1+x)e^{-1/2-2x/\varepsilon-x/2}\right\} \tag{40}$$

This approximation is comparable with the perturbation iteration solution if further

$$e^{-x/2} \cong 1 - \frac{x}{2} \tag{41}$$

is taken in (40).

Example Problem 2

Consider the non-homogenous linear boundary value problem

$$\varepsilon y'' + 2y' = x \quad y(0) = \alpha, \quad y(1) = \beta \tag{42}$$

The Outer Solution

To find the outer solution, the original equation is used

$$F(y', y'', \varepsilon) = \varepsilon y'' + 2y' - x = 0 \tag{43}$$

The boundary layer is located near $x \approx 0$. Therefore, the outer solution is not expected to satisfy the condition at $x = 0$. Substituting (43) into (3) and rearranging, the iteration equation is

$$2y_n' - x + 2\varepsilon(y_c')_n + \varepsilon y_n'' = 0, \quad n = 0, 1, 2, \dots \tag{44}$$

Choose a simple initial guess compatible with the boundary condition at the right

$$y_0=\beta \tag{45}$$

to start the iteration. The first iteration solution satisfying the boundary condition at the right is

$$y_1 = \beta + \frac{x^2 - 1}{4} \tag{46}$$

This solution is not expected to satisfy the boundary condition at the left hand side i.e. $y_1(0)\neq 0$.

The Inner Solution

The inner solution is valid within a small region called the boundary layer where the solution has a sharp turn. It turns out that the boundary layer variable by stretching the coordinate inside this layer is

$$\xi = \frac{x}{\varepsilon} \tag{47}$$

and the transformed equation is

$$\frac{d^2Y}{d\xi^2} + 2\frac{dY}{d\xi} - \varepsilon^2\xi = 0 \tag{48}$$

Defining

$$F(Y', Y'', \varepsilon) = Y'' + 2Y' - \varepsilon^2\xi = 0 \tag{49}$$

the iteration Eq. (3) takes the form

$$Y_n'' + 2Y_n' + 2\varepsilon(Y_c')_n + \varepsilon(Y_c'')_n = 0, \quad n = 0, 1, 2, \ldots \tag{50}$$

Starting from a simple guess satisfying the boundary condition at the left

$$Y_0=\alpha \tag{51}$$

the first iteration is

$$Y_1 = \alpha + \varepsilon c_3(1 - e^{-2\xi}) \tag{52}$$

Note that this solution is valid in the neighborhood of $x = 0$ and therefore is not expected to satisfy the boundary condition at the right hand side. The undetermined coefficient will be determined by the matching conditions.

Matching

The matching of solutions requires

$$(y_1)^i = (Y_1)^0 \tag{53}$$

in the overlapping region. In accordance, the outer expansion is written in terms of the inner variable, approximated and equated to the inner expansion written in terms of the outer variable and approximated. Hence

$$(y_1)^i = \beta + \frac{\varepsilon^2\xi^2 - 1}{4} \cong \beta - \frac{1}{4} \tag{54}$$

The inner expansion is represented in terms of the outer variable x and approximated

$$(Y_1)^0 = \alpha + \varepsilon c_3(1 - e^{-2x/\varepsilon}) \cong \alpha + \varepsilon c_3 \tag{55}$$

Note that for x kept fixed and not negligibly small, $e^{-2x/\varepsilon}$ is an exponentially small term which can be neglected. Equating (54) and (55) gives the undetermined constant

$$\varepsilon c_3 = \beta - \frac{1}{4} - \alpha \tag{56}$$

Hence the inner solution in terms of the original variable is

$$Y_1 = \beta - \frac{1}{4} + \left(\alpha - \beta + \frac{1}{4}\right)e^{-2x/\varepsilon} \tag{57}$$

The composite solution is

$$y = Y_1 + y_1 - (y_1)^i \tag{58}$$

or

$$y = \beta + \frac{x^2 - 1}{4} + \left(\alpha - \beta + \frac{1}{4}\right)e^{-2x/\varepsilon} \tag{59}$$

which is a valid approximation throughout the whole domain. This solution satisfies both of the boundary conditions and can also be retrieved by the method of matched asymptotic expansions. The exact solution of the problem is

$$y = \alpha + \frac{\alpha - \beta + \frac{1}{4} - \frac{1}{4}\varepsilon}{1 - e^{-2/\varepsilon}}\left(e^{-2x/\varepsilon} - 1\right) + \frac{1}{4}x^2 - \frac{1}{4}\varepsilon x \tag{60}$$

Neglecting the exponentially small term $e^{-2/\varepsilon}$ $e^{-2/\varepsilon}$ and $O(\varepsilon)$ terms, solution (59) is obtained, hence the solution is a valid approximation of the exact solution.

APPLICATIONS TO NON-LINEAR DIFFERENTIAL EQUATIONS

A first order and a second order nonlinear boundary value problems are solved in this section using the method.

Example Problem 3

Consider the first order non-linear boundary value problem

$$\varepsilon y' + y^2 = 4 \quad y(0) = 1 \tag{61}$$

The outer solution

For the outer solution, the original equation is used

$$F(y, y', \varepsilon) = \varepsilon y' + y^2 - 4 = 0 \tag{62}$$

The boundary layer is assumed to be located near $x\approx 0$. If there arises any inconsistency in the solutions, the assumption may be revised. The outer solution is not expected to satisfy the condition at $x = 0$. Substituting (62) into (3) and re-arranging, the iteration equation is

$$y_n^2 - 4 + 2y_n\varepsilon(y_c)_n + \varepsilon y_n' = 0, \quad n = 0, 1, 2, \ldots \tag{63}$$

Since there is no boundary condition at the right hand side, the initial guess might be selected so that the iteration equation takes the simplest form. Hence

$$y_0=2 \tag{64}$$

The first iteration solution is

$$y_1=2 \tag{65}$$

which is the same solution with the starting value. Therefore the outer expansion seems to be converged.

The Inner Solution

The boundary layer variable by stretching the coordinate inside this layer is

$$\xi = \frac{x}{\varepsilon} \tag{66}$$

and the transformed equation is

$$\frac{dY}{d\xi} + Y^2 - 4 = 0 \tag{67}$$

The iteration Eq. (3) takes the form

$$Y_n' + Y_n^2 - 4 + 2Y_n\varepsilon(Y_c)_n + \varepsilon(Y_c')_n = 0, \quad n = 0, 1, 2, \ldots \tag{68}$$

Starting from a simple guess

$$Y_0 = 2 \tag{69}$$

which makes the most simplification in the iteration equation, the successive iterations are

$$Y_1 = 2 - e^{-4\xi} \tag{70}$$

$$Y_2 = 2 - \frac{5}{4}e^{-4\xi} + \frac{1}{4}e^{-8\xi} \tag{71}$$

Matching

The matching of solutions requires

$$(y_2)^i = (Y_2)^0 \tag{72}$$

in the overlapping region. Since the outer solution is constant, the expansion and approximation in the inner variable is the same

$$(y_2)^i = 2 \tag{73}$$

The inner expansion is represented in terms of the outer variable x and approximated

$$(Y_2)^0 = 2 - \frac{5}{4}e^{-4x/\varepsilon} + \frac{1}{4}e^{-8x/\varepsilon} \cong 2 \tag{74}$$

The solutions are in agreement in the overlapping region. The composite expansion valid throughout the domain is

$$y = 2 - \frac{5}{4}e^{-4x/\varepsilon} + \frac{1}{4}e^{-8x/\varepsilon} \tag{75}$$

The exact solution is

$$y = 2\tanh\left[\tanh^{-1}\left(\frac{1}{2}\right) + \frac{2x}{\varepsilon}\right] \tag{76}$$

Solution (75) is a very good approximation of the exact solution (76). In Fig. 1, $\varepsilon = 0.1$, and in Fig. 2, $\varepsilon = 100$ are taken. For small or large perturbation parameters, the agreement is excellent.

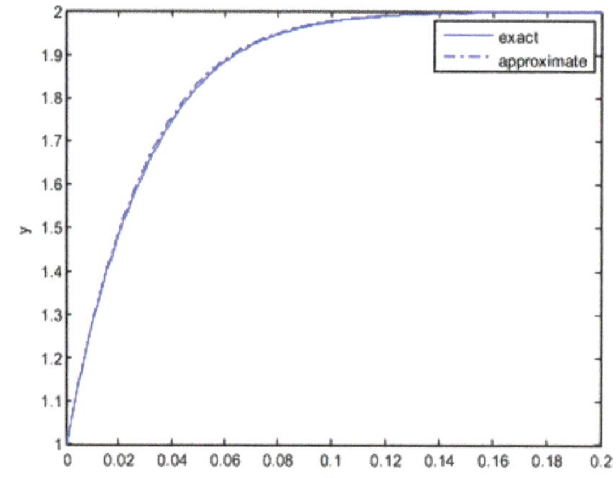

Figure 1: Comparison of the exact and approximate solutions for $\varepsilon = 0.1$ (Example Problem 3).

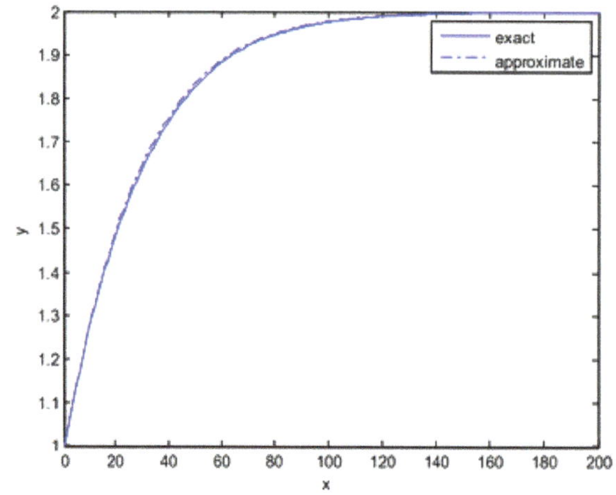

Figure 2: Comparison of the exact and approximate solutions for $\varepsilon = 100$ (Example Problem 3).

Example Problem 4

Consider the second order non-linear boundary value problem (Cole 1968)

$$\varepsilon y'' + yy' - y = 0 \quad y(0) = 0, \quad y(1) = 3 \tag{77}$$

The Outer Solution

For the outer solution, the original equation is used

$$F(y, y', y'', \varepsilon) = \varepsilon y'' + yy' - y = 0 \tag{78}$$

The boundary layer is located near $x \approx 0$. The outer solution is not expected to satisfy the condition at $x = 0$. Substituting (78) into (3) and rearranging, the iteration equation is

$$y_n y'_n - y_n + (y'_n - 1)\varepsilon(y_c)_n + y_n \varepsilon(y'_c)_n + \varepsilon y''_n = 0, \quad n = 0, 1, 2, \ldots \tag{79}$$

$y_0 = 3$ which is inspired from the right hand side boundary condition would not be a good initial guess. Rather, one has to look for a trivial function which most simplifies the iteration equation for $n = 0$. Such an initial guess is

$$y_0 = x \tag{80}$$

The first and second iteration solutions satisfying the right hand side condition is

$$y_1 = x + 2 \tag{81}$$

$$y_2 = x + 2 \tag{82}$$

which shows that the outer solution converged.

The Inner Solution

The boundary layer variable by stretching the coordinate inside this layer is

$$\xi = \frac{x}{\varepsilon} \tag{83}$$

and the transformed equation is

$$\frac{d^2Y}{d\xi^2} + Y\frac{dY}{d\xi} - \varepsilon Y = 0 \tag{84}$$

The iteration Eq. (3) takes the form

$$Y''_n + Y_n Y'_n + Y'_n \varepsilon(Y_c)_n + Y_n \varepsilon(Y'_c)_n + \varepsilon(Y''_c)_n - \varepsilon Y_n = 0, \quad n = 0, 1, 2, \ldots \tag{85}$$

Starting from a simple guess

$$Y_0 = 2 \tag{86}$$

which makes the most simplification in the iteration equation, the first iteration is

$$Y_1 = (2 + c_1)(1 - e^{-2\xi}) + \varepsilon\xi \tag{87}$$

Matching this solution with the first iteration of the outer solution makes a simplification and the constant c_1 turns out to be zero. Hence, the final first iteration is

$$Y_1 = 2(1 - e^{-2\xi}) + \varepsilon\xi \tag{88}$$

The second iteration yields

$$Y_2 = 2\left(1 - e^{-2\xi}\right) + \varepsilon\xi\left(1 + e^{-2\xi}\right) + e^{-4\xi} + \varepsilon c_2 + \left(\varepsilon\xi^2 - 1 - \varepsilon c_2\right)e^{-2\xi} \tag{89}$$

Matching

The matching of solutions requires

$$(y_2)^i = (Y_2)^0 \tag{90}$$

in the overlapping region. The outer solution is written in the inner variable and approximated including first order terms yield the same result

$$(y_2)^i = 2 + \varepsilon\xi = 2 + x \tag{91}$$

Then the inner expansion is represented in terms of the outer variable x and approximated

$$(Y_2)^0 = 2\left(1 - e^{-2x/\varepsilon}\right) + x\left(1 + e^{-2x/\varepsilon}\right) + e^{-4x/\varepsilon} + \varepsilon c_2 + \left(\frac{x^2}{\varepsilon} - 1 - \varepsilon c_2\right)e^{-2x/\varepsilon} \cong 2 + x + \varepsilon c_2 \tag{92}$$

since terms of $e^{-2x/\varepsilon}$,$e^{-4x/\varepsilon}$ are exponentially small in the overlapping region. Comparing (92) with (91), it is found that $c_2 = 0$. The inner expansion in the final form is

$$Y_2 = 2\left(1 - e^{-2\xi}\right) + \varepsilon\xi\left(1 + e^{-2\xi}\right) + e^{-4\xi} + \left(\varepsilon\xi^2 - 1\right)e^{-2\xi} \tag{93}$$

The composite expansion is calculated as a final step

$$y = Y_2 + y_2 - (y_2)^i \tag{94}$$

or

$$y = 2 - 3e^{-2x/\varepsilon} + x\left(1 + e^{-2x/\varepsilon}\right) + e^{-4x/\varepsilon} + \frac{x^2}{\varepsilon}e^{-2x/\varepsilon} \tag{95}$$

Solution (95) is contrasted with the numerical solution of the original equation obtained by an adaptive step size Runge–Kutta method combined with shooting. In Fig. 3, $\varepsilon = 0.1$ is taken and the match is excellent. In Fig. 4, a larger perturbation parameter, namely $\varepsilon = 1$ is taken and there is a reasonably good agreement between the results.

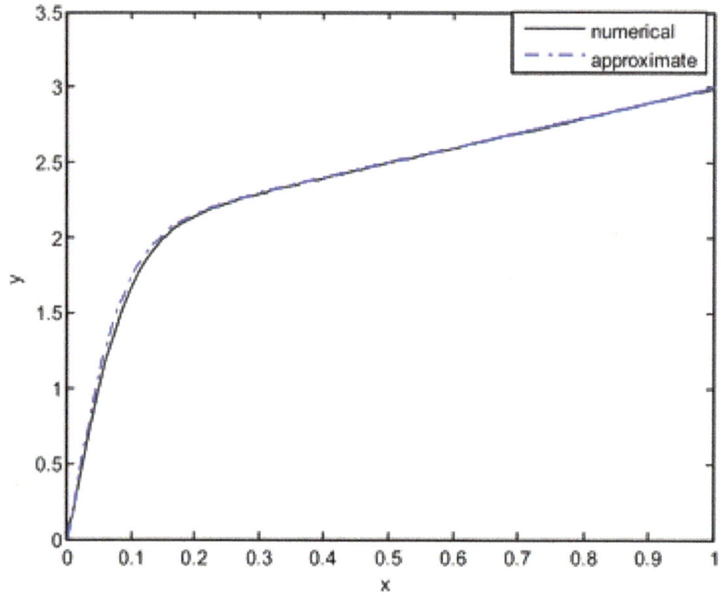

Figure 3: Comparison of the exact and approximate solutions for $\varepsilon = 0.1$ (Example Problem 4).

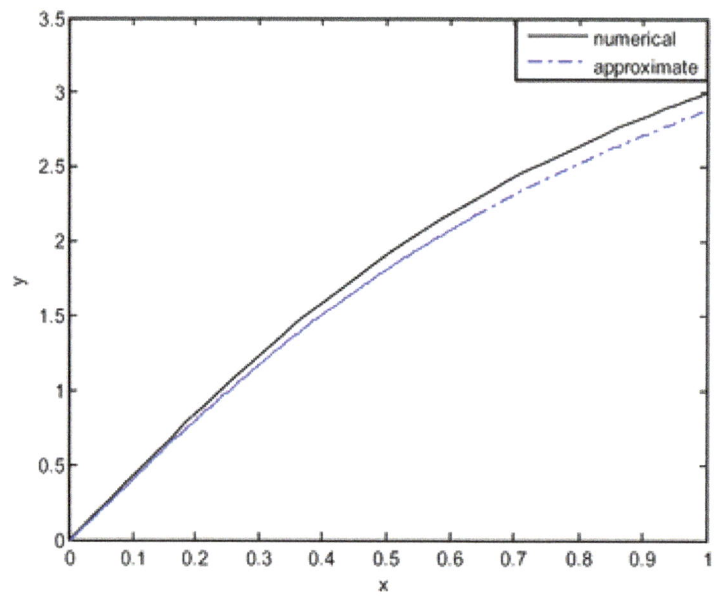

Figure 4: Comparison of the exact and approximate solutions for $\varepsilon = 1$ (Example Problem 4).

CONCLUDING REMARKS

Boundary layer type problems are treated by the perturbation iteration method for the first time. The simplest algorithm of PIA(1,1) in which there is only one correction term in the perturbation expansion and at most first order derivatives in the Taylor series expansions is considered. For PIA(1,1), due to the reduction of differential order of the iterated equations, an outer and an inner expansion must be found independently via the appropriate iteration procedures. The solutions are then matched with each other to construct a composite expansion valid throughout the whole domain. Singular linear and nonlinear problems are treated with the method and results are compared with the available analytical or numerical solutions. It is shown that the method can be effectively applied to singular problems producing admissible analytical approximate solutions.

ACKNOWLEDGEMENTS

The support of the Turkish Academy of Sciences (TÜBA) to cover the publication charges is highly appreciated.

Competing Interests

The author declares that he has no competing interests

REFERENCES

1. Aksoy Y, Pakdemirli M (2010) New perturbation–iteration solutions for Bratu-type equations. Comput Math Appl 59:2802–2808
2. Aksoy Y, Pakdemirli M, Abbasbandy S, Boyacı H (2012) New perturbation–iteration solutions for nonlinear heat transfer equations. Int J Numer Methods Heat Fluid Flow 22:814–828
3. Bluman GW, Kumei S (1989) Symmetries and differential equations. Springer, New York
4. Cole JD (1968) Perturbation methods in applied mathematics. Blaisdell, Waltham
5. Dolapci IT, Şenol M, Pakdemirli M (2013) New perturbation iteration solutions for Fredholm and Volterra integral equations. J Appl Math. doi:10.1155/2013/682537
6. He JH (2001) Iteration perturbation method for strongly nonlinear oscillators. J Vib Control 7:631–642
7. Mickens RE (1987) Iteration procedure for determining approximate solutions to nonlinear oscillator equations. J Sound Vib 116:185–187
8. Mickens RE (2005) A generalized iteration procedure for calculating approximations to periodic solutions of "Truly Nonlinear Oscillators". J Sound Vib 287:1045–1052
9. Mickens RE (2006) Iteration method solutions for conservative and limit-cycle force oscillators. J Sound Vib 292:964–968
10. Nayfeh AH (1981) Introduction to perturbation techniques. Wiley, New York
11. Pakdemirli M (2013) Review of the new perturbation iteration method. Math Comput Appl 18:139–151
12. Pakdemirli M (2015) Perturbation–iteration method for strongly nonlinear vibrations. J Vib Control. doi:10.1177/1077546315586647
13. Pakdemirli M, Boyacı H (2007) Generation of root finding algorithms via perturbation theory and some formulas. Appl Math Comput 184:783–788
14. Pakdemirli M, Boyacı H, Yurtsever HA (2007) Perturbative derivation and comparisons of root-finding algorithms with fourth order derivatives. Math Comput Appl 12:117–124
15. Pakdemirli M, Boyacı H, Yurtsever HA (2008) A root finding algorithm with fifth order derivatives. Math Comput Appl 13:123–128
16. Pakdemirli M, Aksoy Y, Boyacı H (2011) A new perturbation–iteration approach for first order differential equations. Math Comput Appl 16:890–899
17. Şenol M, Dolapci IT, Aksoy Y, Pakdemirli M (2013) Perturbation–iteration method for first-order differential equations and systems. Abstr Appl Anal. doi:10.1155/2013/704137
18. Van Dyke M (1975) Perturbation methods in fluid mechanics. The Parabolic Press, Stanford

CHAPTER 7

A UNIFORMLY VALID APPROXIMATION ALGORITHM FOR NONLINEAR ORDINARY SINGULAR PERTURBATION PROBLEMS WITH BOUNDARY LAYER SOLUTIONS

Süleyman Cengizci, Mehmet Tarık Atay and Aytekin Eryılmaz

BACKGROUND

Nonlinear problems have always been more attractive than linear ones for scientists. The main reason for this is that almost all natural phenomena in nature lead us to nonlinear models to describe them. In these models, finding exact solutions are quite difficult or sometimes impossible. Because of this reason, it is needed to get at least an approximate solution to these types of problems by certain methods.

Singularly perturbed problems occurs frequently in electrical systems, celestial mechanics, particle physics, quantum mechanics, (semi/super) conductor systems, fluid mechanics, thermal processes and in chemical/biochemical reactions (Kumar 2011). These problems are characterized by the presence of a very small positive parameter $0 < \varepsilon \ll 1$ that multiplies the highest order derivative term in the differential equation. This small parameter is known as *singular perturbation parameter*. In the case of $\varepsilon = 0$, problem is called as *reduced problem* since the order of the differential equation reduces. Consider the general form of a singularly perturbed nonlinear second-order ordinary differential equation

$$\varepsilon y''(x) = f(x, y(x), y'(x)) \quad 1$$

subject to boundary condions

$$y(0) = \alpha,\ y(1) = \beta, \quad (\alpha, \beta \in \mathbb{R}). \quad 2$$

It is a well established fact that a nonlinear second-order two-point boundary value problem (1–2) with small parameter plays a critical role in nonlinear physics. Also it is of great practical interest to study this nonlinear phenomena. Singular perturbation problems and the methods used to tackle them are very important concepts because of their mathematical properties, physical meanings and applications in engineering sciences. We refer the reader to consult the reference Kumar (2011) for more detailed information and some significant examples.

The first study in the perturbation theory was presented by Prandtl (1905). But, the term *singular perturbation* was used for the first time by Friedrichs and Wasow (1946). Scientists have paid great attention for this theory for more than a century. After the first studies, a number of excellent books were published such as O'Malley (1974), Bender and Orszag (1978), Kevorkian and Cole (1981), Eckhaus (1973), Eckhaus (1979), Lagerstrom (1988), Hinch (1991), Van Dyke (1975), Johnson (2006), Verhulst (2006), Holmes (1995) and Roos et al. (1996). Thanks to these great books and the other works, today we have certain*traditional asymptotic methods*. Some of them are the Method of Matched Asymptotic Expansions (MMAE), the Method of Multiple-scale Analysis, the Periodic Averaging Method, the Method of Wentzel–Kramers–Brillouin (WKB) Approximation and the Method of Strained Coordinates.

Towards the end of 1980s, various methods, apart from the traditional asymptotic methods, began to appear. In those years Kadalbajoo, Reddy, Jiwari et al. conducted so many significant studies such as Kadalbajoo et al. (1987), Kadalbajoo and Reddy (1987), Chawla and Katti (1982), Mo (1993), Kadalbajoo and Patidar (2003), Kadalbajoo and Gupta (2010), Mittal and Jiwari (2011), Sharma et al. (2012a) and Sharma et al. (2012b). In 2003, Kadalbajoo and Patidar made a detailed survey of

Citation: Süleyman Cengizci, Mehmet Tarık Atay, Aytekin Eryılmaz, A uniformly valid approximation algorithm for nonlinear ordinary singular perturbation problems with boundary layer solutions, DOI: 10.1186/s40064-016-1865-6.

singular perturbation problems in partial differential equations (PDEs) (Kadalbajoo and Patidar 2003). In 2010, Kadalbajoo and Gupta in their study Kadalbajoo and Gupta (2010) made a great survey on the numerical methods for singularly perturbed problems. In 2011, Parul studied the traditional methods to solve this kind of problems and gave important examples occuring in engineering and science (Kumar 2011, 2011). In 2012, Roos made a survey, particularly of singularly perturbed convection–reaction–diffusion problems covering the years 2008–2012 (Roos 2012). In the mean time, a number of intriguing numerical methods were presented such as Reproducing Kernel Method (Cui and Geng 2007; Geng and Cui 2007; Geng 2011; Geng and Cui 2011; Li et al. 2012), Variational Iteration Method (Kumar and Mishra 2014), Haar Wavelet Approach (Pandit and Kumar 2014). We must state that there are so many various methods and the above-mentioned methods are just some of them.

In this paper, we study on an efficient asymptotic method called SCEM that generates uniformly valid approximations (UVA) to the solution of singularly perturbed nonlinear boundary value problems. Applying the present method, we are able to get rid of tedious matching procedure of MMAE. We propose a UVA at the first step and then seek appropriate approximations called outer approximation and complementary approximations such that the resulting successive approximations satisfy the boundary conditions exactly.

The paper is organized as follows: "About asymptotic expansions" section gives a brief description of the asymptotic expansions. The overview of SCEM is given in "Successive Complementary Expansion Method" section. In "Numerical examples" section, we consider four numerical problems for comparison with existing methods. The conclusion is given in the last section.

ABOUT ASYMPTOTIC EXPANSIONS

In this section, we briefly review the basic concepts of asymptotic approximation theory.

The limit process is fundamental tool of mathematical analysis. It is defined for a given real continuous function f: given any $\delta > 0$ there exists a number $n_0(\delta)$ such that $|f(x) - f| < \delta$ for any $n \geqslant n_0(\delta)$. This definition gives information about the behavior of the function f as $x \to a$, but not about how. Therefore, asymptotic approximation theory requires another definition that enables us to describe the behavior of functions and compare them in a more precise way under the limit process: Bachmann–Landau notations. Consider real and continuous functions $f(\varepsilon)$ and $g(\varepsilon)$, where $0 < \varepsilon \le \varepsilon_0 \ll 1$.$f(\varepsilon) = O(g(\varepsilon))$, $\varepsilon \to 0$ if there exist positive constants K and ε_0 such that $|f(\varepsilon)| \le K|g(\varepsilon)|$ for $\varepsilon \to 0$ in $(0, \varepsilon_0]$.$f(\varepsilon) = o(g(\varepsilon))$, for $\varepsilon \to 0$ if lim$\varepsilon \to 0$f(ε)g(ε)=0. $f(\varepsilon)$ and $g(\varepsilon)$ are said to be asymptotically equivalent, $f(\varepsilon) \approx g(\varepsilon)$ as $\varepsilon \to 0$ if $f(\varepsilon) = O(g(\varepsilon))$ and $g(\varepsilon) = O(f(\varepsilon))$for $\varepsilon \to 0$. $f(\varepsilon) = O_S(g(\varepsilon))$ if $f(\varepsilon) = O(g(\varepsilon))$ and $f(\varepsilon) \neq o(g(\varepsilon))$ for $\varepsilon \to 0$. Here the subscript S denotes *sharp estimate*. Now consider a sequence of functions $\{\phi_{n+1}\}$, $n = 0, 1, \dots$. Such a sequence is an *asymptotic sequence* if $\phi_{n+1}(\varepsilon) = o(\phi_n(\varepsilon))$ for $\varepsilon \to 0$ and $n = 0, 1, \dots$ Let $y(x, \varepsilon)$ be defined in some domain Ω of x and some neighborhood of $\varepsilon = 0$. The series

$$y_a(x,\varepsilon) = \sum_{i=0}^{n} \phi_i(\varepsilon) y_i(x) \quad 3$$

is called *regular asymptotic expansion (Poincaré expansion)* of $y(x, \varepsilon)$ as $\varepsilon \to 0$ if the condition

$$y(x,\varepsilon) - \sum_{i=0}^{n} \phi_i(\varepsilon) y_i(x) = o(\phi_i(\varepsilon))$$

is satisfied. And *generalized asymptotic expansions* are defined as

$$y(x,\varepsilon) = \sum_{i=0}^{n} \phi_i(\varepsilon) y_i(x,\varepsilon) = O_S(\phi_{i+1}(\varepsilon)), \quad \varepsilon \to 0.$$

Interesting cases occur when the function $y(x, \varepsilon)$ is not regular in Ω, so expansion (3) is uniformly valid only in a restricted region $\Omega_0 \in \Omega$ that is called *outer region*. In this case, the asymptotic expansion that is valid in outer region is often called *outer expansion* and (3) can be given as

$$y_a(x,\varepsilon) = E_0\phi = \sum_{i=0}^{n} \phi_i^{(0)}(\varepsilon) y_i^{(0)}(x), \quad 4$$

where $\phi_i^{(0)}(\varepsilon)$ is an asymptotic sequence and the special operator E_0 called the *outer expansion operator* at a given order $\phi(\varepsilon)$.

On the other hand, we encounter with a singular perturbation problem and we must introduce boundary layer domains. Here, in the simplest case (we assume that the problem has boundary layer near the origin $x = 0$), we introduce an *inner region* which can be formally denoted $\Omega_1 = \Omega - \Omega_n$ and located near the origin. The boundary layer variable is $\bar{x} = \frac{x}{\xi(\varepsilon)}, \xi(\varepsilon)$ being the order of thickness of this boundary layer. If a regular expansion can be constructed in Ω_1, we can write

$$y_a(x,\varepsilon) = E_1\phi = \sum_{i=0}^{n} \phi_i^{(1)}(\varepsilon) y_i^{(1)}(\bar{x}). \quad 5$$

This *inner expansion operator*E_1 is defined in Ω_1 at the same order $\phi(\varepsilon)$ as the outer expansion operator E_0;thus, $y(x, \varepsilon) - E_1 y(x, \varepsilon) = o(\phi(\varepsilon))$ and so

$y_a(x, \varepsilon) = E_0 y(x, \varepsilon) + E_1 y(x, \varepsilon) - E_1 E_0 y(x, \varepsilon)$

is clearly uniformly valid approximation to order $\delta(\varepsilon)$ satisfying the *modified Van Dyke principle (MVDP)* $E_1 E_0 y(x, \varepsilon) = E_0 E_1 y(x, \varepsilon)$. This is the main idea underlying the method of matched asymptotic expansions (MMAE). We know that MMAE which has been designed for finding uniformly valid approximations to singularly perturbed boundary value problems is a powerful mathematical technique. It is based on finding two different approximations for different two regions, which are called as inner region (where the solution exhibits rapid changes) and outer region (which is far from the inner region). In the last step, these two approximations are matched using the limit process to obtain a uniformly valid approximation. For more details we refer the reader to O'Malley (1974), Bender and Orszag (1978), Kevorkian and Cole (1981), Eckhaus (1973), Eckhaus (1979), Lagerstrom (1988), Hinch (1991), Van Dyke (1975), Johnson (2006),

Verhulst (2006), Mauss and Cousteix (2002), Cousteix and Mauss (2004), Cousteix and Mauss (2009), Cathalifaud et al. (2010) and Nayfeh (1973).

SUCCESSIVE COMPLEMENTARY EXPANSION METHOD

Sometimes the matching procedure in MMAE can be tedious or impossible. Therefore, we wish to present and examine an efficient asymptotic method named as *successive complementary expansion method*, which is designed by French scientists J. Mauss and J. Cousteix in order to obtain uniformly valid approximations to the boundary layer problems occur in fluid mechanics in Mauss and Cousteix (2002). In SCEM, instead of finding two different approximations to match later, a uniformly valid approximation that exactly satisfies the boundary conditions is proposed at the first step. So, thanks to the SCEM we will not be in need of any matching procedure. We cannot ignore the fact that SCEM is not the first method which does not require any matching procedure. For instance, the WKB method and the Method of Multiple-scale Analysis also do not require any matching procedure (Cousteix and Mauss 2004). But their applicabilities are restricted to some certain problems. The uniformly valid SCEM approximation is in the regular form given as follows

$$y_n^{scem}(x,\bar{x},\varepsilon)=\sum_{i=0}^{n}\delta_i(\varepsilon)[y_i(x)+\Psi_i(\bar{x})], \qquad 6$$

where $\delta_i(\varepsilon)$ is an asymptotic sequence and $\Psi_i(\bar{x})$ are the complementary approximation functions that depends on $\bar{x}$. Functions $y_i(x)$ are the outer approximations that have been found by MMAE and they only depend on x, not also on ε. In its regular form, SCEM is equivalent to MMAE. If the functions $y_i(x)$ and $\Psi_i(\bar{x})$ also depend on ε, the uniformly valid SCEM approximation is named as *generalized SCEM approximation* and given in the following form (Cousteix and Mauss 2007; Mauss and Cousteix 2002; Cousteix and Mauss 2009)

$$y_{ng}^{scem}(x,\bar{x},\varepsilon)=\sum_{i=0}^{n}\bar{\delta}_i(\varepsilon)\left[\bar{y}_i(x,\varepsilon)+\bar{\Psi}_i(\bar{x},\varepsilon)\right] \qquad 7$$

with

$$y_{ng}^{scem}(x,\bar{x},\varepsilon)=y_n^{scem}(x,\bar{x},\varepsilon)+o(\delta_{n+1}(\varepsilon)).$$

The sequence of order functions $\delta_i(\varepsilon)$ may or not be same with $\bar{\delta}_i(\varepsilon)$. If only one-term SCEM approximation is desired, then one seeks a uniformly valid SCEM approximation in the form of

$$y_0^{scem}(x,\bar{x},\varepsilon)=y_0(x,\varepsilon)+\Psi_0(\bar{x},\varepsilon).$$

To improve the accuracy of SCEM approximation, the first SCEM approximation can be iterated using (7). It means that successive complementary terms will be add to the approximation. To this end, second SCEM approximation will be sought in the form of

$$y_1^{scem}(x,\bar{x},\varepsilon)=y_0(x,\varepsilon)+\Psi_0(\bar{x},\varepsilon)+\varepsilon\left(y_1(x,\varepsilon)+\Psi_1(\bar{x},\varepsilon)\right)$$

In Cousteix and Mauss (2007), error estimates for first and second SCEM approximations are given as $\left|y-y_0^{scem}\right|<\varepsilon K_0$ and $\left|y-y_1^{scem}\right|<\varepsilon^2K_1$, where K_0 and K_1 are positive constants independent of ε and y is the exact solution of the problem.

NUMERICAL EXAMPLES

In this section, we present four numerical experiments to show the efficiency and the robustness of the proposed method. All the numerical calculations are performed using Shampine et al. (2000).

Example 1

Consider the singular perturbation problem

$$\varepsilon y''(x)+2y'(x)+e^{y(x)}=0, \quad x\in(0,1) \qquad 8$$

subject to boundary conditions

$$y(0)=0,\ y(1)=0. \qquad 9$$

This problem has a boundary layer near the point $x = 0$ and uniformly valid asymptotic approximation is given as follows in Bender and Orszag (1978)

$$y(x)=\ln\left(\frac{2}{x+1}\right)-(\ln 2)e^{\frac{-2x}{\varepsilon}}. \qquad 10$$

To obtain SCEM approximation for $n = 0$ (first-term approximation) of (8–9), we seek an approximation in the form of

$$y_0^{scem}(x,\bar{x},\varepsilon)=y_0(x,\varepsilon)+\Psi_0(\bar{x},\varepsilon). \qquad 11$$

For $\varepsilon = 0$, Eq. (8) yields

$$2y'(x)+e^{y(x)}=0 \qquad 12$$

and we have the outer solution as follows

$$y_0(x,\varepsilon)=\ln\left(\frac{2}{x+1}\right). \qquad 13$$

Now we look for a uniformly valid SCEM approximation in the form of

$$y_{ng}^{scem}(x,\bar{x},\varepsilon)=\ln\left(\frac{2}{x+1}\right)+\Psi_0(\bar{x},\varepsilon)+\sum_{i=1}^{n}\varepsilon^n[y_i(x,\varepsilon)+\Psi_i(\bar{x},\varepsilon)]; \qquad 14$$

where $\bar{x}=\frac{x}{\varepsilon}$. Substituting (14) into the Eq. (8) and balancing the terms with respect to the powers of parameter ε yields

$$\varepsilon\left(\Psi_0+\ln\left(\frac{2}{x+1}\right)\right)''+2\left(\Psi_0+\ln\left(\frac{2}{x+1}\right)\right)'+e^{\left(\Psi_0+\ln\left(\frac{2}{x+1}\right)\right)}=0 \qquad 15$$

and consequently, applying the balancing procedure we have

$$\Psi_0''(\bar{x},\varepsilon)+2\Psi_0'(\bar{x},\varepsilon)=0 \qquad 16$$

with the boundary conditions

$$\Psi_0(0,\varepsilon)=-\ln(2),\ \Psi_0\left(\frac{1}{\varepsilon},\varepsilon\right)=0. \qquad 17$$

Then,

$$y_{0g}^{scem}(x, \bar{x}, \varepsilon) = \left[\ln\left(\frac{2}{x+1}\right) + \Psi_0(\bar{x}, \varepsilon) \right] \quad 18$$

is the first term of the SCEM approximation.

The numerical results of the problem are shown in Tables 1, 2 and in Figs. 1, 2. We deliberately choose the node points as possible as near the boundary layer (for this problem, near the point $x = 0$) in order to see efficiency of the method.

Table 1: Numerical results of Example 1 for $\varepsilon = 0.1$

x	UVA solution (Bender and Orszag 1978)	SCEM approximation	Absolute error
0.000	0.000000000000	0.000000000000	0.000000000000
0.001	0.012725733435	0.012725733463	2.828981493e(−11)
0.005	0.060974133872	0.060974134008	1.359571344e(−10)
0.010	0.115695936573	0.115695936832	2.589762846e(−10)
0.100	0.504029730749	0.504029731985	1.2353316147e(−9)
0.200	0.498130190310	0.498130191712	1.4025156036e(−9)
0.300	0.429064776009	0.429064777435	1.4251414492e(−9)
0.400	0.356442418964	0.356442420392	1.4282036109e(−9)
0.500	0.287650603618	0.287650605047	1.4286180016e(−9)
0.600	0.223139292470	0.223139293899	1.4286740679e(−9)
0.700	0.162518353125	0.162518354554	1.4286816729e(−9)
0.800	0.105360437654	0.105360439083	1.4286826582e(−9)
0.900	0.051293283830	0.051293285259	1.4286828040e(−9)
1.000	−0.000000001428	0.000000000000	1.4286828220e(−9)

Table 2: Numerical results of Example 1 for $\varepsilon = 0.001$

x	UVA solution (Bender and Orszag 1978)	SCEM approximation	Absolute error
0.000	0.000000000000	0.000000000000	0.000000000000
0.001	0.598340410221	0.598340410221	0.000000000000
0.005	0.688128170215	0.688128170215	0.000000000000
0.010	0.683196848278	0.683196848278	0.000000000000
0.100	0.597837000755	0.597837000755	0.000000000000
0.200	0.510825623765	0.510825623765	0.000000000000
0.300	0.430782916092	0.430782916092	0.000000000000
0.400	0.356674943938	0.356674943938	0.000000000000
0.500	0.287682072451	0.287682072451	0.000000000000
0.600	0.223143551314	0.223143551314	0.000000000000
0.700	0.162518929497	0.162518929497	0.000000000000
0.800	0.105360515657	0.105360515657	0.000000000000
0.900	0.051293294387	0.051293294387	0.000000000000
1.000	0.000000000000	0.000000000000	0.000000000000

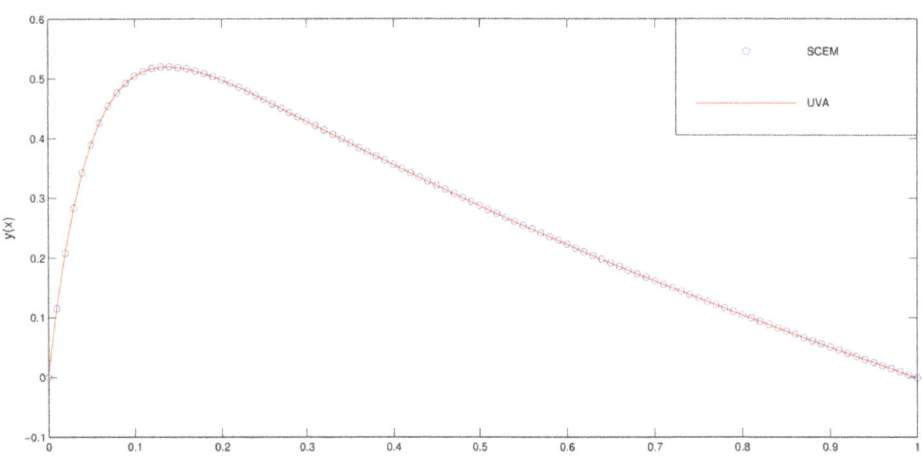

Figure 1: Comparison of SCEM and UVA solutions of Example 1 for $\varepsilon = 0.1$

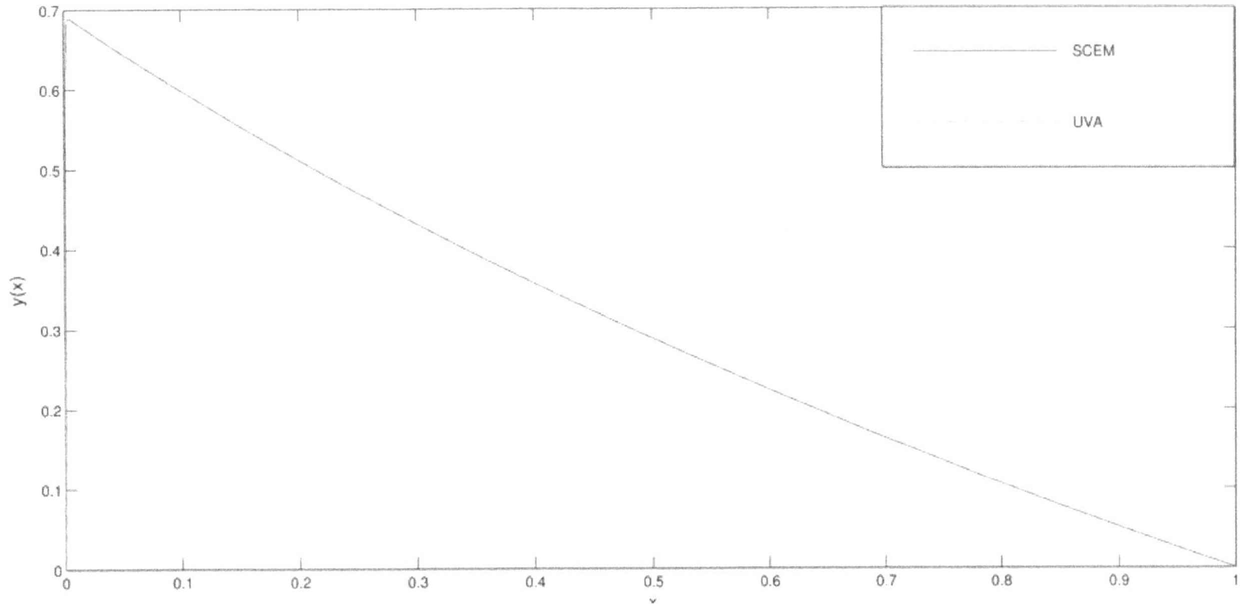

Figure 2: Comparison of SCEM and UVA solutions of Example 1 for $\varepsilon = 0.001$

Example 2

Consider the second singular perturbation problem given in O'Malley (1974)

$$\varepsilon y''(x) + y(x)y'(x) = 0, \quad x \in (-1, 1) \qquad 19$$

subject to boundary conditions

$$y(-1) = 0, \; y(1) = -1, \qquad 20$$

and with the exact solution

$$y(x,\varepsilon) = -\frac{1 - e^{\frac{-(x+1)}{\varepsilon}}}{1 + e^{\frac{-(x+1)}{\varepsilon}}}. \qquad 21$$

This problem has boundary layer near the point $x = -1$.

In Table 3 and Figs. 3, 4 we deliberately choose the node points as possible as near the boundary layer (for this problem, near the point $x = -1$) in order to see efficiency of the method.

Table 3: Numerical results of Example 2 for $\varepsilon = 0.001$

x	Exact solution	SCEM approximation	Absolute error
−1.000	0.000000000000	0.000000000000	0.000000000000
−0.995	−0.986614298151	−0.986614102977	1.95174316e(−7)
−0.990	−0.999909204262	−0.999909184974	1.92884420e(−8)
−0.980	−0.999999995877	−0.999999995802	7.55888684e(−11)
−0.400	−1.000000000000	−1.000000000000	0.000000000000
−0.200	−1.000000000000	−1.000000000000	0.000000000000
0.000	−1.000000000000	−1.000000000000	0.000000000000
0.200	−1.000000000000	−1.000000000000	0.000000000000
0.400	−1.000000000000	−1.000000000000	0.000000000000
0.600	−1.000000000000	−1.000000000000	0.000000000000
0.800	−1.000000000000	−1.000000000000	0.000000000000
1.000	−1.000000000000	−1.000000000000	0.000000000000

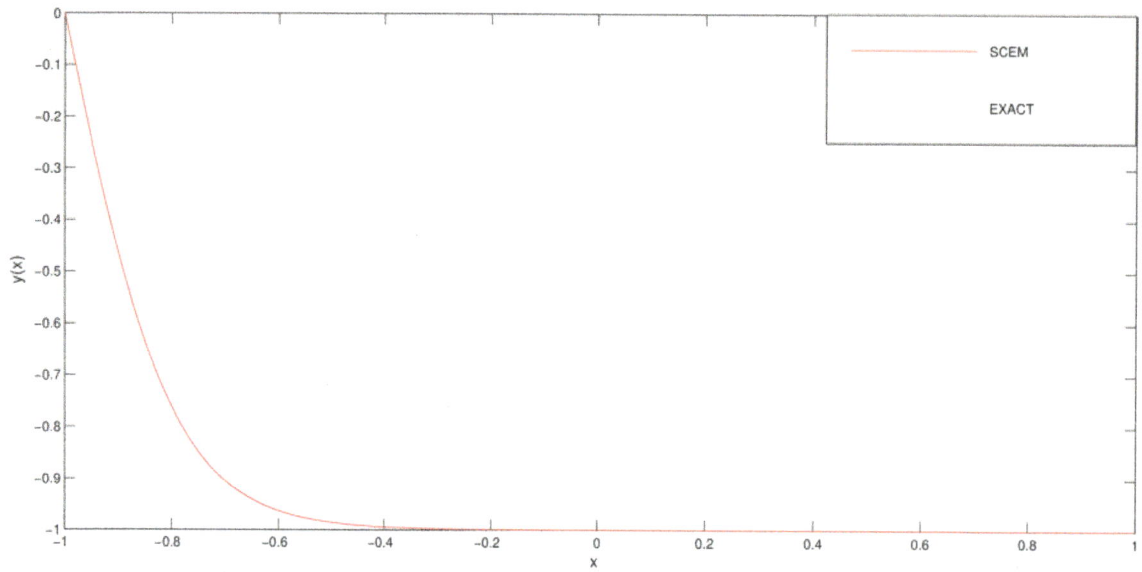

Figure 3: Comparison of SCEM and exact solutions of Example 2 for $\varepsilon = 0.1$

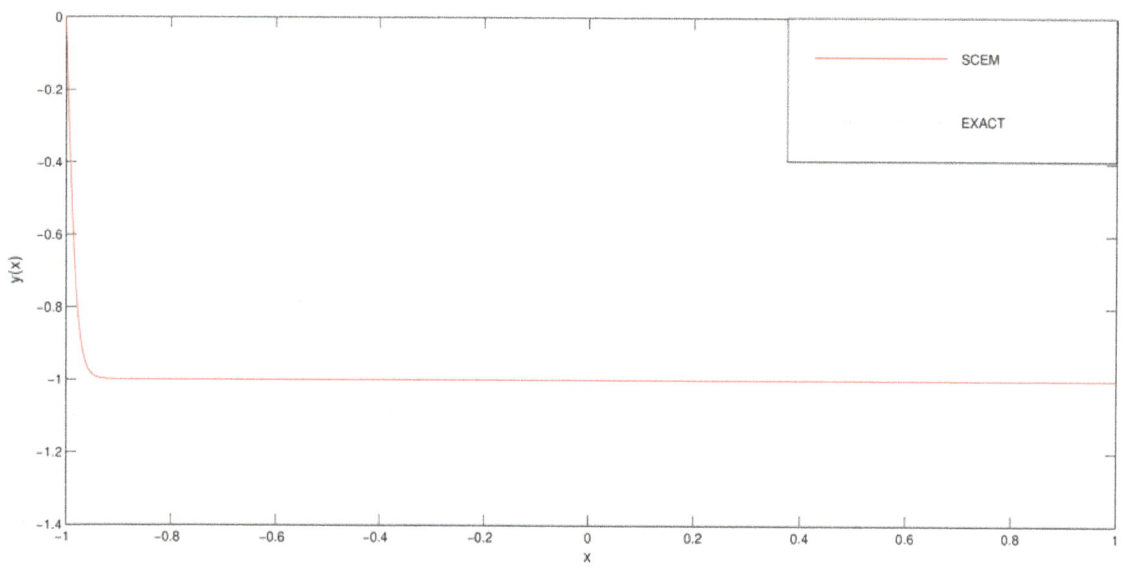

Figure 4: Comparison of SCEM and exact solutions of Example 2 for $\varepsilon = 0.01$

Example 3

Consider the following singular perturbation problem given in O'Malley (1974)

$$\varepsilon y''(x) + e^{y(x)}y'(x) - \frac{\pi}{2}\sin\left(\frac{\pi x}{2}\right)e^{2y(x)} = 0, \quad x \in (0, 1), \tag{22}$$

with the boundary conditions

$y(0) = 0, \ y(1) = 0$ 23

and with the asymptotic solution

$$y = -\ln\left[\left(1 + \cos\frac{\pi x}{2}\right)\left(1 - \frac{1}{2}e^{-\frac{x}{2\varepsilon}}\right)\right] + O(\varepsilon). \tag{24}$$

This problem has boundary layer near the point $x = 0$.

In Tables 4, 5 and Figs. 5, 6, 7 we deliberately choose the node points as possible as near the boundary layer (for this problem, near the point $x = 0$) in order to see efficiency of the method.

Table 4: Numerical results of Example 3 for $\varepsilon = 0.01$

x	Asymptotic solution (O'Malley 1974)	SCEM approximation	Absolute error
0.000	0.0000000000	0.0000000000	0.0000000000
0.001	−0.0476179808	−0.0476179344	4.633630e(−8)
0.005	−0.1998179193	−0.1998178062	1.130920e(−7)
0.010	−0.3317348800	−0.3317346943	1.857260e(−7)
0.050	−0.6496961369	−0.6496952868	8.501800e(−7)
0.100	−0.6835976643	−0.6835975658	9.854120e(−8)
0.200	−0.6683483288	−0.6683483281	6.84303e(−10)
0.300	−0.6371090868	−0.6371090868	4.61752e(−12)
0.400	−0.5927835996	−0.5927835996	3.10862e(−14)
0.500	−0.5347999967	−0.5347999967	2.22044e(−16)
0.600	−0.4623401221	−0.4623401221	1.11022e(−16)
0.700	−0.3743118452	−0.3743118452	0.0000000000
0.800	−0.2692764695	−0.2692764695	0.0000000000
0.900	−0.1453415344	−0.1453415344	0.0000000000
1.000	0.0000000000	0.0000000000	0.0000000000

Table 5: Numerical results of Example 3 for $\varepsilon = 0.001$

x	Asymptotic solution (O'Malley 1974)	SCEM approximation	Absolute error
0.000	0.0000000000	0.0000000000	0.0000000000
0.001	−0.3317959489	−0.3317935405	2.40830e(−06)
0.005	−0.6512232379	−0.6512204646	2.77327e(−06)
0.010	−0.6897108336	−0.6897104246	4.08988e(−07)
0.050	−0.6916046583	−0.6916046583	2.66675e(−13)
0.100	−0.6869723256	−0.6869723256	0.0000000000
0.200	−0.6683710290	−0.6683710290	0.0000000000
0.300	−0.6371092397	−0.6371092397	0.0000000000
0.400	−0.5927836007	−0.5927836007	0.0000000000
0.500	−0.5347999967	−0.5347999967	0.0000000000
0.600	−0.4623401221	−0.4623401221	0.0000000000
0.700	−0.3743118452	−0.3743118452	0.0000000000
0.800	−0.2692764695	−0.2692764695	0.0000000000
0.900	−0.1453415344	−0.1453415344	0.0000000000
1.000	0.0000000000	0.0000000000	0.0000000000

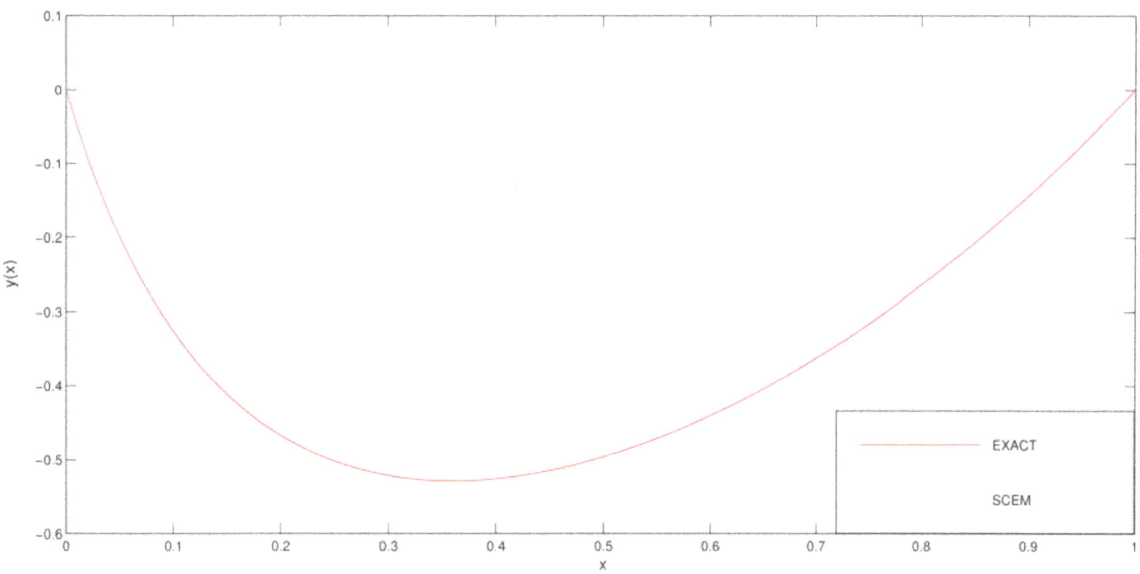

Figure 5: Comparison of SCEM and asymptotic solutions of Example 3 for ε= 0.1

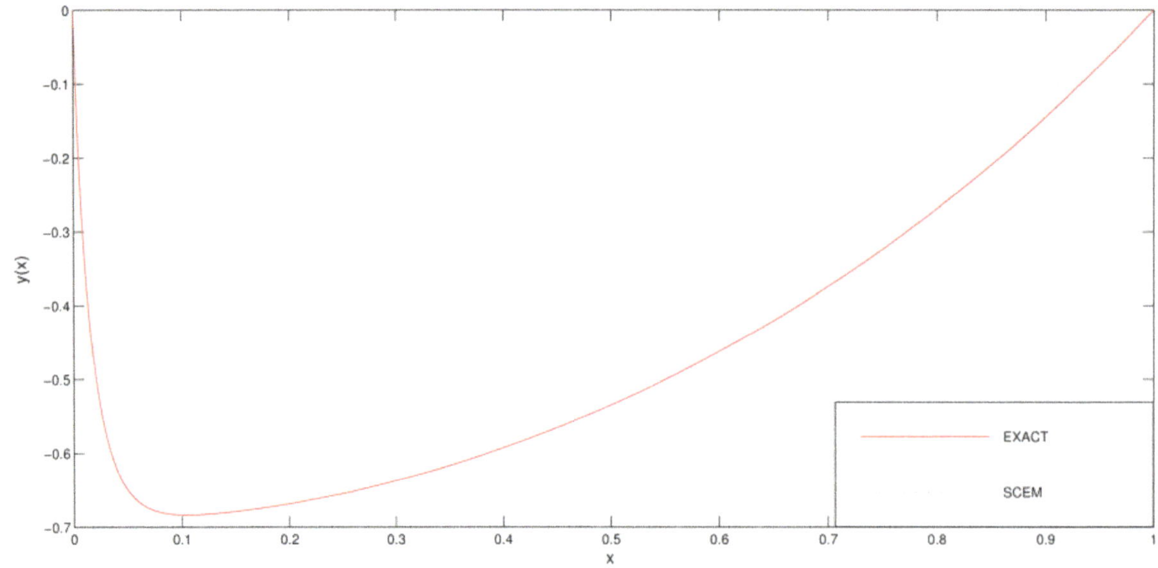

Figure 6: Comparison of SCEM and asymptotic solutions of Example 3 for $\varepsilon = 0.01$

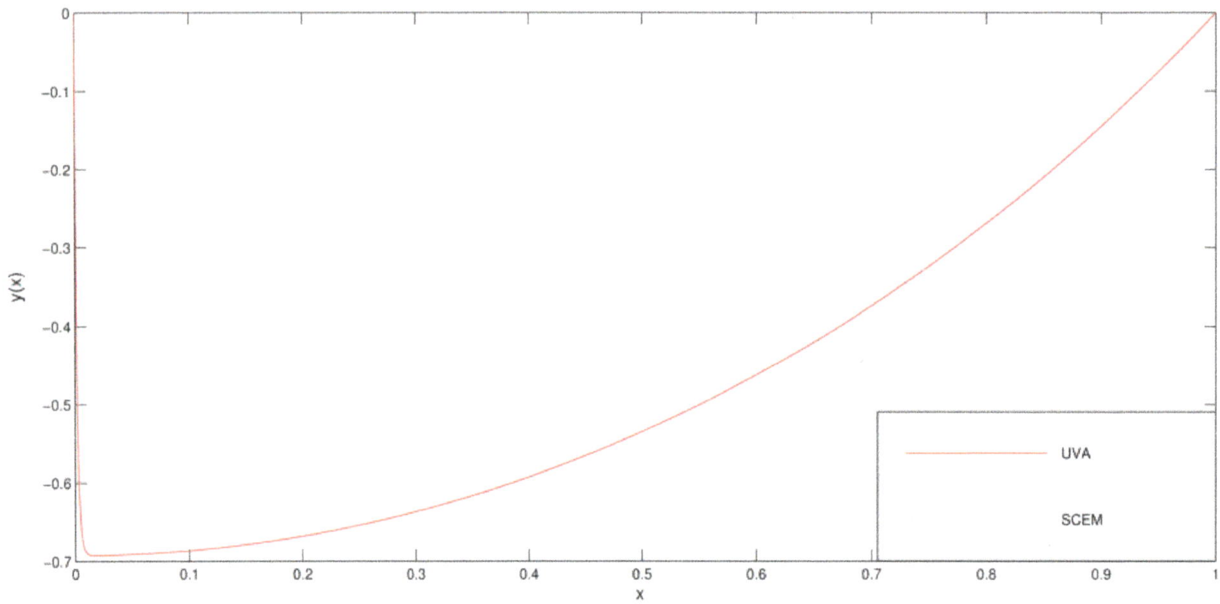

Figure 7: Comparison of SCEM and asymptotic solutions of Example 3 for $\varepsilon = 0.001$

Example 4

As the last example, let us consider the singular perturbation problem given in Kevorkian and Cole (1981)

$$\varepsilon y''(x) + y(x)y'(x) - y(x) = 0, \; x \in [0, 1] \tag{25}$$

with the boundary conditions

$$y(0) = -1, \; y(1) = 3.9995. \tag{26}$$

This problem has boundary layer near the point $x = 0$ and the uniformly valid approximation is given in Kevorkian and Cole (1981) as

$$y = x + c_1 \tanh\left(\left(\frac{c_1}{2}\right)\left(\frac{x}{\varepsilon} + c_2\right)\right), \tag{27}$$

where $c_1 = 2.9995$ and $c_2 = \frac{1}{c_1} \ln\left[\frac{c_1 - 1}{c_1 + 1}\right]$.

In Tables 6, 7 and Figs. 8, 9 we deliberately choose the node points as possible as near the boundary layer (for this problem, near the point $x = 0$) in order to see efficiency of the method.

Table 6: Numerical results of Example 4 for $\varepsilon = 0.01$

x	UVA solution	SCEM approximation	Absolute error
0.000	−1.0000000000000	−1.0000000000000	0.0000000000000
0.001	−0.5813960926361	−0.5791385346450	2.2575579e(−3)
0.005	1.1529592608432	1.1618243705484	8.8651097e(−3)
0.010	2.4659396723180	2.4727365388982	6.7968665e(−3)
0.050	3.0494963201391	3.0494968235068	5.0336773e(−7)
0.070	3.0694999908694	3.0694999930517	2.1823303e(−9)
0.100	3.0994999999988	3.0994999999993	4.689582e(−13)
0.200	3.1995000000000	3.1995000000000	0.000000000000
0.400	3.3995000000000	3.3995000000000	0.000000000000
0.600	3.5995000000000	3.5995000000000	0.000000000000
0.800	3.7995000000000	3.7995000000000	0.000000000000
1.000	3.9995000000000	3.9995000000000	0.000000000000

Table 7: Numerical results of Example 4 for $\varepsilon = 0.001$

x	UVA solution	SCEM approximation	Absolute error
0.000	−1.0000000000000	−1.0000000000000	0.000000000000
0.001	2.4569396723180	2.4576445075493	7.0483523e(−4)
0.005	3.0044963201391	3.0044963670615	4.6922397e(−8)
0.010	3.0094999999988	3.0094999981336	1.8652617e(−9)
0.050	3.0495000000000	3.0495000000101	1.01216812e(−11)
0.070	3.0695000000000	3.0694999930517	7.23865412e(−13)
0.100	3.0995000000000	3.0995000000000	1.02140518e(−14)
0.200	3.1995000000000	3.1995000000000	0.000000000000
0.400	3.3995000000000	3.3995000000000	0.000000000000
0.600	3.5995000000000	3.5995000000000	0.000000000000
0.800	3.7995000000000	3.7995000000000	0.000000000000
1.000	3.9995000000000	3.9995000000000	0.000000000000

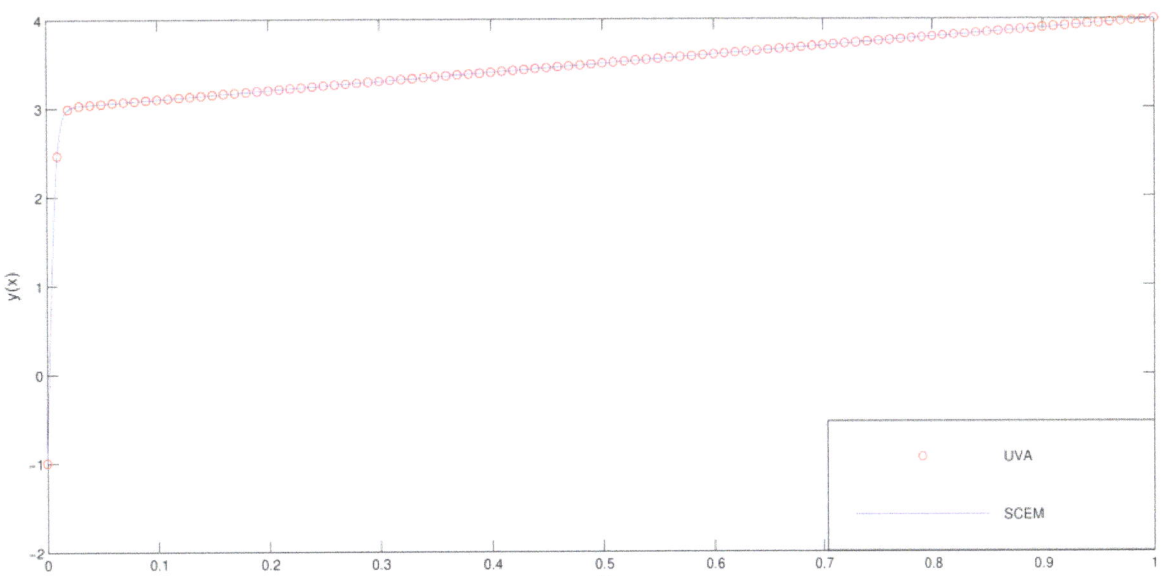

Figure 8: Comparison of SCEM and UVA solutions of Example 4 for $\varepsilon = 0.01$

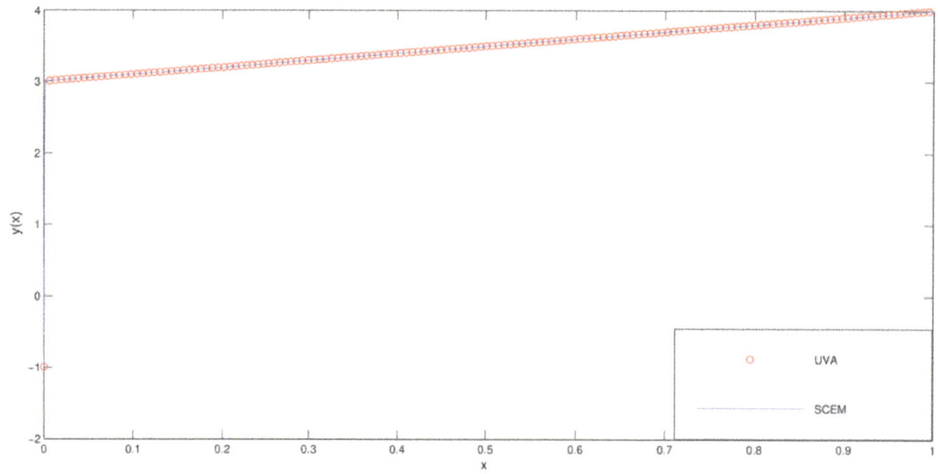

Figure 9: Comparison of SCEM and UVA solutions of Example 4 for $\varepsilon = 0.001$

CONCLUSION

In this paper, an efficient method so-called SCEM has been presented for singularly perturbed two-point second order nonlinear boundary value problems in ordinary differential equations and then the results have been compared with those which are previously obtained by various methods in literature. SCEM is very easy to implement using a mathematical software. As a result of our study, even though only one-term SCEM approximations are used in numerical examples, we obtain highly accurate approximations. As one can see in the numerical examples, SCEM does not require any matching procedures. Moreover, the boundary conditions are satisfied exactly, not asymptotically. Consequently, the present method is well-suited for solving nonlinear singular perturbation problems.

ACKNOWLEDGEMENTS

The authors would like to express their sincere thanks to the editor in chief, editor and anonymous reviewers for their valuable suggestions to revise the manuscript.

Competing interests

The authors declare that they have no competing interests.

REFERENCES

1. Bender CM, Orszag SA (1978) Advanced mathematical methods for scientists and engineers. McGraw-Hill, New York
2. Cathalifaud P, Mauss J, Cousteix J (2010) Nonlinear aspects of high Reynolds number channel flows. Eur J Mech B/Fluids 29:295–304
3. Chawla MM, Katti CP (1982) Finite difference methods and their convergence for a class of singular two-point boundary value problems. Numer Math 39:341–350
4. Cousteix J, Mauss J (2004) Approximations of the Navier–Stokes equations for high Reynolds number flows past a solid wall. J Comput Appl Math 166(1):101–122
5. Cousteix J, Mauss J (2007) Asymptotic analysis and boundary layers. Scientific computation, vol XVIII. Springer, Berlin
6. Cousteix J, Mauss J (2009) Interactive boundary layer models for channel flow. Eur J Mech B Fluids 28:72–87
7. Cui MG, Geng FZ (2007) Solving singular two-point boundary value problem in reproducing kernel space. J Comput Appl Math 205:6–15
8. Eckhaus W (1973) Matched asymptotic expansions and singular perturbation. North Holland Publishing Company, Amsterdam
9. Eckhaus W (1979) Asymptotic analysis of singular perturbations, studies in mathematics and its applications, vol 9. North-Holland Publishing Co., Amsterdam
10. Friedrichs KO, Wasow W (1946) Singular perturbations of nonlinear oscillations. Duke Math J 13:367–381
11. Geng FZ (2011) A novel method for solving a class of singularly perturbed boundary value problems based on reproducing kernel method. Appl Math Comput 218(8):4211–4215
12. Geng FZ, Cui MG (2007) Solving singular nonlinear second-order periodic boundary value problems in the reproducing kernel space. Appl Math Comput 192:389–398
13. Geng FZ, Cui MG (2011) A novel method for nonlinear two-point boundary value problems: combination of ADM and RKM. Appl Math Comput 217:4676–4681
14. Hinch EJ (1991) Perturbation methods. Cambridge University Press, Cambridge
15. Holmes MH (1995) Introduction to perturbation methods, vol 20 of texts in applied mathematics. Springer, New York
16. Johnson RS (2006) Singular perturbation theory: mathematical and analytical techniques with applications to engineering. Springer, Berlin
17. Kadalbajoo MK, Gupta V (2010) A brief survey on numerical methods for solving singularly perturbed problems. Appl Math Comput 217:3641–3716

18. Kadalbajoo MK, Mohan K, Reddy YN (1987) Numerical treatment of singularly perturbed two point boundary value problems. Appl Math Comput 21(2):93–110
19. Kadalbajoo MK, Patidar KC (2003) Singularly perturbed problems in partial differential equations: a survey. Appl Math Comput 134(2):371–429
20. Kadalbajoo MK, Reddy YN (1987) Approximate method for the numerical solution of singular perturbation problems. Appl Math Comput 21(3):185–199
21. Kevorkian J, Cole JD (1981) Perturbation methods in applied mathematics. Springer, New York
22. Kumar M (2011) Parul: methods for solving singular perturbation problems arising in science and engineering. Math Comput Model 54:556–575
23. Kumar M (2011) Parul: a recent development of computer methods for solving singularly perturbed boundary value problems. Int J Differ Equ 404276:1–32
24. Kumar M, Mishra HK (2014) Variational iteration method for a singular perturbation boundary value problems. Am J Numer Anal 2(4):102–106
25. Lagerstrom PA (1988) Matched asymptotic expansions: ideas and techniques, vol 76 of applied mathematical sciences. Springer, New York
26. Li Z et al (2012) The solution of a class of singularly perturbed two-point boundary value problems by the iterative reproducing kernel method. Abst Appl Anal 2012:1–7. doi:10.1155/2012/984057
27. Mauss J, Cousteix J (2002) Uniformly valid approximation for singular perturbation problems and matching principle. CR Méc 330(10):697–702
28. Mittal RC, Jiwari R (2011) A numerical scheme for singularly perturbed Burger–Huxley equation. J Appl Math Inform 29(3–4):813–829
29. Mo JQ (1993) A singularly perturbed nonlinear boundary value problem. J Math Anal Appl 178(1):289–293
30. Nayfeh AH (1973) Perturbation methods, pure and applied mathematics. Wiley, New York
31. O'Malley RE Jr (1974) Introduction to singular perturbations. Academic Press, New York
32. Pandit S, Kumar M (2014) Haar wavelet approach for numerical solution of two parameters singularly perturbed boundary value problems. Appl Math Inf Sci 6:2965–2974
33. Prandtl L (1905) Uber Flussigkeits-bewegung bei kleiner reibung, verhandlungen, III. Int. Math. Kongresses, Tuebner, Leipzig, pp 484–491
34. Roos HG, Stynes M, Tobiska L (1996) Numerical methods for singularly perturbed differential equations, convection-diffusion and flow problems. Springer, Berlin
35. Roos H (2012) Robust numerical methods for singularly perturbed differential equations: a survey covering 2008–2012. ISRN Appl Math 379547
36. Shampine LF, Kierzenka J, Reichelt MW (2000) Solving boundary value problems for ordinary differential equations in MATLAB with bvp4c. Tutorial notes, pp 437–448.http://www.mathworks.com/bvp_tutorial
37. Sharma D, Jiwari R, Kumar S (2012) Numerical solution of two point boundary value problems using Galerkin–Finite element method. Int J Nonlinear Sci 13(2):204–210
38. Sharma D, Jiwari R, Kumar S (2012) A comparative study of modal matrix and finite elements methods for two point boundary value problems. Int J Appl Math Mech 8(13):29–45
39. Van Dyke M (1975) Perturbation methods in fluid mechanics. Parabolic Press, Stantord
40. Verhulst F (2006) Methods and applications of singular perturbations: boundary layers and multiple timescale dynamics, vol 50. Springer, Berlin

CHAPTER 8

LAPLACE TRANSFORM HOMOTOPY PERTURBATION METHOD FOR THE APPROXIMATION OF VARIATIONAL PROBLEMS

U. Filobello Nino, H. Vazquez Leal, M. M. Rashidi, H. M. Sedighi, A. Perez Sesma, M. Sandoval Hernandez, A. Sarmiento Reyes, A. D. Contreras Hernandez, D. Pereyra Diaz, C. Hoyos Reyes, V. M. Jimenez Fernandez, J. Huerta Chua, F. CastroG onzalez and J. R. Laguna Camacho

Keywords: Homotopy perturbation method, Nonlinear differential equation, Approximate solutions, Laplace transform, Laplace transform homotopy perturbation method, Variational calculus, Euler equation

BACKGROUND

The calculus of variations is a powerful branch of the analysis with many applications in both pure and practical mathematics (Lanczos 1986). It has also been found that even, the laws of physics can be expressed in a compact and elegant way through variational principles, as occurs with the Lagrange equations of mechanics, which can be deduced from the variational principle of Hamilton. Unlike elementary calculus problems, which seeks to find the points at which a function reaches its maximum and minimum values, the variational calculus considers the problem of some magnitude, whose values depend on a entire curve, throughout an integral (for instance: surface area or descent time). In fact the aim is to find the curve that extremizes the aforementioned quantity in question.

As will be seen in the next section, the procedure that we will follow consist in finding the differential equation for a function, that leads an integral to take an extreme value.

Although some variational problems were solved by using special methods (like Bernoulli's solution for the brachistochrone problem (see "Basic idea of variational calculus" section) or so-called isoperimetric problem, proposed and solved by the ancient Greeks), it was Euler who presented the variational calculus as a coherent branch of the analysis by discovering the basic differential equation for an extremization curve.

As discussed in the next section, the Euler equation is a nonlinear equation, quite complicated and impossible to solve it in general, and although some applications lead to specific cases which are soluble, it is clear the importance of searching for methods to find approximate solutions to these equations. In fact, nonlinear problems arise in many phenomena, practical or theoretical. For the same reason, several methods focused to find approximate solutions to nonlinear differential equations, algebraic and transcendental equations and integro-differential equations have been reported, such as those based on based on tanh method (Evans and Raslan 2005), exp-function (Xu 2007; Mahmoudi et al. 2008), Adomian's decomposition method (ADM) (Adomian and Rach 1985, Adomian 1988; Babolian and Biazar 2002; Rach 2012; Fatoorehchi and Abolghasemi 2012, 2013; Fatoorehchi et al. 2014, 2015a, b, c, d; Fatoorehchi and Abolghasemi 2015e; Hashim 2006), parameter expansion (Zhang and Xu 2007), homotopy perturbation method (HPM) (He1998, 1999, 2003, 2008; Filobello-Nino et al. 2012, 2013c, 2014a, b, 2015b;Ghaderi 2012; Aminikhah and Hemmatnezhad 2012; Aminikhah 2011, 2012; Vazquez-Leal

Citation: Filobello-Nino, U., Vazquez-Leal, H., Rashidi, M.M. et al., Laplace transform homotopy perturbation method for the approximation of variational problems, doi:10.1186/s40064-016-1755-y.

et al. 2012a, b; Fatoorehchi and Abolghasemi2011), homotopy analysis method (HAM) (Rashidi et al. 2012a, 2012b; Lakestani and Kazemian 2013), series method (Filobello-Nino et al. 2015a), Lie algebra method (Shang 2012, 2013, 2015; Olver 1993; Casas 1996) and perturbation method (PM) (Filobello-Nino et al. 2013a, b) among many others.

A particularly relevant job, which deals with a combined version of the Laplace transform method (LTM) and Adomian's decomposition method (ADM) (named the H2LTM), it is described in reference (Fatoorehchi and Abolghasemi 2015e). This work introduced a technique for extending the LTM to solve nonlinear differential equations obtaining analytical and accurate solutions.

The rest of this work is organized as follows. "Basic idea of variational calculus" section presents a basic idea of the main results of variational calculus required for this work. In "Standard HPM" section, we introduce the basic idea of standard HPM method. "LT-HPM and some of its modified versions" section presents an introduction of LT-HPM and some of its modified versions useful for this work. Additionally, "Cases study" section presents four cases study of variational problems. In addition a wide discussion on the results is presented in "Discussion" section. Finally, a brief conclusion is given in "Conclusions" section.

BASIC IDEA OF VARIATIONAL CALCULUS

This paper deals with the problem of finding a function $y = y(x)$ with continues second derivatives (Elsgolts1983; Levi 1980; Simmons 1983; Zelikin 2005; Boyce 1998), in order to find an approximate solution to the variational problem of extreme an integral of the form

$$I(y) = \int_{x_1}^{x_2} f(x, y, y')dx, \qquad 1$$

so that its graph join two points, namely (x_1, y_1) and (x_2, y_2).

From variational calculus, we know that if $y(x)$ is a function that satisfies the above requirements, then satisfies the Euler equation.

$$\frac{d}{dx}\left(\frac{\partial f}{\partial y'}\right) - \frac{\partial f}{\partial y} = 0, \text{ where } y' = dy/dx. \qquad 2$$

Some interesting cases of the integral (1) are (Simmons 1983; Filobello-Nino et al. 2013d):

- The problem of finding the shortest curve joining two given points, requires to minimize:

$$\int_{x_1}^{x_2} \sqrt{1+y'^2}dx. \qquad 3$$

- In the problem of finding the curve that generates the surface of revolution of smallest area when revolves about the axis x, we have to minimize:

$$\int_{x_1}^{x_2} 2\pi y\sqrt{1+y'^2}dx. \qquad 4$$

- The problem of brachistochrone, search for the vertical curve without friction that joins two fixed points, through which a particle slides in the shortest possible time (Connor and Robertson 1997, 1998; Filobello-Nino et al. 2013d). This variational problem for the curve of quickest descent, requires minimize:

$$\int_{x_1}^{x_2} \frac{\sqrt{1+y'^2}}{\sqrt{2gy}}dx \qquad 5$$

When the variable y does not appear in the function f, Euler's equation (2) is reduced immediately to the form

$$\frac{d}{dx}\left(\frac{\partial f}{\partial y'}\right) = 0, \qquad 6$$

integrating (6), we see that $\partial f/\partial y'$ is conserved in according with differential equation

$$\frac{\partial f}{\partial y'} = c, \qquad 7$$

where c is a constant.

On the other side if f is not an explicit function of x, then Euler's equation can be expressed in the form

$$h(x, y, y') = c, \qquad 8$$

where

$$h(x,y,y') = y'\frac{\partial f}{\partial y'} - f. \qquad 9$$

It is customary to refer to the solutions of the Euler equation as stationary functions, and the corresponding value of the integral (1) as stationary value. Besides, the solutions to the above mentioned Euler equation, which are unrestricted by the boundary conditions, are called extremals (Simmons 1983).

STANDARD HPM

The standard homotopy perturbation method (HPM) was proposed by Ji Huan He, it was introduced to approach various kinds of nonlinear problems. The HPM is considered as a combination of the classical perturbation technique and the homotopy (whose origin is in the topology), but not restricted to small parameters as occur with traditional perturbation methods (He 1998, 1999).

To figure out how HPM works, consider a general nonlinear differential equation in the form

$$A(u) - f(x) = 0, \ x \in \Omega, \qquad 10$$

with the following boundary conditions

$$B(u, \partial u/\partial n) = 0, \ x \in \Gamma, \qquad 11$$

where A is a general differential operator, B is a boundary operator, $f(x)$ a known analytical function and Γ is the domain boundary for Ω. A can be divided into two operators L and N, where L is linear and N nonlinear; so that (10) can be rewritten as

$$L(u) + N(u) - f(x) = 0. \qquad 12$$

Generally, a homotopy can be constructed as (He 1998, 1999).

$$H(U,p) = (1-p)[L(U) - L(u_0)] + p[L(U) + N(U) - f(x)] = 0, \ p \in [0, 1], \ x \in \Omega \qquad 13$$

or

$$H(U, p) = L(U) - L(u_0) + p[L(u_0) + N(U) - f(x)] = 0, \ p \in [0, 1], \ x \in \Omega \qquad 14$$

where p is a homotopy parameter, whose values are within range of 0 and 1, u_0 is the first approximation for the solution of (12) that satisfies the boundary conditions.

Assuming that solution for (13) or (14) can be written as a power series of p as

$$U = v_0 + v_1 p + v_2 p^2 + \cdots . \qquad 15$$

Substituting (15) into (14) and equating identical powers of p terms, there can be found values for the sequence $v_0, v_1, v_2, \ldots$

When $p \rightarrow 1$, it yields the approximate solution for (1) in the form

$$U = v_0 + v_1 + v_2 + v_3 \cdots . \qquad 16$$

LT-HPM AND SOME OF ITS MODIFIED VERSIONS

Basic idea of the LT-HPM Method

LT-HPM follows the same steps of standard HPM until (14), next we apply Laplace transform on both sides of homotopy equation (14), to obtain (Aminikhah and Hemmatnezhad 2012; Aminikhah 2011, 2012; Filobello-Nino et al. 2013c, 2014a).

$$\Im\{L(U) - L(u_0) + p[L(u_0) + N(U) - f(r)]\} = 0, \qquad 17$$

(employing the freedom of homotopy formulation, it is possible that LT-HPM substitutes $L(u_0)$ for an arbitrary function $z(x)$ provided with some unknown parameters $A, B, C, \ldots$ to be adequately determined [see cases study 3 and 4)].

Using the differential property (170) of LT, we have (Spiegel 1998):

$$s^n \Im\{U\} - s^{n-1}U(0) - s^{n-2}U'(0) - \cdots - U^{(n-1)}(0) = \Im\{L(u_0) - pL(u_0) + p[-N(U) + f(r)]\} \qquad 18$$

or

$$\Im(U) = \left(\frac{1}{s^n}\right)\left\{s^{n-1}U(0) + s^{n-2}U'(0) + \cdots + U^{(n-1)}(0)\right\} + \left(\frac{1}{s^n}\right)\Im\{L(u_0) - pL(u_0) + p[-N(U) + f(r)]\}, \qquad 19$$

applying inverse Laplace transform to both sides of (19), we obtain

$$U = \Im^{-1}\left\{\left(\frac{1}{s^n}\right)\left\{s^{n-1}U(0) + s^{n-2}U'(0) + \cdots + U^{(n-1)}(0)\right\} + \left(\frac{1}{s^n}\right)\Im\{L(u_0) - pL(u_0) + p[-N(U) + f(r)]\}\right\}. \qquad 20$$

Assuming that the solutions of (12) can be expressed as a power series of p

$$U = \sum_{n=0}^{\infty} p^n v_n. \qquad 21$$

Then substituting (21) into (20), we get

$$\sum_{n=0}^{\infty} p^n v_n = \Im^{-1}\left\{\begin{array}{l}\left(\frac{1}{s^n}\right)\left\{s^{n-1}U(0) + s^{n-2}U'(0) + \cdots + U^{(n-1)}(0)\right\} \\ + \left(\frac{1}{s^n}\right)\Im\left\{L(u_0) - pL(u_0) + p\left[-N(\sum_{n=0}^{\infty} p^n v_n) + f(r)\right]\right\}\end{array}\right\}, \qquad 22$$

comparing coefficients of p, with the same power leads to

$$p^0 : v_0 = \Im^{-1}\left\{\left(\frac{1}{s^n}\right)\left(s^{n-1}U(0) + s^{n-2}U'(0) + \cdots + U^{(n-1)}(0)) + \Im\{L(u_0)\}\right)\right\},$$

$$p^1 : v_1 = \Im^{-1}\left\{\left(\frac{1}{s^n}\right)(\Im\{-N_0(v_0) - L(u_0) + f(r)\})\right\},$$

$$p^2 : v_2 = \Im^{-1}\left\{\left(\frac{1}{s^n}\right)\Im\{-N_1(v_0, v_1)\}\right\},$$

$$p^3 : v_3 = \Im^{-1}\left\{\left(\frac{1}{s^n}\right)\Im\{-N_2(v_0, v_1, v_2)\}\right\},$$

$$\ldots,$$

$$p^j : v_j = \Im^{-1}\left\{\left(\frac{1}{s^n}\right)\Im\{-N_{j-1}(v_0, v_1, v_2, \ldots, v_{j-1})\}\right\},$$

$$\ldots. \qquad 23$$

Assuming that $U(0) = u_0 = \alpha_0$, $U(0) = \alpha_1, \ldots, U^{n-1}(0) = \alpha_{n-1}$; the exact solution may be obtained as follows

$$u = \lim_{p \rightarrow 1} U = v_0 + v_1 + v_2 + \cdots . \qquad 24$$

The above notation has emphasized the fact that the nonlinear operator N can be expanded in formal series with respect to the embedding parameter p. The same criterion is used throughout "LT-HPM and some of its modified versions" section (Marinca and Herisanu 2011).

Basic idea of Non-Linearities Distribution Laplace Transform-Homotopy Perturbation Method (NDLT-HPM)

Vazquez-Leal et al. (2012a) introduced a modified version of homotopy perturbation method, the nonlinearities distribution homotopy perturbation method (NDHPM), which sometimes eases the solutions searching process for (12) and reduces the complexity of solving differential equations in terms of power series.

As first step, we introduced the following homotopy (Vazquez-Leal et al. 2012a).

$$H(U,p) = (1-p)[L(U) - L(u_0)] + p[L(U) + N(U,p) - f(x,p)] = 0 \quad 25$$

or

$$H(U,p) = L(U) - L(u_0) + p[L(u_0) + N(U,p) - f(x,p)] = 0,\ p \in [0,1]\ x \in \Omega \quad 26$$

It can be noticed that the homotopy function (25) is essentially the same as (13), except for the nonlinear operator N and the non homogeneous function f, which contain embedded the homotopy parameter p. Thus, a way to introducenon-linearities distribution Laplace transform-homotopy perturbation method (NDLT-HPM) is apply directly, Laplace transform on both sides of homotopy equation (36), to obtain (Filobello-Nino et al. 2014b).

$$\Im\{L(U) - L(u_0) + p[L(u_0) + N(U,p) - f(x,p)\} = 0, \quad 27$$

(more generally, one could substitute $f(x,p)$ in (25) by another function $g(x,p)$ such that, $g(x,p) \rightarrow f(x)$ when$p \rightarrow 1$, see cases study above).

Using the differential property of LT, we have (Spiegel 1998).

$$s^n\Im\{U\} - s^{n-1}U(0) - s^{n-2}U'(0) - \cdots - U^{(n-1)}(0) = \Im\{L(u_0) - pL(u_0) + p[-N(U,p) + f(x,p)]\}, \quad 28$$

or

$$\Im(U) = \left(\frac{1}{s^n}\right)\Big\{s^{n-1}U(0) + s^{n-2}U'(0) + \cdots + U^{(n-1)}(0) + \Im\{L(u_0) - pL(u_0) + p[-N(U,p) + f(x,p)]\}\Big\}, \quad 29$$

applying inverse Laplace transform to both sides of (29), we obtain

$$U = \Im^{-1}\left\{\left(\frac{1}{s^n}\right)\Big\{s^{n-1}U(0) + s^{n-2}U'(0) + \cdots + U^{(n-1)}(0) + \Im\{L(u_0) - pL(u_0) + p[-N(U,p) + f(x,p)]\}\Big\}\right\}. \quad 30$$

Assuming that the solutions of (12) and $f(x,p)$ can be expressed as a power series of p

$$U = \sum_{n=0}^{\infty} p^n v_n(x), \quad 31$$

and

$$f = \sum_{m=0}^{\infty} p^m f_m(x), \quad 32$$

respectively, where $f_m(x)$ are usually obtained by applying Taylor series method to $f(x,p)$.

Then substituting (32) and (31) into (30), we get

$$\sum_{n=0}^{\infty} p^n v_n = \Im^{-1}\left\{\begin{array}{l}\left(\frac{1}{s^n}\right)\{s^{n-1}U(0) + s^{n-2}U'(0) + \cdots + U^{(n-1)}(0)\} \\ + \left(\frac{1}{s^n}\right)\Im\left\{L(u_0) - pL(u_0) + p\left[-N\left(\sum_{n=0}^{\infty} p^n v_n, p\right) + \sum_{m=0}^{\infty} p^m f_m(x)\right]\right\}\end{array}\right\}, \quad 33$$

comparing coefficients of p, with the same power leads to

$$p^0 : v_0 = \Im^{-1}\left\{\left(\frac{1}{s^n}\right)\left(s^{n-1}U(0) + s^{n-2}U'(0) + \cdots + U^{(n-1)}(0)) + \Im\{L(u_0)\}\right)\right\},$$

$$p^1 : v_1 = \Im^{-1}\left\{\left(\frac{1}{s^n}\right)\left(\Im\{N_0(v_0) - L(u_0) + f_0(x)\}\right)\right\},$$

$$p^2 : v_2 = \Im^{-1}\left\{\left(\frac{1}{s^n}\right)\Im\{N_1(v_0, v_1, f_0, f_1)\}\right\},$$

$$p^3 : v_3 = \Im^{-1}\left\{\left(\frac{1}{s^n}\right)\Im\{N_2(v_0, v_1, v_2, f_0, f_1, f_2)\}\right\},$$

$$\ldots$$

$$p^j : v_j = \Im^{-1}\left\{\left(\frac{1}{s^n}\right)\Im\{N_{j-1}(v_0, v_1, v_2, \ldots, v_{j-1}, f_0, f_1, f_2, \ldots f_{j-1})\}\right\}$$

$$\ldots. \qquad 34$$

Assuming that the initial approximation has the form: $U(0) = u_0 = \alpha_0$, $U(0) = \alpha_1$, ..., $U^{n-1}(0) = \alpha_{n-1}$;therefore the exact solution may be obtained as follows

$$u = \lim_{p\to 1} U = v_0 + v_1 + v_2 + \cdots \qquad 35$$

Basic idea of Laplace Transform Homotopy Perturbation Method with Variable Coefficients (CVLT-HPM)

To obtain (18), we assumed that the coefficient of $L(U)$ is one. In this section we will assume the mentioned coefficient is a positive whole power of x, thus we rewrite explicitly (17) for this case as (Filobello-Nino et al.2015b).

$$\Im\{(x^m L(U) - x^m z(x)) + p[x^m z(x) + N(U) - f(x)]\} = 0, \qquad 36$$

where m is a positive integer, and employing the versatility and freedom of HPM we have substituted $L(u_0)$for an arbitrary function $z(x)$. We will choose as a trial function $z(x)$, a polynomial provided with some unknown parameters A, B, C, ... to be determined [see (44)].

Using the properties (170) and (171), we have (Spiegel 1998).

$$(-1)^m \frac{d^m\left[s^n\Im\{U\} - s^{n-1}U(0) - s^{n-2}U'(0) - \cdots - U^{(n-1)}(0)\right]}{ds^m}$$
$$= \Im\{x^m z(x) + p[-z(x)x^m - N(U) + f(x)]\}, \qquad 37$$

after integrating m times, we obtain

$$s^n\Im\{U\} - s^{n-1}U(0) - s^{n-2}U'(0) - \cdots - U^{(n-1)}(0)$$
$$= (-1)^m \int\int\cdots\int \Im\{x^m z(x) + p[-z(x)x^m - N(U) + f(x)]\}dsds'\cdots ds''$$
$$(m\text{-times}), \qquad 38$$

or

$$U = \Im^{-1}\left\{\frac{1}{s^n}\left\{s^{n-1}U(0) + s^{n-2}U'(0) + \cdots + U^{(n-1)}(0)\right.\right.$$
$$\left.\left. + (-1)^m \int\int\cdots\int \Im\{r^m z(x) + p[-z(x)x^m - N(U) + f(x)]\}dsds'\cdots ds''\right\}\right\}. \qquad 39$$

Assuming also, that the solutions of the ODE to solve, can be expressed as a power series of p

$$U = \sum_{n=0}^{\infty} p^n v_n. \qquad 40$$

Then substituting (40) into (39), we get

$$\sum_{n=0}^{\infty} p^n v_n = \Im^{-1}\left\{\begin{array}{l}\left(\frac{1}{s^n}\right)\left\{s^{n-1}U(0) + s^{n-2}U'(0) + \cdots + U^{(n-1)}(0)\right\} \\ + \left(\frac{(-1)^m}{s^n}\right)\int\int\cdots\int \Im\left\{x^m z(x) + p\left[-x^m z(x) - N\left(\sum_{n=0}^{\infty} p^n v_n\right) + f(x)\right]\right\}dsds'\cdots ds''\end{array}\right\}, \qquad 41$$

comparing coefficients of p, with the same power leads to

$$p^0 : v_0 = \Im^{-1}\left\{\left(\frac{1}{s^n}\right)\left(s^{n-1}U(0) + s^{n-2}U'(0) + \cdots + U^{(n-1)}(0)) + (-1)^m \int\int\cdots\int \Im\{x^m z(x)\}dsds'\cdots ds''\right)\right\}$$

$$p^1 : v_1 = \Im^{-1}\left\{\left(\frac{1}{s^n}\right)(-1)^m \int\int\cdots\int (\Im\{-N_0(v_0) - x^m z(x) + f(x)\})dsds'\cdots ds''\right\}$$

$$p^2 : v_2 = \Im^{-1}\left\{\left(\frac{1}{s^n}\right)(-1)^m \int\int\cdots\int (\Im\{-N_1(v_0, v_1)\})dsds'\cdots ds''\right\}$$

$$p^3: v_3 = \Im^{-1}\left\{\left(\frac{1}{s^n}\right)(-1)^m \int\int\cdot\cdot\int (\Im\{-N_2(v_0, v_1, v_2)\})ds ds'\cdot\cdot ds''\right\}$$

$\ldots,$

$$p^j: v_j = \Im^{-1}\left\{\left(\frac{1}{s^n}\right)(-1)^m \int\int\cdot\cdot\int \left(\Im\{-N_{j-1}(v_0, v_1, v_2, \ldots v_{j-1})\}\right)ds ds'\cdot\cdot ds''\right\},$$

$\ldots.$ 42

Assuming that $U(0) = u_0 = \alpha_0$, $U(0) = \alpha_1$, …, $U^{n-1}(0) = \alpha_{n-1}$; then an approximate solution may be obtained as follows

$$u = \lim_{p\to 1} U = v_0 + v_1 + v_2 + \cdots.$$ 43

For boundary value problems it is expected that some of the initial conditions above mentioned, be initially unknown; therefore (43) can be expressed as

$u = u(x, A, B, C, \ldots, \alpha_i)$. 44

In order to calculate adequately the values for $A, B, C, \ldots, \alpha_i$, we will deduce an algebraic system for them, in the following way.

1. Re quiring that (44) satisfies the boundary conditions at the endpoint of the interval.
2. In order to find the values of the total number of parameters, we require add more algebraic equations to those found in (1), up to get as many equations as parameters to determine. If are required, say *j* additional equations, then a convenient possibility is incorporating the following *j*-equations (Marinca and Herisanu 2011).

$R(x_1, A, B, C, \ldots, \alpha_i) = R(x_2, A, B, C, \ldots, \alpha_i) = \cdots = R(x_j, A, B, C, \ldots, \alpha_i) = 0,$

where the residual is defined, by substituting of (44) into (12), to obtain(see 176).

$R(x, A, B, C, \ldots, \alpha_i) = L(u(x, A, B, C, \ldots, \alpha_i)) + N(u(x, A, B, C, \ldots, \alpha_i)) - f(x)$. The above points $x_1, x_2, x_3, \ldots, x_j$ belong to the interest interval, and it is assumed that $u(x, A, B, C, \ldots, \alpha_i)$ is the approximate solution of (12) given by (33).

For the case of second order ODES, where $y'(0) = A \neq 0$; it is possible to show the term containing $y'(x)$ could give rise to an inappropriate term $\Im^{-1}\left\{\frac{Ln(s)}{s^2}\right\}$. To avoid this potential problem, we should perform, as a work rule, an adequate transformation in order to express our differential equation in its normal form (Simmons 1983). Such as it is known, the normal form does not contain the problematic term $y'(x)$, and therefore allows applying VCLT-HPM algorithm in accordance with the above procedure (Filobello-Nino et al. 2015b).

CASES STUDY

This section, applies LT-HPM and some of their modifications aforementioned, in order to find analytical approximate expressions for the stationary functions, solutions of the Euler equations, which turn out from the following cases study.

Case 1

This case study uses basic version of LT-HPM.

We will get an approximate solution for the problem of finding a function $y(x)$ that extremizes the value of the integral.

$$\int_{x_1}^{x_2}\left(\frac{1}{2}y'^2 - y^3 - x^2y^2\right)dx.$$ 45

After employing Euler equation (2), for $f = ½y'^2 - y^3 - x^2y^2$ we obtain

$y''(x) + 3y^2(x) + 2x^2y(x) = 0.$

As a case study, we consider the following boundary conditions $y(0) = 0, y(1) = 1/2$ so that we can express formally our problem in the following way

$y''(x) + 3y^2(x) + 2x^2y(x) = 0,\ y(0) = 0,\ y(1) = ½$ 46

Unfortunately there is not a systematic way to obtain an exact solution for this boundary value problem.

To obtain an approximate solution for (46), we identify

$L(y) = y''(x),$ 47

$N(y) = 3y^2 + 2x^2y(x).$ 48

We construct the following homotopy in accordance with (13)

$(1-p)(y''-y''0)+p[y''+3y^2+2x^2y]=0,$ 49

or

$y''=y''0+p[-y''0-3y^2-2x^2y].$ 50

Applying Laplace transform to (50) we get

$I\{y''\}=I\{y''0+p[-y''0-3y^2-2x^2y]\}.$

As it is explained in (Spiegel 1998), it is possible to rewrite the above equation, as

$$s^2Y(s) - sy(0) - y'(0) = \Im\left(y_0'' + p\left(-y_0'' - 3y^2 - 2x^2y\right)\right),$$ 51

where we have defined $Y(s) = \Im(y(x))$.

After applying the initial condition $y(0) = 0$, the last expression can be simplified as follows

$$s^2Y(s) - A = \Im\left(y_0'' + p\left(-y_0'' - 3y^2 - 2x^2y\right)\right),$$ 52

where, we defined $A = y'(0)$.

Solving for $Y(s)$ and applying Laplace inverse transform $\Im^{-1}$

$$y(x) = \Im^{-1}\left\{\frac{A}{s^2} + \frac{1}{s^2}\left(\Im\left(y_0'' + p\left(-y_0'' - 3y^2 - 2x^2y\right)\right)\right)\right\}$$

53

Next, we assume the solution for (46) has the form

$$y(x) = \sum_{n=0}^{\infty} p^n v_n(x),$$ 54

and we choose

$$y_0(x) = Ax, \tag{55}$$

as the first approximation for the solution of (46) that satisfies the condition $y(0) = 0$.

After substituting (54) and (55) into (53), we get

$$\sum_{n=0}^{\infty} p^n v_n = \Im^{-1}\left\{\frac{A}{s^2} + \frac{p}{s^2}\Im\left(-3\left(v_0 + pv_1 + p^2v_2 + \cdots\right)^2 - 2x^2\left(v_0 + pv_1 + p^2v_2 + \cdots\right)\right)\right\}, \tag{56}$$

Equating terms with identical powers of p, we obtain

$$p^0 : v_0(x) = \Im^{-1}\left\{\frac{A}{s^2}\right\}, \tag{57}$$

$$p^1 : v_1(x) = \Im^{-1}\left\{\left(\frac{1}{s^2}\right)\Im\left(-3v_0^2 - 2x^2v_0\right)\right\}, \tag{58}$$

$$p^2 : v_2(x) = \Im^{-1}\left\{\left(\frac{1}{s^2}\right)\Im\left(-6v_0v_1 - 2x^2v_1\right)\right\}, \tag{59}$$

Solving the above equations for $v_0(x)$, $v_1(x)$, $v_2(x)$, … we obtain

$$p^0 : v_0(x) = Ax, \tag{60}$$

$$p^1 : v_1(x) = \frac{-A^2}{4}x^4 - \frac{A}{10}x^5, \tag{61}$$

$$p^2 : v_2(x) = \frac{A^3}{28}x^7 + \frac{11A^2}{560}x^8 + \frac{A}{360}x^9, \tag{62}$$

and so on.

By substituting (60)–(62) into (54) and calculating the limit when $p \to 1$, results in a second order approximation

$$y(x) = Ax - \frac{A^2}{4}x^4 - \frac{A}{10}x^5 + \frac{A^3}{28}x^7 + \frac{11A^2}{560}x^8 + \frac{A}{360}x^9. \tag{63}$$

In order to calculate the value of A, we require that (63) satisfies the boundary condition $y(1) = 1/2$, so that we obtain

$$A = 0.651098543. \tag{64}$$

Thus, substituting (64) into (63), we obtain a handy LT-HPM solution.

$$y(x) = 0.651098543x - 0.1059823282x^4 - 0.06510985430x^5 + 0.009857848493x^7 + 0.008327182928x^8 + 0.001808607064x^9. \tag{65}$$

Case 2

This case study employs CVLT-HPM method.

We will obtain the exact solution for the problem of finding a function that extremizes the value of the integral.

$$\int_{x_1}^{x_2}\left(\frac{1}{2}x^2y'^2 + y^2 + 8x^2y\right)dx. \tag{66}$$

After employing Euler equation (2), for $f = ½x^2y'^2 + y^2 + 8x^2y$ we obtain

$$x^2y''(x) + 2xy'(x) - 2y(x) - 8x^2 = 0. \tag{67}$$

Considering the boundary conditions $y(0) = 0, y(1) = 1$, our purpose will be to get an approximate solution for the following non homogeneous variable coefficients linear equation.

$$x^2y''(x) + 2xy'(x) - 2y(x) - 8x^2 = 0,\ \ y(0) = 0,\ \ y(1) = 1. \tag{68}$$

Since it is not known in advance the value of $y'(0)$ and (67) has a term containing $y'(x)$ [see "Basic idea of Laplace transform homotopy perturbation method with variable coefficients (CVLT-HPM)" section (2)], then, it is better to write (67) in a simpler way by using the following substitution (Simmons 1983)

$$xy(x) = u(x), \tag{69}$$

this transforms an equation in the standard form as (67), in its normal form given by

$$u''(x) + q(x)u(x) - 8x = 0, \tag{70}$$

where

$$q(x) = \frac{-2}{x^2}. \tag{71}$$

Besides (69) implies the following boundary conditions for u:

$$u(0) = 0,\ \ u(1) = 1, \tag{72}$$

since $y(0) = 0$ and $y(1) = 1$.

Therefore, we can express formally our problem in terms of u as follows

$$u''(x) - \frac{2}{x^2}u(x) - 8x = 0,$$
$$u(0) = 0,\ \ u(1) = 1. \tag{73}$$

In the same way, we note that

$$u'(x) = xy'(x) + y(x),$$

implies the condition

$$u'(0) = 0. \tag{74}$$

With the purpose of applying CVLT-HPM, we will rewrite (70) as follows

$$xu''(x) - \frac{2}{x}u(x) - 8x^2 = 0. \tag{75}$$

In accordance with "Basic idea of Laplace transform homotopy perturbation method with variable coefficients (CVLT-HPM)" section, it is convenient to construct the following homotopy.

$$(1-p)(xu'' - xD) + p\left[xu'' - \frac{2u}{x} - 8x^2\right] = 0, \tag{76}$$

where we have chosen as trial function, the constant function $w(x) = D$.

In order to apply CVLT-HPM, we simplify (76) as follows

$$xu'' = xD + p\left[-xD + \frac{2u}{x} + 8x^2\right]. \tag{77}$$

After applying Laplace transform to (77) we get

$$\Im\{xu''\} = \Im\left\{xD + p\left[-xD + \frac{2u}{x} + 8x^2\right]\right\}.$$

As it is explained in (Spiegel 1998) it is possible to rewrite the above equation, as

$$-\frac{d}{ds}\left(s^2U(s)\right)=\Im\left(xD+p\left(-xD+\frac{2u}{x}+8x^2\right)\right),$$

78

where we have defined $U(s) = \Im(u(x))$ and employed the initial conditions $u(0) = 0$, and $u'(0) = 0$ [see (73) and (74)].

After integrating (78), we obtain

$$s^2U(s)=-\int\Im\left\{xD+p\left[-xD+\frac{2u}{x}+8x^2\right]\right\}ds. \quad 79$$

Applying $\Im^{-1}$ to this last integral expression, yields in

$$u(x)=-\Im^{-1}\left\{\left(\frac{1}{s^2}\right)\int\Im\left\{xD+p\left[-xD+\frac{2u}{x}+8x^2\right]\right\}ds\right\}. \quad 80$$

Next, we assume a series solution for $y(x)$, in the form

$$u(x)=\sum_{n=0}^{\infty}p^n v_n, \quad 81$$

Substituting (81) into (80), we get

$$\sum_{n=0}^{\infty}p^n v_n=\Im^{-1}\left\{\left(\frac{1}{s^2}\right)\int\Im\left\{-xD+p\left[xD-\frac{2\sum_{n=0}^{\infty}p^n v_n}{x}-8x^2\right]\right\}ds\right\}.$$

82

On comparing the coefficients with identical powers of p, and perform the indicated operations, we have

$$p^0: v_0=\Im^{-1}\left\{\left(\frac{1}{s^2}\right)\int\Im\{-xD\}ds\right\}, \quad 83$$

$$p^1: v_1=\Im^{-1}\left\{\left(\frac{1}{s^2}\right)\int\Im\left\{xD-\frac{2v_0}{x}-8x^2\right\}ds\right\}, \quad 84$$

$$p^2: v_2=\Im^{-1}\left\{\left(\frac{1}{s^2}\right)\int\Im\left\{\frac{-2v_1}{x}\right\}ds\right\}, \quad 85$$

$$p^3: v_3=\Im^{-1}\left\{\left(\frac{1}{s^2}\right)\int\Im\left\{\frac{-2v_2}{x}\right\}ds\right\}, \quad 86$$

$$p^4: v_4=\Im^{-1}\left\{\left(\frac{1}{s^2}\right)\int\Im\left\{\frac{-2v_3}{x}\right\}ds\right\}, \quad 87$$

Solving the above operations for $v_0(x)$, $v_1(x)$, $v_2(x)$… we obtain

$$p^0: v_0(x)=\frac{D}{2}x^2, \quad 88$$

$$p^1: v_1(x)=\frac{4x^3}{3}, \quad 89$$

$$p^2: v_2(x)=\frac{4x^3}{(3)(3)}, \quad 90$$

$$p^3: v_3(x)=\frac{4x^3}{(3)(3)(3)}, \quad 91$$

$$p^4: v_4(x)=\frac{4x^3}{(3)(3)(3)(3)}, \quad 92$$

$$p^n: v_n(x)=\frac{4x^3}{(3)(3)(3)(3)\ldots(3)(3)}=\frac{4x^3}{(3)^n}, \quad 93$$

and so on.

By substituting solutions (88)–(93) into (81) and calculating the limit value $p \to 1$, results in.

$$u(x)=\frac{D}{2}x^2+\frac{4}{3}\left[1+\frac{1}{3}+\left(\frac{1}{3}\right)^2+\left(\frac{1}{3}\right)^3+\cdots+\left(\frac{1}{3}\right)^n\right]x^3. \quad 94$$

Clearly, the second term of the right hand side of (94) is a geometric series of ratio $r = 1/3$. Since $|r| < 1$, it is well known that the mentioned series converges and the sum of the first n terms is given by $\frac{1-(1/3)^n}{1-1/3}$ (Simmons 1983); therefore (94) becomes in

$$u(x)=\frac{D}{2}x^2+2\left[1-(1/3)^n\right]x^3. \quad 95$$

Assuming that $n \to \infty$, we get

$$u(x)=\frac{D}{2}x^2+2x^3. \quad 96$$

From the boundary condition $y(1) = 1$ we get $D = -2$, therefore

$$u(x) = -x^2 + 2x^3. \quad 97$$

Finally, from (69) we obtain the solution for the boundary value problem (68)

$$y(x) = -x + 2x^2. \quad 98$$

Thus, for this case study CVLT-HPM provides an exact solution for (68).

Case 3

Next, we will analyze the case of

$$\int_{x_1}^{x_2}\frac{\sqrt{1+y'^2}}{y^{1/n}}dx. \quad 99$$

After employing Euler equation (8), we get

$$y^{2/n}(x)(1+y'^2(x))=c \quad 100$$

where c is a constant.

Later, we will require to express (100) in such a way that it does not involves c. Therefore we differentiate (100) respect to x.

$$\frac{d\left(y^{2/n}(x)(1+y'^2(x))\right)}{dx}=0. \quad 101$$

We rewrite (100) as follows

$$y^2y'^2=cy^{2\left(1-\frac{1}{n}\right)}-y^2, \quad 102$$

or

$$\left(\frac{dy^2}{dx}\right)^2=4\left[cy^{2\left(1-\frac{1}{n}\right)}-y^2\right], \quad 103$$

so that

$$y^{2'}=2\sqrt{c}\,y^{\left(\frac{n-1}{n}\right)}\left[1-\frac{y^{2/n}}{c}\right]^{1/2}, \quad 104$$

where we employed the notation

$$y^{2'}=\left(\frac{dy^2}{dx}\right). \quad 105$$

Next, we propose the following substitution

$y = u^{n/n-1},\ n \neq 1.$ 106

Thus, (104) adopts the form

$$\frac{du^{\frac{2n}{n-1}}}{dx} = 2\sqrt{cu}\left[1 - \frac{u^{2/n-1}}{c}\right]^{1/2}.\ n \neq 1.$$ 107

We will consider as a case study $n = 3$. It corresponds to the equation (see 100) (Filobello-Nino et al. 2013d).

$y^{2/3}(x)(1 + y'^2(x)) = c.$ 108

We note, it is possible to rewrite (108) as

$$x = \int \frac{y^{1/3}dy}{\sqrt{c - y^{2/3}}} + c.$$ 109

Although (109) can be solved performing the algebraic substitution $y = v^3$, the dependent variable $y(x)$remains implicit.

In fact, we propose the following boundary value problem.

$$y^{2/3}(x)(1 + y'^2(x)) = c,$$
$$y(0) = 0, \quad y(20) = 1.$$ 110

In order to get an explicit analytical approximate solution for (110), we will consider (106) and (107) for $n= 3$.

$y(x) = u^{3/2}(x),$ 111

$$\frac{du^3}{dx} = 2\sqrt{cu}\left[1 - \frac{u}{c}\right]^{1/2}.$$ 112

Next, we use Newton's binomial to transform (112) into the following approximate equation

$$\frac{du^3}{dx} - 2\sqrt{cu} + \frac{u^2}{\sqrt{c}} + \frac{u^3}{4c^{3/2}} = 0.$$ 113

Such as it will be explained in discussion section, although LT-HPM is an appropriate formulation for this case study, it is required a reformulation because the term $2c\sqrt{u}$, gives rise to an approximate solution inconsistent with the proposed boundary condition $u(0) = 0$(from condition (111) we see that $y(0) = 0 \rightarrow u(0) = 0$). In order to delay the effect of the above mention term in the iterative process, we begin proposing the following homotopy, which is an original contribution of this work (see also, "Discussion" section).

$$(1 - p^n)(du^3/dx - w) + p^n\left[du^3/dx + u^3/4c^{3/2} + u^2\sqrt{c} - 2\sqrt{cu}\right] = 0,$$ 114

where we have introduced a trial function w which may depend, for our purposes of

$w = w(x, u, p),$ 115

and n is an integer number $n > 1$.

We rewrite the above equation as follows.

$$\frac{du^3}{dx} = w + p^n\left[-w - u^3/4c^{3/2} - u^2/\sqrt{c} + 2\sqrt{cu}\right].$$ 116

Next, using NDLT-HPM we propose

$$w = 3Ax^2 - \frac{u^3p}{4c^{3/2}} - \frac{u^2p}{c^{1/2}},$$ 117

so that (116) becomes

$$\frac{du^3}{dx} = 3Ax^2 - u^3p/4c^{3/2} - u^2p/\sqrt{c} + p^n\left[-3Ax^2 + u^3(p-1)/4c^{3/2} + u^2(p-1)/\sqrt{c} + 2\sqrt{cu}\right]$$ 118

It is worthwhile to mention, after evaluating the limit $p \rightarrow 1$, we obtain the equation to solve (116) from (118) as it should be if (118) has to be a valid homotopy.

After considering n values as large as we wish ($n \rightarrow \infty$), it is clear that for practical purposes, (118) is equivalent to

$$\frac{du^3}{dx} = 3Ax^2 - u^3p/4c^{3/2} - u^2p/\sqrt{c}.$$ 119

Thus, (118) let us to delay sufficiently the role of the term $2\sqrt{cu}$ in the iterative process and replacing it for$3Ax^2$; it eliminates the mentioned problem because the iterative process resulting will be consistent with $u(0) = 0$. We will show the calculation of parameter A will allow recover part of lost information for the omission of term $2\sqrt{cu}$, and the substitution of (113) instead of (112).

Following the usual Laplace transform algorithm, applying $\Im$ to (119) we obtain

$$\Im\left\{\frac{du^3}{dx}\right\} = \Im\left\{3Ax^2 - u^3p/4c^{3/2} - u^2p/\sqrt{c}\right\}.$$ 120

As it is explained in (Spiegel 1998) we rewrite (120) as follows

$$s\Im\left\{u^3\right\} = \Im\left\{3Ax^2 - u^3p/4c^{3/2} - u^2p/\sqrt{c}\right\},$$ 121

where we have employed the initial condition $u(0) = 0$.

Applying Laplace inverse transform $\Im^{-1}$ to this last expression, we obtain

$$u^3 = \Im^{-1}\left\{(1/s)\Im\left\{3Ax^2 - u^3p/4c^{3/2} - u^2p/\sqrt{c}\right\}\right\}$$ 122

Next, we assume a series solution for $u(x)$, in the form

$$u(x) = \sum_{n=0}^{\infty} p^n v_n.$$ 123

Substituting (123) into (122), we get

$$\left(\sum_{n=0}^{\infty} p^n v_n\right)^3 = \Im^{-1}\left\{(1/s)\Im\left\{3Ax^2 - \left(\sum_{n=0}^{\infty} p^n v_n\right)^3 p/4c^{3/2} - p\left(\sum_{n=0}^{\infty} p^n v_n\right)\right.\right.$$ 124

After comparing the coefficients of like powers of p, and perform the indicated operations, we have

$$p^0 : v_0^3 = \Im^{-1}\left\{\left(\frac{1}{s}\right)\Im\left\{3Ax^2\right\}\right\}, \quad 125$$

$$p^1 : v_1 v_0^2 + 2v_1 v_0^2 = \Im^{-1}\left\{\left(\frac{1}{s}\right)\Im\left\{\left[-v_0^3/4c^{3/2} - v_0^2/\sqrt{c}\right]\right\}\right\}, \quad 126$$

$$p^2 : 3v_2 v_0^2 + 2v_0 v_1^2 = \Im^{-1}\left\{\left(\frac{1}{s}\right)\Im\left\{\left[-3v_1 v_0^2/4c^{3/2} - 2v_1 v_0/\sqrt{c}\right]\right\}\right\}, \quad 127$$

$$p^3 : v_0^2 v_3 + v_1^3 + 6v_0 v_1 v_2 = \Im^{-1}\left\{\left(\frac{1}{s}\right)\Im\left\{\left[-(3v_2 v_0^2 + 3v_1^2 v_0)/4c^{3/2} - (2v_2 v_0 + v_1^2)/\sqrt{c}\right]\right\}\right\}. \quad 128$$

Solving the above operations for $v_0(x)$, $v_1(x)$, and $v_2(x)$, we obtain

$$p^0 : v_0(x) = A^{1/3}x, \quad 129$$

$$p^1 : v_1(x) = \frac{-A^{1/3}x^2}{48c^{3/2}} - \frac{x}{9\sqrt{c}}, \quad 130$$

$$p^2 : v_2(x) = \frac{13A^{1/3}}{17280c^3}x^3 + \frac{19x^2}{2592c^2} + \frac{4}{243A^{1/3}c}x, \quad 131$$

$$p^3 : v_3(x) = -\frac{10395Ax^4 + 202608x^3cA^{2/3} + 345600x^2c^2A^{1/3} + 102400xc^3}{74649600c^{9/2}A^{2/3}}, \quad 132$$

and so on.

By substituting (129)–(132) into (123), calculating the limit when $p \to 1$, we obtain the following third order approximation.

$$\begin{aligned} u(x) = {} & \frac{13A^{1/3}}{17280c^3}x^3 + \left(\frac{-A^3}{48c^{3/2}} + \frac{19}{2592c^2}\right)x^2 + \left(A^{1/3} - \frac{1}{9\sqrt{c}} + \frac{4}{243A^{1/3}c}\right)x \\ & + \frac{1}{A^{2/3}x^2}\left(\frac{A^{1/3}x^2}{48c^{3/2}} - \frac{x}{9\sqrt{c}}\right)^3 - \frac{37}{14400c^{7/2}}x^3 \\ & - \frac{7}{1296c^{5/2}A^{1/3}}x^2 - \frac{41A^{1/3}}{276480c^{9/2}}x^4. \end{aligned} \quad 133$$

From (111) and (133) we obtain

$$\begin{aligned} y(x) = {} & \left(\frac{13A^{1/3}}{17280c^3}x^3 + \left(\frac{-A^3}{48c^{3/2}} + \frac{19}{2592c^2}\right)x^2 + \left(A^{1/3} - \frac{1}{9\sqrt{c}} + \frac{4}{243A^{1/3}c}\right)x \right. \\ & \left. + \frac{1}{A^{2/3}x^2}\left(\frac{A^{1/3}x^2}{48c^{3/2}} - \frac{x}{9\sqrt{c}}\right)^3 - \frac{37}{14400c^{7/2}}x^3 - \frac{7}{1296c^{5/2}A^{1/3}}x^2 - \frac{41A^{1/3}}{276480c^{9/2}}x^4\right)^{3/2}. \end{aligned} \quad 134$$

In order to calculate the values of A and c, we will require that (134) satisfies the proposed boundary condition $y(20) = 1$. Also, proceeding as it is indicated in "Basic idea of Laplace transform homotopy perturbation method with variable coefficients (CVLT-HPM)" section, we substitute (134) into (101) for $n = 3$ and evaluating the expression so obtained for some value x. Choosing $x = 19.8$, which lies into the interval of interest, $0 \leq x \leq 20$(see "Discussion" section below), we obtain an system of algebraic equations for A and c, whose solution is given by:

$$A = 0.001806042527, \quad c = 3.428590549. \quad 135$$

After substituting (135) into (134) we obtain the following approximation

$$y(x) = \left(\begin{aligned} & -0.00003215665052x^3 - 0.001414069432x^2 + 0.1011974436x \\ & + \frac{(67.42920175)(0.0003996329765x^2 - 0.06000669099x)^3}{x^2} - 7.057905179 \times 10^{-8}x^4 \end{aligned}\right)^{3/2} \quad 136$$

Case 4

This case is studied, by using a combination of LT-HPM and NDLT-HPM.

Finally we will consider the problem of extremizing an integral of the form

$$\int_{x_1}^{x_2} \frac{\sqrt{1+y'^2}}{g(x)}dx, \quad 137$$

where, $g(x)$ is an arbitrary function, except having Taylor series.

Defining $f(x) = 1/g(x)$, it is clear that (f1) is formally equivalent to the case

$$\int_{x_1}^{x_2} f(x)\sqrt{1+y'^2}dx. \quad 138$$

After employing Euler equation (7), for $f = \frac{\sqrt{1+y'^2}}{g(x)}$, we obtain the following differential equation

$$y'(x) = \frac{cg(x)}{\sqrt{1-c^2g^2(x)}}, \quad 139$$

where c is a constant.

Separating variables, we rewrite (139) in terms of

$$y = \int \frac{cg(x)dx}{\sqrt{c-c^2g^2(x)}} + C. \quad 140$$

Unfortunately the above integral hardly can be solved in terms of elementary functions for many $g(x)$ functions.

In principle it is possible to provide an scheme in order to obtain analytical approximate solutions for problems like (139), by using LT-HPM.

After multiplying by $y(x)$, it is possible to rewrite (139) as

$$\frac{dy^2}{dx} = \frac{2cgy}{\sqrt{1-c^2g^2}} = 2cgy\left(1-c^2g^2\right)^{-1/2}. \quad 141$$

We employ Newton's binomial to transform (141) into the following approximate equation

$$\frac{dy^2}{dx} = 2cgy\left(1+\frac{1}{2}c^2g^2+\frac{3}{8}c^4g^4\right). \quad 142$$

Following "Basic idea of non-linearities distribution Laplace transform-homotopy perturbation method (NDLT-HPM)" and "Basic idea of laplace transform homotopy perturbation method with variable coefficients (CVLT-HPM)" sections, it is convenient to construct a homotopy as follows

$$(1-p)(dy^2/dx - 2Ax) + p[dy^2/dx - 2cg(x,p)y - c^3g^3(x,p)y - 3c^5g^5(x,p)y/4] = 0, \quad 143$$

where in one hand we have chosen as a polynomial trial function $w(x) = 2Ax$, (A is a parameter to be determined) and on the other hand we emphasized the possibility of embedded the homotopy parameter $p$into the function $g(x)$ (see "Discussion" section).

Next we rewrite the above equation as

$$\frac{dy^2}{dx} = 2Ax + p\left[-2Ax + 2cyg(x,p) + c^3g^3(x,p)y + 3c^5g^5(x,p)y/4\right]. \quad 144$$

Following Laplace transform algorithm, we apply $\Im$

$$\Im\left\{\frac{dy^2}{dx}\right\} = \Im\left\{2Ax + p\left[-2Ax + 2cyg(x,p) + c^3g^3(x,p)y + 3c^5g^5(x,p)y/4\right]\right\}. \quad 145$$

Such as it was explained in (Spiegel 1998) we rewrite (145) as follows

$$s\Im\{y^2\} - y(0) = \Im\{2Ax + p[-2Ax + 2cyg(x,p) + c^3g^3(x,p)y + 3c^5g^5(x,p)y/4]\}. \quad 146$$

Applying Laplace inverse transform $\Im^{-1}$ to the above expression we get

$$y^2 = \Im^{-1}\left\{\frac{y(0)}{s} + \frac{\Im}{s}\left\{2Ax + p\left[-2Ax + 2cyg(x,p) + c^3g^3(x,p)y + 3c^5g^5(x,p)y/4\right]\right\}\right\}. \quad 147$$

Next, in order to show how to work the methodology indicated by (147), we propose as cases: $g(x) = \exp(x)$;with two different boundary conditions $y(0) = 0$, $y(1) = 1/2$ and $y(0) = 0, y(1) = 1$ in order to show the flexibility of the proposed methodology. So that for instance, we can express formally (139) for our first case as follows

$$y'(x) = \frac{c\exp(x)}{\sqrt{1-c^2\exp(2x)}} \quad y(0) = 0, \;\; y(1) = 1/2, \quad 148$$

in the same manner after, we will consider the case $y(0) = 0, y(1) = 1$.

Later, we will require expressing (148) in such a way that it does not involves c. For that purpose we obtain the following equivalent expression for (148)

$$\frac{y'(x)}{\exp(x)\sqrt{1+y'(x)^2}} = c,$$

differentiating respect to x, in order to eliminate c we get

$$\frac{d\left(\frac{y'(x)}{\exp(x)\sqrt{1+y'(x)^2}}\right)}{dx} = 0. \quad 149$$

Thus, the proposed analytical approximate solution (147), adopts the form

$$y^2 = \Im^{-1}\left\{\frac{\Im}{s}\left\{2Ax + p\left[-2Ax + 2cy\exp(px) + c^3 y\exp(3px) + 3c^5 y\exp(5px)/4\right]\right\}\right\}. \quad 150$$

Next, we assume a series solution for $y(x)$, in the form

$$y(x) = \sum_{n=0}^{\infty} p^n v_n. \quad 151$$

In order to keep control of the different p order approximations from (150), we introduce the following Taylor series expansions.

$$\exp(px) = 1 + px + \frac{p^2x^2}{2} + \cdots$$

$$\exp(3px) = 1 + 3px + \frac{9p^2x^2}{2} + \cdots,$$

$$\exp(5px) = 1 + 5px + \frac{25p^2x^2}{2} + \cdots. \quad 152$$

Substituting (151) and (152) into (150), we get

$$\left(\sum_{n=0}^{\infty} p^n v_n\right)^2 = \Im^{-1}\left\{\frac{\Im}{s}\left\{2Ax + p\left[-2Ax + 2c\left(\sum_{n=0}^{\infty} p^n v_n\right)\left(1 + px + \frac{p^2x^2}{2} + \cdots\right)\right.\right.\right.$$
$$+ c^3\left(\sum_{n=0}^{\infty} p^n v_n\right)\left(1 + 3px + \frac{9p^2x^2}{2} + \cdots\right)$$
$$\left.\left.\left. + 3c^5\left(\sum_{n=0}^{\infty} p^n v_n\right)\left(1 + 5px + \frac{25p^2x^2}{2} + \cdots\right)/4\right]\right\}\right\}. \quad 153$$

On comparing the coefficients of like powers of p, we have

$$p^0 : v_0^2 = \Im^{-1}\left\{\frac{1}{s}\Im(2Ax)\right\}, \quad 154$$

$$p^1 : 2v_0v_1 = \Im^{-1}\left\{\left(\frac{1}{s}\right)\Im\left\{\left[-2Ax + 2cv_0 + c^3 v_0 + 3c^5 v_0/4\right]\right\}\right\}, \quad 155$$

$$p^2 : v_1^2 + 2v_0v_2 = \Im^{-1}\left\{\left(\frac{1}{s}\right)\Im\left\{v_1\left(2c + c^3 + 3c^5/4\right) + v_0 x\left(2c + 3c^3 + 15c^5/4\right)\right\}\right\}, \quad 156$$

$$p^3 : 2v_0v_3 + 2v_1v_2 = \Im^{-1}\left\{\left(\frac{1}{s}\right)\Im\left\{v_2\left(2c + c^3 + \frac{3}{4}c^5\right) + v_1 x\left(2c + 3c^3 + \frac{15c^5}{4}\right)\right.\right.$$
$$\left.\left. + v_0 x^2\left(c + \frac{9c^3}{2} + \frac{75c^5}{8}\right)\right\}\right\}, \quad 157$$

After solving the above operations for $v_0(x)$, $v_1(x)$, and $v_2(x)$ we obtain

$$p^0 : v_0(x) = \sqrt{Ax}, \quad 158$$

$$p^1 : v_1(x) = \left[-\frac{\sqrt{A}}{2} + \frac{c}{2} + \frac{c^3}{4} + \frac{3c^5}{16}\right]x, \tag{159}$$

$$p^2 : v_2(x) = \left[-\frac{\sqrt{A}}{2} + \frac{c}{2} + \frac{c^3}{4} + \frac{3c^5}{16}\right]\left[\frac{\sqrt{A}}{2} + \frac{c}{2} + \frac{c^3}{4} + \frac{3c^5}{16}\right]x + \frac{1}{6}\left(2c + 3c^3 + \frac{15}{4}c^5\right)x^2, \tag{160}$$

$$p^3 : v_3(x) = \frac{1}{8}\left(c + \frac{9c^3}{2} + \frac{75c^5}{8}\right)x^3 + \frac{1}{6\sqrt{A}}\left[2c + 3c^3 + \frac{15c^5}{4}\right]\left[\frac{c}{3} + \frac{c^3}{6} + \frac{c^5}{8}\right]x^2$$
$$+ \left(\frac{-\sqrt{A}}{2} + \frac{c}{2} + \frac{c^3}{4} + \frac{3c^5}{16}\right)\left(\frac{\sqrt{A}}{2} + \frac{c}{2} + \frac{c^3}{4} + \frac{3c^5}{16}\right)\frac{x}{2}, \tag{161}$$

and so on.

By substituting (158)–(161) into (151) and calculating the limit when $p \to 1$, results in a handy third order approximation

$$y(x) = \frac{1}{8}\left(c + \frac{9c^3}{2} + \frac{75c^5}{8}\right)x^3 + \frac{1}{6}\left[2c + 3c^3 + \frac{15c^5}{4}\right]\left[1 + \frac{1}{\sqrt{A}}\left(\frac{c}{3} + \frac{c^3}{6} + \frac{c^5}{8}\right)\right]x^2$$
$$+ \left[\sqrt{A} + \left(\frac{-\sqrt{A}}{2} + \frac{c}{2} + \frac{c^3}{4} + \frac{3c^5}{16}\right)\left(1 + \frac{3}{2}\left(\frac{\sqrt{A}}{2} + \frac{c}{2} + \frac{c^3}{4} + \frac{3c^5}{16}\right)\right)\right]x. \tag{162}$$

In order to calculate the values of A and c, we will require that (162) satisfies the boundary condition $y(1) = 1/2$. Also, following "Basic idea of Laplace transform homotopy perturbation method with variable coefficients (CVLT-HPM)" section, we substitute (162) into (149) and evaluate the expression so obtained for some value, let us say $x = 1/5$, which lies into the interval of interest, $0 \leq x \leq 1$ (see "Discussion" below). After performing the above substitution, we obtain a system of algebraic equations for A and c, whose solution provides

$$A = 1.889510047, \quad c = 0.3720801807 \tag{163}$$

Thus, substituting (163) into (162), we obtain a handy approximate solution for (158)

$$y(x) = 0.2474371491x + 0.1687199868x^2 + 0.08384286388x^3. \tag{164}$$

Proceeding the same manner for the conditions, $y(0) = 0$, $y(1) = 0.7$, we obtain the values $A = 0.002029581087$, $c = 0.3131923646$. Finally, substituting them into (162), we get

$$y(x) = 0.2267590836x + 0.4132800611x^2 + 0.05996085479x^3. \tag{165}$$

DISCUSSION

In order to use a pure numerical solution as reference, we utilized the scheme based on trapezoid combined with Richardson extrapolation (Ascher et al. 1995; Ascher and Petzold 1998) from the built-in numerical routines provided by Maple 15. Moreover, the command was setup with an absolute error (AE) tolerance of 1×10^{-12}.

This work proposed LT-HPM and some its modifications, in order to find analytical approximate expressions for stationary functions solutions of the Euler equations, which turned out from four cases studied. Such as it was already mention, research in this area of knowledge is relevant because calculus of variations is a branch of the analysis with applications for both, pure and applied mathematics and expressing the basic principles of physics with simplicity and elegance among others.

We noted Euler equation (2) which describes these problems is of nonlinear nature and therefore the task of finding analytical solutions for this equation is in general a very difficult task. As a matter of fact, although some cases admit exact solutions, many variational problems admit just numerical solutions (Razzaghi and Elnagar 1994; Razzaghi and Yousefi 2001). Nevertheless this article showed one case study for which was not possible to find a numerical solution. Unlike (Lakestani and Kazemian 2013) where just one particular case of variational problem was solved, the goal of this paper was expose how LT-HPM and some of its modifications are employed in order to analyze a wide variety of cases study. Indeed, cases study 3 and 4 were presented in the most general way possible and to achieve it, we employed some modified versions of LT-HPM.

Thus, case study 1 proposed the basic version of LT-HPM ("Basic idea of the LT-HPM method" section) in order to find an analytical approximate solution for (46); essentially we used the procedure followed in (Filobello-Nino et al. 2013c). The proposed method expresses the problem of solving a nonlinear differential equation in terms of solving an algebraic equation for some unknown initial condition [see (64)]. This formulation has shown dependable performance for the case of differential equations without the presence of singular points and this case satisfies these requirements (Filobello-Nino et al. 2013c). Figure 1 shows the comparison between numerical solution of (46) and LT-HPM second order approximation (65). From this figure is clear the proposed solution provides a good approximation. In more precise terms, to verify the precision of (63) and (65) we calculate the square residual error (SRE) of (65), which is explained in "Appendix 2". SRE is in general terms a positive number representative of the total error by using an approximate solution for a given problem (Marinca and Herisanu 2011). The resulting value for our case study was 0.001918936920, which confirms the accuracy of the proposed solution. If more accuracy has to be required, one must consider higher order approximations.

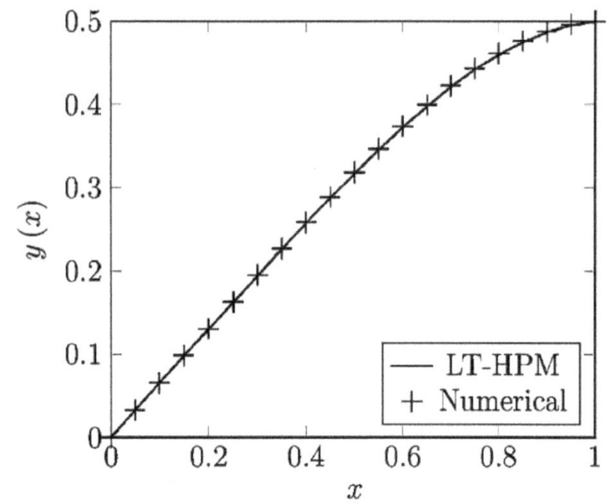

Figure 1: Comparison between numerical solution of (46) (*cross*) and LT-HPM approximation (65) (*line*).

Our second variational problem (66) gave rise to the second order non homogeneous linear differential equation with variable coefficients (68). Since the aforementioned equation is something long, and following the indications of "Basic idea of Laplace transform homotopy perturbation method with variable coefficients (CVLT-HPM)" section, we ease the application of the proposed methodology, transforming (67) in its normal form (70), by using the substitution (69) (Simmons 1983). The equivalent equation (73) is also of variable coefficients; therefore we employed the modified version of LT-HPM explained in "Basic idea of Laplace transform homotopy perturbation method with variable coefficients (CVLT-HPM)" section (Filobello-Nino et al. 2015b). CVLT-HPM expresses the problem to be solved in terms of a differential equation for Laplace transform $u(s)$, although it easily solvable [see (78)–(80)].

Once $u(s)$ was expressed in terms of $u(x)$, we assumed a series solution in the form (81). CVLT-HPM method, calculated the different approximate orders, expressing the n-th iterative process in terms of integrals of the lower order approximations, previously calculated [see (83)–(87)].

Unlike basic LT-HPM algorithm ("Basic idea of the LT-HPM method" section), CVLT-HPM employed its homotopy freedom in order to substitute $L(u_0)$ as it was originally defined by HPM method ("Standard HPM" section), for the polynomial trial function $w(x) = D$ (constant), whose value was finally determined from the boundary condition $y(1) = 1$. It is remarkable to note we obtained an exact solution for (98), through the convergent geometric series (94), which shows that in those cases where a problem has an exact solution, CVLT-HPM method is able to provide it.

Case study 3 was presented in the most general possible manner (99). This case is relevant because, such as it was explained in "Background" section, some values of n corresponds to known cases. Thus,$n = 1/2$ defines the problem of brachistochrone (Connor and Robertson 1997, 1998; Filobello-Nino et al. 2013d) and $n = -1$the problem of finding the curve that generates the surface of revolution of smallest area (Simmons 1983).

The corresponding Euler equation (102), presents an additional challenge; the presence of a highly nonlinear term that corresponds to the multiplication of y^2 by y'^2 which difficult the application of LT-HPM algorithm. To overcome these difficulties, we proposed the substitution (106) in order to obtain (107). Despite of this equation involved the derivative of a power of dependent variable $u^{2n/n-1}$, the methodology explained following Eqs. (114)–(119), turned out adequate to obtain an accurate analytical approximate solution for the case study $n = 3$ [see (110), (111), and (112)]. A relevant point of the proposed method is, although (108) is easily expressed in terms of separated variables, and the resulting integral (109) can be solved performing the algebraic substitution $y = v^3$, the dependent variable $y(x)$ remains implicit, and therefore Newton–Raphson method would be required to obtain values of $y(x)$, given some values of x. In the sequel, this case study not even admits a numerical solution.

On the other hand, such as we mentioned, the formulation of LT-HPM would turn out to be inappropriate for this case, because the term $2\sqrt{c}u$, would give rise to an approximate solution inconsistent with the proposed boundary condition $u(0) = 0$. As a matter of fact, if trial function was chosen as $z(x) = 0$ then $v_0(x) = 0$, and the rest of the iterations would vanish. If trial function is $z(x) = A$(constant), then the subsequent iterative process would give rise to fractional powers of x(in fact $v_1(x)= \sqrt[3]{Ax^{1/3}}$). The problem of solutions containing these powers is that, its derivatives becomes infinite at $x = 0$, unlike the initial conditions of the problem. It is not difficult to verify that a similar problem would occur if the trial function were chosen as one of the following: $z(x) = Ax$, $z(x) = Ax^3$, $z(x) = Ax^4$, ... and so on. Only the case $z(x) = Ax^2$, provides a whole powers polynomial function consistent with $u(0) = 0$. With the purpose of delay the effect of the problematic term $2\sqrt{c}u$, and introduce in the iterative process the presence of term Ax^2, we proposed employ the homotopy equation (119). In order to motivate it, we proposed the homotopy (114), which allows introduce a trial function, even depending of the homotopy parameter p, $w = w(x, u, p)$ in order to reach our goals. Nevertheless, we note (114) is a valid homotopy because (113) is recovered in the limit $p \to 1$, as expected. In particular by choosing w in accordance with (117), let us employed for practical purposes, to Eq. (119) instead of (113) in the limit $n \to \infty$. The above is due to the homotopy freedom of (114), which allowed delay sufficiently the role of the term $2\sqrt{c}u$, and essentially substituted it for $3Ax^2$; thus the iterative process resulted to be consistent with $u(0) = 0$. The proposed homotopy (114), which is employed in order to delay and substitute a problematic term of a differential equation, including the introduction of a trial function depending of parameter p [see (115) and (117)] is an original contribution of this work. A relevant fact of the proposed method is the determination of parameter A allowed recovers part of lost information for the omission of term $2\sqrt{c}u$, and the substitution of (113) instead of (112). Then LT-HPM provided a third order approximation for u (133), and through Eq. (111) an approximate solution (134) for $y(x)$. In order to calculate the values of A and c we required one hand, (134) satisfied the proposed boundary condition $y(20) = 1$, and on the other hand we deduced another equation by substituting (134) into (101) for $n = 3$ and then, evaluating the expression so obtained for some value which lied into the interval of interest $0 \leq x \leq 20$. It is important to notice that unlike the others cases study, this example proposed a nonlinear differential equation defined in an interval relatively large in order to show the flexibility of the proposed method. Further research is required to propose a systematic procedure to obtain suitable cancelling points. For instance, the cancelling point for this case study was arbitrarily selected as $x = 19.8$ after some trial and error attempts. This strategy was effective to obtain a low residual error. The resulting system of algebraic equations was solved to obtain the values of A and c and finally an approximate solution for $y(x)$ [see (136)].

It is worth to note the value of the SRE of (136) is just 0.06334882582; from this we depict the accuracy of our approximate solution (see Fig. 2). If more accuracy has to be required, one must considerer higher order approximations. An outstanding point is that, numerical routines from Maple 15 failed in this case in order to provide a numerical solution as reference. Since the numerical solutions are often the only manner to obtain information from a nonlinear problem, LT-HPM method indeed, is a useful strategy capable of supporting approximate methods, especially for cases like this.

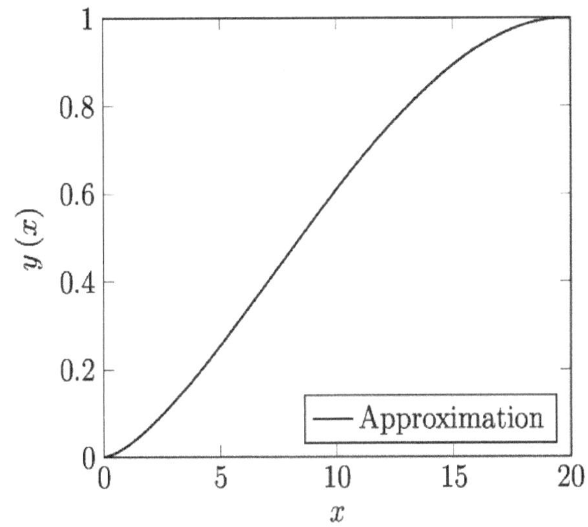

Figure 2: Combination of NDLT-HPM and LT-HPM to obtain approximation (136).

Finally, our case study 4, proposed the problem of extremizing (137), which was also introduced in the most general way. Such as it was said $g(x)$ represents an arbitrary function, although having Taylor series. In the sequel, the mathematical steps (137)–(139) and (141)–(147) were perform without assuming a particular $g(x)$.

We note the proposed homotopy (143) chose the polynomial trial function $z(x) = 2Ax$ and embedded the homotopy parameter p into $g(x)$(see "Basic idea of non-linearities distribution Laplace transform-homotopy perturbation method (NDLT-HPM)" section). (Filobello-Nino et al. 2014b) employed this technique, provided for the freedom of homotopy formulations in order to redistribute the nonlinearities (Vazquez-Leal et al. 2012a) and non-polynomial nonhomogeneous terms, between the successive iterations of LT-HPM method, and thus to ease the search for an approximate solution. The main reason is that NDLT-HPM distributes the contributions of the above mentioned functions along the iterations for the successive parameter powers p that allows to NDLT-HPM method being simpler at the different stages of iterations. As a matter of fact, we proposed a combination between LT-HPM and NDLT-HPM method, in order to obtain better results. As a case study, we considered $g(x) = \exp(x)$ (although for this choosing of $g(x)$ (140) is integrable, it is expected the method works even for cases where (140) cannot be expressed in terms of elementary functions). Its Taylor expansions (152) in terms of the homotopy parameter and further substitution in the proposed approximate solution (150) show the manner which a function is included in the iterative process; in the same way we note the trial function $z(x) = 2Ax$, was chosen in order to get a whole power series solution, with good precision. With the end of calculating the values of A and c we required that (162) satisfies the boundary condition $y(1) = 1/2$. Also, following "Basic idea of Laplace transform homotopy perturbation method with variable coefficients (CVLT-HPM)" section, we substituted (162) into (149) and evaluate the expression so obtained for some value, we chose $x = 1/5$, which lies into $0 \leq x \leq 1$. After performing the above substitution, we obtain a system of algebraic equations for A and c, whose solution provided the approximate solution (164). As a matter of fact, proceeding at the same manner, for the proposed second case study given by the conditions $y(0) = 0$, $y(1) = 0.7$, we obtained the approximate solution (165). For both cases, we obtained handy accurate approximate solutions; the SRE of (164) and (165) were 0.002934863992 and 0.05009724353 respectively (see Figs. 3,4) from which we infer that the proposed method is potentially useful in the resolution of the kind of problems as (139). It is possible to obtain more accurate solutions, considering higher order approximations, although could be lost the handy character of our approximation. Worth mentioning even in these case studies, the value of $x = 1/5$, was chosen by trial and error attempts in order to find a good approximate solution.

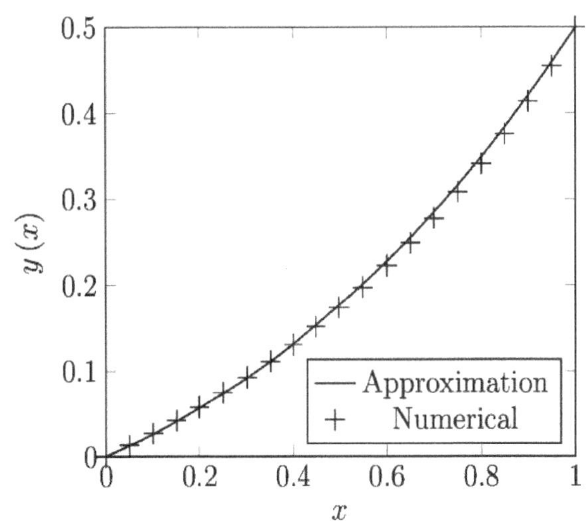

Figure 3: Combination of LT-HPM and NDLT-HPM to obtain approximation (164) (*line*) and numerical solution (*cross*).

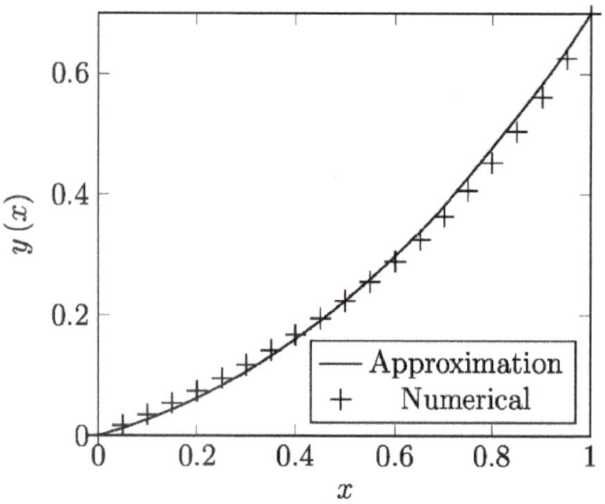

Figure 4: Combination of LT-HPM and NDLT-HPM to obtain approximation (165) (*line*) and numerical solution (*cross*).

It should be mentioned that, variational problems have been successfully studied for other authors, although the most of them have proposed just numerical approximations (Razzaghi and Elnagar 1994; Razzaghi and Yousefi 2001). On the other hand, HAM was employed to obtain an approximate solution for brachistocrone problem (Lakestani and Kazemian 2013). This work obtained precise solutions but long. As it is well known, HAM is accurate and powerful, but sometimes its approximate solutions turn out to be long and cumbersome, and for the same reason, they are not adequate for practical applications; on the other hand we obtained handy accurate approximate solutions with a reduced number of terms. Such as we have already mentioned, unlike (Lakestani and Kazemian 2013) that just studied one variational problem, the goal of this paper was expose in a general manner, how to find analytical solutions for several interesting cases of (2). On the other hand, (Ghaderi 2012) employed homotopy perturbation method (HPM) and Green's function method in order to find approximate solutions to variational calculus. Nevertheless unlike our work, the mentioned article, dealt with moving boundary and isoperimetric problems. The main advantage of (Ghaderi2012) is that the mathematical properties of the Green's function are well established, however its study does not belong to the elementary mathematics; unlike the above, LT-HPM is an iterative method which is based on elementary Laplace transform properties (Spiegel 1998) and simple algebraic steps, making it an ideal technique for practical applications. Finally, future research will be required to improve the present proposal to be applied to infinite intervals.

CONCLUSIONS

This work introduced LT-HPM and some of its modifications in order to find analytical approximate solutions for some Euler's ordinary differential equations, whose solutions extremizing the value of integrals of the form (1). As it is well known, the calculus of variations is a powerful branch of the analysis that can be applied to several problems of pure mathematics, also to express the basic principles of the mathematical physics in an elegant and simple way.

The relevance of this work consisted mainly on two points. On one hand it showed that the proposed methodology is potentially efficient in order to find solutions to the difficult linear and nonlinear problems arising from variational problems (even CVLT-HPM was able to provide exact solutions, see case study 2). On the other hand the article suggested for some cases, appropriate mathematical manipulations of the aforementioned equations, in order to transform them into a more accessible form, to increase the possibility of success (see cases study 3 and 4). The above is even more important, considering that many times even the numerical methods fail to solve these complicated equations, such as happened with our case study 3, and the numerical solutions and qualitative methods are usually the only manner to obtain information from a nonlinear problem. Finally, an outstanding issue for a future work is improving the methodology used to calculate the parameters introduced for our proposed solutions. A forthcoming paper should investigate a systematic manner to find the optimum values for the aforementioned parameters, without using the method of trial and error attempts in its determination.

AUTHORS' CONTRIBUTIONS

All authors contributed extensively in the development and completion of this article. All authors read and approved the final manuscript.

ACKNOWLEDGEMENTS

We gratefully acknowledge the financial support from the National Council for Science and Technology of Mexico (CONACyT) through grant CB-2010-01 #157024. The authors would like to thank Rogelio-Alejandro Callejas-Molina, and Roberto Ruiz-Gomez for their contribution to this project.

Competing interests

The authors declare that they have no competing interests.

APPENDIX 1

Laplace transform of $F(t)$ is denoted by $\Im\{F(t)\}$ and is defined by the integral (Connor and Robertson 1997).

$$\Im\{F(t)\} = f(s) = \int_0^\infty e^{-st}F(t)dt. \qquad 166$$

Linearity of Laplace transform is an important property, which is expressed as

$$\Im\{c_1F_1(t) + c_2F_2(T)\} = c_1f_1(s) + c_2f_2(s), \qquad 167$$

where c_1 and c_2 are constants and we have denoted: $\Im\{F_1(t)\} = f_1(s)$, $\Im\{F_2(t)\} = f_2(s)$.

Some known properties of LT, widely employed in this study are

- $\Im\{1\} = \frac{1}{s} \quad (s > 0),$ 168
- $\Im\{t^n\} = \frac{n!}{s^{n+1}} \quad (s > 0),$ 169
- $\Im\{F^{(n)}(t)\} = s^n f(s) - s^{n-1}F(0) - s^{n-2}F(0) - \cdots - F^{(n-1)}(0),$ 170

where $F^{(n)}(t)$ denotes the nth derivative of $F(t)$ and $\Im\{F(t)\} = f(s)$.

$$\Im\{t^nF(t)\} = (-1)^n \frac{d^nf(s)}{dt^n},$$,n denotes a positive integer, 171

where $\Im\{F(t)\} = f(s)$.

If Laplace transform of $F(t)$ is $f(s)$, then $F(t)$ is called the inverse Laplace transform of $f(s)$ and is expressed by $F(t) = \Im^{-1}\{f(s)\}$, where $\Im^{-1}$ is called the inverse Laplace transform operator.

From Eqs. (168) and (169) is clear that

$$1 = \Im^{-1}\left(\frac{1}{s}\right), \quad 172$$

$$t^n = \Im^{-1}\left(\frac{n!}{s^{n+1}}\right), \quad 173$$

and so on.

The following important result is obtained from (167) and denotes the linearity property of $\Im^{-1}$

$$\Im^{-1}\{c_1f_1(s) + c_2f_2(s)\} = c_1F_1(t) + c_2F_2(t). \quad 174$$

Appendix 2

The square residual error (SRE) of an approximate solution is given by

$$\int_a^b R^2(u(r))dr, \quad 175$$

a and b are the end points, the residual $R(u(r))$ is defined as

$$R(x) = L(u(x)) + N(u(x)) - f(x), \quad 176$$

where $u(r)$ is an approximate solution for the equation to be solved, which generally may depend of some parameters (See the end of "LT-HPM and some of its modified versions" section).

The square residual error is in general terms a positive number, which is representative of the total error committed, by using the approximate solution $u(r)$ SRE, would be zero only for the case where $u(r)$ turns out to be the exact solution of the differential equation under study.

REFERENCES

1. Adomian George (1988) A review of the decomposition method in applied mathematics. J Math Anal Appl 135(2):501–544
2. Adomian G, Rach R (1985) On the solution of algebraic equations by the decomposition method. J Math Anal Appl 105(1):141–166
3. Aminikhah H (2011) Analytical approximation to the solution of nonlinear Blasius viscous flow equation by LTNHPM. ISRN Math Anal. doi:10.5402/2012/957473
4. Aminikhah H (2012) The combined Laplace transform and new homotopy perturbation method for stiff systems of ODE s. Appl Math Model 36:3638–3644
5. Aminikhah H, Hemmatnezhad M (2012) A novel effective approach for solving nonlinear heat transfer equations. Heat Transf Asian Res 41(6):459–467
6. Ascher U, Petzold L (1998) Computer methods for ordinary differential equations and differential-algebraic equations. SIAM, Philadelphia
7. Ascher UM, Mattheij RM, Russell RD (1995) Numerical solution of boundary value problems for ordinary differential equations. Classics in applied mathematics 13. Society for Industrial and Applied Mathematics. ISBN: 978-0-89871-354-1
8. Babolian E, Biazar J (2002) On the order of convergence of Adomian method. Appl Math Comput 130(2):383–387. doi:10.1016/S0096-3003(01)00103-5
9. Boyce DP (1998) Ecuaciones Diferenciales y Problemas con Valores a la Frontera, Cuarta Edición. Limusa Noriega Editores, México
10. Casas F (1996) Solution of linear partial differential equations. J Comput Appl Math 76:159–170
11. Connor J and Robertson E (1997) Brachistochrone. Available via DIALOG http://www-history.mcs.st-and.ac.uk/HistTopics/Brachistochrone.html. Accessed July 2015
12. Connor J and Robertson E (1998) Jacob Bernoulli available via DIALOG http://www-history.mcs.standrew.ac.vk/history/References/Bernoulli_Jacob.html. Accessed July 2015
13. Elsgolts L (1983) Differential equations and the calculus of variations. Mir Publishers, Moscow
14. Evans DJ, Raslan KR (2005) The Tanhfunction method for solving some important nonlinear partial differential. Int J Comput Math 82:897–905. doi:10.1080/00207160412331336026
15. Fatoorehchi H, Abolghasemi H (2011) Analytical solution to intra-phase mass transfer in falling film contactors via homotopy perturbation method. Int Math Forum 6(67):3315–3321
16. Fatoorehchi H, Abolghasemi Hossein (2012) Investigation of nonlinear problems of heat conduction in tapered cooling fins via symbolic programming. Appl Appl Math 7:717–734
17. Fatoorehchi Hooman, Abolghasemi Hossein (2013) A more realistic approach toward the differential equation governing the glass transition phenomenon. Intermetallics 32:35–38
18. Fatoorehchi Hooman, Abolghasemi Hossein (2015) Series solution of nonlinear differential equations by a novel extension of the Laplace transform method. Int J Comput Math 2015:1–31
19. Fatoorehchi H, Abolghasemi Hossein, Rach Randolph (2014) An accurate explicit form of the Hankinson–Thomas–Phillips correlation for prediction of the natural gas compressibility factor. J Petrol Sci Eng 117:46–53
20. Fatoorehchi H, Abolghasemi H, Zarghami R (2015a) Analytical approximate solutions for a general nonlinear resistor–nonlinear capacitor circuit model. Appl Math Model 39(19):6021–6031
21. Fatoorehchi H, Abolghasemi H, Zarghami R, Rach R (2015b) Feedback control strategies for a cerium–catalyzed Belousov–Zhabotinsky chemical reaction system. Can J Chem Eng 93(7):1212–1221

22. Fatoorehchi H, Zarghami R, Abolghasemi H, Rach R (2015c) Chaos control in the cerium-catalyzed Belousov–Zhabotinsky reaction using recurrence quantification analysis measures. Chaos Solitons Fractals 76:121–129

23. Fatoorehchi H, Abolghasemi H, Zarghami R, Rach R, von Freeden S (2015d) A novel and computationally efficient algorithm for stability analysis of multi input-multi output process control systems. Korean J Chem Eng 2015:1–11

24. Filobello-Nino U, Vazquez-Leal H, Sarmiento-Reyes A, Perez-Sesma A, Hernandez-Martinez L, Herrera-May A, Jimenez-Fernandez VM, Marin-Hernandez A, Pereyra-Diaz D, Diaz-Sanchez A (2013a) The study of heat transfer phenomena using PM for approximate solution with Dirichlet and mixed boundary conditions. Appl Comput Math 2(6):143–148. doi:10.11648/j.acm.20130206.16 **(ISSN: 2328-5605)**

25. Filobello-Nino U, Vazquez-Leal H, Khan Y, Yildirim A, Jimenez-Fernandez VM, Herrera-May AL, Castaneda-Sheissa R, Cervantes-Perez J (2013b) Using perturbation method and Laplace-Padé approximation to solve nonlinear problems. Miskolc Math Notes 14(1):89–101

26. Filobello-Nino U, Vazquez-Leal H, Khan Y, Perez-Sesma A, Diaz-Sanchez A, Jimenez-Fernandez VM, Herrera-May A, Pereyra-Diaz D, Mendez-Perez JM, Sanchez-Orea J (2013c) Laplace transform-homotopy perturbation method as a powerful tool to solve nonlinear problems with boundary conditions defined on finite intervals. Comput Appl Math 34(1):1–16. doi:10.1007/s40314-013-0073-z

27. Filobello-Nino U, Vazquez-Leal H, Pereyra-Diaz D, Yildirim A, Perez-Sesma A, Castaneda-Sheissa R, Sanchez-Orea J, Hoyos-Reyes C (2013d) A generalization of the Bernoulli's method applied to brachistochrone-like problems. Appl Math Comput 219(12):6707–6718

28. Filobello-Nino U, Vazquez-Leal H, Cervantes-Perez J, Benhammouda B, Perez-Sesma A, Hernandez-Martinez L, Jimenez-Fernandez VM, Herrera-May AL, Pereyra-Diaz D, Marin-Hernandez A, Huerta-Chua J (2014a) A handy approximate solution for a squeezing flow between two infinite plates by using of Laplace transform-homotopy perturbation method. SpringerPlus 3(421):1–10. doi:10.1186/2193-1801-3-421

29. Filobello-Nino U, Vazquez-Leal H, Benhammouda B, Hernandez-Martinez L, Hoyos-Reyes C, Perez-Sesma JAA, Jimenez-Fernandez V, Manuel Pereyra-Diaz D, Marin-Hernandez A, Diaz-Sanchez A, Huerta-Chua J, Cervantes-Perez J (2014b) Nonlinearities distribution Laplace transform-homotopy perturbation method. SpringerPlus 3:594. doi:10.1186/2193-1801-3-594

30. Filobello-Nino U, Vazquez-Leal H, Benhammouda B, Perez-Sesma A, Jimenez-Fernandez VM, Cervantes-Perez J, Sarmiento-Reyes A, Huerta-Chua J, Morales-Mendoza LJ, Gonzalez-Lee M, Diaz-Sanchez A, Pereyra-Díaz D, López-Martínez R (2015a) Analytical solutions for systems of singular partial differential-algebraic equations. Hindawi Publishing Corporation Discrete Dyn Nat Soc. doi:10.1155/2015/752523

31. Filobello-Nino U, Vazquez-Leal H, Sandoval-Hernandez M, Perez-Sesma A, Sarmiento-Reyes A, Benhammouda B, Jimenez-Fernandez VM, Huerta-Chua J, Hernandez-Machuca SF, Mendez-Perez JM, Morales-Mendoza LJ, Gonzalez-Lee M (2015b) Extension of Laplace transform-homotopy perturbation method to solve nonlinear differential equations with variable coefficients defined with Robin boundary conditions. Neural Comput Appl. doi:10.1007/s00521-015-2080-z

32. Filobello-Niño U, Vazquez-Leal H, Castañeda-Sheissa R, Yildirim A, Hernandez Martinez L, PereyraDíaz D, Pérez Sesma A, Hoyos Reyes C (2012) An approximate solution of Blasius equation by using HPM method. Asian J Math Stat. doi:10.3923/ajms.2012 **(ISSN 1994-5418)**

33. Ghaderi S (2012) Homotopy perturbation method for solving moving boundary and isoperimetric problems. Appl Math 3:403–409. doi:10.4236/am.2012.35062

34. Hashim Ishak (2006) Adomian decomposition method for solving BVPs for fourth-order integro-differential equations. J Comput Appl Math 193(2):658–664

35. He JH (1998) A coupling method of a homotopy technique and a perturbation technique for nonlinear problems. Int J Non-Linear Mech 351:37–43. doi:10.1016/S0020-7462(98)00085-7

36. He JH (1999) Homotopy perturbation technique. Comput Methods Appl Mech Eng 178:257–262. doi:10.1016/S0045-7825(99)00018-3

37. He Ji-Huan (2003) Homotopy perturbation mcthod: a new nonlinear analytical technique. Appl Math Comput 135(1):73–79

38. He Ji-Huan (2008) Recent development of the homotopy perturbation method. Topol Methods Nonlinear Anal 31(2):205–209

39. Lakestani M, Kazemian M (2013) Solving Brachistochrone problem using homotopy analysis method. Acta Universitatis Apulensis 33(2013):165–175

40. Lanczos C (1986) The variational principles of mechanics, 4th edn. Dover publications Inc, New York

41. Levi E (1980) Theories and methods of applied mathematics. National Autonomous University of Mexico, Mexico

42. Mahmoudi J, Tolou N, Khatami I, Barari A, Ganji DD (2008) Explicit solution of nonlinear ZK-BBM wave equation using Exp-function method. J Appl Sci 8:358–363. doi:10.3923/jas.2008.358.363

43. Marinca V, Herisanu N (2011) Nonlinear dynamical systems in engineering, 1st edn. Springer, Berlin Heidelberg

44. Olver PJ (1993) Applications of Lie groups to differential equations, 2nd edn. Springer, New York

45. Rach R (2012) A bibliography of the theory and applications of the Adomian decomposition method, 1961–2011. Kybernetes 41(7/8). doi:10.1108/k.2012.06741gaa.007

46. Rashidi MM, Rastegari MT, Asadi M, Bg OA (2012a) A study of non-newtonian flow and heat transfer over

a non-isothermal wedge using the homotopy analysis method. Chem Eng Commun 199:231–256

47. Rashidi M, Pour SM, Hayat T, Obaidat S (2012b) Analytic approximate solutions for steady flow over a rotating disk in porous medium with heat transfer by homotopy analysis method. Comput Fluids 54:1–9
48. Razzaghi M, Elnagar GN (1994) A pseudospectral collocation method for the brachistochrone problem. Math Comput Simul 36(3):241–246. doi:10.1016/0378-4754(94)90008-6
49. Razzaghi M, Yousefi S (2001) Legendre wavelets method for the solution of nonlinear problems in the calculus of variations. Math Comput Model 34(1–2):45–54. doi:10.1016/S0895-7177(01)00048-6
50. Shang Yilun (2012) A Lie algebra approach to susceptible-infected-susceptible epidemics. Electron J Differ Equ 33:1–7
51. Shang Yilun (2013) Lie algebra method for solving biological population model. J Theor Appl Phys 7(1):1–4
52. Shang Y (2015) Analytical solution for an in-host viral infection model with time-inhomogeneous rates. Acta Phys Pol B 46(8):1567–1577. doi:10.5506/APhysPolB.46.1567
53. Simmons G (1983) Differential equations with applications and historical notes. McGraw-Hill, New York
54. Spiegel MR (1998) Teoría y Problemas de Transformadas de Laplace, primera edición. Serie de compendios Schaum, McGraw-Hill, México
55. Vazquez-Leal H, Filobello-Nino U, Castañeda-Sheissa R, Hernandez-Martınez L, Sarmiento-Reyes A (2012a) Modified HPMs inspired by homotopy continuation methods. Math Probl Eng Hindawi Publishing Corporation. doi:10.1155/2012/309123
56. Vazquez-Leal H, Sarmiento-Reyes A, Khan Y, Filobello-Nino U, Diaz-Sanchez A (2012b) Rational biparameter homotopy perturbation method and Laplace-Padé coupled version. J Appl Math. doi:10.1155/2012/923975
57. Xu F (2007) A generalized soliton solution of the Konopelchenko–Dubrovsky equation using exp-function method. Z Naturforschung A 62(12):685–688
58. Zelikin M (2005) Control Óptimo y Cálculo Variacional. Editorial URSS, Rusia
59. Zhang L-N, Xu L (2007) Determination of the limit cycle by He's parameter expansion for oscillators in a potential. Z Naturforschung A 62(7–8):396–398

CHAPTER 9

AN ANALYTICAL COUPLED TECHNIQUE FOR SOLVING NONLINEAR LARGE-AMPLITUDE OSCILLATION OF A CONSERVATIVE SYSTEM WITH INERTIA AND STATIC NON-LINEARITY

Md. Abdur Razzak and Md. Shamsul Alam

Keywords: Homotopy perturbation method (HPM) and variational approach, Non-linear oscillation, Large amplitude, Cantilever beam

BACKGROUND

Nonlinear oscillation problems are important in physical sciences, mechanical structures and engineering structures. Nonlinear vibration of oscillation systems are modeled by nonlinear differential equations. It is almost difficult to get exact solution for such nonlinear differential equations. Several methods have been used to solve weakly (small parameters, so-called perturbation parameters) nonlinear differential equations. Among all, most widely used technique is perturbation method (Marion 1970; Krylov and Bogoliubov 1947; Bogoliubov and Mitropolskii 1961; Nayfeh 1973; Nayfeh and Mook 1979; Nayfeh 1981). The perturbation method is not applied when a small parameter is absent in a nonlinear problem. Nonlinear of planar, large-amplitude free vibrations of a slender, inextensible cantilever beam carrying a lumped mass with rotary inertia at an intermediate position along its span is one of the problems that does not contain small parameter. In general, such problem is not always possible to get exact solution because of their complexity and thus the analytical approximate techniques must be needed to solve such problem. Moreover, there have been many strongly nonlinear problems arising in both science and engineering. To eliminate the limitations of classical perturbation technique, many analytical techniques such as variational iterative method (He et al. 2010; Herisanu and Marinca 2010a, b), variational method (He 2007; Kaya et al. 2010; Khan et al. 2011), energy balance method (EBM) (He 2002, 2006), homotopy analysis method (Liao 2003) used to solve strongly nonlinear problems. Recently, Khan et al. (2013) generalized the standard homotopy analysis method to solve nonlinear oscillators with rational terms. Moreover, Khan and Mirzabeigy (2014) has been improved He's energy balance method, especially the second-order approximation is considered here.

The homotopy perturbation method (HPM) (He 1999, 2004; Rafei et al. 2007; Ganji and Sadighi 2006; Ghorbani and Nadjafi 2007) is another effective technique for solving strongly nonlinear problems. The homotopy perturbation method was first introduced by He (1999). Generally, it is a method which is a combination of the classical perturbation method and the homotopy method in topology (He 2000). The solution procedure of HPM is very simple, only a few iteration steps lead to accurate approximations. Recently, some authors (Wang et al. 2012; Khan et al. 2014; Aminikhaha and Hemmatnezhad 2011; Akindeinde 2015; Suleman and Wu 2015) have been improved and modified the homotopy perturbation method. Moreover, another modified version of HPM named as optimal homotopy perturbation method (OHPM) (Marinca and Herisanu 2010, 2011;

Citation: Md. Abdur Razzak and Md. Shamsul Alam, An analytical coupled technique for solving nonlinear large-amplitude oscillation of a conservative system with inertia and static non-linearity, DOI 10.1186/s40064-016-2089-5.

Herisanu and Marinca 2012) have also been used for solving strongly nonlinear systems. Furthermore, some authors (Akbarzade 2010; Khan et al. 2012; Akbarzade and Khan 2012) have developed an analytical approximate technique coupling of the homotopy perturbation method and variational method in order to get high accuracy. Khan et al. (2012) obtained fourth-order approximations of strongly nonlinear problems, but the solution procedures of third and fourth-order approximations are very laborious process. On the contrary, some authors (Akbarzade and Khan 2012; Hamdan and Dado 1997; Wu et al. 2003; Herisanu and Marinca 2010a, b) have investigated the free vibration of a conservative oscillator having inertia and static non-linearity. In the previous articles (Hamdan and Dado 1997; Wu et al. 2003), the second-order approximate frequencies as well as the corresponding periodic solutions of such nonlinear oscillator were determined by using harmonic balance method. The results of Hamdan and Dado (1997) and Wu et al. (2003) are valid for weak nonlinearities and small amplitudes of oscillation. Moreover, in terms of large amplitudes, the results of the researchers (Hamdan and Dado (1997) and Wu et al. (2003) do not provide better outcome. On the other hand, the solution procedure of the article (Herisanu and Marinca 2010a, b) is very laborious. One of the shortcomings of the articles (Akbarzade 2010; Akbarzade and Khan 2012) is their trial functions do not satisfy the initial conditions.

In this paper, an analytical coupled method [a combination of homotopy perturbation method (He 2004) and variational method (He 2007)], along with a new trial function, has been presented to obtain the approximate frequency and the corresponding periodic solution of the strongly nonlinear oscillation of a conservative oscillator having inertia and static non-linearities (Akbarzade and Khan 2012; Hamdan and Dado 1997; Wu et al. 2003; Herisanu and Marinca 2010a, b). The new trial function of the present paper has satisfied the initial conditions. The results obtained in this paper (first-order approximate frequencies) are much better result for large values of amplitude than other existing results (Hamdan and Dado 1997; Wu et al. 2003). The method is very easy and straightforward.

FORMULATION AND SOLUTION METHOD

Consider the nonlinear oscillator (Hamdan and Dado 1997; Wu et al. 2003; Herisanu and Marinca 2010a, b)

$$\frac{d^2u}{dt^2} + u + \alpha u^2 \frac{d^2u}{dt^2} + \alpha u \left(\frac{du}{dt}\right)^2 + \beta u^3 = 0, \tag{1}$$

subject to the initial conditions

$$u(0) = A, \quad \frac{du}{dt}(0) = 0. \tag{2}$$

By considering the nonlinear oscillator, Eq. (1), the following homotopy can be constructed:

$$u'' + \omega^2 u + p[\alpha u^2 u'' + \alpha u u'^2 + \beta u^3 + (1 - \omega^2)u] = 0, \tag{3}$$

where $p \in [0, 1]$ and ω is an unknown angular frequency of the nonlinear oscillator which is further to be determined. When $p = 0$, Eq. (3) becomes the linearized equation, $u'' + \omega^2 u = 0$. When $p = 1$, it turns out to be the original one.

Let us consider that the periodic solution to Eq. (1) may be written as a power series in p:

$$u = u_0 + pu_1 + p^2 u_2 + \cdots . \tag{4}$$

Substituting Eq. (4) into Eq. (3) and equating the coefficients of p^0 and p^1, we obtain

$$u_0'' + \omega^2 u_0 = 0, \quad u_0(0) = A, \quad u_0'(0) = 0, \tag{5}$$

and

$$u_1'' + \omega^2 u_1 + \alpha u_0^2 u_0'' + \alpha u_0 u_0'^2 + \beta u_0^3 + (1 - \omega^2) u_0 = 0, \quad u_1(0) = 0, \quad u_1'(0) = 0. \tag{6}$$

The solution of Eq. (5) is $u_0 = A\cos\omega t$, where ω will be determined from the variational formulation for u_1, which reads:

$$J(u_1) = \int_0^T \left\{ -\frac{1}{2} u_1'^2 u_1'' + \frac{1}{2}\omega^2 u_1^2 + \alpha u_0^2 u_0'' u_1 + \alpha u_0 u_0'^2 u_1 + \beta u_0^3 u_1 + (1 - \omega^2) u_0 u_1 \right\} dt,$$

$$T = \frac{2\pi}{\omega}. \tag{7}$$

In previous article (Akbarzade and Khan 2012), a trial function was chosen in the following form:

$$u_1(t) = B\left(\cos\omega t - \frac{1}{3}\cos 5\omega t\right) \tag{8}$$

The accuracy of the first-order approximate solution, Akbarzade and Khan (2012) was chosen the trial function in the following form:

$$u_1(t) = B_1\left(\cos\omega t - \frac{1}{3}\cos 3\omega t\right) + B_3\left(\frac{1}{3}\cos 3\omega t - \frac{3}{5}\cos 5\omega t + \frac{5}{7}\cos 7\omega t\right) \tag{9}$$

Here, we observe that the trial functions Eqs. (8)–(9) are not satisfied the initial conditions $u_1(0)=0, u'_1(0)=0$ when substitutes $t = 0$ in the Eqs. (8)–(9). It is the main shortcomings of the article Akbarzade and Khan (2012).

In this paper, the limitation of the article Akbarzade and Khan (2012) has been removed by choosing a simple new trial function in the following form:

$$u_1(t) = B\left(\cos\omega t - \frac{1}{3}\cos 3\omega t - \frac{2}{3}\cos 5\omega t\right). \tag{10}$$

The new trial function given in Eq. (10) is satisfied the initial conditions $u_1(0)=0$, $u'_1(0)=0$. The trial function given in Eq. (10) makes the solution rapidly converges; furthermore, the determination of first-order approximation is very easy.

Substituting u_1 into functional Eq. (7), we obtain the following result:

$$J(A,B,\omega) = \frac{B\pi\left(9A + 6A^3\beta - (9A + 52B + 3\alpha A^3)\omega^2\right)}{9\omega}. \quad 11$$

Setting:

$$\frac{\partial J}{\partial B} = 0, \quad \frac{\partial J}{\partial \omega} = 0. \quad 12$$

Solving Eq. (12), we obtain the first approximate frequency as a function of amplitude as

$$\omega = \omega_0 = \sqrt{\frac{3 + 2A^2\beta}{3 + \alpha A^2}}, \quad 13$$

where ω_0 is the first-order analytical approximate frequency.

Therefore, the first-order approximate solution of Eq. (1) becomes

$$u(t) = A\cos\omega\, t, \quad 14$$

where ω is given in Eq. (13).

Thus, the determination of first-order approximation is very easy and straightforward. On the other hand, the determination of second-order approximation of the article (Herisanu and Marinca 2010a, b) is very laborious process; thus, seems to be complex.

RESULTS AND DISCUSSION

An analytical coupled technique [combining of the homotopy perturbation method (He 2004) and variational method (He 2007)], along with a simple new trial function, has been presented to determine the approximate frequency and the corresponding periodic solution of the above-mentioned nonlinear oscillator given by Eq. (1). Recently, some authors (Akbarzade and Khan 2012; Hamdan and Dado 1997; Wu et al. 2003; Herisanu and Marinca 2010a, b) have determined the approximate frequencies and the periodic solutions of such nonlinear oscillator. They (Akbarzade and Khan 2012; Hamdan and Dado 1997; Wu et al. 2003; Herisanu and Marinca 2010a, b) were obtained second-order approximation because their first-order approximation did not provide better result. On the other hand, the solution procedures of the article (Herisanu and Marinca 2010a, b) are not easy and it is very laborious process also. In this situation, the first-order approximation of the present paper gives significantly better result than other existing second-order approximations (Hamdan and Dado 1997; Wu et al. 2003).

To verify the efficiency and accuracy of the present method, the approximate frequencies have been obtained for several amplitudes when $\alpha = \beta = 1$ and $\alpha = \beta = 2$ and have been compared those results with other existing harmonic balance methods (Hamdan and Dado 1997; Wu et al. 2003). All results are shown respectively in Tables 1 and 2. The absolute relative errors of the present paper (first-order frequencies) have been compared with the numerical frequency and give less than 0.00 % in the limit as $A \to \infty$ whereas the absolute relative errors of the first and second-order analytical approximations (obtained by Hamdan and Dado 1997) give less than 13.40 and 10.33 %, respectively. On the other hand, the absolute relative errors of the first and second-order analytical approximations (obtained by Wu et al. 2003) give less than 13.40 and 3.72 %, respectively. Thus, the convergent rate of the present method is very faster than Hamdan and Dado (1997); Wu et al. (2003).

Table 1: Comparison between the numerical frequency ω, the approximate frequency obtained by present method (given in Eq. 13) and other existing frequencies (Hamdan and Dado 1997; Wu et al. 2003) for $\alpha = \beta = 1$ as well as several large amplitudes

A	Numerical frequency ω_e	Hamdan and Dado (1997) (error%)		Wu et al. (2003) (error%)		Present method (error%) ω_0
		ω_0	ω_1	ω_0	ω_1	
5	1.34288	1.20953	1.24841	1.20953	1.32217	1.37581
		9.93	7.04	9.93	1.54	2.45
10	1.38928	1.22074	1.26285	1.22074	1.35084	1.40388
		12.13	9.10	12.13	2.77	1.05
15	1.40138	1.22295	1.26570	1.22295	1.35672	1.40955
		12.73	9.68	12.73	3.19	0.58
20	1.40632	1.22373	1.26671	1.22373	1.35883	1.41158
		12.98	9.92	12.98	3.38	0.38
25	1.40883	1.22409	1.26718	1.22409	1.35981	1.41252
		13.11	10.05	13.11	3.48	0.26
30	1.41029	1.22429	1.26743	1.22429	1.36035	1.41304
		13.19	10.13	13.19	3.54	0.20
50	1.41261	1.22458	1.26781	1.22458	1.36113	1.41379
		13.31	10.25	13.31	3.65	0.08
100	1.41375	1.22470	1.26797	1.22470	1.36147	1.41411
		13.37	10.31	13.37	3.70	0.03
200	1.41408	1.22473	1.26801	1.22473	1.36155	1.41419
		13.39	10.33	13.39	3.72	0.00
500	1.41419	1.22474	1.26802	1.22474	1.36157	1.41421
		13.40	10.33	13.40	3.72	0.00
1000	1.41421	1.22474	1.26802	1.22474	1.36157	1.41421
		13.40	10.33	13.40	3.72	0.00

Table 2: Comparison between the numerical frequency ω, the approximate frequency obtained by present method (given in Eq. 13) and other existing frequency (Hamdan and Dado 1997; Wu et al. 2003) for $\alpha = \beta = 2$ as well as several large amplitudes.

A	Numerical frequency ω_e	Hamdan and Dado (1997) (error%)		Wu et al. (2003) (error%)		Present method (error%) ω_0
		ω_0	ω_1	ω_0	ω_1	
5	1.37132	1.21687	1.25786	1.21687	1.34073	1.39406
		11.26	8.27	11.26	2.23	1.66
10	1.40006	1.22272	1.26541	1.22272	1.35613	1.40898
		12.67	9.61	12.67	3.14	0.64
15	1.40707	1.22384	1.26685	1.22384	1.35913	1.41187
		13.02	9.96	13.02	3.41	0.34
20	1.40986	1.22424	1.26736	1.22424	1.36020	1.41289
		13.17	10.11	13.17	3.52	0.22
25	1.41127	1.22442	1.26760	1.22442	1.36069	1.41337
		13.24	10.18	13.24	3.59	0.15
30	1.41207	1.22452	1.26773	1.22452	1.36096	1.41363
		13.28	10.22	13.28	3.62	0.11
50	1.41335	1.22466	1.26791	1.22466	1.36135	1.41400
		13.35	10.29	13.35	3.68	0.05
100	1.41397	1.22472	1.26799	1.22472	1.36152	1.41416
		13.39	10.32	13.39	3.71	0.01
200	1.41414	1.22474	1.26801	1.22474	1.36156	1.41420
		13.40	10.33	13.40	3.72	0.00
500	1.41420	1.22474	1.26802	1.22474	1.36157	1.41421
		13.40	10.33	13.40	3.72	0.00
1000	1.41421	1.22474	1.26802	1.22474	1.36158	1.41421
		13.40	10.34	13.40	3.72	0.00

Next, the approximate solution of Eq. (1) has been determined by using present method and harmonic balance method (Wu et al. 2003) for $\alpha = \beta = 1$, $A = 10$ and shown in Fig. 1. Finally, the approximate solution of Eq. (1) has been determined by using present method and harmonic balance method (Wu et al. 2003) for $\alpha = \beta = 2$, $A = 10$ and shown in Fig. 2. All figures include numerical solution obtained by fourth order Runge–Kutta method.

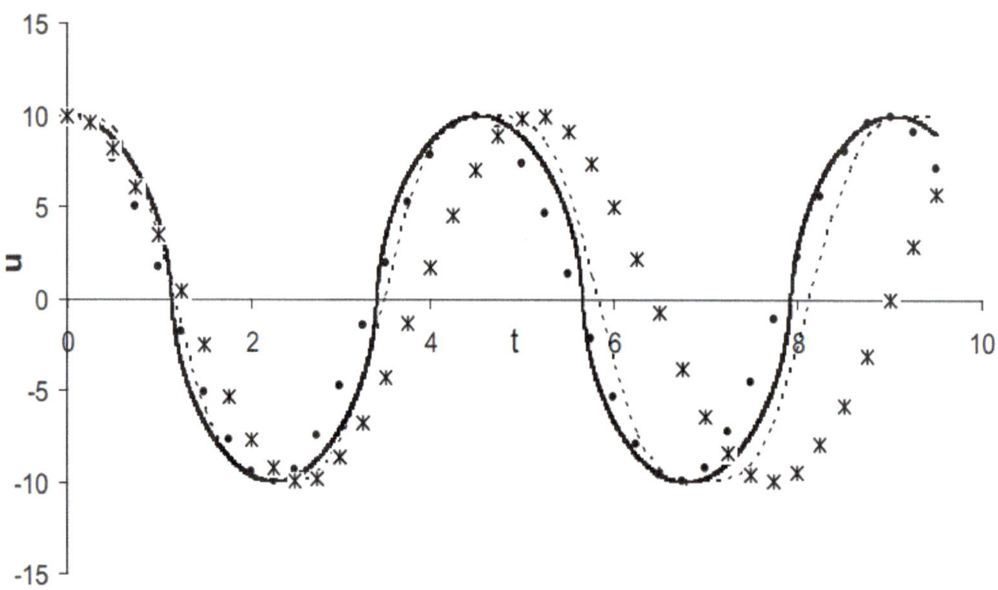

Figure 1: Comparison of the analytical approximate periodic solution obtained by present method (denoting by circles line) with numerical solution obtained by fourth order Runge–Kutta method (denoted by solid line) and also with the first-order (denoted by cross lines) as well as second-order (denoted by dash line) approximations obtained by harmonic balance method (Wu et al. 2003) for $\alpha = \beta = 1$, $A = 10$.

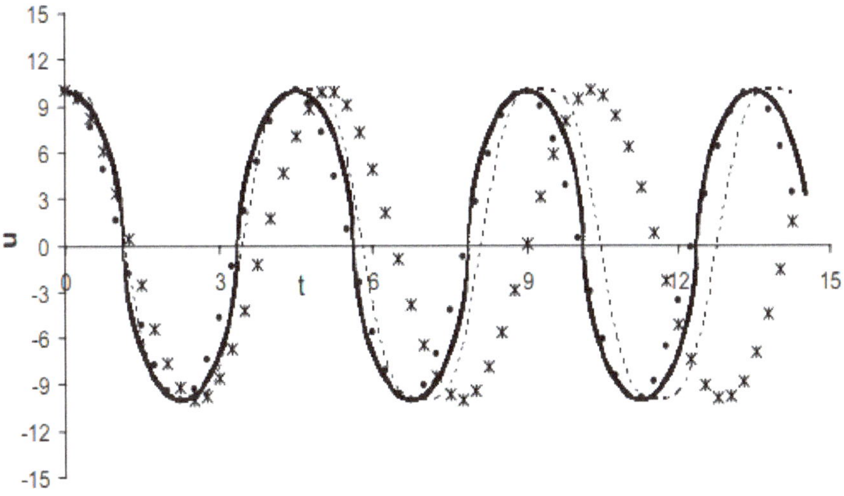

Figure 2: Comparison of the analytical approximate periodic solution obtained by present method (denoting by circles line) with numerical solution obtained by fourth order Runge–Kutta method (denoted by solid line) and also with the first-order (denoted by cross lines) as well as second-order (denoted by dash line) approximations obtained by harmonic balance method (Wu et al. 2003) for $\alpha = \beta = 2$, $A = 10$.

From all the figures, we see that the first-order approximate solution obtained by harmonic balance method deviates from numerical solution. Moreover, the second-order approximate solution obtained by harmonic balance method does not better agreement with the corresponding numerical solution. On the other hand, the first-order approximate solution obtained by present method gives excellent agreement with the corresponding numerical solution. Therefore, the present method is suitable for solving Eq. (1) than Akbarzade and Khan (2012); Hamdan and Dado (1997); Wu et al. (2003); Herisanu and Marinca (2010a, b) for strong nonlinearity as well as large amplitudes of oscillation.

CONCLUSION

In this paper, a simple analytical technique has been presented to solve of nonlinear oscillations of planar, flexural large amplitudes free vibration of a slender, inextensible cantilever beam carrying a lumped mass with rotary inertia at an intermediate position along its span. Generally, the first-order approximation is considered in this paper. The first-order approximation gives rapidly converges to the corresponding numerical solution. The present method gives better result than other existing results for large amplitudes of oscillation. It has been proved that the present method is very effective and convenient and provides more accurate result for solving strongly nonlinear oscillators.

ACKNOWLEDGEMENTS

The authors are grateful to the reviewers for their helpful comments/suggestions in improving the manuscript. The authors are also grateful to Mr. Harun Or Rashid, Lecturer (English), for his help in language.

Competing interests

The author declares that they have no competing interests.

REFERENCES

1. Akbarzade M (2010) Coupled method of homotopy perturbation method and variational approach for solution to nonlinear cubic-quintic duffing oscillator. Adv Theor Appl Mech 3(7):329–337
2. Akbarzade M, Khan Y (2012) Dynamic model of large amplitude non-linear oscillations arising in the structural engineering: analytical solutions. Math Comput Model 55:480–489
3. Akindeinde SO (2015) Homotopy perturbation method for the strongly nonlinear Darcy-Forscheimer model. Math Theor Model 5(9):78–84
4. Aminikhaha H, Hemmatnezhad M (2011) An effective modification of the homotopy perturbation method for stiff systems of ordinary differential equations. Appl Math Lett 24:1502–1508
5. Bogoliubov NN, Mitropolskii YA (1961) Asymptotic methods in the theory of nonlinear oscillations. Gordan and Breach, New York
6. Ganji DD, Sadighi A (2006) Application of He's homotopy-perturbation method to nonlinear coupled systems of reaction-diffusion equations. Int J Non Sci Num Simul 7:411–418
7. Ghorbani A, Nadjafi JS (2007) He's homotopy perturbation method for calculating adomian polynomials. Int J Non Sci Num Simul 8:229–232
8. Hamdan MN, Dado MHF (1997) Large amplitude free vibrations of a uniform cantilever beam carrying an intermediate lumped mass and rotary inertia. J Sound and Vib 206:151–168
9. He JH (1999) Homotopy perturbation technique. Comput Methods Appl Mech Eng 3:257–262
10. He JH (2000) New perturbation technique which is also valid for large parameters. J Sound Vib 229:1257–1263
11. He JH (2002) Preliminary report on the energy balance for nonlinear oscillations. Mech Res Commun 29:107–111
12. He JH (2004) The homotopy perturbation method for nonlinear oscillators with discontinuous. Appl Math Comput 151:287–292

13. He JH (2006) Some asymptotic methods for strongly nonlinear equations. Int J Mod Phys B 20:1141–1199
14. He JH (2007) Variational approach for nonlinear oscillators. Chaos Solitions Fractals 34(5):1430–1439
15. He JH, Wu GC, Austin F (2010) The variational iterative method which should be followed. Nonlinear Sci Lett A 1(1):1–30
16. Herisanu N, Marinca V (2010a) A modified variational iterative method for strongly nonlinear oscillators. Nonlinear Sci Lett A 1(2):183–192
17. Herisanu N, Marinca V (2010b) Explicit analytical approximation to large-amplitude non-linear oscillations of a uniform cantilever beam carrying an intermediate lumped mass and rotary inertia. Meccanica 45:847–855
18. Herisanu N, Marinca V (2012) Optimal homotopy perturbation method for non-conservative dynamical system of a rotating electrical machine. Zeitschrift fur Naturforschung A 67:509–516
19. Kaya MO, Durmaz S, Demirbag SA (2010) He's variational approach to multiple coupled nonlinear oscillators. Int J Non Sci Num Simul 11(10):859–865
20. Khan Y, Mirzabeigy A (2014) Improved accuracy of He's energy balance method for analysis of conservative nonlinear oscillator. Neural Comput Appl 25:889–895
21. Khan Naj A, Ara A, Khan Nad A (2014) On solutions of the nonlinear oscillators by modified homotopy perturbation method. Math Sci Lett 3(3):229–236
22. Khan Y, Faraz N, Yildirim A (2011) New soliton solutions of the generalized Zakharov equations using He's variational approach. Appl Math Lett 24:965–968
23. Khan Y, Akbarzadeb M, Kargar A (2012) Coupling of homotopy and the variational approach for a conservative oscillator with strong odd-nonlinearity. Sci Iran A 19(3):417–422
24. Khan Y, Fardib M, Boroujenib FH (2013) Nonlinear oscillators with rational terms: a new semi-analytical technique. Sci Iran A 20(4):1153–1160
25. Krylov NN, Bogoliubov NN (1947) Introduction to nonlinear mechanics. Princeton University Press, New Jersey
26. Liao SJ (2003) Beyond perturbation: introduction to homotopy analysis method. CRC Press, Boca Raton
27. Marinca V, Herisanu N (2010) Optimal homotopy perturbation method for strongly nonlinear differential equations. Non Sci Lett A 1:273–280
28. Marinca V, Herisanu N (2011) Nonlinear dynamic analysis of an electrical machine rotor-bearing system by the optimal homotopy perturbation method. Comput Math Appl 61:2019–2024
29. Marion JB (1970) Classical dynamics of particles and system. Harcourt Brace Jovanovich, San Diego
30. Nayfeh AH (1973) Perturbation methods. Wiley, New York
31. Nayfeh AH (1981) Introduction to perturbation techniques. Wiley, New York
32. Nayfeh AH, Mook DT (1979) Nonlinear oscillations. Wiley, New York
33. Rafei M, Ganji DD, Daniali H (2007) Solution of the epidemic model by homotopy perturbation method. Appl Math Comput 187:1056–1062
34. Suleman M, Wu Q (2015) Comparative solution of nonlinear quintic cubic oscillator using modified homotopy perturbation method. Adv Math Phys. doi:10.1155/2015/932905 (Article ID 932905)
35. Wang F, Li W, Zhang H (2012) A new extended homotopy perturbation method for nonlinear differential equations. Math Comput Model 55:1471–1477
36. Wu BS, Lim CW, Ma YF (2003) Analytical approximation to large-amplitude oscillation of a non-linear conservative system. Int J Nonlinear Mech 38:1037–1043

CHAPTER 10

HOMOTOPY PERTURBATION METHOD

S. Narayanamoorthy and S. P. Sathiyapriya

Keywords: Homotopy perturbation method, Fuzzy Volterra integral equations, Numerical technique and algorithm, Approximate solutions, Error analysis

BACKGROUND

Integral equations find special applicability within many scientific and mathematical disciplines. Indeed modeling physical problems using integral equations with the exact parameters is not always easy but also impossible in the real problems. For this purpose, one way is using uncertainty measures for handling such lack of information. One of the most recent approaches is using fuzzy concept (Zadeh 1965). So instead of using deterministic models of integral equations, we can go in for fuzzy integral equations. Hence there occurs a need to develop mathematical models and numerical procedures that would appropriately treat general fuzzy integral equations and solve them. In this paper, we apply homotopy perturbation method (HPM) to solve both linear and nonlinear fuzzy Volterra integral equations of the second kind (FVIE-2). Many research papers dealing with fuzzy integral equations exists in open literatures and some of them are reviewed and cited here for better understanding of the present analysis. We know that solving fuzzy integral equations requires appropriate definitions of fuzzy function and fuzzy integral of a fuzzy function. We refer to the reader (Dubois and Prade 1978) for basic arithmetic operations on fuzzy numbers and he also presented the elementary fuzzy calculus (Dubois and Prade 1982) based on extension principle. Alternative approaches were later suggested by others (Goetschel and Voxman 1986; Kaleva 1987). Various methods and applications of linear and nonlinear integral equations were also reported (Wazwaz 2011). The fuzzy integral equations (Friedman et al. 1999; Subrahmanyam and Sudarsanam 1996) is one of the most important fields of fuzzy set theory. The concepts of fuzzy integral equations and fuzzy integro-differential equations have motivated a large amount of research works (Mirzaee et al.2012; Hussain and Ali 2013a, b; Mosleh and Otadi 2013; Allahviranloo et al. 2014; Otadi and Mosleh 2014) in last decades because of its applications in scientific phenomena. Therefore investigating fuzzy integral equations by finding accurate and efficient methods for solving these equations has become an active research undertaking.

Homotopy perturbation method (He 1999) is a perturbation technique coupled with the homotopy technique was developed by He JH and was further improved by him (He 2000,2003, 2004). In homotopy perturbation method, a complicated problem under study is continuously deformed into a simple problem which is easy to solve to obtain an approximate solution. HPM yields a very rapid convergence of the solution series in most cases, usually only few iterations lead to very accurate solutions. This method has been applied to many problems (Chun 2007; Filobello-Nino et al. 2014b) and also in particular to integral equations (Abbasbandy 2007) and fuzzy integral equations (Matinfar and Saeidy 2010). In recent past, various analytical and numerical methods were used such as Adomian decomposition method, direct computation method, series solution method, successive approximation method and conversion to integral equations. However, these methods are not easy to use and require tedious calculations. But HPM was proved (Saberi-Nadjafi and Ghorbani 2009) to be an effective and reliable tool for handling most of the linear and nonlinear differential, ordinary and partial, as well as linear and nonlinear integral equations. The main reason consists in the fact that this gives flexibility in the choice of basis functions for the solution and does not involve linear inversion operators (as compared to the Adomian decomposition method), while still retaining a simplicity that makes the method easily understandable from the standpoint of general perturbation methods.

Citation: S. Narayanamoorthy, S. P. Sathiyapriya, Homotopy perturbation method, DOI: 10.1186/s40064-016-2038-3.

Fuzzy integral equations are important in studying and solving a large proportion of problems in various topics in applied mathematics and it also arise in many industrial fields, such as electromagnetic fields and thermal problems. It is well-known that a large class of initial and boundary value problems can be converted to Volterra integral equations and also the problem of heat conduction with a variable heat transfer coefficient is reduced to the solution of Volterra integral equations of the second kind. Several authors applied HPM to solve linear and nonlinear partial differential equations of fractional order (Momani and Odibat 2007a, b), Volterra integral equations (Grover and Tomer 2011), singularly perturbed Volterra integral equations (Alnasr and Momani 2008) and n-th order fuzzy linear differential equations (Tapaswini and Chakraverty 2013). The reader is also referred to two recent papers where the authors (Filobello-Nino et al. 2014a; Vazquez-Leal and Sarmiento-Reyes 2015) make use of HPM in their application oriented works. In the very recent studies, it has been found that homotopy perturbation method was also used to solve fractional fisher's equation (Hamdi Cherif et al. 2016) and a system of nonlinear chemistry problems (Ramesh Rao 2016). Taking into account all these specifications, the proposed work is mainly dealt with fuzzy Volterra integral equations. Up to authors' knowledge no research has been carried out so far by constructing a numerical algorithm using HPM for solving fuzzy Volterra integral equations of the second kind and our work is aimed at these.

The paper is organized as follows: the fuzzy Volterra integral equations are recalled with the required theorem and the basic concept of HPM is provided. Our main results are stated in the numerical technique, which presents the detailed description of the proposed method. Besides we introduce the algorithm for solving FVIE-2. Additionally test problems are included in the section 'Illustrative examples' and concluding results are made over them. Finally, a brief conclusion is provided.

Remark

Some of the basic notations and definitions that are not described in this paper are standard and usual. One can refer to the authors (Dubois and Prade 1978; Goetschel and Voxman 1983,1986; Puri and Ralescu 1983, 1986; Kaleva 1987) for the notion of fuzzy numbers and its arithmetic operations, Hausdorff distance between fuzzy numbers, continuity and existence of definite integral of fuzzy functions.

FUZZY VOLTERRA INTEGRAL EQUATIONS

The integral equations which are discussed in this section are the Volterra integral equations of the second kind. The Fredholm integral equation of the second kind (Hochstadt 1973) is given by

$$F(x) = f(x) + \lambda \int_a^b k(x,t)F(t)dt \tag{1}$$

where $\lambda > 0$, a and b are constants, k(x, t) is an arbitrary kernel function over the square $a \le x,t \le b$ and f(x) is a function of $a \le x \le b$.

If the kernel satisfies k(x, t) = 0, x > t,

we obtain the Volterra integral equation of the second kind and is given by

$$F(x) = f(x) + \lambda \int_a^x k(x,t)F(t)dt \tag{2}$$

In the above equation a refers to constant and x is a variable. If f(x) is a crisp function, then the solutions of the following equations

$$\left(\underline{\int_a^b f(x;\alpha)dx}\right) = \int_a^b \underline{f}(x;\alpha)dx \tag{3}$$

$$\left(\overline{\int_a^b f(x;\alpha)dx}\right) = \int_a^b \overline{f}(x;\alpha)dx \tag{4}$$

are crisp as well, where $(\underline{f}(x;\alpha), \overline{f}(x;\alpha))$ is the parametric form of f(x). However, if f(x) is a fuzzy function these equations possess fuzzy solutions.

Let $F: I \rightarrow E^1$ be a fuzzy function. The integral of F over I, denoted by $\int_I F(x)dx$ or $\int_a^b F(x)dx$, is defined levelwise by

$$\left[\int_I F(x)dx\right]_\alpha = \int_I F_\alpha(x)dx = \left\{\int_I f(x)dx | f: I \rightarrow R \text{ is a measurable function for } F_\alpha\right\}$$

for each $0 \le \alpha \le 1$.

Now the parametric form of fuzzy Volterra integral equation of the second kind (FVIE-2) are as follows (Ghanbari 2010)

Let $(\underline{f}(x,\alpha), \overline{f}(x;\alpha))$ and $(\underline{F}(x;\alpha), \overline{F}(x;\alpha))$, be the parametric forms of f(x) and F(x) respectively, $0 \le \alpha \le 1$, $a \le x \le b$. Then the parametric forms of FVIE-2 are given by

$$\underline{F}(x,\alpha) = \underline{f}(x,\alpha) + \lambda \int_a^x \underline{k(x,t)F(t,\alpha)}dt \tag{5}$$

$$\overline{F}(x,\alpha) = \overline{f}(x,\alpha) + \lambda \int_a^x \overline{k(x,t)F(t,\alpha)}dt \tag{6}$$

for $0 \le \alpha \le 1$, where

$$\underline{k(x,t)F(t,\alpha)} = \begin{cases} k(x,t)\underline{F}(t,\alpha) & k(x,t) \ge 0 \\ k(x,t)\overline{F}(t,\alpha) & k(x,t) < 0 \end{cases} \tag{7}$$

and

$$\overline{k(x,t)F(t,\alpha)} = \begin{cases} k(x,t)\overline{F}(t,\alpha) & k(x,t) \geq 0 \\ k(x,t)\underline{F}(t,\alpha) & k(x,t) < 0 \end{cases} \tag{8}$$

The following theorem provides sufficient conditions for the existence of a unique solution to Eq. (1), where f(x) is a fuzzy function.

Theorem

(Wu and Ma 1990) Let k(x,t) be continuous for $a \leq x, t \leq b$, $\lambda > 0$, and f(x) is a fuzzy continuous function of x, $a \leq x \leq b$. If $\lambda < \frac{1}{M(b-a)}$, where $M = \max_{(a \leq x, t \leq b)} |k(x,t)|$, then the iterative procedure

$$F_0(x) = f(x),$$

$$F_k(x) = f(x) + \lambda \int_a^b k(x,t)F_{k-1}(t)dt, \quad k \geq 1, \tag{9}$$

converges to the unique solution of Eq. (1). Specifically,

$$\sup_{(a \leq x \leq b)} D(F(x), F_k(x)) \leq \frac{L^k}{1-L} \sup_{(a \leq x \leq b)} D(F_0(x), F_1(x)), \tag{10}$$

where $L = \lambda M(b-a)$. This refers that $F_k(x)$ converges uniformly in x to F(x), i.e., given arbitrary $\varepsilon > 0$ we can find N such that

$$D(F(x), F_k(x)) < \varepsilon, \quad a \leq x \leq b, \ k > N.$$

The proof this theorem can be easily extended for fuzzy Volterra integral equations of the second kind, i.e., for Eq. (2) where f(x) is a fuzzy function as well.

Basic Concept of Homotopy Perturbation Method and its Application to Volterra Integral Equations

For convenience of the reader, we present the review of the HPM. The essential idea of this method is to introduce a homotopy parameter, say p, which takes the values from 0 to 1. When p = 0, the system of equation usually reduces to a sufficiently simplified form, which normally admits a rather simple solution. As p gradually increases to 1, the system goes through a sequence of deformation, the solution of each of which is close to that at the previous stage of deformation. Eventually at p = 1, the system takes the original form of the equation and final stage of deformation gives the desired solution.

To figure out how HPM works, consider the nonlinear integral equation

$$L(u)+N(u)=f(r), \ r\epsilon\Omega \tag{11}$$

with boundary conditions $B(u, \frac{\partial u}{\partial n}) = 0, r\epsilon\Gamma$,

where L is a linear operator, N is a nonlinear operator, B is a boundary operator, Γ is the boundary of the domain Ω and f(r) is a known analytic function. In order to use the HPM, a suitable construction of homotopy is of vital importance. If L(u) = 0 with some possible unknown parameter can best describe the original nonlinear system. Generally, a homotopy can be constructed as (He 1999, 2000)

$$H(U,p) = (1-p)[L(U) - L(u_0)] + p[L(U) + N(U) - f(r)] = 0 \tag{12}$$

or

$$H(U,p) = L(U) - L(u_0) + p[L(u_0) + N(U) - f(r)] = 0, \tag{13}$$

where $p\epsilon[0,1]$ and $r\epsilon\Omega$.

The nonlinear Volterra integral equation is given by

$$\gamma(x) = f(x) + \int_a^x k(x,t)\{R(\gamma(t)) + N(\gamma(t))\}dt, \quad a \leq x, \ t \leq b \tag{14}$$

where $\gamma(x)$ is an unknown function that will be determined, k(x, t) is the kernel of the integral equation, f(x) is an analytic function, $R(\gamma)$ and $N(\gamma)$ are linear and nonlinear function of γ respectively.

To illustrate the HPM, we reconstitute Eq. (14) as

$$L(u) = u(x) - f(x) - \int_a^x k(x,t)\{R(u(t)) + N(u(t))\}dt, \tag{15}$$

with the solution $u(x) = \gamma(x)$ and we define the homotopy H(U, p) by

$$H(U,0) = F(u), \quad H(U,1) = L(u), \tag{16}$$

where F(u) is a functional operator with solution, say u_0, which can be obtained easily. We can choose a homotopy

$$H(u,p) = (1-p)F(u) + pL(u) = 0, \tag{17}$$

and continuously trace an implicitly defined curve from a starting point $H(u_0,0)$ to a solution function $H(\gamma, 1)$. The embedding parameter p monotonically increases from 0 to 1 as the trivial problem F(u) = 0 continuously deformed to the original problem L(u) = 0. The embedded parameter $p\epsilon[0,1]$ can be considered as an expanding parameter.

$$u = \sum_{n=0}^{\infty} p^n u_n \tag{18}$$

When $p \to 1$, Eq. (17) corresponds to Eq. (15) and Eq. (18) becomes the approximate solution of Eq. (14), i.e.,

$$\gamma(x) = \lim_{p \to 1} u = \sum_{n=0}^{\infty} u_n \tag{19}$$

The above series is convergent in most of the cases and also the rate of convergence depends on L(u) (He 2000).

In other words, it is very natural to assume that the solution of [Eqs. (16) and (17)] can be expressed as

$$\gamma = u_0 + p^1 u_1 + p^2 u_2 + \cdots \tag{20}$$

Equating the terms with identical powers of p, we get

$$p^0 : u_0(x) = f(x)$$

$$p^1 : u_1(x) = \int_a^x k(x,t) u_0(t) dt$$

$$\vdots$$

$$p^n : u_n(x) = \int_a^x k(x,t) u_{n-1}(t) dt \text{ and so on.} \tag{21}$$

Therefore when p = 1, the approximate solution of above equation can be readily obtained as follows

$$\gamma = \lim_{p \to 1} u_0 + u_1 + u_2 + \cdots \tag{22}$$

The above series is convergent for most cases.

PROPOSED NUMERICAL TECHNIQUE FOR SOLVING FVIE-2 USING HPM

As HPM is considered as a combination of the classical perturbation technique and the homotopy (whose origin is in the topology), but not restricted to small parameters as occur with traditional perturbation methods, because HPM neither requires small parameter nor linearization, but only few iterations to obtain highly accurate solution. Hence we introduce the recursive scheme for solving FVIE-2 as follows

In view of homotopy, we can define the following convex homotopy,

$$\begin{cases} (1-p)\left[\underline{V}(x,\alpha) - \underline{u}_0(x,\alpha)\right] + p\left[\underline{V}(x,\alpha) - \underline{f}(x,\alpha) - \int_a^x k(x,t)\underline{V}(x,\alpha)dt\right] = 0 \\ (1-p)\left[\bar{V}(x,\alpha) - \bar{u}_0(x,\alpha)\right] + p\left[\bar{V}(x,\alpha) - \bar{f}(x,\alpha) - \int_a^x k(x,t)\bar{V}(x,\alpha)dt\right] = 0 \end{cases} \tag{23}$$

where α is the fuzzy parameter $(0 \le \alpha \le 1)$.

The solution of Eq. (23) is assumed in the following form

$$\begin{cases} \underline{V}(x,\alpha) = \sum_{i=0}^{\infty} p^i \underline{V}_i(x,\alpha) \\ \bar{V}(x,\alpha) = \sum_{i=0}^{\infty} p^i \bar{V}_i(x,\alpha) \end{cases} \tag{24}$$

where $(\underline{V}_i, \bar{V}_i)\forall i$ are unknown functions to be determined.

The initial approximation is taken as

$$\begin{cases} \underline{V}(0,\alpha) = \underline{u}_0(0,\alpha) = \underline{f}(x,\alpha) \\ \bar{V}(0,\alpha) = \bar{u}_0(0,\alpha) = \bar{f}(x,\alpha) \end{cases} \tag{25}$$

Substituting Eqs. (24) and (25) in Eq. (23) reduces to

$$\begin{cases} \underline{V}(x,\alpha) = \underline{u}_0(x,\alpha) + p\int_a^x k(x,t)\underline{V}(x,\alpha)dt \\ \bar{V}(x,\alpha) = \bar{u}_0(x,\alpha) + p\int_a^x k(x,t)\bar{V}(x,\alpha)dt \end{cases} \tag{26}$$

Equating the coefficients with like powers of p, we yield the following iterations

$$p^0 : \begin{cases} \underline{V}(0,\alpha) = \underline{f}(x,\alpha) \\ \bar{V}(0,\alpha) = \bar{f}(x,\alpha) \end{cases} \tag{27}$$

$$p^1 : \begin{cases} \underline{V}_1(x,\alpha) = \int_a^x k(x,t)\underline{f}(t,\alpha)dt \\ \bar{V}_1(x,\alpha) = \int_a^x k(x,t)\bar{f}(t,\alpha)dt \end{cases} \tag{28}$$

$$p^2 : \begin{cases} \underline{V}_2(x,\alpha) = \int_a^x k(x,t)\underline{V}_1(t,\alpha)dt \\ \bar{V}_2(x,\alpha) = \int_a^x k(x,t)\bar{V}_1(t,\alpha)dt \end{cases}$$

$$\vdots$$

and so on. (29)

Finally, the solution of FVIE-2 is given by

$$\begin{cases} \underline{u}(x,\alpha) = \lim_{p\to 1} \underline{V}(x,\alpha) = \sum_{i=0}^{\infty} \underline{V}_i(x,\alpha) \\ \bar{u}(x,\alpha) = \lim_{p\to 1} \bar{V}(x,\alpha) = \sum_{i=0}^{\infty} \bar{V}_i(x,\alpha) \end{cases} \tag{30}$$

In general, we obtain the following iteration formulae

$$p^{i+1} : \begin{cases} \underline{V}_{i+1}(x,\alpha) = \int_a^x k(x,t)\underline{V}_i(t,\alpha)dt \\ \bar{V}_{i+1}(x,\alpha) = \int_a^x k(x,t)\bar{V}_i(t,\alpha)dt \end{cases} \tag{31}$$

These relations [Eq. (31)] will enable us to determine the components $\underline{u}(x,\alpha)$ and $\bar{u}(x,\alpha)$ recursively for $i \ge 0$.

ALGORITHM OF THE APPROACH

On the basis of our proposed results using HPM, which are discussed in the above section, we convert to a numerical HPM algorithm that is one of the main results of this paper. Hence, in an algorithmic form, HPM can be expressed and implemented as follows.

The Eqs. (5) and (6) with the iteration index $i \ge 0$, and a appropriate value for the tolerance (here Tol=10−5Tol=10−5).

Main steps

Step 1: For reckoning the initial data, let

$\underline{V}_0 = \underline{V}$ and $\bar{V}_0 = \bar{V}$ where $x = 0$

Step 2: Use the recursive relations [Eq. (31)] to compute the values of $\underline{V}_{i+1}$' by using $\underline{V}_i(x)$ and $\bar{V}_{i+1}(x)$ by using $\bar{V}_i(x)$

Step 3: If $\left|\underline{V}_{i+1}(x) - \underline{V}_i(x)\right| < Tol, max\left|\bar{V}_{i+1}(x) - \bar{V}_i(x)\right| < Tol\}$, , then go to step 4, else set i = i + 1 and go to step 2

Step 4: Print $\underline{u}(x) = \sum_{i=0}^{\infty} \underline{V}_i$ and $\bar{u}(x) = \sum_{i=0}^{\infty} \bar{V}_i$ as the approximates of the exact solutions

Illustrative Examples

To give a clear overview and to demonstrate the efficiency of the homotopy perturbation method, we implement the proposed numerical technique by solving a linear FVIE-2 and a nonlinear FVIE-2 with known exact solutions.

LINEAR FUZZY VOLTERRA INTEGRAL EQUATIONS OF THE SECOND KIND

Example 1

We consider the linear FVIE-2 given by

$$F(x) = f(x) + \lambda \int_a^x k(x,t)F(t)dt, \quad 0 \le x \le 1, 0 \le t \le x$$

where $\lambda = 1$, a = 0, k(x, t) = sinhx and

$$f(x) = \left(\underline{f}(x,\alpha), \bar{f}(x,\alpha)\right)$$
$$= \left\{\left(\cosh x + 1 - \cosh^2 x\right)\left(\alpha^2 + \alpha\right), \left(\cosh x + 1 - \cosh^2 x\right)\left(4 - \alpha^3 - \alpha\right)\right\}, \quad 0 \le \alpha \le 1$$

The exact solution in this case is given by

$$\underline{F}(x,\alpha) = \underline{u}(x,\alpha) = coshx\left(\alpha^2 + \alpha\right)$$

$$\bar{F}(x,\alpha) = \bar{u}(x,\alpha) = coshx\left(4 - \alpha^3 - \alpha\right)$$

Here we have

$$\underline{k(x,t)F(t,\alpha)} = (sinhx)\underline{F}(t,\alpha)$$

$$\overline{k(x,t)F(t,\alpha)} = (sinhx)\bar{F}(t,\alpha)$$

To solve the given equation by HPM we construct a convex homotopy as follows

$$H(\underline{V},p,\alpha) = \underline{V}(x,\alpha) - \left(coshx + 1 - cosh^2x\right)\left(\alpha^2 + \alpha\right) + p\int_a^x k(x,t)\underline{V}(t,\alpha)dt$$

$$H(\bar{V},p,\alpha) = \bar{V}(x,\alpha) - \left(coshx + 1 - cosh^2x\right)\left(4 - \alpha^3 - \alpha\right) + p\int_a^x k(x,t)\bar{V}(t,\alpha)dt$$

Now we make use of the iteration formulae [Eq. (31)] and applying the numerical algorithm, the HPM solution series with few iterative terms are as follows

$p^0 : (\underline{V}_0(x,\alpha), \bar{V}_0(x,\alpha))$

is the initial fuzzy approximations,where

$$\underline{V}_0(x,\alpha) = \underline{f}(x,\alpha) = \left(coshx + 1 - cosh^2x\right)\left(\alpha^2 + \alpha\right)$$

$$\bar{V}_0(x,\alpha) = \bar{f}(x,\alpha) = \left(coshx + 1 - cosh^2x\right)\left(4 - \alpha^3 - \alpha\right)$$

$p^1 : (\underline{V}_1(x,\alpha), \bar{V}_1(x,\alpha)),$

where

$$\underline{V}_1(x,\alpha) = \frac{-1}{4}\alpha(1+\alpha)sinhx(-2x - 4\,sinhx + sinh2x)$$

$$\bar{V}_1(x,\alpha) = \frac{1}{4}\left(-4 + \alpha + \alpha^3\right)sinhx(-2x - 4\,sinhx + sinh2x)$$

In the same way, the iterations $p^2 : \left(\underline{V}_2(x,\alpha), \bar{V}_2(x,\alpha)\right)$ and $p^3 : \left(\underline{V}_3(x,\alpha), \bar{V}_3(x,\alpha)\right)$ were also computed for calculating fuzzy approximate solutions which could not be stated here as the iterations contain very long expressions.

We approximate $\underline{F}(x,\alpha)$ and $\bar{F}(x,\alpha)$ by setting p = 1 and obtain the following fuzzy solutions, which are given by

$$\underline{F}(x,\alpha) = \underline{u}(x,\alpha) = \sum_{i=0}^{3} \underline{V}_i(x,\alpha)$$
$$= \left(coshx + 1 - cosh^2x\right)\left(\alpha^2 + \alpha\right)$$
$$+ \left(\frac{-1}{4}\alpha(1+\alpha)sinhx(-2x - 4\,sinhx + sinh2x)\right) + \cdots$$

and

$$\bar{F}(x,\alpha) = \bar{u}(x,\alpha) = \sum_{i=0}^{3} \bar{V}_i(x,\alpha)$$
$$= \left(coshx + 1 - cosh^2x\right)\left(4 - \alpha^3 - \alpha\right)$$
$$+ \left(\frac{1}{4}\left(-4 + \alpha + \alpha^3\right)sinhx(-2x - 4\,sinhx + sinh2x)\right) + \cdots$$

Error Analysis

Absolute errors are computed as

$$\underline{E}(x,\alpha) = \left|coshx\left(\alpha^2 + \alpha\right) - \underline{u}(x,\alpha)\right|$$

$$\bar{E}(x,\alpha) = \left|coshx\left(4 - \alpha^3 - \alpha\right) - \bar{u}(x,\alpha)\right|$$

The components of iteration formulae [Eq. (31)] were obtained by the Mathematica program (Mathematica package version 5.2) according to the numerical algorithm where the tolerance is given a suitable positive value. The numerical results of the obtained approximate solutions are compared with the exact solutions for different α-values and are presented in Table 1. Moreover exact and approximate solutions are shown graphically in Fig. 1.

Table 1: Exact and approximate solutions with four iterations at x = 0.5

α	Exact solution		Approximate solution		Error	
	$\underline{u}(x,\alpha)$	$\bar{u}(x,\alpha)$	$\underline{u}(x,\alpha)$	$\bar{u}(x,\alpha)$	$\underline{E}(x,\alpha)$	$\bar{E}(x,\alpha)$
0	0	4.51050	0	4.51034	0	0.00016
0.1	0.124039	4.39661	0.124034	4.39645	0.000005	0.00016
0.2	0.270630	4.27596	0.270620	4.27580	0.000010	0.00016
0.3	0.439774	4.14177	0.439758	4.14162	0.000016	0.00015
0.4	0.631471	3.98729	0.631447	3.98714	0.000024	0.00015
0.5	0.845719	3.80574	0.845688	3.80560	0.000031	0.00014
0.6	1.082520	3.59036	1.082480	3.59023	0.000040	0.00013
0.7	1.341870	3.33439	1.341820	3.33427	0.000050	0.00012
0.8	1.623780	3.03106	1.623720	3.03095	0.000060	0.00011
0.9	1.928240	2.67360	1.928170	2.67350	0.000070	0.00010
1	2.255250	2.25525	2.255170	2.25517	0.000080	0.00008

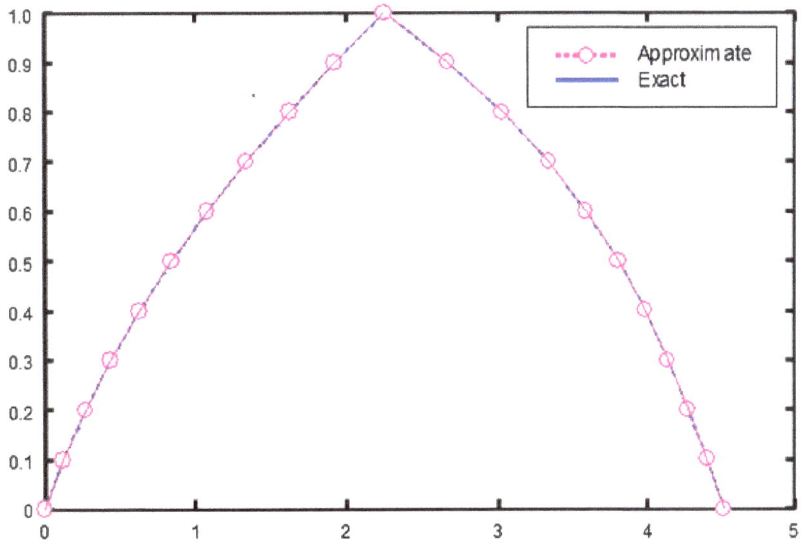

Figure 1: Plots of exact and approximate solutions of Example 1.

NONLINEAR FUZZY VOLTERRA INTEGRAL EQUATIONS OF THE SECOND KIND

Example 2

We consider the nonlinear FVIE-2 given by

$$F(x) = f(x) + \lambda \int_a^a k(x,t)F^2(t)dt, \quad 0 \le x \le 1,\ 0 \le t \le x$$

where $\lambda = 1$, a = 0, k(x, t) = $x^2(1 - 2t)$ and

$$\begin{aligned} f(x) &= \left(\underline{f}(x,\alpha), \bar{f}(x,\alpha)\right) \\ &= \Bigg\{ \left((2-\alpha)^2\left(\frac{1}{2}x^6 + x^5 - x^3 + \frac{11}{32}x^2\right) - \frac{11}{32}x^2 + \alpha x + \alpha \right), \\ &\quad \left(\alpha^2\left(\frac{1}{2}x^6 + x^5 - x^3 + \frac{11}{32}x^2\right) + (2-\alpha)\left(\frac{-11}{32}(2-\alpha)x^2 + x + 1\right) \right) \Bigg\}, \quad 0 \le \alpha \le 1 \end{aligned}$$

The exact solution in this case is given by

$$\underline{F}(x,\alpha) = \underline{u}(x,\alpha) = \alpha(x+1)$$

$$\bar{F}(x,\alpha) = \bar{u}(x,\alpha) = (2-\alpha)(x+1)$$

Here we have

$$\underline{k(x,t)F^2(t,\alpha)} = x^2(1-2t)\underline{F}^2(t,\alpha)$$

$$\overline{k(x,t)F^2(t,\alpha)} = x^2(1-2t)\bar{F}^2(t,\alpha)$$

To solve the given equation by HPM we construct a convex homotopy as follows

$$\begin{aligned} H(\underline{V},p,\alpha) &= \underline{V}(x,\alpha) - \left((2-\alpha)^2\left(\frac{1}{2}x^6 + x^5 - x^3 + \frac{11}{32}x^2\right) - \frac{11}{32}x^2 + \alpha x + \alpha \right) \\ &\quad + p\int_a^x k(x,t)\underline{V}^2(t,\alpha)dt \end{aligned}$$

$$\begin{aligned} H(\bar{V},p,\alpha) &= \bar{V}(x,\alpha) - \left(\alpha^2\left(\frac{1}{2}x^6 + x^5 - x^3 + \frac{11}{32}x^2\right) + (2-\alpha)\left(\frac{-11}{32}(2-\alpha)x^2 + x + 1\right) \right) \\ &\quad + p\int_a^x k(x,t)\bar{V}^2(t,\alpha)dt \end{aligned}$$

Now we make use of the iteration formulae [Eq. (31)] and applying the numerical algorithm, the HPM solution series with few iterative terms are as follows

$p^0 : (\underline{V}_0(x,\alpha), \bar{V}_0(x,\alpha))$ is the initial fuzzy approximations,

where

$$\underline{V}_0(x,\alpha) = \underline{f}(x,\alpha) = (2-\alpha)^2\left(\frac{1}{2}x^6 + x^5 - x^3 + \frac{11}{32}x^2\right) - \frac{11}{32}x^2 + \alpha x + \alpha$$

$$\bar{V}_0(x,\alpha) = \bar{f}(x,\alpha) = \alpha^2\left(\frac{1}{2}x^6 + x^5 - x^3 + \frac{11}{32}x^2\right) + (2-\alpha)\left(\frac{-11}{32}(2-\alpha)x^2 + x + 1\right)$$

$$p^1 : (\underline{V}_1(x,\alpha), \bar{V}_1(x,\alpha),)$$

where

$$\begin{aligned}
&\underline{V}_1(x,\alpha)\\
&= x^2\left(\alpha^2 x + \frac{1}{48}\alpha(33 + \alpha(-92 + 11\alpha))x^3\right.\\
&\quad - \frac{1}{64}\alpha(161 + \alpha(-140 + 43\alpha))x^4\\
&\quad + \frac{1}{5120}(1089 + \alpha(1064 + \alpha(102 + \alpha(-328 + 121\alpha))))x^5\\
&\quad - \frac{1}{3072}(5313 + \alpha(-25048 + \alpha(23046 + \alpha(-6856 + 473\alpha))))x^6\\
&\quad + \frac{1}{56}(-2+\alpha)^2(-1+\alpha)(-65 + 19\alpha)x^7\\
&\quad - \frac{1}{128}(-2+\alpha)^2(95 + \alpha(-4 + 21\alpha))x^8\\
&\quad - \frac{1}{288}(-2+\alpha)^2(355 + \alpha(-324 + 97\alpha))x^9\\
&\quad + \frac{1}{160}(-2+\alpha)^2(159 + 37(-4+\alpha)\alpha)x^{10} + \frac{3}{11}(-2+\alpha)^4 x^{11}\\
&\quad \left. - \frac{1}{12}(-2+\alpha)^4 x^{12} - \frac{7}{52}(-2+\alpha)^4 x^{13} - \frac{1}{28}(-2+\alpha)^4 x^{14}\right)
\end{aligned}$$

$$\begin{aligned}
&\bar{V}_1(x,\alpha)\\
&= \frac{-1}{64}x^2\left(-64(2+\alpha)^2 x + \frac{16}{3}(2+\alpha)(-35 + 23\alpha)x^3\right.\\
&\quad - 4(-2+\alpha)(5 + \alpha(3 + 8\alpha))x^4\\
&\quad + \frac{1}{5}(-825 + \alpha(1298 + \alpha(-729 + 128\alpha)))x^5\\
&\quad + \frac{1}{3}(121 + \alpha(-242 + \alpha(-351 + 280\alpha)))x^6 - \frac{3}{27}(-1+\alpha)\alpha^2(15 + 2\alpha)x^7\\
&\quad + 2\alpha^2(51 + \alpha(-31 + 8\alpha))x^8 + \frac{8}{9}\alpha^2(-1 + \alpha(17 + 16\alpha))x^9\\
&\quad \left. - \frac{8}{5}\alpha^2(11 + \alpha(-11 + 12\alpha))x^{10} - \frac{192}{11}\alpha^4 x^{11} + \frac{16}{3}\alpha^4 x^{12} + \frac{112}{13}\alpha^4 x^{13} + \frac{16}{7}\alpha^4 x^{14}\right)
\end{aligned}$$

In the same way, the iterations $p^2 : (\underline{V}_2(x,\alpha), \bar{V}_2(x,\alpha))$ and $p^3 : (\underline{V}_3(x,\alpha), \bar{V}_3(x,\alpha))$ were also computed for calculating fuzzy approximate solutions which could not be stated here as the iterations contain very long expressions.

We approximate $\underline{F}(x,\alpha)$ and $\bar{F}(x,\alpha)$ by setting p = 1 and the obtained fuzzy solutions are given by

$$\begin{aligned}
\underline{F}(x,\alpha) &= \underline{u}(x,\alpha) = \sum_{i=0}^{3} \underline{V}_i(x,\alpha)\\
&= \left((2-\alpha)^2\left(\frac{1}{2}x^6 + x^5 - x^3 + \frac{11}{32}x^2\right) - \frac{11}{32}x^2 + \alpha x + \alpha\right)\\
&\quad + x^2\left(\alpha^2 x + \frac{1}{48}\alpha\left(33 + \alpha(-92 + 11\alpha)x^3 - \frac{1}{64}\alpha(161 + \alpha(-140 + 43\alpha))x^4 + \cdots\right.\right.
\end{aligned}$$

and

$$\begin{aligned}
\bar{F}(x,\alpha) &= \bar{u}(x,\alpha) = \sum_{i=0}^{3} \bar{V}_i(x,\alpha)\\
&= \left(\alpha^2\left(\frac{1}{2}x^6 + x^5 - x^3 + \frac{11}{32}x^2\right) + (2-\alpha)\left(\frac{-11}{32}(2-\alpha)x^2 + x + 1\right)\right)\\
&\quad - \frac{1}{64}x^2\left(-64(2+\alpha)^2 x + \frac{16}{3}(2+\alpha)(-35 + 23\alpha)x^3 + \cdots\right.
\end{aligned}$$

Error Analysis

Absolute errors are computed as

$$\underline{E}(x,\alpha) = |\alpha(x+1) - \underline{u}(x,\alpha)|$$

$$\bar{E}(x,\alpha) = |(2-\alpha)(x+1) - \bar{u}(x,\alpha)|$$

By the same way of the previous example, we obtained the components of iteration formulae [Eq. (31)] by using the Mathematica program (Mathematica package version 5.2) according to the numerical algorithm where the tolerance is given a suitable positive value. The numerical results of the obtained approximate solutions are compared with the exact solutions for different α-values and are presented in Table 2. Moreover exact and approximate solutions are shown graphically in Fig. 2.

Table 2: Exact and approximate solutions with four iterations at x = 0.5

α	Exact solution		Approximate solution		Error	
	$\underline{u}(x,\alpha)$	$\bar{u}(x,\alpha)$	$\underline{u}(x,\alpha)$	$\bar{u}(x,\alpha)$	$\underline{E}(x,\alpha)$	$\bar{E}(x,\alpha)$
0	0	3.00000	0	2.98221	0	0.01779
0.1	0.150000	2.85000	0.148744	2.83473	0.001256	0.01527
0.2	0.300000	2.70000	0.297259	2.68701	0.002741	0.01299
0.3	0.450000	2.55000	0.445361	2.53908	0.004639	0.01092
0.4	0.600000	2.40000	0.599574	2.39095	0.000426	0.00905
0.5	0.750000	2.25000	0.749317	2.24261	0.000683	0.00739
0.6	0.900000	2.10000	0.899174	2.09409	0.000826	0.00591
0.7	1.050000	1.95000	1.049622	1.94538	0.000378	0.00462
0.8	1.200000	1.80000	1.199686	1.79650	0.000314	0.00350
0.9	1.350000	1.65000	1.349564	1.64746	0.000436	0.00254
1	1.500000	1.50000	1.499300	1.49827	0.000700	0.00173

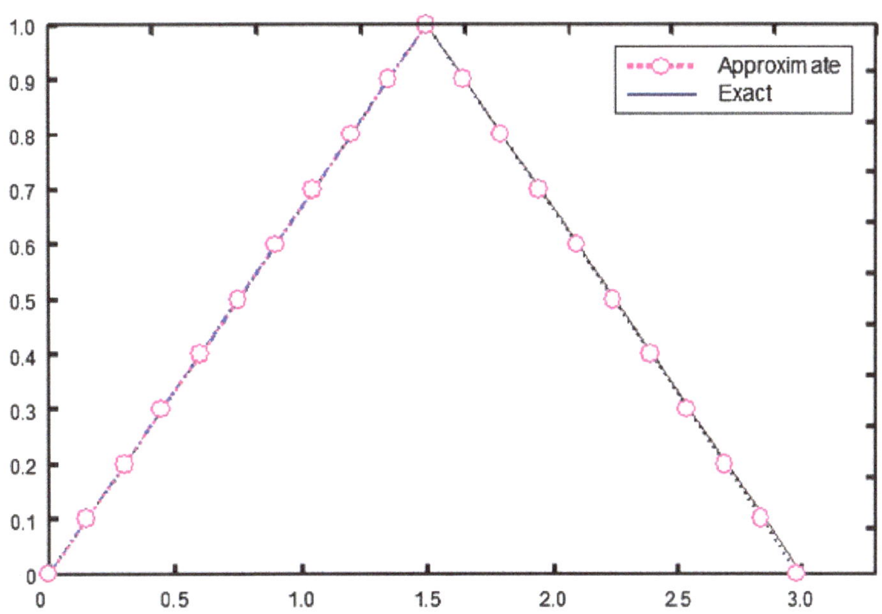

Figure 2: Plots of exact and approximate solutions of Example 2.

CONCLUDING REMARKS ON NUMERICAL RESULTS

By looking into the computed values presented above (Tables 1, 2), one may conclude that the obtained approximate solutions by applying the proposed method (HPM) are in high agreement with the exact solutions. Furthermore, from the graphical representation (Figs. 1, 2), it is apparent that the plot of exact and approximate solutions merely coincide due to the occurrence of very minimum amount of errors (E−(x,α)E_(x,α) and E¯(x,α)E¯(x,α)). The fuzzy approximate solutions are calculated at four iterations for the numerical examples considered. The results presented clearly shows that the iterative steps number 3–4 in the numerical algorithm accompany for the convergence of the fuzzy solution so that the computations are minimized and the accuracy of the approximate values is also achieved with exceptionally few iteration and overall errors are reduced.

CONCLUSIONS

In this paper, we developed a simple yet versatile numerical technique using HPM and successfully applied it by means of the proposed algorithm for solving the linear and nonlinear fuzzy Volterra integral equations of the second kind. Mathematica has been used for computations in this paper. The examples analyzed illustrates that the answers are trusty and reveals the effectiveness of the proposed method. The reliability of HPM due to its precise results and the reduction in computation gives HPM a wider applicability. Moreover, it avoids the tedious work needed by the traditional numerical methods. Advantage of the proposed method lies in the free selection of initial approximation in a straightforward manner and also it overcomes the drawbacks of handling larger equations. This method does not involve any discretization of variables and hence it is free from rounding off errors. One can apply this method to higher order equations also.

Further research can be focused on studying the existence and convergence of HPM for linear and nonlinear FVIE-2. Finally we conclude that this work provides the applicable computational technique will aid the practical applications already used in the design of various fuzzy dynamical systems and it is an issue of considerable importance at this current trend due to its versatile nature.

ACKNOWLEDGEMENTS

The authors would like to thank the Editor and anonymous referees for their helpful suggestions and valuable comments.

Competing interests

The authors declare that they have no competing interests.

REFERENCES

1. Abbasbandy S (2007) Application of He's homotopy perturbation method to functional integral equations. Chaos, Solitons Fractals 31(5):1243–1247
2. Allahviranloo T et al (2014) An application of a semi-analytical method on linear fuzzy Volterra integral equations. J Fuzzy Set Valued Anal 2014:1–15
3. Alnasr MH, Momani Shaher (2008) Application of homotopy perturbation method to singularly perturbed Volterra integral equations. J Appl Sci 8(6):1073–1078
4. Chun C (2007) Integration using He's homotopy perturbation method. Chaos, Solitons Fractals 34(4):1130–1134
5. Dubois D, Prade H (1978) Operations on fuzzy numbers. J Syst Sci 9:613–626
6. Dubois D, Prade H (1982) Towards fuzzy differential calculus. Fuzzy Sets Syst 8:1–7
7. Filobello-Nino U et al (2014a) A handy approximate solution for a squeezing flow between two infinite plates by using of laplace transform-homotopy perturbation method. SpringerPlus 3:421
8. Filobello-Nino U et al (2014b) Nonlinearities distribution laplace transform- homotopy perturbation method. SpringerPlus 3:594
9. Friedman M, Ma M, Kandel A (1999) Numerical solutions of fuzzy differential and integral equations. Fuzzy Sets Syst 106:35–48
10. Ghanbari M (2010) Numerical solution of fuzzy linear Volterra integral equations of the second kind by homotopy analysis method. Int J Ind Math 2(2):73–87
11. Goetschel R, Voxman W (1983) Topological properties of fuzzy numbers. Fuzzy Sets Syst 10:87–99
12. Goetschel R, Voxman W (1986) Elementary fuzzy calculus. Fuzzy Sets Syst 18:31–43
13. Grover M, Tomer A (2011) A new approach to evaluate second kind of Volterra integral equation with homotopy perturbation method and variational iteration method. IRACST-Eng Sci Technol: Int J 1(1):24–29
14. Hamdi Cherif M et al (2016) Homotopy perturbation method for solving the fractional fisher's equation. Int J Anal Appl 10(1):9–16
15. He JH (1999) Homotopy perturbation technique. Comput Methods Appl Mech Eng 178(3–4):257–262
16. He JH (2000) A coupling method of a homotopy technique and a perturbation technique for non-linear problems. Int J Non-linear Mech 35(1):37–43
17. He JH (2003) Homotopy perturbation method: a new nonlinear analytical technique. Appl Math Comput 135(1):73–79
18. He JH (2004) Comparison of homotopy perturbation method and homotopy analysis method. Appl Math Comput 156(2):527–539
19. Hochstadt H (1973) Integral equations. Wiley, Newyork
20. Hussain Eman A, Ali Ayad W (2013a) Homotopy analysis method for solving fuzzy integro-differential equations. Mod Appl Sci 7(3):15–25
21. Hussain Eman A, Ali Ayad W (2013b) Homotopy analysis method for solving nonlinear fuzzy integral equations. Int J Appl Math 28(1):1177–1189
22. Kaleva O (1987) Fuzzy differential equations. Fuzzy Sets Syst 24(2):301–317
23. Matinfar M, Saeidy M (2010) Application of homotopy perturbation method for fuzzy integral equations. J Math Comput Sci 1(4):377–385
24. Mirzaee F et al (2012) Numerical solution of Fredholm fuzzy integral equations of the second kind via direct method using triangular functions. J Hyperstruct 1(2):46–60
25. Momani S, Odibat Z (2007a) Comparison between homotopy perturbation method and the variational iteration method for linear fractional partial differential equations. Comput Math Appl 54(7–8):910–919
26. Momani S, Odibat Z (2007b) Homotopy perturbation method for nonlinear partial differential equations of fractional order. Phys Lett A 365(5–6):345–350
27. Mosleh M, Otadi M (2013) Solution of fuzzy Volterra integral equations in a Bernstein polynomial basis. J Adv Inf Technol 4(3):148–155
28. Otadi M, Mosleh M (2014) Numerical solutions of fuzzy nonlinear integral equations of the second kind. Iran J Fuzzy Syst 1(1):135–145

29. Puri ML, Ralescu D (1983) Differential for fuzzy function. J Math Anal Appl 91:552–558
30. Puri ML, Ralescu D (1986) Fuzzy random variables. J Math Anal Appl 114:409–422
31. Ramesh Rao TR (2016) A comparison between the differential transform method and homotopy perturbation method for a system of non linear chemistry problems. Indian J Appl Res 6(2):171–180
32. Saberi-Nadjafi J, Ghorbani A (2009) He's homotopy perturbation method: an effective tool for solving integral and integro-differential equations. Comput Math Appl 58:2379–2390
33. Subrahmanyam PV, Sudarsanam SK (1996) A note on fuzzy Volterra intergral equations. Fuzzy Sets Syst 81(2):188–191
34. Tapaswini S, Chakraverty S (2013) Numerical solution of n-th order fuzzy linear differential equations by homotopy perturbation method. Int J Comput Appl 64(6):5–10
35. Vazquez-Leal H, Sarmiento-Reyes A (2015) GHM method for obtaining rational solutions of nonlinear differential equations. SpringerPlus 4:241. doi:10.1186/s40064-015-1011-x
36. Wazwaz AM (2011) Linear and nonlinear integral equations: methods and applications. Springer, Berlin, Heidelberg, Newyork
37. Wu C, Ma M (1990) On the integrals, series and integral equations of fuzzy set valued functions. J Harbin Inst Technol 21:11–19
38. Zadeh LA (1965) Fuzzy sets. Inf Control 8:338–353

CHAPTER 11

PERTURBATION TECHNIQUE FOR A CLASS OF NONLINEAR IMPLICIT SEMILINEAR IMPULSIVE INTEGRO-DIFFERENTIAL EQUATIONS OF MIXED TYPE WITH NONCOMPACTNESS MEASURE

Heng-you Lan and Yi-shun Cui

Keywords: Nonlinear first-order implicit semilinear impulsive integro-differential equation, Monotone iterative technique, Monotone condition and noncompactness measure condition, Lower and upper solution, Existence and uniqueness

INTRODUCTION

The theory of impulsive differential equations has become an important area of investigation in recent years stimulated by their numerous applications to problems arising in mechanics, electrical engineering, medicine, biology, ecology, *etc*. Various evolutionary processes undergo abrupt changes of states at certain moments of time; between intervals of continuous evolution such changes can be well approximated as being instantaneous changes at state, or in the form of impulses. These process are modeled by impulse differential equations and have been the most important research directions and connections for impulsive differential equations; see, for example, [1, 2, 3, 4,5, 6, 7] and the references therein. Subsequently, many authors have investigated the existence of solutions to impulsive differential equations or (implicit) impulsive integro-differential equations with their strong applications in Banach spaces; see [1, 2, 3, 4, 5, 6, 7, 8, 9, 10, 11, 12, 13, 14, 15, 16, 17, 18, 19, 20, 21, 22, 23, 24, 25, 26, 27] and the references therein.

Recently, Lan and Cui [15] studied a class of initial value problems of nonlinear first-order implicit impulsive integro-differential equations in Banach space. By using the Mönch fixed point theorem, they obtained some new existence theorems of solutions for this class of nonlinear first-order implicit impulsive integro-differential equations in Banach spaces under some weaker conditions. Furthermore, some (implicit) impulsive differential equations under various initial and boundary conditions has also been studied by several authors; see, for example, [11, 13, 22, 24, 28] and the references therein. By using a monotone iterative technique in the presence of lower and upper solutions, Lan [23] discussed the existence of solutions for a new class of nonlinear first-order implicit impulsive integro-differential equations in Banach spaces. Under wide monotone conditions and the noncompactness measure conditions, he also obtained the existence of extremal solutions and a unique solution between lower and upper solutions. In [25], Chen and Li introduced and studied a class of semilinear impulsive evolution equations in Banach spaces by using a mixed monotone iterative technique. The presented results improved and extended some relevant results in ordinary differential equations and partial differential equations. For related works, see [9, 26, 27, 29, 30] and the references therein.

On the other hand, the monotone iterative technique, which is one of the approximation methods for finding solutions of a

Citation: Heng-you Lan, Yi-shun Cui, Perturbation technique for a class of nonlinear implicit semilinear impulsive integro-differential equations of mixed type with noncompactness measure, DOI: 10.1186/s13662-014-0329-y.

comparatively large class of impulsive differential equations, can be applied in practice easily; see, for example, [14, 16, 17, 23, 25, 26, 27, 29]. Further, some nice examples of the monotone iterative technique can be found in [20, 21]. As a matter of fact, Li and Liu [16] pointed out that 'the monotone iterative technique in the presence of lower and upper solutions is an important method for seeking solutions of differential equations in abstract spaces'. Moreover, Li and Liu [16] used a monotone iterative technique in the presence of lower and upper solutions to discuss the existence of solutions for the initial value problem of the impulsive integro-differential equation of Volterra type in a Banach space. Under monotone conditions and the noncompactness measure condition of the nonlinearity function *f*, the authors also obtained the existence of extremal solutions and a unique solution between lower and upper solutions. In [14], by using a monotone iterative technique in the presence of lower and upper solutions, we discussed the existence of solutions for a new system of nonlinear mixed type implicit impulsive integro-differential equations in Banach spaces. Under some monotonicity conditions and the noncompactness measure conditions, they also obtained the existence of extremal solutions and a unique solution between lower and upper solutions.

Motivated and inspired by the above works, by using the Arzela-Ascoli theorem, the Bellman inequality, and the monotone iterative technique in the presence of lower and upper solutions, we discuss the existence of mild solutions for the following nonlinear first-order implicit semilinear impulsive differential equation problem in Banach space BB: Find u:J→B such that

$$\begin{cases} u'(t) = Au(t) + f(t, u(t), Tu(t), u'(t)), & t \neq t_k, \\ \Delta u|_{t=t_k} = I_k(u(t_k)), & k = 1, 2, \ldots, m, \\ u(t_0) = u_0, \end{cases} \tag{1.1}$$

where $J=[t_0,t_0+a]\subset R=(-\infty,+\infty)$ is a compact interval, the operator A is the infinitesimal generator of a positive C_0-semigroup $\{G(t), t \geq t_0\}$ on $\mathbb{B}, f \in C(J \times \mathbb{B} \times \mathbb{B} \times \mathbb{B}, \mathbb{B})$ is a nonlinear continuous operator, $t_0 < t_1 < \cdots < t_m < t_0 + a < +\infty$, $u_0 \in \mathbb{B}$ is a given element, $\hbar \in C(D, \mathbb{R}^+)$, $D = \{(t,s) \mid s,t \in J, t \geq s\}$, $\mathbb{R}^+ = [0, +\infty)$,

$$Tu(t) = \int_{t_0}^{t} \hbar(t,s)u(s)\,ds,$$

and for $k = 1, 2, \ldots, m$, $I_k \in C[\mathbb{B}, \mathbb{B}]$ is an impulsive function, $\Delta u|_{t=t_k}$ denotes the jump of $u(t)$ at $t = t_k$, i.e., $\Delta u|_{t=t_k} = u(t_k^+) - u(t_k^-)$, $u(t_k^-)$ and $u(t_k^+)$ represent the left and right limits of u(t) at $t=t_k$, respectively. Further, under wide monotone conditions and the noncompactness measure conditions, we obtain the existence of extremal solutions and a unique mild solution between lower and upper solutions.

PRELIMINARIES

Throughout this paper, let $\mathbb{B}$ be an ordered Banach space with the norm $\|\cdot\|$ and partial order $\leq$, whose positive cone $P = \{x \in \mathbb{B} \mid x \geq 0\}$ is normal with normal constant N, and $A : \mathrm{dom}(A) \subset \mathbb{B} \to \mathbb{B}$ be a closed linear operator and generate a C_0-semigroup G(t) $(t \geq t_0)$ in $\mathbb{B}$. Let $J = [t_0, t_0 + a]$, $t_0 < t_1 < \cdots < t_m < t_0 + a < +\infty$, $J_0 = [t_0, t_1], J_1 = (t_1, t_2], \ldots, J_k = (t_k, t_{k+1}], \ldots, J_m = (t_m, t_0 + a]$, and

$$PC(J, \mathbb{B}) = \{x : J \to \mathbb{B} \mid x(t) \text{ is continuous at } t \neq t_k, \text{ and left continuous at } t = t_k, \text{ and } x(t_k^+) \text{ exists}, k = 1, 2, \ldots, m\}.$$

Evidently, $PC(J, \mathbb{B})$ is a Banach space with norm $\|x\|_{PC} = \sup_{t \in J} x(t)$. Let $J' = J \setminus \{t_1, t_2, \ldots, t_m\}$, and $\mathbb{B}_*$ the Banach space generated by dom(A)dom (A) with norm $\|\cdot\|_* = \|\cdot\| + \|A\cdot\|$. An abstract function $x \in PC(J, \mathbb{B}) \cap C^1(J', \mathbb{B}) \cap C(J', \mathbb{B}_*)$ is called a solution of problem (1.1) if x(t)x(t) satisfies all the equalities of (1.1).

Let

$$PC^1(J, \mathbb{B}) = \{x \in PC(J, \mathbb{B}) \cap C^1(J', \mathbb{B}) \cap C(J', \mathbb{B}_*) \mid x'(t_k^+), x'(t_k^-) \text{ exist}, k = 1, 2, \ldots, m\},$$

where $x'(t_k^+)$ and $x'(t_k^-)$ represent the right and left derivatives of x(t) at $t=t_k$, respectively. For $x \in PC^1(J, \mathbb{B})$, by virtue of the mean value theorem

$$x(t_k) - x(t_k - \tau) \in \tau\overline{co}\{x'(t) : t_k - \tau < t < t_k\} \quad (\tau > 0),$$

it is easy to see that the left derivative x'_(tk) exists and

$$x'_-(t_k) = \lim_{h \to 0^+} \tau^{-1}[x(t_k) - x(t_k - \tau)] = x'(t_k^-).$$

In the sequel, $x'(t_k)$ is understood as $x'_-(t_k)$, then $x' \in PC^1(J, \mathbb{B})$. If $x \in PC(J, \mathbb{B}) \cap C^1(J', \mathbb{B}) \cap C(J', \mathbb{B}_*)$ is a solution of problem (1.1), then by the continuity of *f* and the closed linearity of *A*, we know $x \in PC^1(J, \mathbb{B})$. Evidently, $PC^1(J, \mathbb{B})$ is a Banach space with norm $\|x\|_{PC^1} = \max\{\sup_{t \in J} \|x(t)\|, \sup_{t \in J} \|x'(t)\|\}$.

A mapping $F : J \to \mathbb{B}$ is differentiable at t∈J if there exists a $F'(t) \in \mathbb{B}$ such that the limits

$$\lim_{\tau\to 0^+}\frac{F(t+\tau)-F(t)}{\tau} \quad \text{and} \quad \lim_{\tau\to 0^+}\frac{F(t)-F(t-\tau)}{\tau}$$

exist and are equal to F'(t). Here the limits are taken in $\mathbb{B}$. At the endpoints of *J*, we consider the one-sided derivatives.

By the well-known result [31], we know that there exist C>0 and σ∈R such that $\|G(t)\| \le Ce^{\sigma t}$. Letting

$$\delta_0 := \inf\{\sigma \in \mathbb{B} \mid \exists C>0, \|G(t)\| \le Ce^{\sigma t}\},$$

then δ_0 is called the increasing index of G(t). It follows from the properties of the C_0-semigroup that the C_0-semigroup $G(t)$ $(t \ge t_0)$ is exponentially stable if and only if $\delta_0<0$.

Let $C^1(J,\mathbb{B})$ denote the Banach space of all continuous differentiable $\mathbb{B}$-value functions on interval *J* with norm $\|x\|_{C^1} = \max_{t\in J}\|x'(t)\|$. Let $\alpha(\cdot)$ denote the Kuratowski measure of noncompactness of the bounded set. For the details of the definition and properties of the measure of noncompactness, see [32]. For any $E \subset C^1(J,\mathbb{B})$ and $t \in J$, set $E(t) = \{x(t) \mid x \in E\} \subset \mathbb{B}$. If *E* is bounded in $C^1(J,\mathbb{B})$, then E(t) is bounded in $\mathbb{B}$, and $\alpha(E(t)) \le \alpha(E)$.

Lemma 2.1

Assume that the C_0-semigroup G(t) $(t\ge t_0)$ is exponentially stable, i.e., $\delta_0<0$. Then for any $p \in PC^1(J,\mathbb{B})$ and $v_k, u_0 \in \mathbb{B}$, $k=1,2,\ldots,m$, the initial value problem of linear impulsive evolution equation in $\mathbb{B}$

$$\begin{cases} u'(t) = Au(t) + p(t), & t \in J', \\ \Delta u|_{t=t_k} = v_k, & k = 1,2,\ldots,m, \\ u(t_0) = u_0, \end{cases} \tag{2.1}$$

has a unique mild solution $u \in PC^1(J,\mathbb{B})$ expressed by

$$u(t) = G(t-t_0)u_0 + \int_{t_0}^{t} G(t-s)p(s)\,ds + \sum_{t_0<t_k<t} G(t-t_k)v_k.$$

Proof

It follows from Theorem 2.9 of [31, Chapter 4] and Lemma 2.2 in [26] that this conclusion follows directly.

Lemma 2.2 [33] *If H is a bounded subset of $PC^1(J,\mathbb{B})$, the element of* H' *is equicontinuous at* J_k *for all* k=0,1,2,…,m *then*

$$\alpha(H) = \max\left\{\sup_{t\in J}\alpha(H(t)), \sup_{t\in J}\alpha(H'(t))\right\},$$

where H'(t)={x'(t):x∈H}.

Lemma 2.3 [33] *Let $E \subset C(J,\mathbb{B})$ be bounded and equicontinuous. Then* α(E(t)) *is continuous on J, and*

$$\alpha\left(\left\{\int_J x(t)\,dt \,\middle|\, x \in E\right\}\right) \le \int_J \alpha(E(t))\,dt.$$

Lemma 2.4 [34, Corollary 3.1(b)] *Let $E = \{x_n\} \subset PC(J,\mathbb{B})$ be a bounded and countable set. Then* α(E(t)) *is Lebesgue integral on J, and*

$$\alpha\left(\left\{\int_J x_n(t)\,dt\right\}\right) \le 2\int_J \alpha(E(t))\,dt.$$

EXISTENCE AND UNIQUENESS THEOREMS

In this section, we will prove our main results concerning the mild solutions of the nonlinear first-order implicit impulsive integro-differential equation (1.1) in Banach spaces.

Definition 3.1

If a function $y \in PC^1(J,\mathbb{B})$ satisfies

$$\begin{cases} y'(t) \le Au(t) + f(t,u(t),Tu(t),u'(t)), & t \ne t_k, \\ \Delta y|_{t=t_k} \le I_k(u(t_k)), & k=1,2,\ldots,m, \\ y(t_0) \le u_0, \end{cases} \tag{3.1}$$

then we call it a lower solution of problem (1.1); if all the inequalities of (3.1) are inverse, then we call it an upper solution of problem (1.1).

Definition 3.2

A C_0-semigroup G(t) $(t\ge t_0)$ in $\mathbb{B}$ is said to be positive, if the order inequality G(t)x≥θ holds for every x≥θ, x∈$\mathbb{B}$, and $t\ge t_0$.

It is easy to see that for any $M \ge 0$, $A - MI$ also generates a C_0-semigroup $\Gamma(t) = e^{-M(t)}G(t)$ $(t \ge t_0)$ in $\mathbb{B}$. Γ(t) is a positive C_0-semigroup if G(t) is a positive C_0-semigroup for all $t\ge t_0$.

Now, let us first list the following assumptions for convenience:

(H_1): Problem (1.1) has a lower solution $y_0 \in PC^1(J,\mathbb{B})$ and an upper solution $x_0 \in PC^1(J,\mathbb{B})$ with $y_0 \le x_0$, and there exist constants M∈(0,1) such that

$$f(t,u_2,v_2,w_2) - f(t,u_1,v_1,w_1) \ge -M(u_2-u_1)$$

for all $t \in J$ and $y_0(t) \le u_1 \le u_2 \le x_0(t)$, $Ty_0(t) \le v_1 \le v_2 \le Tx_0(t)$, and $y_0'(t) \le w_1 \le w_2 \le x_0'(t)$.

(H_2): I_k(x) is increasing on the order interval $[y_0(t), x_0(t)]$ for $t \in J$, $k=1,2,\ldots,m$.

(H_3): There exist 0<2L<1−M such that

$$\alpha(\{f(t,u_n(t),v_n(t),w_n(t))\}) \le L[\alpha(\{u_n(t)\}) + \alpha(\{v_n(t)\}) + \alpha(\{w_n(t)\})]$$

for all t∈J, and increasing or decreasing monotonic sequences $\{u_n\} \subset [y_0(t), x_0(t)]$, $\{v_n\} \subset [Ty_0(t), Tx_0(t)]$ and $\{w_n\} \subset [y_0'(t), x_0'(t)]$.

In the sequel, we prove the following main results of this paper.

Theorem 3.1

Let $\mathbb{B}$ *be an ordered Banach space, whose positive cone* P *is normal,* A:dom(A) $\subset\mathbb{B}\to\mathbb{B}$ *be a closed linear operator, the positive* C_0*-semigroup* G(t) $(t\geq t_0)$ *generated by* A *be compact in* $\mathbb{B}, f\in C(J\times\mathbb{B}\times\mathbb{B}\times\mathbb{B},\mathbb{B})$*, and* $I_k\in C(\mathbb{B},\mathbb{B})$ *for* $k=1,2,\ldots,m$*. Suppose that the conditions* $(H_1)\sim(H_3)$ *hold. Then problem* (1.1) *has minimal and maximal mild solutions between* $[y_0, x_0]$*, which can be obtained by a monotone iterative procedure starting from* y0y0 *and* x0x0*, respectively.*

Proof

Let
$M_1=\sup_{t\in J}\|\Gamma(t)\|$ and $M_2=\sup_{t\in J}\|G'(t)\|$. For any $u\in PC^1(J,\mathbb{B})$, define Fu on J by the equation

$$Fu(t)=\Gamma(t-t_0)u_0+\int_{t_0}^{t}\Gamma(t-s)\left[f(s,u(s),Tu(s),u'(s))+Mu(s)\right]ds$$
$$+\sum_{t_0<t_k<t}\Gamma(t-t_k)I_k(u(t_k)). \tag{3.2}$$

It is easy to see that $F:PC^1(J,\mathbb{B})\to PC^1(J,\mathbb{B})$ is continuous. By Lemma 2.1, we know that the mild solution of problem (1.1) is equivalent to the fixed point of F. Since G(t) $(t\geq t_0)$ is a positive C_0-semigroup, G(0)=I([31]) and it follows from assumptions (H_1) and (H_2) that F is increasing in $[y_0,x_0]$ and maps any bounded set in $[y_0,x_0]$ into a bounded set.

We first show that
$y_0\leq Fy_0,\ Fx_0\leq x_0$. Let $p(t)=y_0'(t)-Ay_0(t)+My_0(t)$, by the definition of lower solution and (2.1), we know that $p\in PC^1(J,\mathbb{B})$ and $p(t)\leq f(t,y_0(t),Ty_0(t),y_0'(t))+My_0(t)$ for $t\in J'$. It follows from Lemma 2.1 that

$$y_0(t)=\Gamma(t-t_0)y_0(t_0)+\int_{t_0}^{t}\Gamma(t-s)g(s)\,ds+\sum_{t_0<t_k<t}\Gamma(t-t_k)\Delta y_0|_{t=t_k}$$
$$\leq\Gamma(t-t_0)u_0+\int_{t_0}^{t}\Gamma(t-s)\left[f(t,y_0(t),Ty_0(t),y_0'(t))+My_0(t)\right]ds$$
$$+\sum_{t_0<t_k<t}\Gamma(t-t_k)I_k(y_0(t_k))$$
$$=Fy_0(t),$$

for all t∈J, *i.e.*, $y_0\leq Fy_0$. Similarly, it can be shown that $Fx_0\leq x_0$. Combining these facts and the increasing property of F in $[y_0,x_0]$, we see that F maps $[y_0,x_0]$ into itself, and $F:[y_0,x_0]\to[y_0,x_0]$ is a continuously increasing operator.

Secondly, we prove that $F:[y_0,x_0]\to[y_0,x_0]$ is completely continuous. Let

$$\Phi u(t)=\int_{t_0}^{t}\Gamma(t-s)\left(f(s,u(s),Tu(s),u'(s))+Mu(s)\right)ds,$$
$$\Psi u(t)=\sum_{t_0<t_k<t}\Gamma(t-t_k)I_k(u(t_k)). \tag{3.3}$$

On the one hand, for all t∈J, we show that $K(t)=\{\Phi_u(t)|u\in[y_0,x_0]\}$ is precompact in $\mathbb{B}$. In fact, for any $\epsilon\in(t_0,t)$ and $u\in[y_0,x_0]$, it follows from (3.3) that

$$\Phi_\epsilon u(t)=\int_{t_0}^{t-\epsilon}\Gamma(t-s)\left(f(s,u(s),Tu(s),u'(s))+Mu(s)\right)ds$$
$$=\Gamma(\epsilon)\int_{t_0}^{t-\epsilon}\Gamma(t-\epsilon-s)\left(f(s,u(s),Tu(s),u'(s))+Mu(s)\right)ds. \tag{3.4}$$

It follows from the condition (H_1) that

$$f(t,y_0(t),Ty_0(t),y_0'(t))+My_0(t)$$
$$\leq f(t,u(t),Tu(t),u'(t))+Mu(t)$$
$$\leq f(t,x_0(t),Tx_0(t),x_0'(t))+Mx_0(t). \tag{3.5}$$

By the normality of the cone P, now we know that there exists a constant $M_3>0$ such that

$$\|f(t,u(t),Tu(t),u'(t))+Mu(t)\|\le M_3,\quad \forall u\in[y_0,x_0].$$

From the compactness of $\Gamma(\epsilon)$, we have $K_\epsilon(t)=\{\Phi_\epsilon u(t)|u\in[y_0,x_0]\}$ is precompact in $\mathbb{B}$. Since

$$\|\Phi u(t)-\Phi_\epsilon u(t)\|\le\int_{t-\epsilon}^{t}\|\Gamma(t-s)\|\cdot\|f(s,u(s),Tu(s),u'(s))+Mu(s)\|\,ds$$
$$\le M_1M_3\epsilon,$$

the set K(t) is totally bounded in $\mathbb{B}$. Moreover, K(t)K(t) is precompact in $\mathbb{B}$.

On the other hand, for all $t_1,t_2\in J$, from (3.3)-(3.5), we have

$$\|\Phi u(t_2)-\Phi u(t_1)\|\le\int_{t_0}^{t_2}(\Gamma(t_2-s)-\Gamma(t_1-s))(\|f(s,u(s),Tu(s),u'(s))+Mu(s)\|)\,ds$$
$$+\int_{t_1}^{t_2}\Gamma(t_2-s)(\|f(s,u(s),Tu(s),u'(s))+Mu(s)\|)\,ds$$
$$\le M_3\int_{t_0}^{t_1}\|\Gamma(t_2-s)-\Gamma(t_1-s)\|\,ds+M_1M_3(t_1-t_1)$$
$$\le M_3\int_{t_0}^{t_0+a}\|\Gamma(t_2-t_1+s)-\Gamma(s)\|\,ds+M_1M_3(t_1-t_1).\tag{3.6}$$

The right side of (3.6) relies on t_2-t_1, but it is independent of u. Since G(t) ($t\ge t_0$) is compact, $\Gamma(t)$ is compact and continuous in the uniform operator topology for all $t\ge t_0$. Thus, the right side of (3.6) tends to 0 as $t_2-t_1\to 0$. Hence, $\Phi([y_0,x_0])$ is an equicontinuous function of the cluster in $C^1(J,\mathbb{B})$.

Similarly, we can prove the compactness of Ψ in (3.3).

For any $t\in J$, since $\{Fu(t)\mid u\in[y_0,x_0]\}=\{\Gamma(t-t_0)+\Phi u(t)+\Psi u(t)\mid u\in[y_0,x_0]\}$, and $Fu(t_0)=u_0$ is precompact in $\mathbb{B}$, we know that $F([y_0,x_0])$ is precompact in $C^1(J_k,\mathbb{B})$ by using the Arzela-Ascoli theorem. Thus, $F:[y_0,x_0]\to[y_0,x_0]$ is completely continuous.

Finally, we show that problem (1.1) has minimal and maximal mild solutions between $[y_0,x_0]$, which can be obtained by a monotone iterative procedure starting from y_0 and x_0, respectively.

It follows from the completely continuity of F that F has minimal and maximal fixed points $\underline{u}$ and $\overline{u}$ in $[y_0, x_0]$, and so they are the minimal and maximal mild solutions of problem (1.1) in $[y_0, x_0]$, respectively.

On the other hand, from the above discussions, we know that $F:[y_0,x_0]\to[y_0,x_0]$ is a continuously increasing operator. Now, we define two sequences $\{y_n\}$ and $\{x_n\}$ in $[y_0,x_0]$ by the iterative scheme

$$y_n=Fy_{n-1},\qquad x_n=Fx_{n-1},\quad n=1,2,\ldots.\tag{3.7}$$

Then it follows from the monotonicity of F that

$$y_0\le y_1\le\cdots\le y_n\le\cdots\le x_n\le\cdots\le x_1\le x_0.\tag{3.8}$$

We prove that $\{y_n\}$ and $\{x_n\}$ are uniformly convergent in J.

For convenience, let $E=\{y_n\mid n\in\mathbb{N}\}$ and $E_0=\{y_{n-1}\mid n\in\mathbb{N}\}$. Since $E=F(E_0)$, by (3.2) and the boundedness of E_0, we easily see that E is equicontinuous in every interval J'_k, where $J'_1=[t_0,t_1]$ and $J'_k=(t_{k-1},t_k]$, $k=2,3,\ldots,m$. From $E_0=E\cup\{y_0\}$ and Lemma 2.2, it follows that $\alpha(E_0(t))=\alpha(E(t))$ and

$$\alpha^1(E(t))=\max\{\alpha(E(t)),\alpha(E'(t))\}$$

for all $t\in J$. Letting

$$\phi(t)=\alpha^1(E(t))=\alpha^1(E_0(t)),\quad t\in J,$$

by Lemma 2.3, we know that $\phi\in PC^1(J,\mathbb{R}^+)$. Going from J'_1 to J'_{m+1} interval by interval, we show that $\phi(t)\equiv 0$ in J.

In fact, for $t\in J$, there exists a J'_k such that $t\in J'_k$. By Lemma 2.3, we have

$$\begin{aligned}
\alpha\big(T(E_0(t))\big) &= \alpha\left(\left\{\int_{t_0}^{t} \hbar(t,s)y_{n-1}(s)\,ds \,\middle|\, n\in\mathbb{N}\right\}\right)\\
&\le \sum_{j=1}^{k-1}\alpha\left(\left\{\int_{t_{j-1}}^{t_j} \hbar(t,s)y_{n-1}(s)\,ds \,\middle|\, n\in\mathbb{N}\right\}\right)\\
&\quad + \alpha\left(\left\{\int_{t_{k-1}}^{t} \hbar(t,s)y_{n-1}(s)\,ds \,\middle|\, n\in\mathbb{N}\right\}\right)\\
&\le \hbar_0\sum_{j=1}^{k-1}\int_{t_{j-1}}^{t_j}\alpha\big(E_0(s)\big)\,ds + \hbar_0\int_{t_{k-1}}^{t}\alpha\big(E_0(s)\big)\,ds\\
&\le \hbar_0\sum_{j=1}^{k-1}\int_{t_{j-1}}^{t_j}\phi(s)\,ds + \hbar_0\int_{t_{k-1}}^{t}\phi(s)\,ds\\
&= \hbar_0\int_{t_0}^{t}\phi(s)\,ds,
\end{aligned}$$

where $\hbar_0 = \max\{|\hbar(t,s)| : (t,s)\in D\}$. Hence,

$$\int_{t_0}^{t}\alpha\big(T(E_0(s))\big)\,ds \le a\hbar_0\int_{t_0}^{t}\phi(s)\,ds. \tag{3.9}$$

It follows from (3.2), Lemma 2.4, assumption (H_3) and (3.9) that, for $t\in J_1'$,

$$\begin{aligned}
\alpha\big(E(t)\big) &= \alpha\big(F(E_0(t))\big)\\
&= \alpha\left(\left\{\int_{t_0}^{t}\Gamma(t-s)\big(f\big(s,y_{n-1}(s),Ty_{n-1}(s),y'_{n-1}(s)\big)+My_{n-1}(s)\big)\,ds \,\middle|\, n\in\mathbb{N}\right\}\right)\\
&\le 2\int_{t_0}^{t}\Gamma(t-s)\alpha\big(\big(\big(f\big(s,y_{n-1}(s),Ty_{n-1}(s),y'_{n-1}(s)\big)+My_{n-1}(s)\big)\mid n\in\mathbb{N}\big)\big)\,ds\\
&\le 2M_1\int_{t_0}^{t}\big\{L\big[\alpha\big(E_0(s)\big)+\alpha\big(T(E_0(s))\big)+\alpha\big(E_0'(s)\big)\big]+M\alpha\big(E_0(s)\big)\big\}\,ds\\
&\le 2M_1\left[(L+M)\int_{t_0}^{t}\alpha\big(E_0(s)\big)\,ds + L\int_{t_0}^{t}\alpha\big(T(E_0(s))\big)\,ds + L\int_{t_0}^{t}\alpha\big(E_0'(s)\big)\,ds\right]\\
&\le 2M_1(M+a\hbar_0L+2L)\int_{t_0}^{t}\phi(s)\,ds,
\end{aligned}$$

$$\begin{aligned}
\alpha\big(E'(t)\big) &= \alpha\big((FE_0)'(t)\big)\\
&= \alpha\left(\frac{d}{dt}\left\{\int_{t_0}^{t}\Gamma(t-s)\big(f\big(s,y_{n-1}(s),Ty_{n-1}(s),y'_{n-1}(s)\big)+My_{n-1}(s)\big)\,ds \,\middle|\, n\in\mathbb{N}\right\}\right)\\
&= \alpha\Bigg(\Bigg\{\int_{t_0}^{t}\Gamma'(t-s)f\big(s,y_{n-1}(s),Ty_{n-1}(s),y'_{n-1}(s)\big)\,ds\\
&\quad + My_{n-1}(t)+\Gamma(0)f\big(t,y_{n-1}(t),Ty_{n-1}(t),y'_{n-1}(t)\big) \,\Bigg|\, n\in\mathbb{N}\Bigg\}\Bigg)\\
&= \alpha\Bigg(\Bigg\{-MF(y_{n-1}(t))\\
&\quad + \int_{t_0}^{t}G'(t-s)e^{-M(t-s)}f\big(s,y_{n-1}(s),Ty_{n-1}(s),y'_{n-1}(s)\big)\,ds\\
&\quad + My_{n-1}(t)+G(0)f\big(t,y_{n-1}(t),Ty_{n-1}(t),y'_{n-1}(t)\big) \,\Bigg|\, n\in\mathbb{N}\Bigg\}\Bigg)\\
&\le -M\alpha\big(F(E_0(t))\big)
\end{aligned}$$

$$+\alpha\left(\int_{t_0}^{t} G'(t-s)e^{-M(t-s)}f(s,y_{n-1}(s),Ty_{n-1}(s),y'_{n-1}(s))\,ds \,\Big|\, n\in\mathbb{N}\right)$$
$$+M\alpha(E_0(t))+\alpha(\{f(t,y_{n-1}(ts),Ty_{n-1}(t),y'_{n-1}(t))\})$$
$$\le -M\alpha(F(E_0(t)))$$
$$+2\int_{t_0}^{t} G'(t-s)e^{-M(t-s)}\alpha(f(s,y_{n-1}(s),Ty_{n-1}(s),y'_{n-1}(s))\mid n\in\mathbb{N})\,ds$$
$$+M\alpha(E_0(t))+\alpha(\{f(t,y_{n-1}(ts),Ty_{n-1}(t),y'_{n-1}(t))\})$$
$$\le -M\alpha(F(E_0(t)))+M\alpha(E_0(t))+\zeta[\alpha(E_0(t))+\alpha(T(E_0)(t))+\alpha(E'_0(t))]$$
$$\le [a\hbar_0\zeta-2M(M+a\hbar_0\zeta+2\zeta)]\int_{t_0}^{t}\phi(s)\,ds+(\zeta+M)\alpha(E_0(t))+L\alpha(E'_0(t))$$
$$\le [a\hbar_0\zeta-2M(M+a\hbar_0\zeta+2\zeta)]\int_{t_0}^{t}\phi(s)\,ds+(M+2\zeta)\alpha^1(E_0(t)),$$

and so

$$\phi(t)\le\Gamma\int_{t_0}^{t}\phi(s)\,ds+(M+2L)\phi(t),$$

i.e.,

$$\phi(t)\le\Theta\int_{t_0}^{t}\phi(s)\,ds,$$

where $\zeta = L(1+2aM_2)$, $\Gamma=\max\{2(M+a\hbar_0 L+2L), a\hbar_0\zeta-2M(M+a\hbar_0\zeta+2\zeta)\}$, and $\Theta=\Gamma/(1-M-2\zeta)$. Hence, by the Bellman inequality, we know that $\phi(t)\equiv 0$ in J'_1. In particular, $\alpha^1(E(t_1))=\alpha^1(E_0(t_1))=\phi(t_1)=0$, and so $\alpha(E(t_1))=\alpha(E_0(t_1))=0$, this means that $E(t_1)$ and $E_0(t_1)$ are precompact in $\mathbb{B}$. Thus $I_1(E_0(t_1))$ is precompact in $\mathbb{B}$, and $\alpha(I_1(E_0(t_1)))=0$.

Now, for $t\in J'_2$, by (3.2) and the above argument for J'_1, we have

$$\alpha(E(t))=\alpha(E(t))=\alpha(F(E_0)(t))$$
$$=\alpha\left(\left\{\Gamma(t-t_0)u_0+\int_{t_0}^{t}\Gamma(t-s)(f(s,y_{n-1}(s),Ty_{n-1}(s),y'_{n-1}(s))+My_{n-1}(s))\,ds\right.\right.$$
$$\left.\left.+\Gamma(t-t_1)I_1(y_{n-1}(t_1))\,\Big|\,n\in\mathbb{N}\right\}\right)$$
$$\le 2(L+M+a\hbar_0L)\int_{t_0}^{t}\phi(s)\,ds+\alpha(I_1(E_0(t_1)))$$
$$\le 2(M+a\hbar_0L+2L)\int_{t_0}^{t}\phi(s)\,ds$$
$$=2(M+a\hbar_0L+2L)\int_{t_1}^{t}\phi(s)\,ds$$

and

$$\phi(t)\le\Theta\int_{t_1}^{t}\phi(s)\,ds.$$

Again by the Bellman inequality, we know that $\phi(t)\equiv 0$ in J'_2, from which we obtain $\alpha(E_0(t_2))=0$ and $\alpha(I_2(E_0(t_2)))=0$.

Continuing such a process interval by interval up to J'_{m+1}, we can prove that $\phi(t)\equiv 0$ in every J'_k, $k=1,2,\dots,m+1$.

For any J_k, if we modify the value of y_n at $t=t_{k-1}$ via $y_n(t_{k-1})=y_n(t^+_{k-1})$, $n\in\mathbb{N}$, then $\{y_n\}\subset C^1(J_k,\mathbb{B})$ and it is equicontinuous. Since $\alpha(\{y_n(t)\})\equiv 0$, $\{y_n(t)\}$ is precompact in $\mathbb{B}$ for every $t\in J_k$. By the Arzela-Ascoli theorem, we know that $\{y_n\}$ is precompact in $C^1(J_k,\mathbb{B})$. Hence, $\{y_n\}$ has a convergent subsequence in $C^1(J_k,\mathbb{B})$. Combining this with the monotonicity (3.8), we easily prove that $\{y_n\}$ itself is convergent in $C^1(J_k,\mathbb{B})$. In particular, $\{y_n(t)\}$ is uniformly convergent in J'_k. Consequently, $\{y_n(t)\}$ is uniformly convergent over the whole of J.

Using a similar argument to that for $\{y_n(t)\}$, we can prove that $\{x_n(t)\}$ is also uniformly convergent in J. Hence, $\{y_n(t)\}$ and $\{x_n(t)\}$ are convergent in $PC^1(J,\mathbb{B})$. Setting

$$\underline{u}=\lim_{n\to\infty}y_n,\qquad \overline{u}=\lim_{n\to\infty}x_n \quad \text{in } PC^1(J,\mathbb{B}) \tag{3.10}$$

and $n\to\infty$ in (3.7) and (3.8), then we have $v_0\le\underline{u}\le\overline{u}\le x_0$ and

$$\underline{u} = F\underline{u}, \qquad \overline{u} = F\overline{u}. \tag{3.11}$$

By the monotonicity of F, it is easy to see that $\underline{u}$ and $\overline{u}$ are the minimal and maximal fixed points of F in $[y_0,x_0]$, and therefore they are the minimal and maximal mild solutions of problem (1.1) in $[y_0,x_0]$, respectively. This completes the proof.

Remark 3.1

In Theorem 3.1, if $\mathbb{B}$ is weakly sequentially complete, the condition (H_3) holds automatically. In fact, by Theorem 2.2 of [35], any monotonic and order-bounded sequence is precompact. Let $\{x_n\}$ and $\{y_n\}$ be two increasing or decreasing sequences obeying condition (H_3), then, by condition (H_1), $\{f(t,x_n, y_n, z_n)+Mx_n\}$is a monotonic and order-bounded sequence. By the property of the measure of noncompactness, we have

$$\alpha(\{Ax_n(t) + f(t,x_n,y_n,z_n) + Mx_n\})$$

$$\leq \alpha(\{Ax_n(t) + f(t,x_n,y_n,z_n) + Mx_n\}) + M\alpha(\{x_n\}) = 0.$$

Hence, condition (H_3) holds.

From Theorem 3.1, we obtain the following result.

Corollary 3.1

Let $\mathbb{B}$ be an ordered and weakly sequentially complete Banach space, whose positive cone P is normal, $f \in C(J \times \mathbb{B} \times \mathbb{B} \times \mathbb{B},\mathbb{B})$, and $I_k \in C(\mathbb{B},\mathbb{B})$, $k = 1,2,\ldots,m$. If the conditions (H_1) and (H_2) are satisfied, then problem (1.1) has minimal and maximal mild solutions between y_0 and x_0, which can be obtained by a monotone iterative procedure starting from y_0 and x_0, respectively.

Next we discuss the uniqueness of the mild solution to problem (1.1) in $[y_0,x_0]$. Assume we replace the assumption (H_3) by the following assumption.

(H_4) There exist positive constants C_i (i=1,2,3) with C3<1 such that

$$f(t,u_2,v_2,w_2) - f(t,u_1,v_1,w_1) \leq C_1(u_2-u_1) + C_2(v_2-v_1) + C_3(w_2-w_1),$$

$$\forall t \in J, \quad y_0(t) \leq u_1 \leq u_2 \leq x_0(t), \qquad Ty_0(t) \leq y_1 \leq y_2 \leq Tx_0(t),$$

For all $t \in J$ and $y_0(t) \leq u_1 \leq u_2 \leq x_0(t)$, $\lambda_1 Ty_0(t) \leq v_1 \leq v_2 \leq \lambda_1 Tx_0(t)$ and $\lambda_2 y_0'(t) \leq w_1 \leq w_2 \leq \lambda_2 x_0'(t)$.
Then we have the following unique existence result.

Theorem 3.2

Let $\mathbb{B}$ be an ordered Banach space, whose positive cone P is normal, $f \in C(J \times \mathbb{B} \times \mathbb{B} \times \mathbb{B},\mathbb{B})$ and $I_k \in C(\mathbb{B},\mathbb{B})$, $k = 1,2,\ldots,m$. If the conditions (H_1), (H_2), and (H_4) hold, then problem (1.1) has a unique mild solution between y_0 and x_0, which can be obtained by a monotone iterative procedure starting from y_0 or x_0.

Proof

We first prove that (H_1) and (H_4) imply (H_3). In fact, for $t \in J$, let $\{u_n\} \subset [y_0(t),x_0(t)]$, $\{v_n\} \subset [Ty_0(t),Tx_0(t)]$, and $\{w_n\} \subset [y_0'(t),x_0'(t)]$ be increasing sequences. For $m,n \in \mathbb{N}$ with m>n, by (H_1) and (H_4),

$$\theta \leq (f(t,u_m,v_m,w_m) - f(t,u_n,v_n,w_n)) + M(u_m-u_n)$$

$$\leq (C_1+M)(u_m-u_n) + C_2(v_m-v_n) + C_3(w_m-w_n).$$

By this inequality and the normality of cone P, we have

$$\|f(t,u_m,v_m,w_m) - f(t,u_n,v_n,w_n)\|$$

$$\leq N\|(C_1+M)(u_m-u_n) + C_2(v_m-v_n) + C_3(w_m-w_n)\| + M\|u_m-u_n\|$$

$$\leq (M+MN+NC_1)\|u_m-u_n\| + NC_2\|v_m-v_n\| + NC_3(w_m-w_n).$$

From this inequality and the definition of the measure of noncompactness, it follows that

$$\alpha\left(\left\{f(t,u_n,v_n,w_n)\right\}\right)$$
$$\leq (M+MN+NC_1)\alpha\left(\{u_n\}\right)+NC_2\alpha\left(\{v_n\}\right)+NC_3\alpha\left(\{w_n\}\right)$$
$$\leq L'\left[\alpha\left(\{u_n\}\right)+\alpha\left(\{v_n\}\right)+\alpha\left(\{w_n\}\right)\right],$$

where $L' = \max\{M+NM+NC_1, NC_2, NC_3\}$. If $\{u_n\}$, $\{v_n\}$, and $\{w_n\}$, are decreasing sequences, the above inequality is also valid. Hence (H_3) holds.

Therefore by Theorem 3.1, problem (1.1) has a minimal solution $\underline{u}$ and a maximal solution $\overline{u}$ in $[y_0,x_0]$. By the proof of Theorem 3.1, (3.7), (3.8), (3.10), and (3.11) are valid. Going from J_1' to J_{m+1}' interval by interval, we show that $\underline{u}(t) \equiv \overline{u}(t)$ in every J_k', $k = 1, 2, \ldots, m+1$.

Indeed, for $t \in J_1'$, by (3.11) and (3.2) and assumption (H_4), we have

$$\theta \leq \overline{u}(t)-\underline{u}(t) = F\overline{u}(t) - F\underline{u}(t)$$
$$= \int_{t_0}^{t} \Gamma(t-s)\left[f\left(s,\overline{u}(s),T\overline{u}(s),\overline{u}'(s)\right)-f\left(s,\underline{u}(s),T\underline{u}(s),\underline{u}'(s)\right)+M\left(\overline{u}(s)-\underline{u}(s)\right)\right]ds$$
$$\leq \int_{t_0}^{t} M_1\left[(M+C_1)\left(\overline{u}(s)-\underline{u}(s)\right)+C_2\left(T\overline{u}(s)-T\underline{u}(s)\right)+C_3\left(\overline{u}'(s)-\underline{u}'(s)\right)\right]ds$$
$$\leq M_1(M+C_1)\int_{t_0}^{t}\left(\overline{u}(s)-\underline{u}(s)\right)ds$$
$$\leq M_1(M+C_1)\int_{t_0}^{t}\left(\overline{u}(s)-\underline{u}(s)\right)ds$$
$$+M_1C_2\hbar_0\int_{t_0}^{t}\int_{t_0}^{s}\left(\overline{u}(\tau)-\underline{u}(\tau)\right)d\tau\, ds+M_1C_3\left(\overline{u}(t)-\underline{u}(t)\right)$$
$$\leq M_1(M+C_1+aC_2\hbar_0)\int_{t_0}^{t}\left(\overline{u}(s)-\underline{u}(s)\right)ds+M_1C_3\left(\overline{u}(t)-\underline{u}(t)\right), \tag{3.12}$$

where $M_1 = \sup_{t\in J}\|\Gamma(t)\|$. It follows from (3.12) and the normality of cone P that

$$\left\|\overline{u}(t)-\underline{u}(t)\right\| \leq M_1N(M+C_1+aC_2\hbar_0)\int_{t_0}^{t}\left\|\overline{u}(s)-\underline{u}(s)\right\|ds+M_1C_3\left\|\overline{u}(t)-\underline{u}(t)\right\|,$$

i.e.,

$$\left\|\overline{u}(t)-\underline{u}(t)\right\| \leq \frac{M_1N(M+C_1+aC_2\hbar_0)}{1-M_1C_3}\int_{t_0}^{t}\left\|\overline{u}(s)-\underline{u}(s)\right\|ds.$$

Thus, by the Bellman inequality, we obtain $\underline{u}(t) \equiv \overline{u}(t)$ in J_1'.

For $t \in J_2'$, since $I_1(\overline{u}(t_1)) = I_1(\underline{u}(t_1))$, using (3.2) and by completely the same argument as above for $t \in J_1'$, we can prove that

$$\left\|\overline{u}(t)-\underline{u}(t)\right\| \leq \frac{M_1N(M+C_1+aC_2\hbar_0)}{1-M_1C_3}\int_{t_0}^{t}\left\|\overline{u}(s)-\underline{u}(s)\right\|ds$$
$$= \frac{M_1N(M+C_1+aC_2\hbar_0)}{1-M_1C_3}\int_{t_1}^{t}\left\|\overline{u}(s)-\underline{u}(s)\right\|ds.$$

Again, by the Bellman inequality, we obtain $\underline{u}(t) \equiv \overline{u}(t)$ in J_2'.

Continuing such a process interval by interval up to J_{m+1}', we see that $\underline{u}(t) \equiv \overline{u}(t)$ over the whole of J. Hence, $u^* := \underline{u} = \overline{u}$ is the unique mild solution of problem (1.1) in $[y_0,x_0]$, which can be obtained by the monotone iterative procedure (3.7) starting from y_0 or x_0.

Remark 3.2

(1) Using the above argument method interval by interval from J_1' to J_{m+1}', we can also improve the main results in [19] and [21], and delete some restrictive conditions there.

(2) In this study, the equicontinuity of the semigroup G(t) ($t \geq t_0$) generated by A is not required.

CONCLUDING REMARKS

In this paper, we introduce and study the following nonlinear first-order implicit impulsive differential equation problem in Banach space $\mathbb{B}$:

Find $u: J \to \mathbb{B} \times \mathbb{B} \times \mathbb{B}$ such that

$$\begin{cases} u'(t) = Au(t) + f(t, u(t), Tu(t), u'(t)), & t \neq t_k, \\ \Delta u|_{t=t_k} = I_k(u(t_k)), & k = 1, 2, \ldots, m, \\ u(t_0) = u_0. \end{cases}$$

By using a monotone iterative technique in the presence of lower and upper solutions, the existence of extremal solutions and a unique mild solution between the lower and upper solutions are obtained under wide monotone conditions and the noncompactness measure conditions. The results presented in this paper improved and generalized some known results concerned with the integro-differential equations and classical (abstract) differential equations.

Moreover, we remark that if the lower solution and the upper solution for problem (1.1) do not exist, then we have the following results.

Theorem 4.1

Let $\mathbb{B}$ be an ordered Banach space, whose positive cone P is normal, A:dom(A) $\subset \mathbb{B} \to \mathbb{B}$ *be a closed linear operator and generate a positive* C_0*-semigroup* G(t) ($t \geq t_0$) *in* $\mathbb{B}$, $f \in C(J \times \mathbb{B} \times \mathbb{B} \times \mathbb{B}, \mathbb{B})$, *and* $I_k \in C(\mathbb{B}, \mathbb{B})$, $k = 1, 2, \ldots, m$.

Assume that there exist

$b > 0$, $x_0 \in \mathrm{dom}(A)$, $x_0 \geq \theta$, $y_k \in \mathrm{dom}(A)$, $y_k \geq \theta$, $k = 1, 2, \ldots, m$, $h \in PC^1(J, \mathbb{B})$ and $h(t) \geq \theta$ *such that*

$$f(t, x, Tx, x') \leq bx + h(t), \qquad I_k(x) \leq y_k, \quad x \geq \theta;$$

$$f(t, x, Tx, x') \geq bx - h(t), \qquad I_k(x) \geq -y_k, \quad x \leq \theta.$$

Then the following results hold:

1. *If the* C_0*-semigroup* G(t) ($t \geq t_0$) *is compact in* $\mathbb{B}$, *and the conditions* (H_1) *and* (H_2) *in Section 3 are satisfied, then problem* (1.1) *has minimal and maximal mild solutions.*
2. *Problem* (1.1) *has minimal and maximal mild solutions when the conditions* (H_1) ~ (H_3) *in Section 3are satisfied.*
3. *If the positive cone P is regular, and the conditions* (H_1) *and* (H_2) *in Section 3 are satisfied, then problem* (1.1) *has minimal and maximal mild solutions.*
4. *Problem* (1.1) *has a unique mild solution when the conditions* (H_1), (H_2), *and* (H_4) *in Section 3 are satisfied.*

Proof

Firstly, we consider the following initial value problem of the linear impulsive evolution equation in $\mathbb{B}$:

$$\begin{cases} u'(t) = Au(t) + bu(t) + h(t), & t \in J', \\ \Delta u|_{t=t_k} = y_k, & k = 1, 2, \ldots, m, \\ u(t_0) = x_0. \end{cases} \tag{4.1}$$

Since (A+bI) generates a positive C_0-semigroup $\Gamma(t) = e^{bt}G(t)$ $(t \geq 0)$ in $\mathbb{B}$, it follows from Theorem 2.9 in [31, Chapter 4] and Lemma 2.1, that problem (4.1) has a unique positive classical solution $\hat{u} \in PC^1(J, E)$. Let $y_0 = -\hat{u}$, $x_0 = \hat{u}$, it is easy to see that y_0 and x_0 are the lower solution and the upper solution of problem (1.1), respectively. So, our conclusions (1)-(4) follow from Theorems 3.1 and 3.2.

DECLARATIONS

Acknowledgements

We thank the referees' valuable comments and suggestions to improve our paper.

This work was partially supported by Sichuan Province Cultivation Fund Project of Academic and Technical Leaders, and the Open Research Fund of Key Laboratory of Higher Education of Sichuan Province for Enterprise Informationalization and Internet of Things (2013WZJ01).

Competing Interests

The authors declare that they have no competing interests.

Authors' Contributions

H-YL conceived of the study and participated in its design and coordination. Y-SC carried out the proof of the corollaries and gave some remarks to show the main results. All authors read and approved the final manuscript.

REFERENCES

1. Samoilenko, AM, Perestyuk, NA: Impulsive Differential Equations. World Scientific, Singapore (1995)
2. Cuevas, C, N'Guérékata, GM, Rabelo, M: Mild solutions for impulsive neutral functional differential equations with state-dependent delay. Semigroup Forum **80**(3), 375-390 (2010)
3. Cuevas, C, Hernández, E, Rabelo, M: The existence of solutions for impulsive neutral functional differential equations. Comput. Math. Appl. **58**(4), 744-757 (2009)
4. Guo, TL, Jiang, W: Impulsive fractional functional differential equations. Comput. Math. Appl. **64**(10), 3414-3424 (2012)
5. Hernández, E, Rabelo, M, Henríquez, H: Existence of solutions for impulsive partial neutral functional differential equations. J. Math. Anal. Appl. **331**, 1135-1158 (2007)
6. Henríquez, H, de Andrade, B, Rabelo, M: Existence of almost periodic solutions for a class of abstract impulsive differential equations. ISRN Math. Anal. **2011**, 1-21 (2011)
7. Hernández, E: Global solutions for abstract impulsive neutral differential equations. Math. Comput. Model. **53**(1-2), 196-204 (2011)
8. Agarwal, RP, O'Regan, D: A multiplicity result for second order impulsive differential equations via the Leggett Williams fixed point theorem. Appl. Math. Comput. **161**, 433-439 (2005)
9. Agarwal, RP, Benchohra, M, Hamani, S, Pinelas, S: Upper and lower solutions method for impulsive differential equations involving the Caputo fractional derivative. Mem. Differ. Equ. Math. Phys. **53**, 1-12 (2011)
10. Akhmet, MU, Turan, M: The differential equation on time scales through impulsive differential equations. Nonlinear Anal. **65**, 2043-2060 (2006)
11. Carl, S, Heikkilä, S: On discontinuous implicit and explicit abstract impulsive boundary value problems. Nonlinear Anal. **41**, 701-723 (2000)
12. Guo, DJ: Multiple positive solutions for first order nonlinear impulsive integro-differential equations in Banach spaces. Appl. Math. Comput. **143**, 233-249 (2003)
13. Lan, HY, Huang, NJ, Kim, JK: First order nonlinear implicit impulsive integro-differential equations in Banach spaces. Dyn. Contin. Discrete Impuls. Syst., Ser. A Math. Anal. **13**(6), 803-813 (2006)
14. Lan, HY: Monotone method for a system of nonlinear mixed type implicit impulsive integro-differential equations in Banach spaces. J. Comput. Appl. Math. **222**(2), 531-543 (2008)
15. Lan, HY, Cui, YS: On the existence of solutions for nonlinear first-order implicit impulsive integro-differential equations. Nonlinear Anal. **71**(5-6), 1670-1677 (2009)
16. Li, YX, Liu, Z: Monotone iterative technique for addressing impulsive integro-differential equations in Banach spaces. Nonlinear Anal. **66**, 83-92 (2007)
17. Ahmad, B, Malar, K, Karthikeyan, K: A study of nonlocal problems of impulsive integrodifferential equations with measure of noncompactness. Adv. Differ. Equ. **2013**, 205 (2013)
18. Li, J, Nieto, JJ, Shen, J: Impulsive periodic boundary value problems of first-order differential equations. J. Math. Anal. Appl.**325**, 226-236 (2007)
19. Liu, LS, Wu, CX, Guo, F: A unique solution of initial value problems for first order impulsive integro-differential equations of mixed type in Banach spaces. J. Math. Anal. Appl. **275**, 369-385 (2002)
20. Nieto, JJ, Rodríguez-López, R: New comparison results for impulsive integro-differential equations and applications. J. Math. Anal. Appl. **328**, 1343-1368 (2007)
21. Sun, JL, Ma, YH: Initial value problems for seconder order mixed monotone type of impulsive integro-differential equations in Banach spaces. J. Math. Anal. Appl. **247**, 506-516 (2000)
22. Huang, NJ, Lan, HY: Existence of the solution for a class of implicit differential equations in Banach spaces. Dyn. Contin. Discrete Impuls. Syst., Ser. A Math. Anal. **13**(1), 27-36 (2006)
23. Lan, HY: Existence and uniqueness results for nonlinear first-order implicit impulsive integro-differential equations with monotone conditions. Dyn. Contin. Discrete Impuls. Syst., Ser. A Math. Anal. **17**(1), 19-30 (2010)
24. Pang, HH, Lu, M, Cai, C: The method of upper and lower solutions to impulsive differential equations with integral boundary conditions. Adv. Differ. Equ. **2014**, 183 (2014)
25. Chen, PY, Li, YX: Mixed monotone iterative technique for a class of semilinear impulsive evolution equations in Banach spaces. Nonlinear Anal. **74**(11), 3578-3588 (2011)
26. Chen, PY, Mu, J: Monotone iterative method for semilinear impulsive evolution equations of mixed type in Banach spaces. Electron. J. Differ. Equ. **2010**, 149 (2010)
27. Chen, PY, Li, YX, Yang, H: Perturbation method for nonlocal impulsive evolution equations. Nonlinear Anal. Hybrid Syst. **8**, 22-30 (2013)
28. Li, DS: Peano's theorem for implicit differential equations. J. Math. Anal. Appl. **258**, 591-616 (2001)
29. Zhang, XP, Sun, YP: Monotone iterative methods of positive solutions for fractional differential equations involving derivatives. Math. Probl. Eng. **2014**, Article ID 254012 (2014)
30. Wang, F, Wang, P: Existence and uniqueness of mild solutions for a class of nonlinear fractional evolution equation. Adv. Differ. Equ. **2014**, 150 (2014)
31. Pazy, A: Semigroups of Linear Operators and Applications to Partial Differential Equations. Applied Mathematical Sciences, vol. 44. Springer, New York (1983)

32. Deimling, K: Nonlinear Functional Analysis. Springer, Berlin (1985)
33. Guo, DJ, Lakshmikantham, V, Liu, XZ: Nonlinear Integral Equations in Abstract Spaces. Kluwer Academic, Dordrecht (1996)
34. Hein, HP: On the behaviour of measure of noncompactness with respect to differentiation and integration of vector-valued functions. Nonlinear Anal. 7, 1351-1371 (1983)
35. Du, Y: Fixed points of increasing operators in ordered Banach spaces and applications. Appl. Anal. **38**, 1-20 (1990)

SECTION III - BIFURCATION ANALYSIS AND CHAOS SYNCHRONIZATION

CHAPTER 12 PARAMETRIC CONTROLLER DESIGN OF HOPF BIFURCATION SYSTEM

Jinbo Lu, Xiaorong Hou and Min Luo

Keywords: School of Energy Science and Engineering, University of Electronic Science and Technology of China, Chengdu, Sichuan 611131, China

INTRODUCTION

Bifurcation is a normal phenomenon of nonlinear system, and the research in bifurcation control has achieved considerable progress in the past 30 years [1]. Hopf bifurcation is an important type of bifurcation. Its practical significance lies in the notion that Hopf bifurcation is a critical state between stable and unstable. There are a lot of related research achievements [2–4], covering a lot of fields, including power system [5, 6], electronic technology [7], network technology [8], new energy [9], and optical technology [10].

Hopf bifurcation research focused on phenomena analysis [11–13] and bifurcation control [14–19]. And the work of Hopf bifurcation control is designing control strategy for Hopf bifurcation system to modify the original system dynamic characteristics [16, 18, 20]. But the general applications of these designed controls are still limited. For enhancing flexibility and universality of Hopf bifurcation controller, some study of applying parametric control to Hopf bifurcation system controller design has emerged [17, 21].

On the basis of nonlinear state feedback theory, [17] gives a relatively common form of Hopf bifurcation control, complete solution, and simulation of controller parameters by example. Its controller is

$$\begin{aligned}
u_q&\left(X, X_1^*, X_2^*, \ldots, X_k^*, \mu\right) \\
&= \sum_{i=1}^{n} A_{qi} \prod_{j=1}^{k}\left(x_i - x_{ij}^*\right) \\
&\quad + \sum_{i=1}^{n}\sum_{j=1}^{k} B_{qij}\left(x_i - x_{ij}^*\right) \prod_{p=1}^{k}\left(x_i - x_{ip}^*\right) \\
&\quad + \sum_{i=1}^{n}\sum_{j=1}^{k} C_{qij}\left(x_i - x_{ij}^*\right)^2 \prod_{p=1}^{k}\left(x_i - x_{ip}^*\right) \\
&\quad + \sum_{i=1}^{n}\sum_{j=1}^{k} D_{qij}\left(x_i - x_{ij}^*\right)^2 \prod_{p=1}^{k}\left(x_i - x_{ip}^*\right)^2 + \cdots \\
&\qquad (q = 1, 2, \ldots, n),
\end{aligned} \tag{1}$$

where X is system state vector, $X_1^*, X_2^*, \ldots, X_k^*$ are k equilibrium points of system, A_{qi}, B_{qij}, C_{qij}, and D_{qij} are the controller parameters, x_i is i state of system, and x_{ij}^* and x_{ip}^* are j and p equilibrium point value of system state , respectively.

As can be seen from controller expression, the controller in [17] contains all of the system equilibrium points. The paper [21] has done a further study on the basis of [17]. The parameters of the controller have been simplified significantly, for a 3-dimensional system with two equilibrium points (the Rössler system). The controller in [21] is

Citation: Jinbo Lu, Xiaorong Hou, and Min Luo, "Parametric Controller Design of Hopf Bifurcation System,", vol. 2015, Article ID 584954, 10 pages, 2015. doi:10.1155/2015/584954.

$$u_3 = \sum_{i=1}^{3} \left[A_i (x_i - x_{i0}^{+}) (x_i - x_{i0}^{-}) \right.$$
$$+ B_i (x_i - x_{i0}^{+})^2 (x_i - x_{i0}^{-})$$
$$\left. + B_{i+3} (x_i - x_{i0}^{+}) (x_i - x_{i0}^{-})^2 \right], \quad (2)$$

where A_i, B_i, and B_{i+3} are the system parameters and x_{i0}^{+} and x_{i0}^{-} are the value of system state x_i on the two equilibrium points. In addition to the controller design work, [21] used plane figure to display the controller constraints and analyzed the bifurcation cases of each connected region.

Although the deep research of Hopf bifurcation control has been done in [17, 21], there are still some limitations. For example, the controller has a complicated form and higher order and includes system equilibrium points, and its parameters constraint solution is cumbersome. Besides, in [17], solution of controller parameters is not sufficient and intuitive. It just expresses several groups of controller parameters by inequality form. Except for the study of [17, 21], [22] did a further application in parametric method on the basis of [21].

In view of the above questions, in this paper, the controller form is improved, and a simple and general parametric controller design method is presented. The designed controller has lower order, and there is no inclusion of system equilibrium points.

The remainder structure of this paper is as follows. Section 2 provides description of Hopf bifurcation system and the method and procedure of parametric controller design. Lorenz system is taken as an example in Section 3; this section designs the related parametric controller and deduces the constraints. In Section 4, numerical simulations are calculated. And in Section 5, the conclusion of this paper is given.

A HOPF BIFURCATION SYSTEM DESCRIPTION AND PARAMETRIC CONTROLLER DESIGN

Hopf Bifurcation System Description

Before discussing how to design a controller for Hopf bifurcation, a general formula of the Hopf bifurcation system is presented. To be more specific, consider the following general nonlinear system:

$$\dot{X} = F(X, \mu) = [f_1(X, \mu), f_2(X, \mu), \ldots, f_n(X, \mu)]^T, \quad (3)$$

where the dot denotes differentiation with respect to time t and $X = (x_1, x_2, \ldots, x_n)^T \in R^n$ is an n-dimensional state vector, while $\mu \in R$ is a scalar parameter, called bifurcation parameter (note: it can be assumed that μ is an -dimensional vector for $m \geq 1$). $F(X, \mu)$ is a matrix of nonlinear polynomial system with Hopf bifurcations, and the following conditions must be satisfied:

(1) Let X^* be an equilibrium (or fixed point) of the system; that is, $F(X^*, \mu) \equiv 0$, for any value of μ.

(2) Suppose that the Jacobian of the system evaluated at the equilibrium X^* has eigenvalues, $\lambda_j(\mu)$, j =1, 2, …,n , which may be real or complex. Assume there exists one pair of the complex conjugates $\lambda_1(\mu)$, $\lambda_2(\mu)$, and all of the others are in left half of complex plane.

(3) Suppose that $\lambda_{1,2}(\mu) = \alpha(\mu) \pm i\beta(\mu)$ cross the imaginary axis at $\mu = \mu^*$. When $\mu = \mu^*$, $\alpha(\mu^*) = 0$, . $d\alpha(\mu^*)/dt \neq 0$, and $\beta(\mu^*) \neq 0$. According to the Hopf theory, a family of limit cycles will bifurcate from the equilibrium X^* at the critical point μ^*.

Generic Parametric Controller Description

For system (3), we design a general formula of nonlinear state feedback control:

$$\dot{X} = F(X, \mu) + u(X, \mu), \quad (4)$$

where

$$u(X, \mu) = [0, 0, \ldots, u_j(X, \mu), 0, \ldots, 0]^T,$$

$$u_j = g(X, \mu) f_j(X, \mu) + \sum_{i=1, i\neq j}^{n} c_{0i} f_i(X, \mu); \quad (5)$$

here

$$g(X, \mu) = c_1 + \sum_{k=1}^{n} \left(c_{k+1} x_k + c_{n+k+1} x_k^2\right), \quad (6)$$

and $\{c_{0i}, c_1, c_{k+1}, c_{n+k+1} \in R\}$ are the parameters of controller. In the controller, the form of $g(X, \mu)$ is simple. $g(X, \mu)$ is the polynomial of system states and it is not including system equilibrium point (generally, the quadratic term is not needed). Comparing with [17], the designed controller is simpler and has lower order and is not including system equilibrium point. Under the control u, in order to keep all the original equilibria unchanged for controlled system (4), the following conditions must be satisfied:

$$u(X^*, \mu) = 0,$$
$$u(X, \mu^*) \neq 0. \quad (7)$$

Remark 1. The control equation given in (5) is not a unique control law. There are many other feasible controllers that may satisfy the system. In (5), the controller consists of system polynomial; $c_{0i} \in R$ also can change for $c_{0i} \in R^n$. That is to say, c_{0i} can be a one-dimensional parameter, which also can be a polynomial composed of system states. So on the premise of satisfying system requests, c_{0i} has a variety of possible expressions, covered by all linear combinations composed of system states, and then it includes the controller expression proposed by [17].

Remark 2. There may not be only one component in $u(X, \mu)$, but usually one component can satisfy the control system requests, so we only design one component here. Besides, as you can see, no matter how many components it has, there will be no inclusion of system equilibrium points value in controller $u(X, \mu)$.

Remark 3. $g(X, \mu)$ form is not fixed. Higher polynomial can be introduced in it. Here, in order to reduce the controller order, $g(X, \mu)$, the highest order is 2, and in most cases, 1 order $g(X, \mu)$ is enough.

Parametric Controller Design Steps

For explaining parametric controller design method and steps, a 3-dimension system controller is designed as an example in this paper. The controller design steps are as follows.

Step 1. According to the parametric controller equation, a 3-dimension system with Hopf bifurcation is given:

$$\begin{aligned} \dot{x}_1 &= f_1(X,\mu), \\ \dot{x}_2 &= f_2(X,\mu), \\ \dot{x}_3 &= f_3(X,\mu). \end{aligned} \tag{8}$$

Here, $X = [x_1, x_2, x_3]^T$. Then, system (8) has parametric controller:

$$\begin{aligned} u(X,\mu) &= [u_1(X,\mu), u_2(X,\mu), u_3(X,\mu)]^T, \\ u_j(X,\mu) &= g(X,\mu) f_j(X,\mu) + \sum_{k=1}^{n} c_{0k} f_k(X,\mu); \end{aligned} \tag{10}$$

here,

$\{u_j(X,\mu) \neq 0, u_k(X,\mu) = 0, k \neq j, j = 1,2,3\}$, $g(X) = c_1 + c_2 x_1 + c_3 x_2 + c_4 x_3$, $\{c_{0k}, c_i \in R, i = 1,2,3,4\}$, and c_{0k} and c_i are controller parameters.

In order to make concise expression, without loss of generality, controller (10) can be written as simplifying form:

$$u_2(X,\mu) = g(X,\mu) \cdot f_2(X,\mu) + c_{01} f_1(X,\mu). \tag{11}$$

Step 2. According to the closed-loop system expression after adding controller, we can get Jacobian matrix including controller parameters. Adding controller (11) on system (8), closed-loop expression can be written as

$$\begin{aligned} \dot{x}_1 &= f_1(X,\mu), \\ \dot{x}_2 &= f_2(X,\mu) + u_2(X,\mu), \\ \dot{x}_3 &= f_3(X,\mu). \end{aligned} \tag{12}$$

At this point, the Jacobian matrix of closed-loop system can be described as:

$$J = \begin{bmatrix} \frac{\partial f_1}{\partial x_1} & \frac{\partial f_1}{\partial x_2} & \frac{\partial f_1}{\partial x_3} \\ j_{21} & j_{22} & j_{23} \\ \frac{\partial f_3}{\partial x_1} & \frac{\partial f_3}{\partial x_2} & \frac{\partial f_3}{\partial x_3} \end{bmatrix}, \tag{13}$$

Where

$$\begin{aligned} j_{21} &= \frac{\partial f_2}{\partial x_1} + \frac{\partial f_2}{\partial x_1} g(X) + c_2 f_2 + c_{01} \frac{\partial f_1}{\partial x_1}, \\ j_{22} &= \frac{\partial f_2}{\partial x_2} + \frac{\partial f_2}{\partial x_2} g(X) + c_3 f_2 + c_{01} \frac{\partial f_1}{\partial x_2}, \\ j_{23} &= \frac{\partial f_2}{\partial x_3} + \frac{\partial f_2}{\partial x_3} g(X) + c_4 f_2 + c_{01} \frac{\partial f_1}{\partial x_3}. \end{aligned} \tag{14}$$

Step 3. In accordance with the gotten Jacobian matrix, we can compute system characteristic equation at the system equilibrium point. Supposing one equilibrium point of system (8) is X_0, then according to the Jacobian matrix, the characteristic equation at the system equilibrium point can be described as follows:

$$D(\lambda) = \det(\lambda I - J)|_{X=X_0} = a_0\lambda^3 + a_1\lambda^2 + a_2\lambda + a_3; \tag{15}$$

here,

$$\begin{aligned} a_0 &= 1, \\ a_1 &= \left[-c_{01}\frac{\partial f_1}{\partial x_2} - \frac{\partial f_1}{\partial x_1} - \frac{\partial f_2}{\partial x_2} g(X) - \frac{\partial f_3}{\partial x_3} \right. \\ &\quad \left. - c_3 f_2 \right]_{X=X_0}, \\ a_2 &= \left[\left(\frac{\partial f_1}{\partial x_1}\frac{\partial f_2}{\partial x_2} - \frac{\partial f_1}{\partial x_2}\frac{\partial f_2}{\partial x_1} - \frac{\partial f_2}{\partial x_3}\frac{\partial f_3}{\partial x_2} + \frac{\partial f_2}{\partial x_2}\frac{\partial f_3}{\partial x_3} \right) \right. \\ &\quad \cdot g(X) - c_2 \frac{\partial f_1}{\partial x_2} f_2 - c_4 \frac{\partial f_3}{\partial x_2} f_2 + c_3 f_2 \left(\frac{\partial f_3}{\partial x_3} + \frac{\partial f_1}{\partial x_1} \right) \\ &\quad + c_{01}\left(\frac{\partial f_3}{\partial x_3}\frac{\partial f_1}{\partial x_2} - \frac{\partial f_1}{\partial x_3}\frac{\partial f_3}{\partial x_2} \right) + \frac{\partial f_1}{\partial x_1}\frac{\partial f_2}{\partial x_2} + \frac{\partial f_1}{\partial x_1}\frac{\partial f_3}{\partial x_3} \\ &\quad \left. - \frac{\partial f_1}{\partial x_2}\frac{\partial f_2}{\partial x_1} - \frac{\partial f_1}{\partial x_3}\frac{\partial f_3}{\partial x_1} - \frac{\partial f_2}{\partial x_3}\frac{\partial f_3}{\partial x_2} + \frac{\partial f_2}{\partial x_2}\frac{\partial f_3}{\partial x_3} \right]_{X=X_0}, \\ a_3 &= \left[\left(-\frac{\partial f_1}{\partial x_1}\frac{\partial f_2}{\partial x_2}\frac{\partial f_3}{\partial x_3} + \frac{\partial f_1}{\partial x_1}\frac{\partial f_2}{\partial x_3}\frac{\partial f_3}{\partial x_2} + \frac{\partial f_1}{\partial x_2}\frac{\partial f_2}{\partial x_1}\frac{\partial f_3}{\partial x_3} \right. \right. \\ &\quad \left. - \frac{\partial f_1}{\partial x_2}\frac{\partial f_2}{\partial x_3}\frac{\partial f_3}{\partial x_1} - \frac{\partial f_1}{\partial x_3}\frac{\partial f_2}{\partial x_1}\frac{\partial f_3}{\partial x_2} + \frac{\partial f_1}{\partial x_3}\frac{\partial f_2}{\partial x_2}\frac{\partial f_3}{\partial x_1} \right) \\ &\quad \cdot g(X) + \left(-c_3\frac{\partial f_1}{\partial x_1}\frac{\partial f_3}{\partial x_3} + c_4\frac{\partial f_1}{\partial x_1}\frac{\partial f_3}{\partial x_2} - c_4\frac{\partial f_1}{\partial x_2}\frac{\partial f_3}{\partial x_1} \right. \\ &\quad \left. + c_2\frac{\partial f_1}{\partial x_2}\frac{\partial f_3}{\partial x_3} + c_3\frac{\partial f_1}{\partial x_3}\frac{\partial f_3}{\partial x_1} - c_2\frac{\partial f_1}{\partial x_3}\frac{\partial f_3}{\partial x_2} \right) f_2 - \frac{\partial f_1}{\partial x_1} \\ &\quad \cdot \frac{\partial f_2}{\partial x_2}\frac{\partial f_3}{\partial x_3} + \frac{\partial f_1}{\partial x_1}\frac{\partial f_2}{\partial x_3}\frac{\partial f_3}{\partial x_2} + \frac{\partial f_1}{\partial x_2}\frac{\partial f_2}{\partial x_1}\frac{\partial f_3}{\partial x_3} - \frac{\partial f_1}{\partial x_2} \\ &\quad \left. \cdot \frac{\partial f_2}{\partial x_3}\frac{\partial f_3}{\partial x_1} - \frac{\partial f_1}{\partial x_3}\frac{\partial f_2}{\partial x_1}\frac{\partial f_3}{\partial x_2} + \frac{\partial f_1}{\partial x_3}\frac{\partial f_2}{\partial x_2}\frac{\partial f_3}{\partial x_1} \right]_{X=X_0}. \end{aligned} \tag{16}$$

Remark 4. X_0 does not require the actual solution. It is only a formal symbol. You can see the specific condition in the subsequent computation example; X_0 in Step 4 is the same as here.

Step 4. Using Hurwitz criterion, we can get the stable constraint of system and simplify it by polynomial division. According to Hurwitz criterion, the constraints including controller parameter are

$$\begin{aligned} a_0 &> 0, \\ a_1 &> 0, \\ a_2 &> 0, \\ a_3 &> 0, \\ g &= a_1 a_2 - a_0 a_3 \geq 0. \end{aligned} \tag{17}$$

Considering $a_0=1$, the system stable constraints can be written as

$$a_1 > 0,$$

$$a_2 > 0,$$

$$a_3 > 0,$$

$$g = a_1 a_2 - a_3 \geq 0. \tag{18}$$

In accordance with polynomial division, for k order real polynomial B(X) and m order real polynomial $A(X), X \in R^n$, $k \geq 1, m \geq 1$, $A(X)$ can be written as

$$A(X) = \overline{A}(X) B(X) + L_{A/B}(X); \tag{19}$$

here, $\overline{A}(X)$ is the integral part of A(X) divided by B(X) and $L_{A/B}(X)$ is the residue of A(X) divided by B(X). If $B(X_0)=0$ at X_0, then $L_{A/B}(X_0) = A(X_0)$. Supposing G(X)=0 is system (8) equilibrium point equation and X_0 is the equilibrium, then the system constraints can be simplified as

$$S: \begin{cases} l_1 = L_{a_1/G}(X_0) > 0 \\ l_2 = L_{a_2/G}(X_0) > 0 \\ l_3 = L_{a_3/G}(X_0) > 0 \\ l_4 = L_{g/G}(X_0) > 0. \end{cases} \tag{20}$$

Step 5. Using computer algebra tools such as symbol computation software and cylindrical algebraic decomposition algorithm to solve semialgebraic set composed of (20) and equilibrium point equation G(X)=0, we can get controller parameter space satisfying system requests.

Remark 5. In constraints l_4, when $l_4=0$, Hopf bifurcation occurs, at which the system is critically stable, and its state characteristic is continuous oscillation. If $l_4>0$, the system is stable and the state is convergent.

LORENZ SYSTEM CONTROLLER DESIGN

It is well known that the Lorenz system widely exists in various research areas [23–26] and it can exhibit complex dynamics, including equilibria, limit cycles, and chaos. This section will explain the controller design and verification process by Lorenz system.

Lorenz System Description

According to [17], Lorenz system can be described as

$$\dot{x} = -a(x - y),$$

$$\dot{y} = bx - xz - y,$$

$$\dot{z} = xy - cz, \tag{21}$$

in which a, b and c are variable parameters. These parameters have direct effect on system characteristics. A little change of them may change the system state and make system trajectory turn from stable to bifurcation and even chaos [17]. According to (21), we can obtain system equilibrium point equation:

$$x_0^3 - (bc - c) x_0 = 0. \tag{22}$$

Controller Design

In accordance with (11), controller can be designed as

$$u_2 = \dot{y}(c_1 + c_2 x + c_3 y + c_4 z) + c_0 \dot{x}, \tag{23}$$

where $c_0 \in R, c_1 \in R, c_2 \in R, c_3 \in R$, and $c_4 \in R$ are controller parameters. According to (21), controller can be rewritten as

$$u_2 = (bx - xz - y)(c_1 + c_2 x + c_3 y + c_4 z) - c_0 a(x - y). \tag{24}$$

Compared with the existing research work [17, 21], (24) has a more concise expression, and it does not include equilibrium point value.

System Stable and Hopf Bifurcation Condition Computation

Lorenz system state characteristics will be directly affected by system parameters. In order to verify controller performance, we select system parameters a=4, b=6, and c=1. When there is no controller, system state trajectory is chaotic. Taking system parameters in Lorenz system (21), we can obtain a new system:

$$\dot{x} = -4(x - y),$$

$$\dot{y} = 16x - xz - y,$$

$$\dot{z} = xy - z. \tag{25}$$

In accordance with (25), the equation at system equilibrium point is

$$x_0^3 - 15x_0 = 0. \tag{26}$$

For simplifying computation process, let $c_0=0$, $c_3=0$, and the controller can be simplified as

$$u_2(x, y, z) = \dot{y}(c_1 + c_2 x + c_4 z). \tag{27}$$

Taking in (27), after computation and arrangement, we can get system constraints:

$$l_1 = c_4 x_0^2 + c_2 x_0 + c_1 + 6, \tag{28}$$

$$l_2 = (5 + 5c_1 + 16c_4) x_0^2 + 16c_2 x_0 - 59c_1 - 55, \tag{29}$$

$$l_3 = (120c_4 + 12c_1 + 12) x_0^2 + 120c_2 x_0 - 60c_1 - 60, \tag{30}$$

$$l_4 = (5c_1^2 + 16c_2^2 + 240c_4^2 + 32c_1c_4 + 23c_1 - 4c_4 + 18) x_0^2 + (32c_1c_2 + 480c_2c_4 - 4c_2) x_0 - 59c_1^2 - 349c_1 - 270. \tag{31}$$

Combined with (26) at equilibrium point, constraints equations (28)~(31) constitute a semialgebra set. Using computer algebra tools, we can simplify the semialgebra set. The arranged system constraints are as follows:

$$\begin{aligned}
&-5 \le c_1 < -1,\\
&c_1 + 15c_4 + 1 > 0,\\
&-15c_2^2 + (c_1 + 15c_4 + 1)^2 > 0,\\
&15c_2^2 + \left(c_1 + 15c_4 - \frac{1}{8}\right)^2 - \frac{1}{64} > 0,\\
&\varphi \ge 0,
\end{aligned} \tag{32}$$

Where

$$\varphi = \left(c_1^2 - 15c_2^2 + 255c_4^2 + 30c_1c_4\right)\left(16c_1^2 - 240c_2^2 + 3600c_4^2 + 480c_1c_4 - 120c_4 - 8c_1 + 1\right). \tag{33}$$

In (32), if $c_1 = -5$ or $\varphi = 0$, the Hopf bifurcation will happen in system. The constraints listed in (32) can use cylindrical algebraic decomposition algorithm to calculate the controller parameter space. But the 3-dimension decomposition result is still not intuitive. In the following numerical simulation, we will assign a certain value to c_1 and then mark out the controller parameter range in 2-dimension space.

SIMULATIONS

Constraints Calculation

Through the above analysis results, we can see $c_1 \in [-5,-1)$. If the system is required to be stable, parameter c_1 should satisfy $c_1 \in (-5,-1)$. Because the system has a bifurcation when $c_1 = -5$, we can choose two different conditions $c_1 = -3$ and $c_1 = -5$ for analysis in numerical simulation.

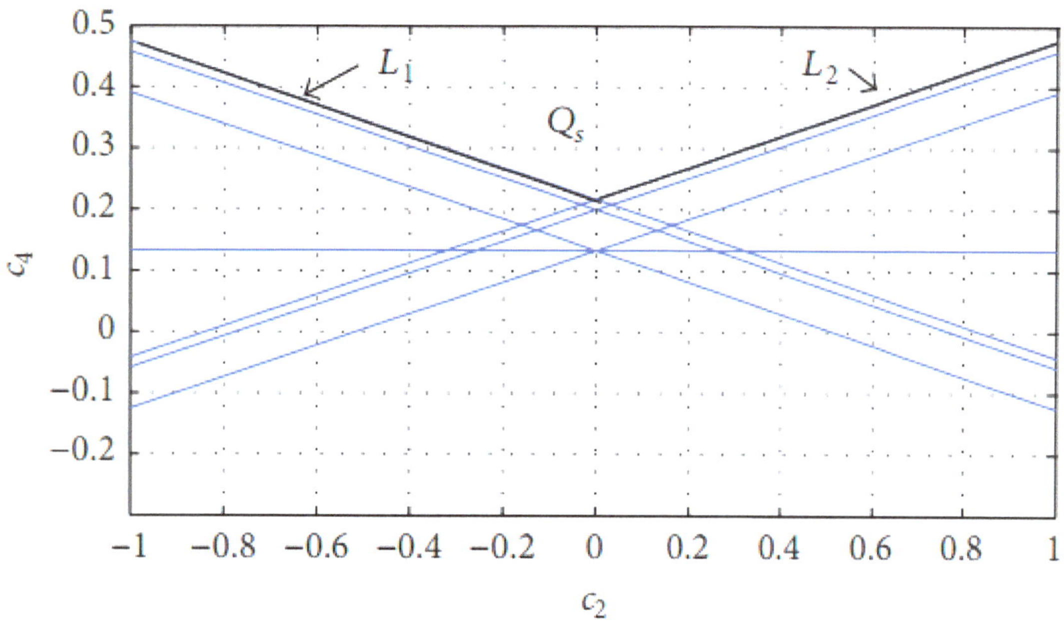

Figure 1: System constraints curves in c_2-c_4 plane when $c_0 = 0$ and $c_1 = -3$.

When $c_1 = -3$, Lorenz system constraints (32) can be rewritten as

$$\begin{aligned}
&c_4 > \frac{2}{15},\\
&-15c_2^2 + 225c_4^2 - 60c_4 + 4 > 0,\\
&15c_2^2 + 225c_4^2 - \frac{375}{4}c_4 + \frac{39}{4} > 0,\\
&\varphi = \left(5c_2^2 - 75c_4^2 + 30c_4 - 3\right)\\
&\quad\cdot\left(240c_2^2 - 3600c_4^2 + 1560c_4 - 169\right) \ge 0.
\end{aligned} \tag{34}$$

When $c_1 = -5$, system constraints (32) can be rewritten as

$$\begin{aligned}
&c_4 > \frac{4}{15},\\
&-15c_2^2 + 225c_4^2 - 120c_4 + 16 > 0,\\
&15c_2^2 + 225c_4^2 - \frac{615}{4}c_4 + \frac{105}{4} > 0,\\
&\varphi = \left(-3c_2^2 + 45c_4^2 - 30c_4 + 5\right)\\
&\quad\cdot\left(-80c_2^2 + 1200c_4^2 - 840c_4 + 147\right) \ge 0.
\end{aligned} \tag{35}$$

Controller Parameter Region

As shown in Figure 1, implicit function curves of constraints (34) equations are drawn in the 2-dimension space. According to cylindrical algebraic decomposition algorithm, we can search the space divided by these function curves; then region Q_s is found. In the region, c_2 and c_4 can stabilize system. The boundary curves of Q_s are L_1 and L_2, and L_1 and L_2 are part of $\varphi = 0$ in (34). L_1 is line $60c_2 + 60\sqrt{15}c_4 - 13\sqrt{15} = 0$ when $c_2 \leq 0$; L_2 is line $-60c_2 + 60\sqrt{15}c_4 - 13\sqrt{15} = 0$ when $c_2 \geq 0$.

In order to provide a convenient explanation, we suppose the three equilibrium points of (26) are E_0, E_-, and E_+, where E_- and E_+ are a pair of symmetric equilibrium points. For parameter space solved by constraints (34), the system state at E_0 is convergent. Choosing different parameters can decide whether there is a bifurcation at system equilibrium points E_- and E_+: (i) there is no existing Hopf bifurcation at two equilibrium points, while the system states are stable, and at the moment the controller parameters c_2 and c_4 value should be in Q_s but not on the boundary; (ii) system states are stable at equilibrium point E_- while at E_+ there is a Hopf bifurcation, and at the moment controller parameter value should on the boundary L_1; (iii) there is a Hopf bifurcation at equilibrium point E_- while system is stable at E_+, and at the moment controller parameter should be on the boundary L_2; (iv) there are Hopf bifurcations at both equilibrium points, and at the moment controller parameter should be on the cross point of L_1 and L_2.

The implicit function curves of constraints (35) equations are shown in Figure 2. Similar to constraints (34), using cylindrical algebraic decomposition algorithm, we can find the parameter space Q_h which satisfies the control requirements. The boundaries of region Q_h are lines L_3 and L_4. They are part of $\varphi = 0$ in (35). L_3 is line $20c_2 + 20\sqrt{15}c_4 - 7\sqrt{15} = 0$ when $c_2 \leq 0$, and L_4 is line $-20c_2 + 20\sqrt{15}c_4 - 7\sqrt{15} = 0$ when $c_2 \geq 0$.

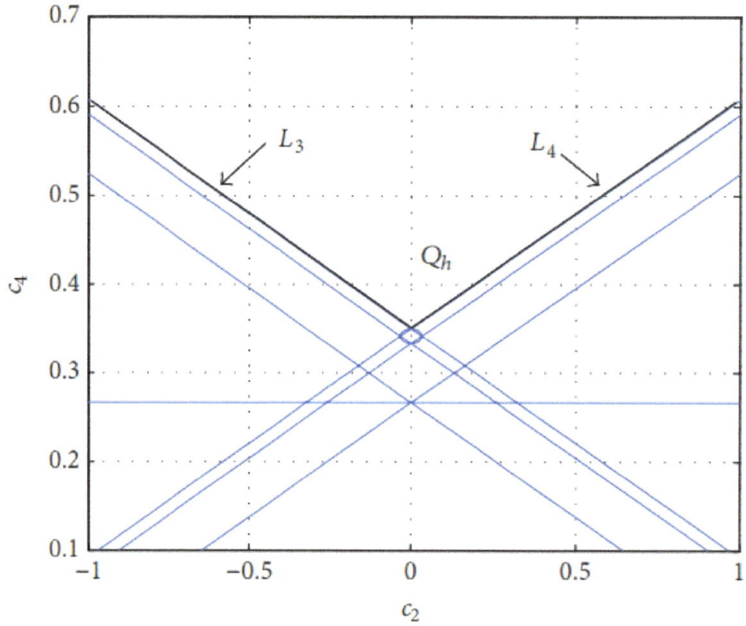

Figure 2: System constraints curves in c_2-c_4 plane when $c_0 = 0$ and $c_1 = -5$.

Unlike constraints (34), there is a Hopf bifurcation at E_0 in (26) by this time. According to the different states at equilibrium points E_- and E_+, the cases are as follows: (i) both equilibria become stable without Hopf bifurcations; at the moment, the controller parameters c_2 and c_4 value should be selected in Q_h and should not be on the boundary; (ii) the system states are stable at equilibrium point E_-, while there is a Hopf bifurcation at E_+, and at the moment controller parameter value should on the boundary L_3; (iii) there is a Hopf bifurcation at equilibrium point E_-, while at E_+, system is stable, and at the moment controller parameter should be on the boundary L_4; (iv) there are Hopf bifurcations at both equilibrium points, and at the moment controller parameter should be on the cross point of L_3 and L_4.

Numeric Simulation Computation Near Equilibria

According to the different conditions at three equilibrium points, we compute the controller parameters below. In order to explain the validity of solved controller parameter range, we conduct the numeric simulation of state stability near equilibrium points for Lorenz system (25). The simulation results are shown in Figures 3–20. When simulating, the initial values of system equilibrium points E_0, E_-, and E_+ are (0, 1, −1), (−4, −4, 16), and (4, 4, 16).

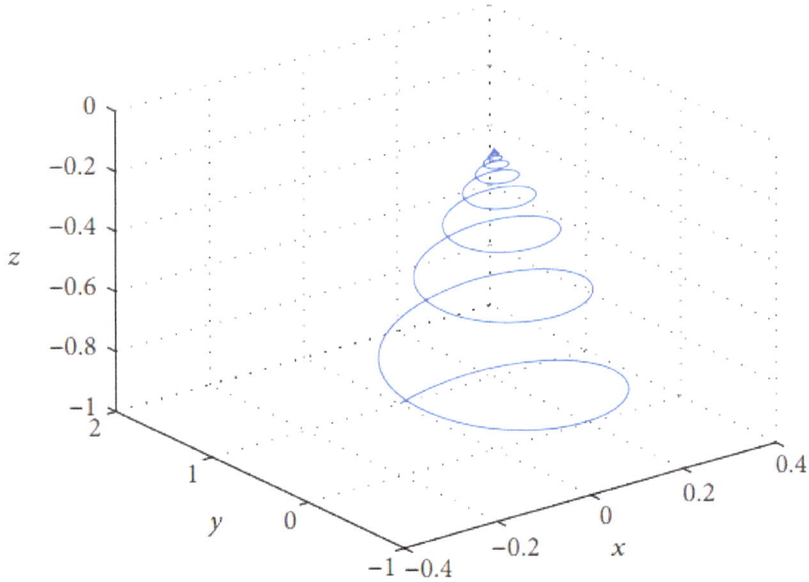

Figure 3: The Lorenz system trajectory is stable at E_0 when $c_1 = -3$.

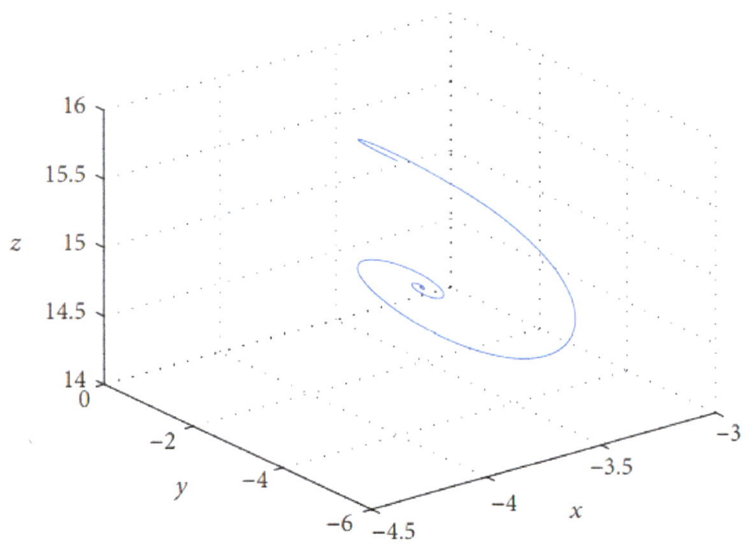

Figure 4: Stable trajectory at E_-, when $c_1 = -3$, for case (i).

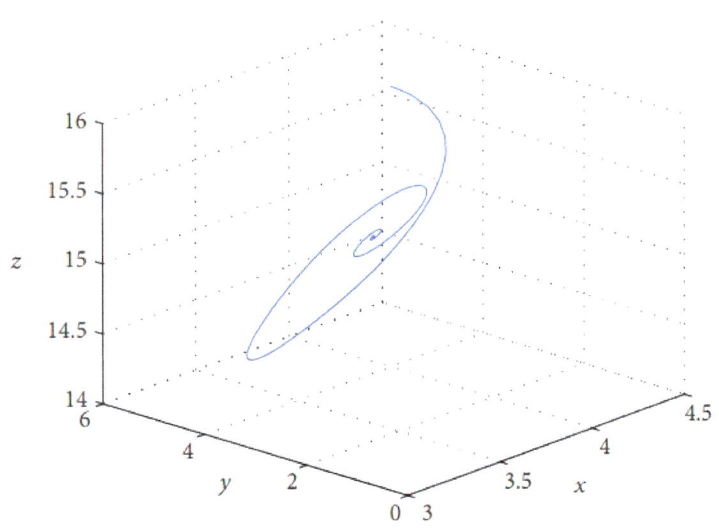

Figure 5: Stable trajectory at E_+, when $c_1 = -3$, for case (i).

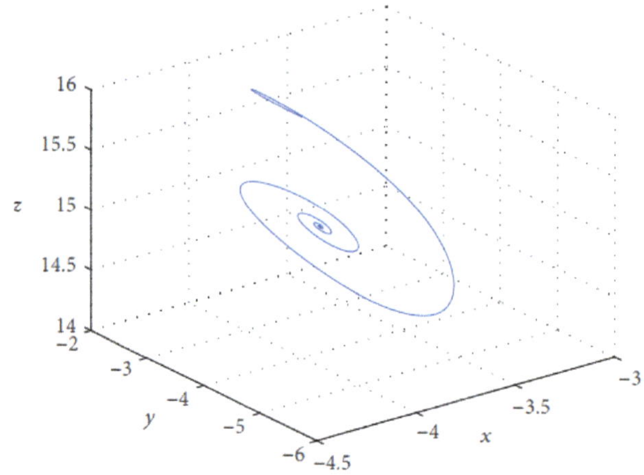

Figure 6: Stable trajectory at E_{-}, when $c1 = -3$,, for case (ii).

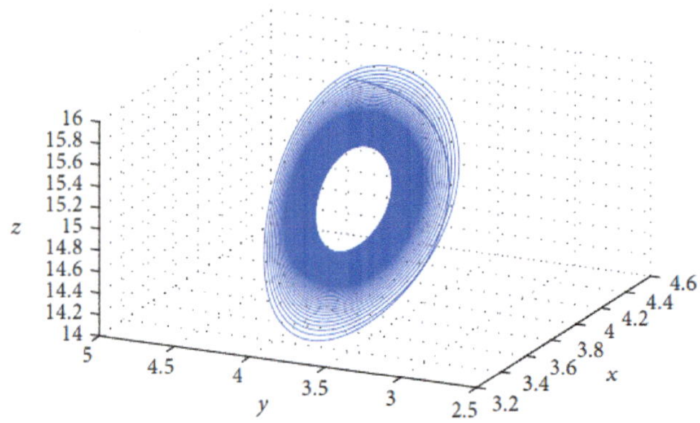

Figure 7: Hopf bifurcation at E_{+}, when $c_1 = -3$, for case (ii).

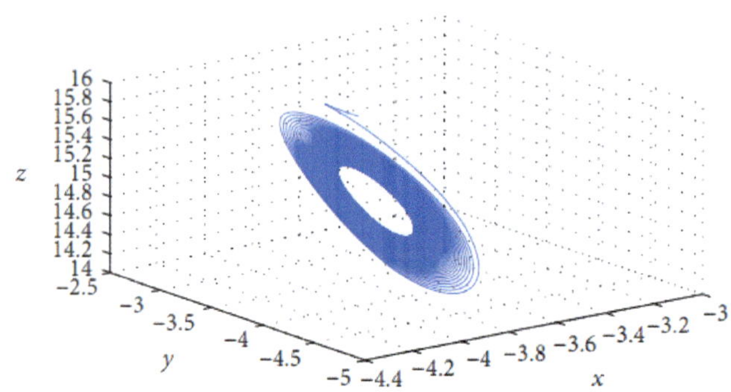

Figure 8: Hopf bifurcation at E_{-}, when $c_1 = -3$, for case (iii).

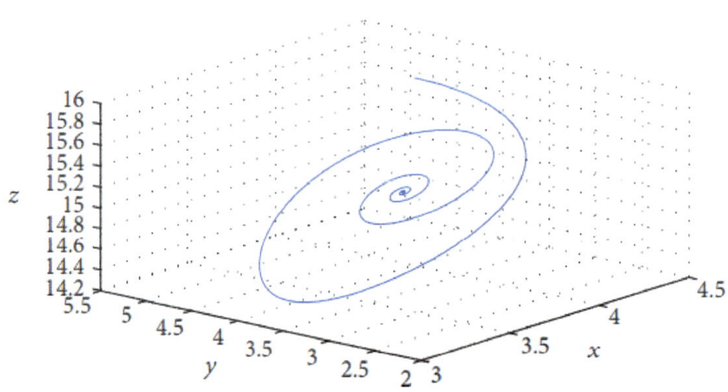

Figure 9: Stable trajectory at E_{+}, when $c_1 = -3$, for case (iii).

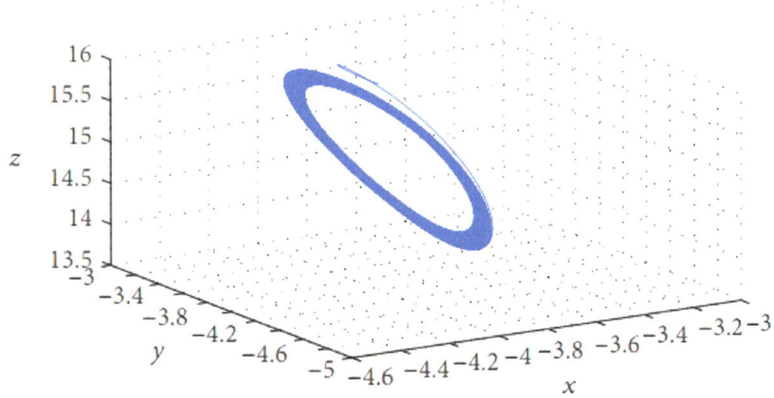

Figure 10: Hopf bifurcation at E_-, when $c_1 = -3$, for case (iv).

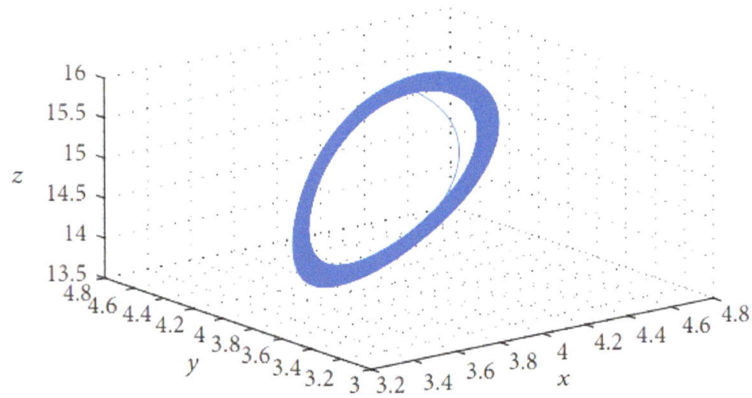

Figure 11: Hopf bifurcation at E_+, when $c_1 = -3$, for case (iv).

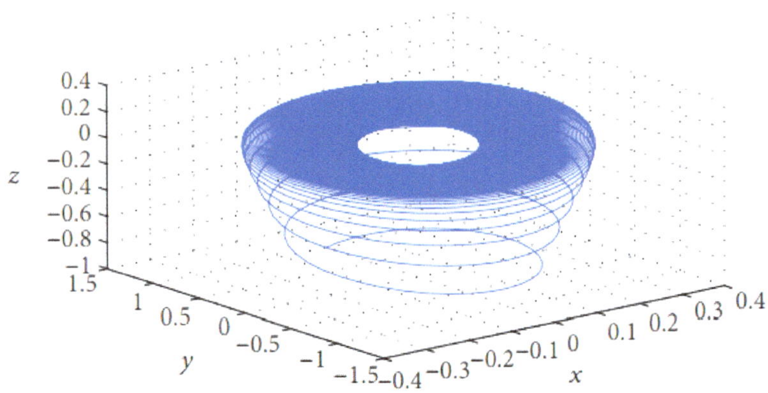

Figure 12: System Hopf bifurcation trajectory at E_0 when $c_1 = -5$.

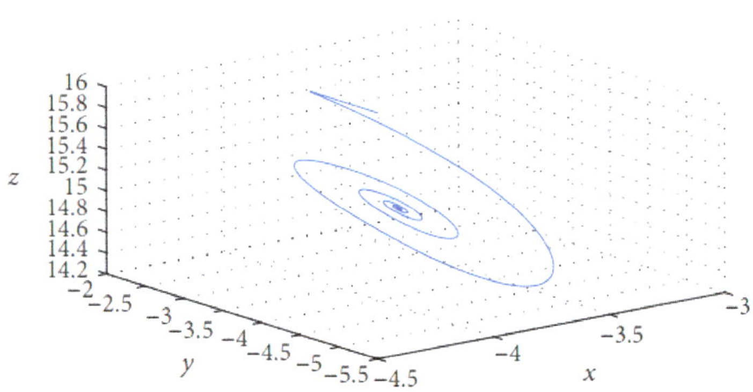

Figure 13: Stable trajectory at E_-, when $c_1 = -5$, for case (i).

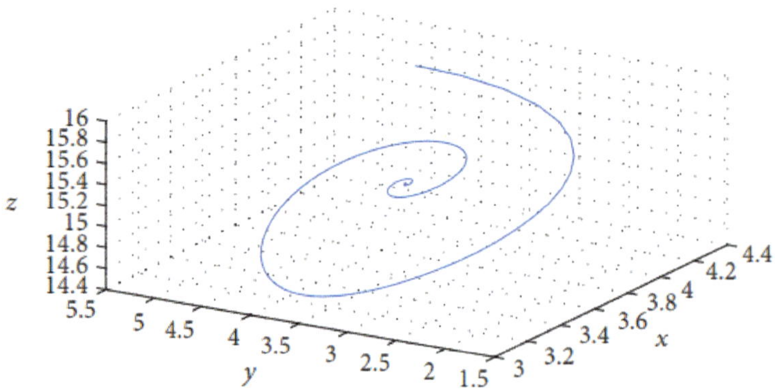

Figure 14: Stable trajectory at E_+, when $c_1 = -5$, for case (i).

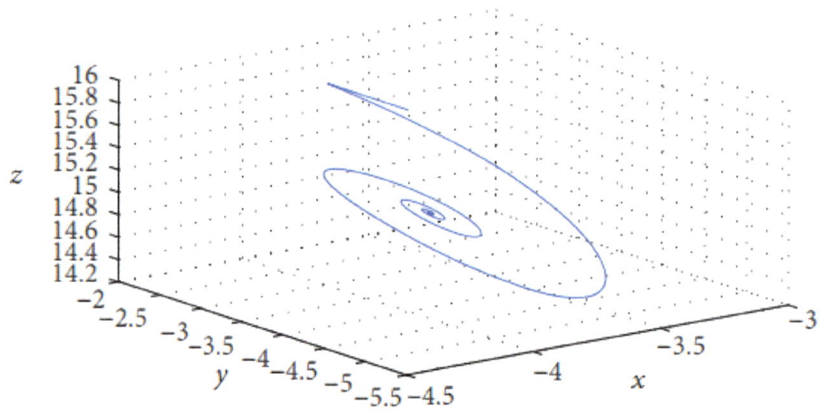

Figure 15: Stable trajectory at E_-, when $c_1 = -5$, for case (ii).

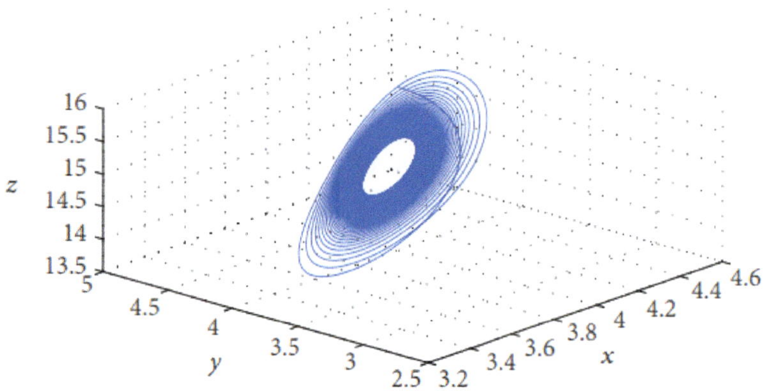

Figure 16: Hopf bifurcation at E_+, when $c_1 = -5$, for case (ii).

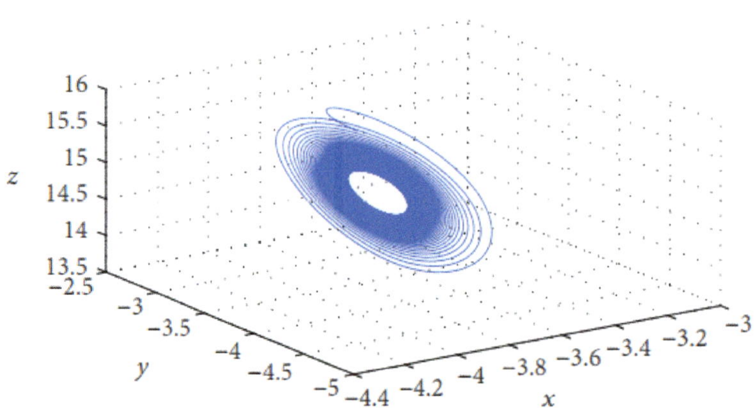

Figure 17: Hopf bifurcation at E_-, when $c_1 = -5$, for case (iii).

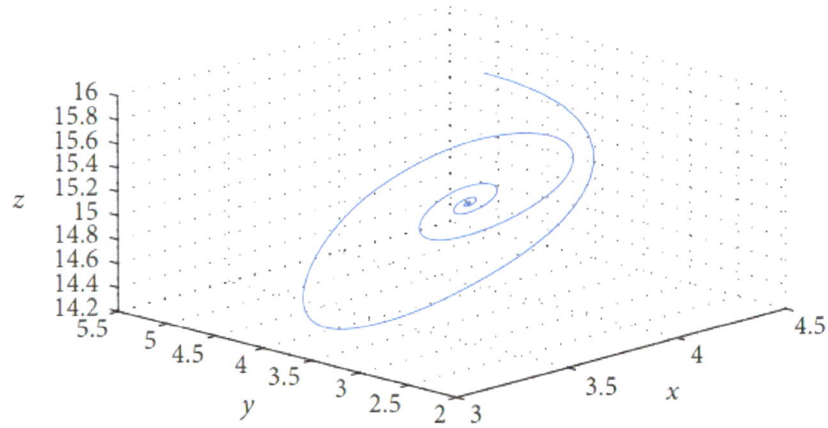

Figure 18: Stable trajectory E_+, when $c_1 = -5$, for case (iii).

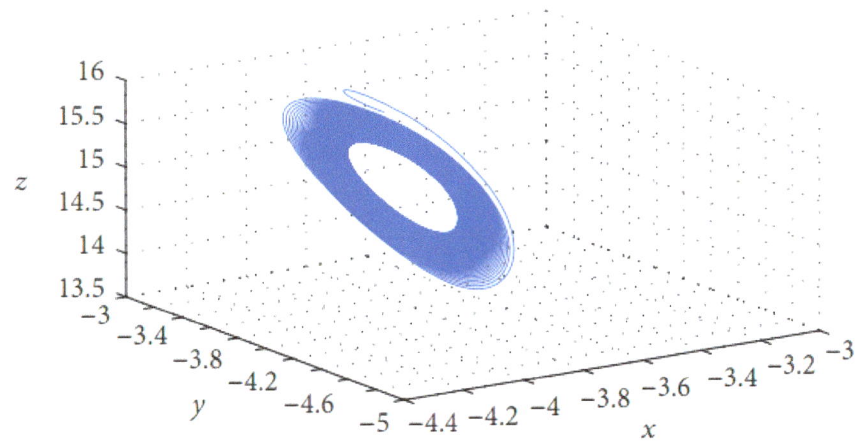

Figure 19: Hopf bifurcation at E_-, when $c_1 = -5$, for case (iv).

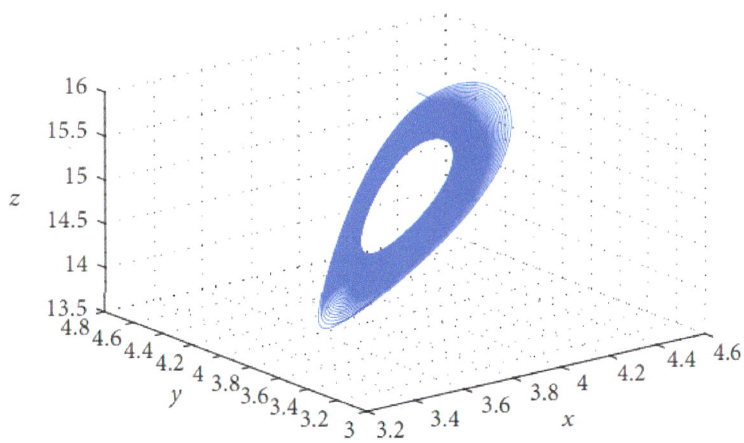

Figure 20: Hopf bifurcation at E_+, when $c_1 = -5$, for case (iv).

When controller parameter $c_1 = -3$, the system trajectory at E_0 is convergent, while according to the different cases of E_- and E_+, parameters c_2 and c_4 can be selected as follows: (i) when the system is stable at two equilibrium points, we can select $c_2 = 1$, $c_4 = 1$; (ii) system is stable at equilibrium point E_-, while at E_+ it has a Hopf bifurcation, and at the moment we can select $c_2 = -1$, $c_4 = (1/60)(13 + 4\sqrt{15}) \approx 0.47487$; (iii) there is a Hopf bifurcation at equilibrium point E_-, while system is stable at E_+, and at the moment we can select $c_2 = 1$, $c_4 = (1/60)(13+4\sqrt{15}) \approx 0.47487$; (iv) both equilibria E_- and E_+ have Hopf bifurcations, and at the moment we can select $c_2 = 0$, $c_4 = 13/60 \approx 0.21667$.

In Figure 3 the Lorenz system trajectory is convergent at E_0 when $c_1=-3$. Figures 4–11 are system simulation trajectories for different cases of E_- and E_+, in which, corresponding to case (i), Figure 4 is the stable convergent trajectory at E_-, and Figure 5 is the stable convergent trajectory at E_+. Corresponding to case (ii), Figure 6 is the stable convergent trajectory at E_-, and Figure 7 is the system trajectory of Hopf bifurcation at E_+. For case (iii), Figure 8 is the Hopf bifurcation trajectory at E_-, and Figure 9 is the stable convergent trajectory at E_+. Figures 10 and 11 are the system trajectories of Hopf bifurcation at E_- and E_+ for case (iv).

When controller parameter $c_1 = -5$, system has Hopf bifurcation at E_0, and its trajectory is a limit cycle. And according to different cases at E_- and E_+, parameters c_2 and c_4 can be chosen as follows: (i) when the system is stable at two equilibrium points, we can select $c_2 = 1$, $c_4 = 1$; (ii) system is stable at equilibrium point E_+, while there is a Hopf bifurcation at , and at the moment we can select $c_2 = -1$, $c_4 = (1/60)(21 + 4\sqrt{15}) \approx 0.60819$; (iii) there is a Hopf bifurcation at equilibrium point E_-, while system is stable at E_+, and at the moment we can select $c_2 = 1$, $c_4 = (1/60)(21+4\sqrt{15}) \approx 0.60819$; (iv) there are Hopf bifurcations at both equilibria, and we can select $c_2 = 0$, $c_4 = 7/20 = 0.35$.

Figure 12 is the system Hopf bifurcation trajectory at E_0 when $c_1 = -5$, and at this moment system converges to a stable limit cycle. Figures 13–20 are system simulation trajectories at E_- and E_+ for different cases, in which, according to case (i), Figures 13 and 14 are the stable trajectories at E_- and E_+. In accordance with case (ii), Figure 15 is the stable trajectory at E_-, and Figure 16 is the Hopf bifurcation trajectory at E_+. For case (iii), Figure 17 is the Hopf bifurcation trajectory at E_-, and Figure 18 is the stable trajectory at E_+. In accordance with condition (iv), Figure 19 is the Hopf bifurcation trajectory at E_-, and Figure 20 is the Hopf bifurcation trajectory at E_+.

Figures 21 and 22 are the system stable trajectories, using the controllers of this paper and [17]. Curve 1 is the system trajectory under the control of [17] and curve 2 is under the control of this paper. From Figures 21 and22, it is easy to see that both controllers can stabilize the Lorenz system at equilibrium, and the controller designed in this paper is more simple and practical.

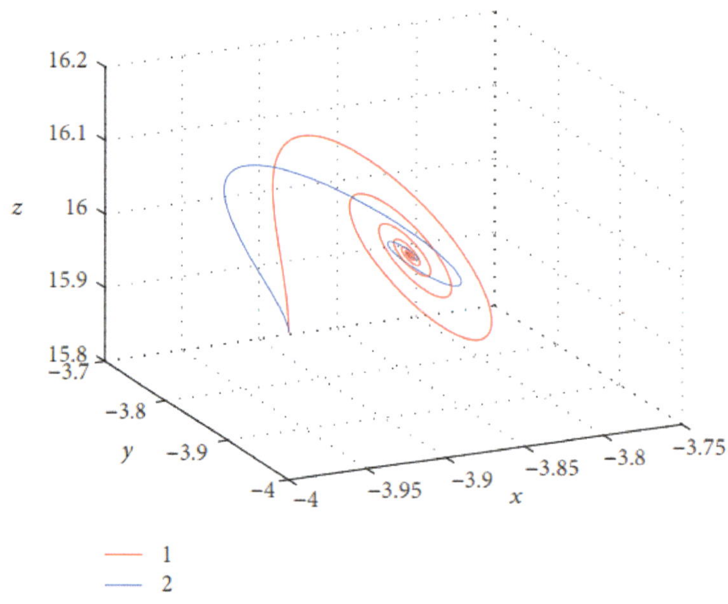

Figure 21: Stable at E_- using controllers in [17] and this paper.

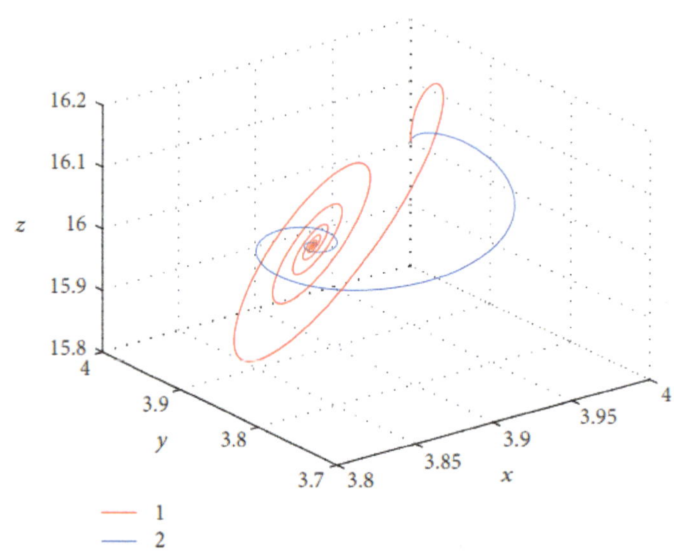

Figure 22: Stable at E_+ using controllers in [17] and this paper.

CONCLUSIONS

Aiming at the limitations of existing controller such as complex controller form, high order, including system equilibrium points in controller, and complicated solving method for parameters constraints, in this paper, a new parametric controller design method is proposed by researching and analyzing present Hopf bifurcation nonlinear system parametric controller design methods. This paper presents the design principle analysis and computation steps of this new parametric controller, designs a corresponding parametric controller for a Lorenz system example, deduces the constraints, and explains the controller method validity by numeric simulation. From simulation, compared with the other methods, it is easy to see that this new parametric controller method has simple form, lower order, not including any equilibrium point value.

Conflict of Interests

The authors declare that there is no conflict of interests regarding the publication of this paper.

ACKNOWLEDGMENT

This work is partially supported by the NSF of China Grants nos. 61374001 and 61074189.

REFERENCES

1. E. H. Abed and J.-H. Fu, "Local feedback stabilization and bifurcation control. I. Hopf bifurcation,"Systems & Control Letters, vol. 7, no. 1, pp. 11–17, 1986. ·
2. G. Chen and K. Yap, "Feedback control of Hopf bifurcations," IEEE Transactions on Circuits and Systems I: Fundamental Theory and Applications, vol. 6, pp. 661–672, 2001.
3. A. Jimenez-Triana, W. K.-S. Tang, G. Chen, and A. Gauthier, "Chaos control in duffing system using impulsive parametric perturbations," IEEE Transactions on Circuits and Systems, vol. 57, no. 4, pp. 305–309, 2010.
4. C. Bi, Y. Xiang, Q. Zhang, and J. Hu, "Simulation study of bifurcation control in synchronous switching converter," in Proceedings of the International Conference on Communications, Circuits and Systems (ICCCAS '13), vol. 2, pp. 463–466, Chengdu, China, November 2013.
5. A. R. Phadke, M. Fozdar, K. R. Niazi, N. Mithulananthan, and R. C. Bansal, "New technique for computation of closest Hopf bifurcation point using real-coded genetic algorithm," IET Generation, Transmission & Distribution, vol. 5, no. 1, pp. 11–18, 2011. ·
6. M. Tomim, B. Lopes, R. Leme, et al., "Modified Hopf bifurcation index for power system stability assessment," IEE Proceedings—Generation, Transmission and Distribution, vol. 152, no. 6, pp. 906–912, 2005.
7. A. Kavitha and G. Uma, "Experimental verification of Hopf bifurcation in DC–DC Luo converter," IEEE Transactions on Power Electronics, vol. 23, no. 6, pp. 2878–2883, 2008.
8. F. Liu, H. O. Wang, and Z.-H. Guan, "Stability analysis and control of bifurcation in a TCP fluid flow model of wireless networks," in Proceedings of the 10th World Congress on Intelligent Control and Automation (WCICA '12), pp. 1026–1030, IEEE, Beijing, China, July 2012.
9. L. Yang, Z. Xu, J. Østergaard, Z. Y. Dong, K. P. Wong, and X. Ma, "Oscillatory stability and eigenvalue sensitivity analysis of a DFIG wind turbine system," IEEE Transactions on Energy Conversion, vol. 26, no. 1, pp. 328–339, 2011.
10. K. E. Chlouverakis, K. M. Al-Aswad, I. D. Henning, and M. J. Adams, "Determining laser linewidth parameter from Hopf bifurcation minimum in lasers subject to optical injection," Electronics Letters, vol. 39, no. 16, pp. 1185–1187, 2003.
11. M. Huang, C. K. Tse, S.-C. Wong, C. Wan, and X. Ruan, "Low-frequency hopf bifurcation and its effects on stability margin in three-phase PFC power supplies connected to non-ideal power grid," IEEE Transactions on Circuits and Systems I: Regular Papers, vol. 60, no. 12, pp. 3328–3340, 2013.
12. P. S. Deivasundari, G. Uma, and R. Santhi, "Experimental verification of Hopf bifurcation in pulse-width modulated inverter fed cage induction motor drive system," IET Power Electronics, vol. 7, no. 2, pp. 340–349, 2014.
13. N. G. Sakellaridis, M. E. Karystianos, and C. D. Vournas, "Homoclinic loop and degenerate hopf bifurcations introduced by load dynamics," IEEE Transactions on Power Systems, vol. 24, no. 4, pp. 1892–1893, 2009.
14. G. Chen, J. L. Moiola, and H. O. Wang, "Bifurcation control:theories,methods,andapplications,"International Journal of Bifurcation and Chaos, vol. 10, no. 3, pp. 511–548, 2000.
15. F. Liu, G. Xiong, Z.-H. Guan, and H. O. Wang, "Stability analysis and control Hopf bifurcation in a FAST TCP model," in Proceedings of the 32nd Chinese Control Conference (CCC '13), pp. 1076–1080, Xi'an, China, July 2013.
16. M. Xiao, W. X. Zheng, and J. Cao, "Hopf bifurcation of an (n+1)-neuron bidirectional associative memory neural network model with delays," IEEE Transactions on Neural Networks and Learning Systems, vol. 24, no. 1, pp. 118–132, 2013.
17. P. Yu and G. Chen, "Hopf bifurcation control using nonlinear feedback with polynomial functions,"International Journal of Bifurcation and Chaos in Applied Sciences and Engineering, vol. 14, no. 5, pp. 1683–1704, 2004. ·
18. T. Yang and X. Chen, "Local L_2 gain of Hopf bifurcation stabilization," in Proceedings of the 47th IEEE Conference on Decision and Control (CDC '08), pp. 4103–4108, IEEE, Cancún, Mexico, December 2008.
19. P. Cai, J.-S. Tang, and Z.-B. Li, "Controlling Hopf bifurcation of a new modified hyperchaotic Lü system,"Mathematical Problems in Engineering, vol. 2015, Article ID 614135, 6 pages, 2015. ·
20. F. Verduzco, J. Alvarez, and A. Carrillo, "Normal form and control of the Hopf bifurcation," inProceedings of

the American Control Conference (ACC '07), pp. 2358–2363, New York, NY, USA, July 2007.

21. Z. Wu and P. Yu, "A method for stability and bifurcation control," IEEE Transactions on Automatic Control, vol. 51, no. 6, pp. 1019–1023, 2006.

22. Z.-Q. Wu, R. Ding, and Y.-S. Chen, "Classification of parametrically constrained bifurcations," Applied Mathematics and Mechanics. English Edition, vol. 31, no. 2, pp. 135–142, 2010.

23. K. Cho and T. Miyano, "Chaotic cryptography using augmented lorenz equations aided by quantum key distribution," IEEE Transactions on Circuits and Systems I: Regular Papers, vol. 62, no. 2, pp. 478–487, 2015.

24. R. Tokunaga, T. Matsumoto, L. O. Chua, and S. Miyama, "The piecewise-linear Lorenz circuit is chaotic in the sense of Shilnikov," IEEE Transactions on Circuits and Systems, vol. 37, no. 6, pp. 766–786, 1990.

25. S. Yu, J. Lu, X. Yu, and G. Chen, "Design and implementation of grid multiwing hyperchaotic Lorenz system family via switching control and constructing super-heteroclinic loops," IEEE Transactions on Circuits and Systems, vol. 59, no. 5, pp. 1015–1028, 2012. ·

26. A. Dwivedi, A. K. Mittal, and S. Dwivedi, "Adaptive synchronisation of diffusionless Lorenz systems and secure communication of digital signals by parameter modulation," IET Communications, vol. 6, no. 13, pp. 2016–2026, 2012.

CHAPTER 13

SIGNAL PROCESSING IN PERIODICALLY FORCED GRADIENT FREQUENCY NEURAL NETWORKS

Ji Chul Kim and Edward W. Large

INTRODUCTION

Neural oscillation is observed throughout the central nervous system and has been suggested to have important functional roles in peripheral, subcortical, and cortical processing (Buzsáki and Draguhn, 2004; Sejnowski and Paulsen, 2006; Koepsell et al., 2010). In the auditory system, oscillatory activities are found at all levels of the pathway, including spontaneous oscillation of haircell bundles (Crawford and Fettiplace, 1985; Martin et al., 2003; Ramunno-Johnson et al., 2009), cochlear dynamics poised near a critical point of oscillatory instability called the Hopf bifurcation (Camalet et al., 2000; Ospeck et al., 2001), chopper and onset cells in the cochlear nucleus mode-locking to periodic stimulation (Laudanski et al., 2010), and the inferior colliculus neurons spontaneously firing at audible frequencies (Schwarz et al., 1993). Mathematical models of non-linear oscillation are used to explain the oscillatory response of these auditory areas to periodic stimulation (Eguíluz et al., 2000; Jülicher et al., 2001; Meddis and O'Mard, 2006; Laudanski et al., 2010; Fredrickson-Hemsing et al., 2012). Forced non-linear oscillator models share important behaviors including synchronization and non-linear compression, but at the same time they exhibit diverse dynamical responses to periodic signals.

To apprehend the full range of possible behaviors, it is important to understand how the dynamical properties of forced non-linear oscillations vary from one parameter regime to another. Here we provide a mathematical analysis of an oscillatory network model that is widely used in auditory modeling. We enumerate the full set of behaviors the model exhibits under periodic forcing. This analysis reveals the signal processing capabilities of a large class of dynamical systems.

Mathematical models of individual neurons and neural populations vary in their degree of physiological detail and mathematical complexity. Biophysically detailed models describe neurophysiological mechanisms with variables representing physical or chemical quantities and can exhibit a range of diverse behaviors observed in the original biological system (e.g., the Hodgkin–Huxley model; Hodgkin and Huxley, 1952). Other models take simpler mathematical forms and capture the local dynamics of a select behavior with fewer variables, thus making the essential dynamics more transparent and amenable to mathematical analysis (Hoppensteadt and Izhikevich, 2001). The canonical model for gradient frequency neural networks (abbr. GrFNNs) is one such simple mathematical model that describes the dynamical properties shared by networks of oscillatory neural populations tuned to a gradient of distinct frequencies, which are found at various stages of the auditory system (Large et al., 2010). The model assumes that each oscillatory neural population (or neural oscillator) in the network is poised near a Hopf

Citation: Kim JC and Large EW (2015), Signal Processing in Periodically Forced Gradient Frequency Neural Networks. Front. Comput. Neurosci. 9:152, doi: 10.3389/fncom.2015.00152.

bifurcation point, which is a transition between quiescence and spontaneous oscillation (Guckenheimer and Holmes, 1983). The canonical model for gradient frequency networks can be considered as an extension of the canonical model for homogeneous (equal or very close) frequency networks of neural oscillators (Hoppensteadt and Izhikevich, 1997) into multi-frequency systems.

When the oscillators are tuned to logarithmically spaced frequencies, the dynamics of the canonical model for gradient frequency neural networks (Large et al., 2010) is described by

$$\tau_i \dot{z}_i = z_i \left(\alpha + i2\pi + (\beta_1 + i\delta_1)|z_i|^2 + \frac{\epsilon(\beta_2 + i\delta_2)|z_i|^4}{1 - \epsilon|z_i|^2} \right) + RT, \quad (1)$$

where z_i is a complex state variable representing the amplitude and phase of synchronized firing of the i th neural population in the network, ϵ is a small real number indicating the degree of non-linearity in the network, a dot over a variable denotes its time derivative, and the roman i denotes the imaginary unit. The right-hand side of Equation (1) consists of the intrinsic terms (all right-hand-side terms except RT), which determine the autonomous behavior of the model, and the input terms (RT), which describe the interaction of the model with the input. The bandwidth of oscillators is constant in logarithmic frequency because the equation is scaled by the time constant τ_i, which is the reciprocal of the natural frequency f_i. The intrinsic parameters α, β_1, and β_2 control the bifurcation of autonomous behavior (see below), and δ_1 and δ_2 determine the dependency of autonomous frequency on amplitude. RT (resonant terms) is a sum of input terms that are potentially resonant to the oscillator's dynamics, which could include both linear and non-linear terms involving external forcing and/or coupling with other oscillators in the network (see Large et al., 2010; Lerud et al., 2014, for possible closed-form expressions of RT). Commonly, models of oscillation near a Hopf bifurcation have intrinsic terms only up to the third order [i.e., the cubic term with β_1 and δ_1 in Equation (1); see Eguíluz et al., 2000; Jülicher et al., 2001 for instance], but the canonical model retains a full series of higher-order terms expressed as a geometric sum (i.e., the term with β_2 and δ_2) to cope with the high-order non-linear input terms in RT.

With the high-order terms governing non-linear interactions of an oscillator with the external signal and also with other oscillators in the network, the canonical model captures the general properties of non-linear dynamics arising in gradient frequency oscillator networks. When driven by an external signal, the oscillators in the canonical model produce non-linear responses containing not only the frequencies in the signal but also non-linear combinations of their natural frequencies and the signal frequencies. Non-linear coupling in the network transforms the signal further by introducing frequencies arising from resonance between oscillators tuned to different frequencies. As a generic model of non-linear, multi-frequency transformation of acoustic signals into neural firing patterns occurring in the auditory system, the canonical model has been used to model auditory processing and music perception. Multi-layer gradient frequency networks were used to model cochlear dynamics by fitting the auditory nerve tuning curves of macaque monkeys (Lerud et al., 2015) and to model the human brainstem frequency-following response to musical intervals by fitting the spectra of auditory evoked potentials (Large and Almonte, 2012; Lerud et al., 2014). Also, both model simulations and analytic predictions were used to explain the perception of musical tonality (Large, 2010; Large et al., in press) and the beat perception in musical rhythm (Large et al., 2015).

Despite its simple mathematical form, the canonical model for gradient frequency neural networks is still difficult to analyze in its entirety because its dynamics is determined by complex interactions among multiple network components. Oscillators in the network are driven by external forcing and at the same time receive input from other oscillators, and both types of interaction may involve linear and/or non-linear coupling which can evolve over time via a generalized form of Hebbian plasticity (Hoppensteadt and Izhikevich, 1996; Large, 2011). Our approach is to analyze individual components of the network separately and attempt to understand its overall dynamics as a combination of its component dynamics. In this paper, we set to analyze and categorize the driven behaviors of canonical oscillators under periodic forcing.

METHODS

We consider the following differential equation describing an oscillator in the canonical model (or simply, a canonical oscillator) driven by sinusoidal forcing of fixed frequency, ω_0, and amplitude, F:

$$\dot{z} = z \left(\alpha + i\omega + \beta_1|z|^2 + \frac{\epsilon\beta_2|z|^4}{1 - \epsilon|z|^2} \right) + Fe^{i\omega_0 t}, \quad (2)$$

where $\omega = 2\pi f$ is the radian natural frequency. To understand the response of a gradient frequency network, our analysis will focus on how the driven state of an oscillator changes as a function of its natural frequency. Since only one oscillator is analyzed, the subscript i in Equation (1) is dropped and the scaling factor is ignored (but see Section 3.6 for frequency scaling of log frequency networks). For the simplicity of analysis, the δ parameters are set to zero, meaning that the intrinsic frequency of the oscillator is not dependent on its amplitude.

The autonomous behavior of the oscillator (i.e., when $F = 0$) is readily seen when it is brought to polar coordinates using $z = re^{i\phi}$. Then, the amplitude and phase dynamics are described by

$$\begin{cases} \dot{r} = \alpha r + \beta_1 r^3 + \frac{\epsilon\beta_2 r^5}{1-\epsilon r^2} \\ \dot{\phi} = \omega. \end{cases}$$

The first equation above defines the amplitude vector field, which shows whether the amplitude increases, decreases, or is stationary over time at a given amplitude value (Figure 1). A fixed point in the vector field, which is obtained by solving $\dot{r} = 0$, represents a steady-state amplitude of the autonomous oscillator. The stability of a fixed point determines if it is an attractor, to which the oscillator returns after small perturbation, or a repeller, from which the oscillator diverges when perturbed. The second equation above shows that the phase ϕ advances at the constant rate of ω.

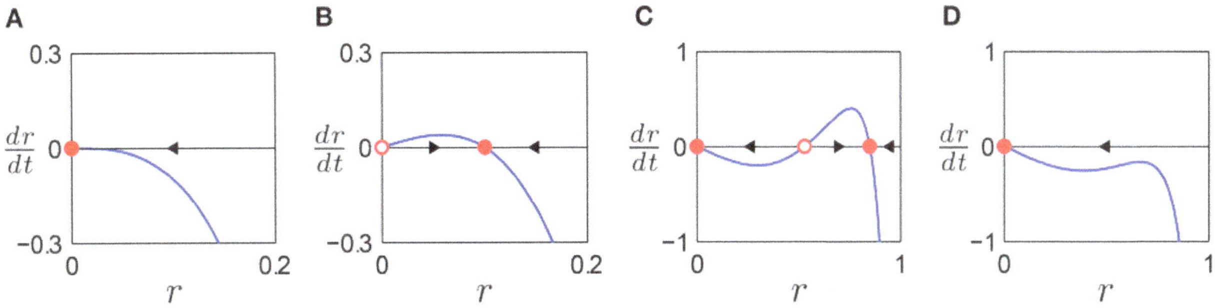

Figure 1. Autonomous behavior of a canonical oscillator in different parameter regimes. Amplitude vector field is shown for **(A)** a critical Hopf regime ($\alpha = 0$, $\beta_1 < 0$, $\beta_2 = 0$), **(B)** a supercritical Hopf regime ($\alpha > 0$, $\beta_1 < 0$, $\beta_2 = 0$), **(C)** a supercritical double limit cycle regime ($\alpha < 0$, $\beta_1 > 0$, $\beta_2 < 0$, local max > 0), and **(D)** a subcritical double limit cycle regime ($\alpha < 0$, $\beta_1 > 0$, $\beta_2 < 0$, local max < 0). Filled circles indicate stable fixed points (attractors) and empty circles unstable fixed points (repellers). Arrows indicate the direction of trajectories in the vector field.

Depending on the values of α, β_1, and β_2, the autonomous amplitude vector field can have one of four distinct topologies. When $\dot{r}$ decreases monotonically as r increases, the origin is the only fixed point which is stable as the arrow indicates (Figure 1A). An oscillator with this type of amplitude vector field decays to zero while oscillating at its natural frequency. A representative parameter regime for this type is the critical point of a supercritical Hopf bifurcation ($\alpha = 0$, $\beta_1 < 0$). (A subcritical Hopf bifurcation occurs when $\alpha = 0$ and $\beta_1 > 0$). When $\dot{r}$ increases from the origin and then decreases after a local maximum, there is a stable non-zero fixed point while the origin is rendered unstable (Figure 1B). An oscillator of this type shows spontaneous oscillation at the amplitude of the stable fixed point (unless the initial condition is zero). The supercritical branch of a supercritical Hopf bifurcation ($\alpha > 0$, $\beta_1 < 0$) is an example. When there are three fixed points with two local extrema, two of the fixed points are stable, indicating bistability between equilibrium at zero and spontaneous oscillation at a non-zero amplitude (Figure 1C). As the local maximum in the vector field moves below the r axis by, say, decreasing β_1, the two non-zero fixed points collide and vanish (Figure 1D). This transition is called a double limit cycle (hereafter, DLC) bifurcation since it involves two limit cycles (closed orbits) in the (r, ϕ) plane, one stable and the other unstable. Thus, we call the regime shown in Figure 1C ($\alpha < 0$, $\beta_1 > 0$, $\beta_2 < 0$, local max > 0)supercritical DLC and the one shown in Figure 1D ($\alpha < 0$, $\beta_1 > 0$, $\beta_2 < 0$, local max < 0) subcritical DLC. The subcritical DLC regime has only one stable fixed point at zero but is different from the critical Hopf regime (Figure 1A) in that it has a local maximum in the vector field. We will show that a subcritical DLC oscillator has different sets of driven behaviors from a critical Hopf oscillator despite their qualitatively identical autonomous behavior.

To examine how a canonical oscillator responds to external forcing, we bring Equation (2) to polar coordinates, again using $z = re^{i\phi}$, and express its dynamics in terms of the relative phase $\psi = \phi - \omega_0 t$ so that a stable fixed point in (r, ψ) indicates a phase-locked state:

$$\begin{cases} \dot{r} = \alpha r + \beta_1 r^3 + \frac{\epsilon \beta_2 r^5}{1-\epsilon r^2} + F\cos\psi \\ \dot{\psi} = \Omega - \frac{F}{r}\sin\psi, \end{cases} \qquad (3)$$

where $\Omega = \omega - \omega_0$ is the frequency difference between the oscillator and the input. We evaluate the stability of fixed point(s) for a range of forcing parameters Ω and F wide enough to encompass all possible qualitatively different driven behaviors of the four regimes of intrinsic parameters introduced above. Stability analysis is crucial for understanding the dynamical responses of the driven oscillator because not all fixed points are stable. The existence of a steady-state solution (i.e., a fixed point) does not guarantee that the oscillator phase-locks to the forcing, since the fixed point could be unstable (Figure 2).

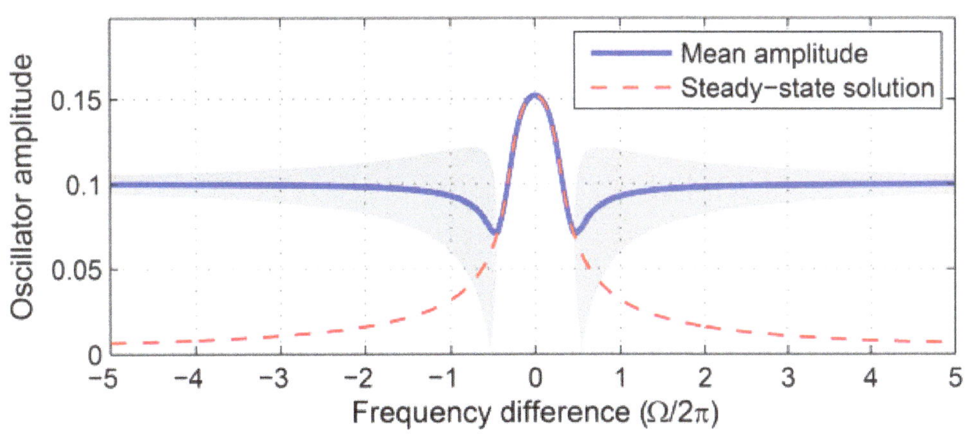

Figure 2. Not all steady-state solutions are stable. Time-averaged amplitude of a canonical oscillator driven by a sinusoidal input obtained from numerical simulations (solid line) is compared with the steady-state solution of Equation 3 (dashed line). Steady-state solutions are stable attractors where they match the mean amplitudes from simulations. The gray shade indicates the range of amplitude fluctuation where the oscillator does not stably phase-lock to the input. Compare with the stability analysis of the same oscillator ($\alpha = 1$, $\beta_1 = -100$, $\beta_2 = 0$, $F = 0.2$) shown in Figure 6.

Fixed points, notated as (r^*, ψ^*), are obtained by solving the steady-state equations $\dot{r} = 0$ and $\dot{\psi} = 0$ simultaneously, and the stability of each fixed point is determined by evaluating the Jacobian matrix, J, at the fixed point:

$$J = \begin{pmatrix} \frac{\partial p}{\partial r} & \frac{\partial p}{\partial \psi} \\ \frac{\partial q}{\partial r} & \frac{\partial q}{\partial \psi} \end{pmatrix}_{(r^*,\psi^*)}$$

$$= \begin{pmatrix} \alpha + 3\beta_1 r^{*2} + \frac{\epsilon\beta_2 r^{*4}(5-3\epsilon r^{*2})}{(1-\epsilon r^{*2})^2} & -F\sin\psi^* \\ 2\delta_1 r^* + \frac{2\epsilon\delta_2 r^{*3}(2-\epsilon r^{*2})}{(1-\epsilon r^{*2})^2} + \frac{F}{r^{*2}}\sin\psi^* & -\frac{F}{r^*}\cos\psi^* \end{pmatrix},$$

where $\dot{r} = p(r, \psi, \cdots)$ and $\dot{\psi} = q(r,\psi,\cdots)\dot{\psi} = q(r,\psi,\cdots)$ in Equation (3). Let T and Δ be the trace and determinant of the Jacobian matrix. Then its eigenvalues are $\lambda_{1,2} = \frac{1}{2}(T \pm \sqrt{T^2 - 4\Delta})$, , and local trajectories near the fixed point have the form $c_1 e^{\lambda_1 t} v_1 + c_2 e^{\lambda_2 t} v_2$ where $\mathbf{v}_{1,2}$ are the eigenvectors of $\lambda_{1,2}$, $c_{1,2}$ are constants determined by initial conditions, and t is time. So the shape of local trajectories near a fixed point, and thus its stability type, is determined by the signs of T, Δ, and $T^2 - 4\Delta$. A fixed point is

- a stable node if $\Delta > 0$, $T^2 - 4\Delta > 0$, and $T < 0$,
- a stable spiral if $\Delta > 0$, $T^2 - 4\Delta < 0$, and $T < 0$,
- an unstable node if $\Delta > 0$, $T^2 - 4\Delta > 0$, and $T > 0$,
- an unstable spiral if $\Delta > 0$, $T^2 - 4\Delta < 0$, and $T > 0$, or
- a saddle point if $\Delta < 0$ (see Strogatz, 1994).

We categorize the driven behavior of a canonical oscillator by examining how the stability of fixed point(s) varies for different values of forcing parameters Ω and F. Stability analysis is done for four regimes of intrinsic behavior using representative parameter settings (see Figure 1): critical Hopf ($\alpha = 0$, $\beta_1 < 0$, $\beta_2 = 0$), supercritical Hopf ($\alpha > 0$, $\beta_1 < 0$, $\beta_2 = 0$), supercritical double limit cycle ($\alpha < 0$, $\beta_1 > 0$, $\beta_2 < 0$, local max > 0), and subcritical double limit cycle regimes ($\alpha < 0$, $\beta_1 > 0$, $\beta_2 < 0$, local max < 0).

RESULTS

Critical Hopf Oscillator

A critical Hopf oscillator is a canonical oscillator poised at the critical point of a supercritical Hopf bifurcation (where $\alpha = 0$ and $\beta_1 < 0$), which means the system is on the verge of spontaneous oscillation. Oscillatory instability at a Hopf bifurcation has recently been shown to underlie non-linear cochlear dynamics characterized by frequency selectivity, sensitivity to weak signals, and non-linear compression (see Hudspeth et al., 2010, for a review), and a bank of oscillators poised at or near Hopf bifurcation points has been used as a model of the cochlea (Jülicher et al., 2001; Duke and Jülicher, 2003; Kern and Stoop, 2003; Magnasco, 2003; Stoop et al., 2005). Here, we choose the simple parameter setting of $\alpha = 0$, $\beta_1 < 0$, and $\beta_2 = 0$ to analyze the behavior of a critical Hopf oscillator under sinusoidal forcing, but there are other parameter regimes that share qualitatively the same driven dynamics (e.g., α, β_1, $\beta_2 < 0$; see Section 3.5 for a classification of parameter regimes by driven behavior).

A stability analysis shows that a critical Hopf oscillator phase-locks to sinusoidal forcing of any frequency and amplitude. For a fixed forcing amplitude, the steady-state amplitude r^* is maximum when the forcing frequency is the same as the natural frequency (i.e., $\Omega = 0$), for which the steady-state relative phase ψ^* is zero indicating in-phase synchronization (Figure 3A). As the natural frequency and the forcing frequency become more different, r^* decreases monotonically and approaches zero while ψ^* approaches $\pm\frac{\pi}{2}$. While the fixed point (r^*, ψ^*) remains stable for all values of Ω, it changes its stability type from a stable node to a stable spiral as $|\Omega|$ increases from 0. It is clearly seen in the (r, ψ) space that the two attractors have distinct local trajectories (Figures3 B, C). The way r and ψ approach their steady-state values in time (monotonic vs. oscillating approach) reflects the difference between a node and a spiral (only the relative phase is shown in Figures 3D,E).

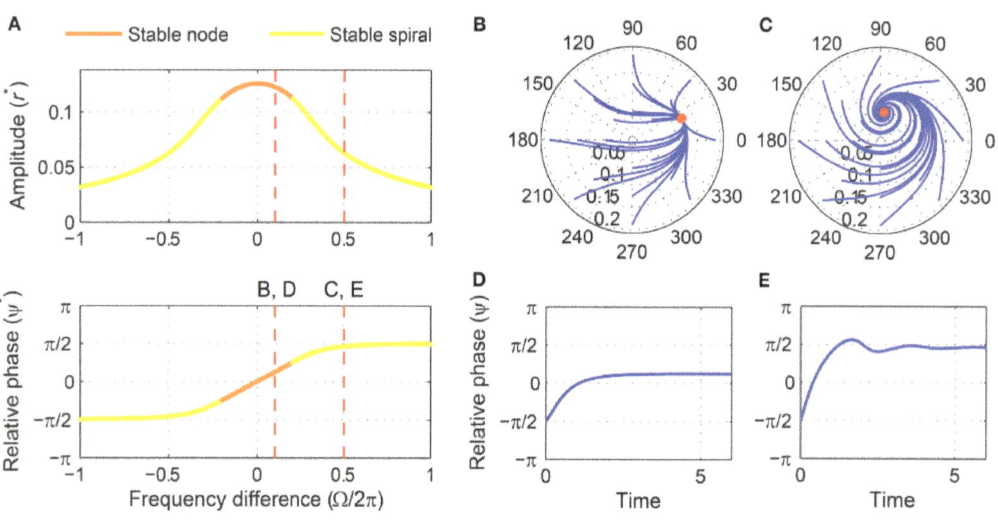

Figure 3. Driven behavior of a critical Hopf oscillator. (A) Steady-state amplitude and relative phase as a function of frequency difference ($\alpha = 0$, $\beta_1 = -100$, $\beta_2 = 0$, $F = 0.2$), with vertical dashed lines indicating the frequency differences used for **(B–E)**, **(B)**trajectories attracted to a stable node in the (r, ψ) plane starting from a set of different initial conditions ($\Omega/2\pi = 0.1$), **(C)** trajectories attracted to a stable spiral ($\Omega/2\pi = 0.5$), **(D)** relative phase plotted over time for a trajectory in **(B)** (phase locking), and **(E)** relative phase plotted over time for a trajectory in**(C)** (phase locking). Filled circles in **(B,C)** indicate stable fixed points.

We can find the boundary between stable nodes and stable spirals by solving $T^2 - 4\Delta = 0$, $\dot{r} = 0$, and $\dot{\psi} = 0$ simultaneously where T and Δ are the trace and determinant of the Jacobian matrix evaluated at a fixed point (see Section 2). We find that the boundary is at

$$|\Omega_c| = \sqrt[3]{-\frac{\beta_1 F^2}{2}},$$

for which

$$(r_c^*, \psi_c^*) = \left(\sqrt[6]{\frac{F^2}{2\beta_1^2}}, \pm\frac{\pi}{4}\right).$$

A critical Hopf oscillator shows the same set of driven behaviors summarized in Figure 3 for all levels of forcing amplitude. With increasing F, r* increases and the node-spiral boundary widens, but no qualitatively different behaviors are introduced as the forcing amplitude changes (Figure 4A, see also Supplementary Video 1).

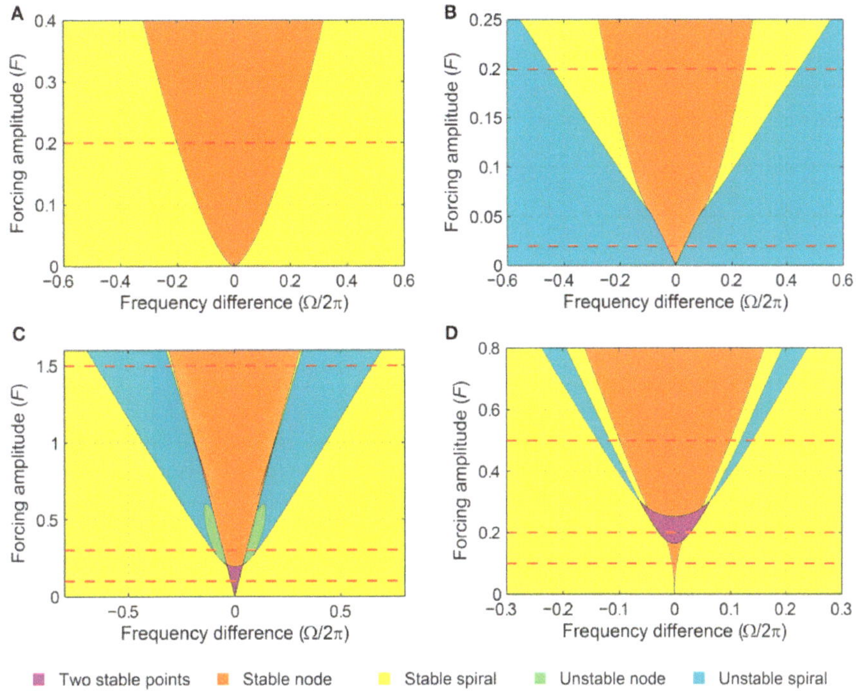

Figure 4. Stability regions for a canonical oscillator under sinusoidal forcing. The stability of driven state (r*, ψ*) is shown as a function of forcing amplitude and frequency difference for **(A)** a critical Hopf oscillator ($\alpha = 0$, $\beta_1 = -100$, $\beta_2 = 0$), **(B)** a supercritical Hopf oscillator ($\alpha = 1$, $\beta_1 = -100$, $\beta_2 = 0$), **(C)** a supercritical double limit cycle oscillator ($\alpha = -1$, $\beta_1 = 4$, $\beta_2 = -1$, $\epsilon = 1$), and **(D)** a subcritical double limit cycle oscillator ($\alpha = -1$, $\beta_1 = 2.5$, $\beta_2 = -1$, $\epsilon = 1$). The color indicates the stability type of a stable fixed point if there is one (purple if there are two). If there is no stable fixed point, the color indicates the stability of an unstable fixed point. Dashed horizontal lines indicate the forcing amplitudes used for Figures 3, 5–11. See also Supplementary Videos 1–4.

Supercritical Hopf Oscillator

A supercritical Hopf oscillator, which is on the supercritical branch of a supercritical Hopf bifurcation ($\alpha > 0$, $\beta_1 < 0$), has a non-zero spontaneous amplitude (Figure 1B) and has been used as a model of spontaneously oscillating systems such as haircell bundles (Fredrickson-Hemsing et al., 2012). (Spontaneous amplitude refers to the steady-state amplitude of an oscillator when no external forcing is applied). A stability analysis shows that it has two distinct sets of driven behaviors depending on the forcing amplitude and that, unlike a critical Hopf oscillator, it does not always phase-lock to sinusoidal forcing.

For weak forcing, there exist three steady-state solutions for small frequency differences, two of which are a saddle-node pair, and just one unstable solution for large frequency differences (Figure5A). As the frequency difference increases from zero for a fixed forcing amplitude, the saddle and node are lost via a saddle-node invariant-circle (SNIC) bifurcation (also called a saddle-node infinite-period or SNIPER bifurcation), which leaves a stable (attracting) limit-cycle orbit with an unstable fixed point inside (Figures 5A–C). The critical frequency difference for which a SNIC bifurcation occurs can be obtained by solving $\Delta = 0$, $\dot{r} = 0$, and $\dot{\psi} = 0$ together. We find the SNIC boundary to be at

$$\Gamma_{SN} = |\Omega_c| = \sqrt{-(\alpha + 3\beta_1 r_c^{*2})(\alpha + \beta_1 r_c^{*2})},$$

where r_c^* is the bigger of the two positive real roots of $2\beta_1^2 r_c^{*6} + 2\alpha\beta_1 r_c^{*4} + F^2 = 0$. For $|\Omega| < \Gamma_{SN}$ the canonical oscillator phase-locks to the forcing, with its driven state attracted to a stable node (Figures 5B,D). For $|\Omega| > \Gamma_{SN}$ for which only one unstable fixed point exists, the relative phase does not converge to a steady-state value but makes full 2π-rotations (i.e., phase slip), meaning the oscillator is not phase-locked to the input, and the amplitude fluctuates near the spontaneous amplitude (Figures 5C,E). (The spontaneous amplitude of the oscillator shown in Figure 5 is $\sqrt{-\alpha/\beta_1} = 0.1)$. Thus, the SNIC bifurcation marks the phase-locking boundary for a supercritical Hopf oscillator under weak forcing. Note that the flow of relative phase is slow near $\frac{\pi}{2}$ (or $-\frac{\pi}{2}$ when $\Omega < 0$) where the saddle-node pair collides and leaves a bottleneck or a "ghost" (Figure 5E).

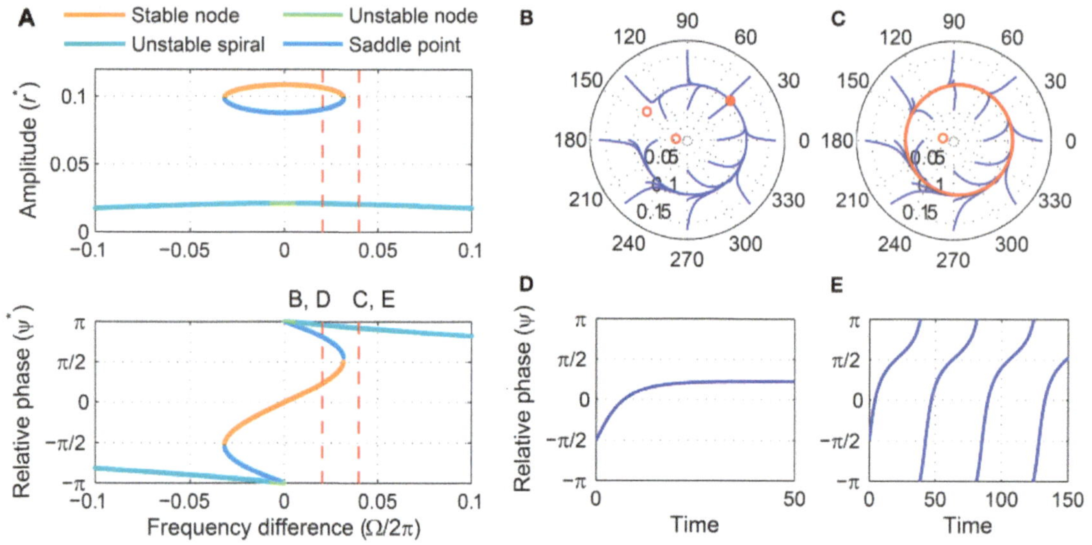

Figure 5. Driven behavior of a supercritical Hopf oscillator under weak forcing. **(A)** Steady-state amplitude and relative phase as a function of frequency difference ($\alpha = 1$, $\beta_1 = -100$, $\beta_2 = 0$, $F = 0.02$), with vertical dashed lines indicating the frequency differences used for **(B–E)**, **(B)** trajectories attracted to a stable node in the (r, ψ) plane ($\Omega/2\pi = 0.02$), **(C)** trajectories drawn to a limit cycle ($\Omega/2\pi = 0.04$), **(D)** relative phase plotted over time for a trajectory in **(B)** (phase locking), and **(E)** relative phase plotted over time for a trajectory in**(C)** (phase slip). In **(B, C)**, filled and empty circles indicate stable and unstable fixed points respectively, and red lines show limit-cycle orbits.

For stronger forcing, only one fixed point exists for all values of frequency difference, but it changes from a stable node to a stable spiral then to an unstable spiral as $|\Omega|$ grows from zero (Figure 6A). Now the phase-locking boundary is at the transition from a stable spiral to an unstable spiral (i.e., a Hopf bifurcation in the (r, ψ) space), which we can find by solving $T = 0$, $\dot{r} = 0$, and $\dot{\psi}=0\dot{\psi}=0$ together. We get

$$\Gamma_H = |\Omega_c| = \sqrt{-\frac{2\beta_1 F^2}{\alpha} - \frac{\alpha^2}{4}},$$

for which

$$r_c^* = \sqrt{-\frac{\alpha}{2\beta_1}} \quad \text{and} \quad \cos\psi_c^* = -\frac{1}{F}\sqrt{-\frac{\alpha^3}{8\beta_1}}.$$

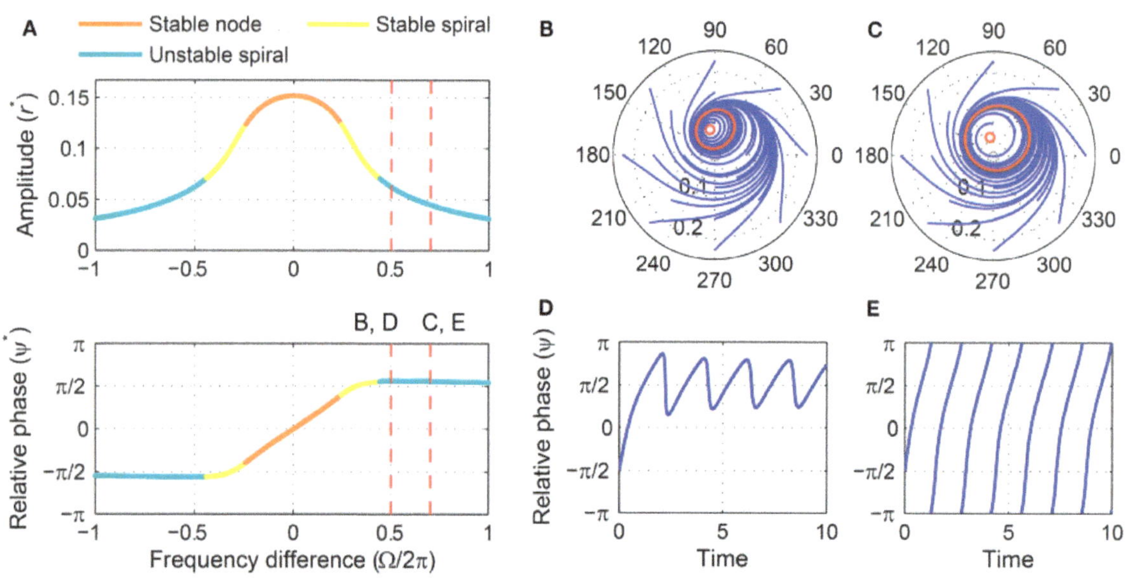

Figure 6. Driven behavior of a supercritical Hopf oscillator under strong forcing. (A) Steady-state amplitude and relative phase as a function of frequency difference ($\alpha = 1$, $\beta_1 = -100$, $\beta_2 = 0$, $F = 0.2$), with vertical dashed lines indicating the frequency differences used for **(B–E)**, **(B)** trajectories attracted to a phase-trapped libration in the (r, ψ) plane ($\Omega/2\pi = 0.5$), **(C)** trajectories attracted to a rotation ($\Omega/2\pi = 0.7$),**(D)** relative phase plotted over time for a trajectory in **(B)** (phase-trapped frequency locking without phase locking), and**(E)** relative phase plotted over time for a trajectory in **(C)** (phase slip). In **(B,C)**, empty circles indicate unstable fixed points, and red lines show limit-cycle orbits.

Note that at the Hopf boundary the steady-state amplitude r_c^* is smaller than the spontaneous amplitude $(r^* = \sqrt{-\alpha/\beta_1}$ when F = 0), and the steady-state relative phase ψ_c^* goes beyond $\pm\frac{\pi}{2}$ since $\cos\psi_c^*$ is negative (see also Figure 6A).

Outside the phase-locking range for strong forcing, the driven behavior of a supercritical Hopf oscillator can be divided into two categories. Just outside the Hopf boundary, the driven state (r, ψ) circles on a stable limit cycle which forms around the unstable spiral and is small enough not to encompass the origin (Figure 6B). In this case, the relative phase changes over time but is bounded and does not traverse the full 2π range (Figure 6D), which is called a libration (as opposed to a rotation, see Strogatz, 1994). When averaged over time, this "phase-trapped" oscillation has the same mean frequency as the input frequency, so it can be described as frequency locking without phase locking (Hoppensteadt and Izhikevich, 1997; Pikovsky et al., 2000, 2001). As |Ω| increases further, the limit cycle around the unstable spiral grows and eventually encompasses the origin (Figure 6C), and the relative phase starts making full rotations (Figure 6E). At this point, the average instantaneous frequency of the oscillator is different from the input frequency and approaches the natural frequency as |Ω| approaches infinity.

The existence of phase-trapped libration (thus, frequency locking) outside the phase-locking boundary is a distinct feature of the Hopf boundary. When crossing the SNIC boundary for weak forcing, the driven state changes directly from phase locking to phase slipping (Figures 5B–E). The same transition of driven behaviors is found for phase models (i.e., oscillators described by their phases only). This is not unexpected since under weak forcing the amplitude of a supercritical Hopf oscillator is effectively constant and does not change much from its spontaneous value.

The SNIC phase-locking boundary and the Hopf boundary exist only for weak and strong forcing levels respectively (Figure 4B; see also Supplementary Video 2 for the transition between Figures5A, 6A), but the two types of phase-locking boundary coexist for a small range of intermediate forcing level. The SNIC boundary exists for forcing amplitudes smaller than

$$F_{SN} = \sqrt{-\frac{8\alpha^3}{27\beta_1}},$$

for which the fixed point at the SNIC bifurcation is

$$(r_c^*, \psi_c^*) = \left(\sqrt{-\frac{2\alpha}{3\beta_1}}, \pm\frac{2\pi}{3}\right).$$

The Hopf boundary, on the other hand, exists for forcing amplitudes greater than

$$F_H = \sqrt{-\frac{\alpha^3}{4\beta_1}},$$

for which the fixed point at the bifurcation is

$$(r_c^*, \psi_c^*) = \left(\sqrt{-\frac{\alpha}{2\beta_1}}, \pm\frac{3\pi}{4}\right).$$

Note that $F_{SN} > F_H$. For forcing amplitudes between the two values, two stable fixed points (a stable node and a stable spiral) coexist for some values of Ω near the locking boundaries. Also, this region of the forcing parameter space (i.e., near the tip of the yellow regions in Figure 4B) contains a complicated but well-studied set of bifurcations. For instance, a Bogdanov–Takens bifurcation is at the lower end of Γ_H, and a cusp point is at the upper end of Γ_{SN}. Detailed analysis of the dynamics around them can be found in Guckenheimer and Holmes (1983), for example. Also, it is worth noting that the same set of bifurcations are found for other periodically driven non-linear oscillators or populations of oscillators such as the forced van der Pol oscillator (Holmes and Rand, 1978) and the forced Kuramoto model (Childs and Strogatz, 2008). However, the canonical model analyzed here, with its simple mathematical form, allows closer analytical examination than is possible for more complex models.

Supercritical Double Limit Cycle Oscillator

As shown in the introduction, a supercritical DLC oscillator ($\alpha < 0$, $\beta_1 > 0$, $\beta_2 < 0$, local max > 0) has two stable autonomous behaviors. Depending on the initial condition, it can be attracted to an equilibrium at zero or oscillate spontaneously with non-zero amplitudes at its natural frequency (Figure 1C). The transition between a supercritical DLC oscillator and a subcritical DLC oscillator, called a double limit cycle bifurcation or a fold limit cycle bifurcation, has been suggested to be involved in several types of bursting neurons (Izhikevich, 2000, 2001). When driven by a sinusoid, a supercritical DLC oscillator shows three distinct sets of behaviors depending on the strength of the forcing, and many of these behaviors involve bistability as well.

Under weak forcing, it has two stable fixed points for small frequency differences and only one for large frequency differences (Figure 7A). The stable fixed point with a small amplitude exists for all values of Ω, but the one with a high amplitude (a stable node) is lost via a SNIC bifurcation and leaves a limit-cycle rotation (phase slip) where it collides with a saddle point (Figures 7B–E). (Due to non-zero β_2 and the higher-order terms it introduces, it is not possible to get a closed-form expression for the frequency difference for which a bifurcation of driven states occurs, as was done for the models with $\beta_2 = 0$ discussed above. However, the location of bifurcations can be obtained using numerical methods.) The transition from phase locking on a stable node to phase slip on a stable limit cycle is identical to what happens at the phase-locking boundary of a supercritical Hopf oscillator under weak forcing (Figure 5), but the presence of a second stable fixed point at low amplitudes differentiates supercritical DLC oscillators from supercritical Hopf oscillators. So, a weakly forced supercritical DLC oscillator shows bistability for all values of frequency difference—phase locking at high or low amplitudes for small values of |Ω|, and phase locking at low amplitudes or phase slip at high amplitudes for large values of |Ω|—and the initial condition determines the driven state to which the oscillator is attracted.

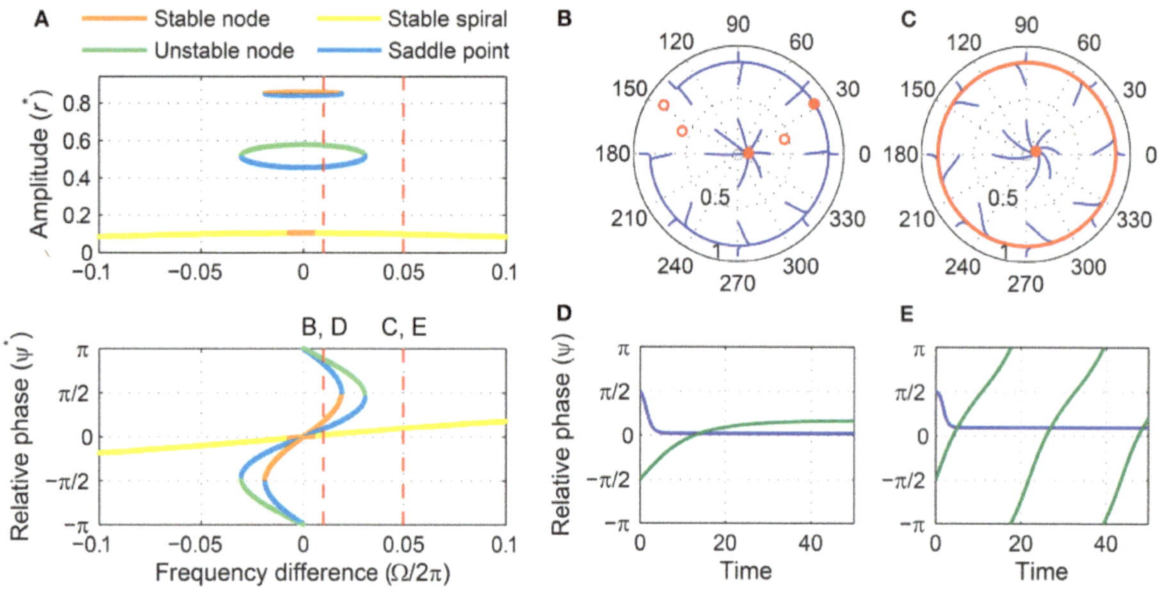

Figure 7. Driven behavior of a supercritical double limit cycle oscillator under weak forcing. **(A)** Steady-state amplitude and relative phase as a function of frequency difference ($\alpha = -1$, $\beta_1 = 4$, $\beta_2 = -1$, $\epsilon = 1$, $F = 0.1$), with vertical dashed lines indicating the frequency differences used for **(B–E)**, **(B)** trajectories attracted to either of two stable fixed points in the (r, ψ) plane ($\Omega/2\pi = 0.01$), **(C)** trajectories drawn to either a stable spiral or a limit-cycle rotation ($\Omega/2\pi = 0.05$), **(D)** relative phase plotted over time for two trajectories in **(B)** (both phase locking), and **(E)** relative phase plotted over time for two trajectories in **(C)** (phase locking in blue, phase slip in green). In **(B,C)**, filled and empty circles indicate stable and unstable fixed points respectively, and red lines show limit-cycle orbits.

For intermediate forcing amplitudes, the saddle-node pair at high amplitudes still exists for small frequency differences, but the stable fixed point at low amplitudes exists only for large frequency differences (Figure 8A). There is a range of intermediate frequency differences for which no stable fixed point exists and all trajectories are attracted to a limit-cycle rotation that the saddle-node pair leaves (Figure 8C). As the frequency difference increases, the fixed point inside the limit cycle changes from an unstable node to an unstable spiral and eventually to a stable spiral (Figures8A,C,D). The emergence of a stable spiral inside a stable (attracting) limit cycle indicates a subcritical Hopf bifurcation. Thus, as the frequency difference increases from zero, the driven state of a supercritical DLC oscillator under intermediate forcing goes from phase locking (a stable node, Figure 8B) to phase slip (a stable limit cycle, Figure 8C) and then to the bistability between phase locking and phase slip (a stable spiral inside a stable limit cycle, Figure 8D).

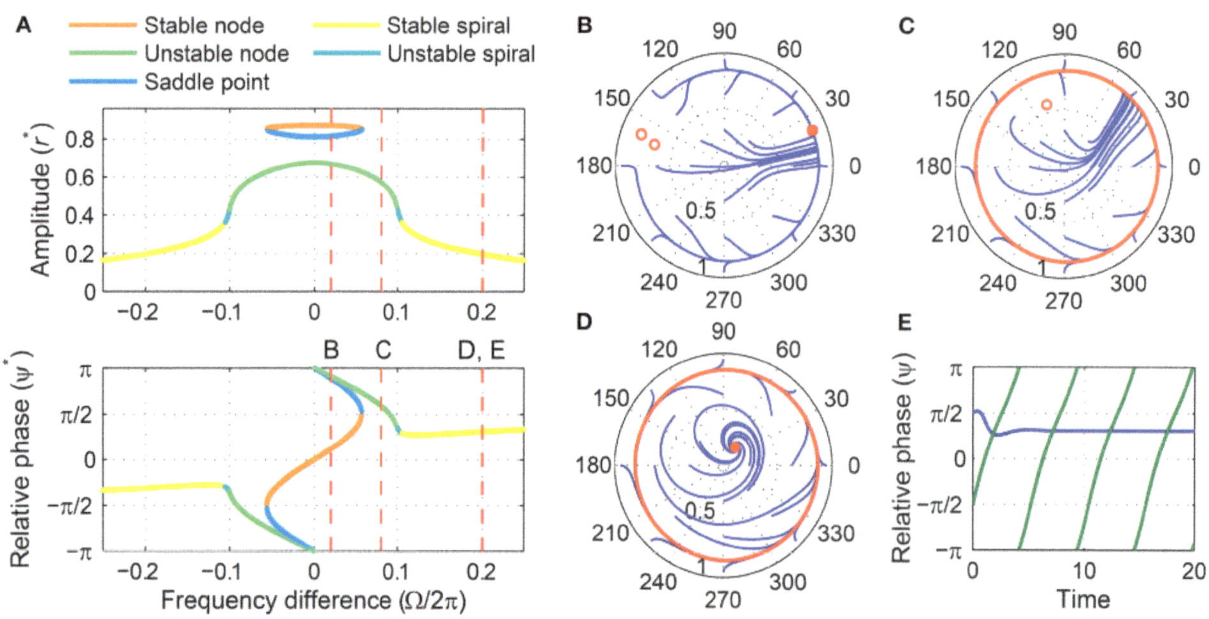

Figure 8. Driven behavior of a supercritical double limit cycle oscillator under intermediate forcing. **(A)** Steady-state amplitude and relative phase as a function of frequency difference ($\alpha = -1$, $\beta_1 = 4$, $\beta_2 = -1$, $\epsilon = 1$, $F = 0.3$), with vertical dashed lines indicating the frequency differences used for **(B–E)**, **(B)** trajectories attracted to a stable node in the (r, ψ) plane ($\Omega/2\pi = 0.02$), **(C)** trajectories drawn to a limit-cycle rotation ($\Omega/2\pi = 0.08$), **(D)** trajectories drawn to either a stable spiral or a limit cycle ($\Omega/2\pi = 0.2$), and **(E)** relative phase plotted over time for two trajectories in **(D)** (phase locking in blue, phase slip in green). In **(B–D)**, filled and empty circles indicate stable and unstable fixed points respectively, and red lines show limit-cycle orbits.

When the forcing amplitude is further increased, only one fixed point exists for any value of frequency difference (Figure 9A). Similar to a strongly forced supercritical Hopf oscillator (Figure6), the driven state (r, ψ) of a strongly forced supercritical DLC oscillator goes through transitions from a stable node (phase locking), a stable spiral (phase locking, Figure 9B), a libration around an unstable spiral (frequency locking without phase locking, Figure 9C), and a rotation around an unstable spiral (phase slip, Figure 9D). In addition to these states, a supercritical DLC oscillator exhibits another driven behavior for even larger frequency differences, bistability between phase locking on a stable spiral and phase slip on a stable limit cycle (Figure 9E). So, a strongly forced supercritical DLC oscillator has two phase-locking boundaries, a supercritical Hopf bifurcation (between Figure 9B and Figure 9C) and a subcritical Hopf bifurcation (between Figure 9D and Figure 9E). Figure 4C and Supplementary Video 3 show how the three sets of driven behaviors shown in Figures 7–9 transition between each other.

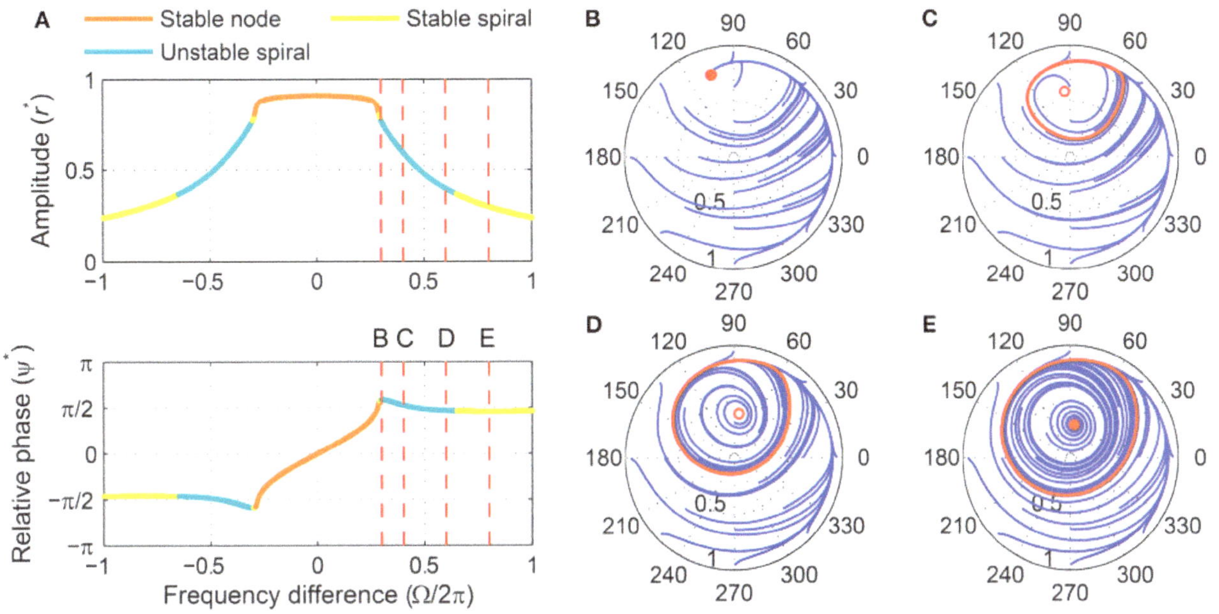

Figure 9. Driven behavior of a supercritical double limit cycle oscillator under strong forcing. **(A)** Steady-state amplitude and relative phase as a function of frequency difference ($\alpha = -1$, $\beta_1 = 4$, $\beta_2 = -1$, $\epsilon = 1$, $F = 1.5$), with vertical dashed lines indicating the frequency differences used for **(B–E)**, **(B)** trajectories attracted to a stable spiral in the (r, ψ) plane ($\Omega/2\pi = 0.3$, phase locking), **(C)** trajectories drawn to a phase-trapped libration ($\Omega/2\pi = 0.4$, frequency locking), **(D)** trajectories drawn to a rotation ($\Omega/2\pi = 0.6$, phase slip), and **(E)** trajectories drawn to either a stable spiral or a rotation ($\Omega/2\pi = 0.8$, phase locking and phase slip). In **(B–E)**, filled and empty circles indicate stable and unstable fixed points respectively, and red lines show limit-cycle orbits.

Subcritical Double Limit Cycle Oscillator

Like a critical Hopf oscillator, a subcritical DLC oscillator is attracted to an equilibrium at zero when it is not driven (Figure 1D). But the presence of a local maximum in the amplitude vector field makes its driven dynamics more varied and interesting than that of a critical Hopf oscillator. Like a supercritical DLC oscillator, a subcritical DLC oscillator exhibits three different sets of driven behaviors depending on the forcing amplitude.

For weak forcing, it behaves like a critical Hopf oscillator, with its driven state attracted to a stable node when $|\Omega|$ is small and to a stable spiral when $|\Omega|$ is large (Figure 10A). For intermediate forcing amplitudes, a pair of fixed points appears at high amplitudes and they are lost via a saddle-node bifurcation at a certain frequency difference (Figures 10B–D). (It is not clearly seen in Figure10B, but the stable fixed point turns back into a stable node just before it collides with the saddle point.) Note that this saddle-node bifurcation does not leave a limit cycle like a SNIC bifurcation (Figure 10D; compare with Figure 7C), making the fixed point at low amplitudes the only (global) attractor. This means that bistability exists only for small frequency differences for a subcritical DLC oscillator under intermediate forcing.

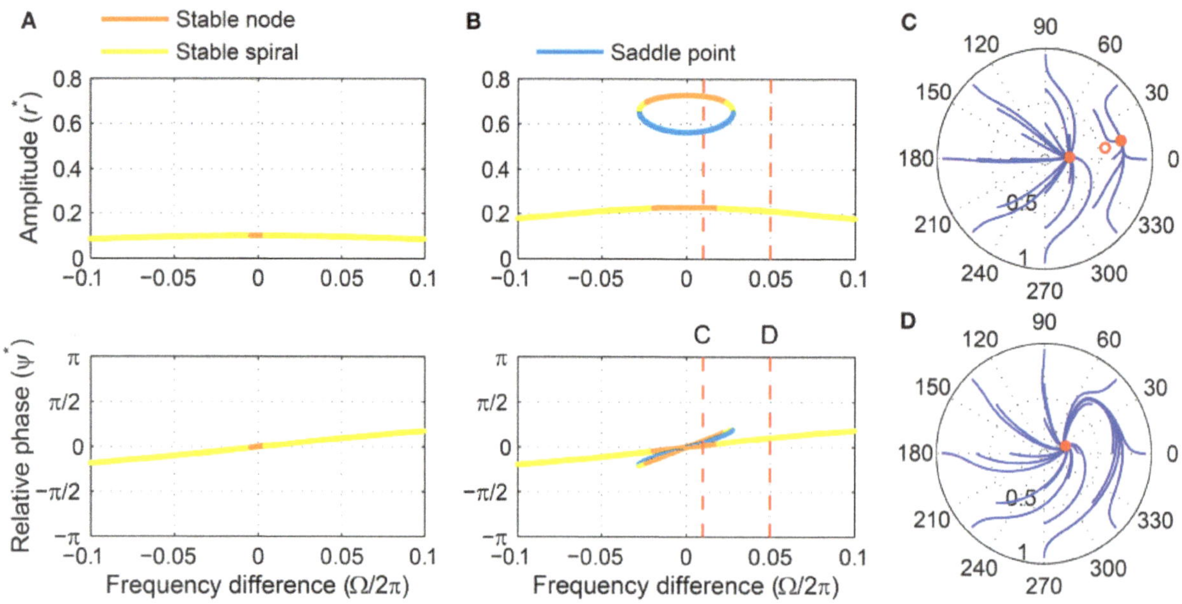

Figure 10. Driven behavior of a subcritical double limit cycle oscillator under weak and intermediate forcing. **(A)** Steady-state amplitude and relative phase as a function of frequency difference for weak forcing ($\alpha = -1$, $\beta_1 = 2.5$, $\beta_2 = -1$, □ = 1, F = 0.1) and**(B)** for intermediate forcing (F = 0.2), with vertical dashed lines indicating the frequency differences used for **(C,D)**, **(C)** trajectories attracted to either of two stable fixed points in the (r, ψ) plane ($\Omega/2\pi = 0.01$, both phase locking), and **(D)** trajectories drawn to a stable spiral ($\Omega/2\pi = 0.05$, phase locking). In **(C,D)**, filled and empty circles indicate stable and unstable fixed points respectively.

When driven strongly, a subcritical DLC oscillator has the same set of fixed points as a supercritical DLC oscillator—a stable node, a stable spiral, an unstable spiral, and a stable spiral as |Ω| increases from zero (compare Figure 9A and Figure 11A). A supercritical Hopf bifurcation occurs at the first phase-locking boundary, where a stable spiral turns unstable and a stable limit cycle grows around it (between Figure 11C and Figure 11D). However, the limit cycle does not grow into a rotation that encompasses the origin, which is the case for a supercritical DLC oscillator. Instead, it shrinks back and turns into a stable spiral via another supercritical Hopf bifurcation (between Figure 11D and Figure 11E). In the absence of a SNIC or subcritical Hopf bifurcation, a strongly driven subcritical DLC oscillator shows no bistability and, since the only non-locked behavior is a libration (Figure11D), it either phase-locks or frequency-locks to the input for all values of Ω. The transitions between different forcing levels are shown in Figure 4D and Supplementary Video 4.

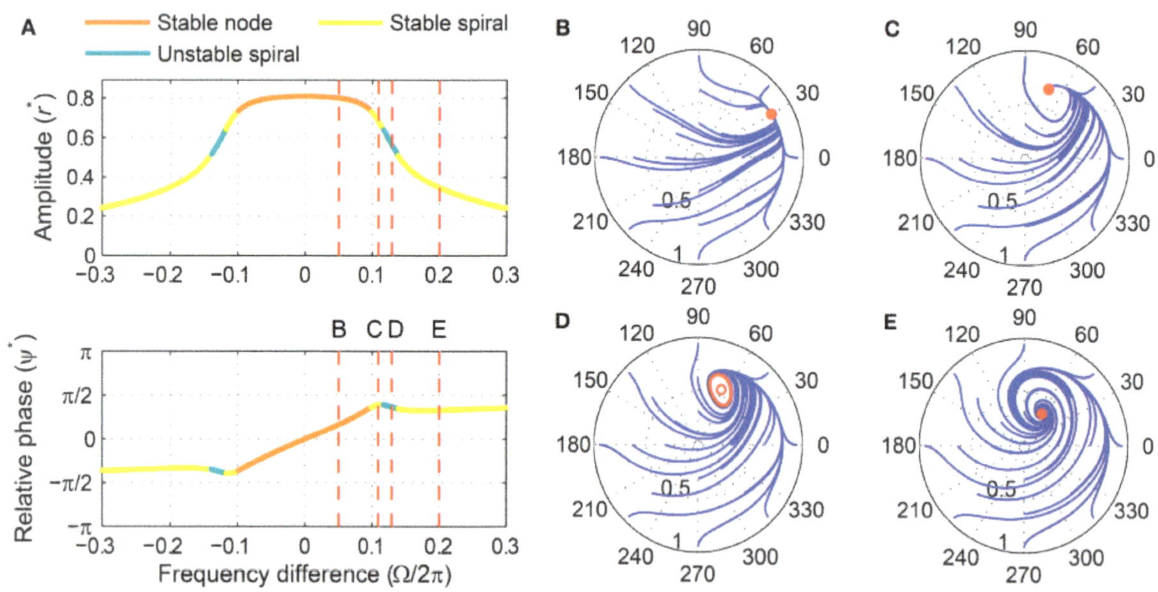

Figure 11. Driven behavior of a subcritical double limit cycle oscillator under strong forcing. **(A)** Steady-state amplitude and relative phase as a function of frequency difference ($\alpha = -1$, $\beta_1 = 2.5$, $\beta_2 = -1$, □ = 1, F = 0.5), with vertical dashed lines indicating the frequency differences used for **(B–E)**, **(B)** trajectories attracted to a stable node in the (r, ψ) plane ($\Omega/2\pi = 0.05$, phase locking), **(C)** trajectories attracted to a stable spiral ($\Omega/2\pi = 0.11$, phase locking), **(D)** trajectories drawn to a phase-trapped libration ($\Omega/2\pi = 0.13$, frequency locking), and **(E)**trajectories drawn to a stable spiral ($\Omega/2\pi = 0.2$, phase locking). In **(B–E)**, filled and empty circles indicate stable and unstable fixed points respectively, and red lines show limit-cycle orbits.

Classification of Parameter Regimes by Driven Behavior

Above we examined the dynamics of periodically forced canonical oscillators in four different intrinsic parameter regimes. We used a representative parameter setting for each of the four distinct driven behaviors, but there are other combinations of intrinsic parameters that show the same sets of behaviors. For instance, the same set of driven behaviors is found for $\alpha > 0$, $\beta_1 < 0$, and $\beta_2 = 0$ (discussed as a supercritical Hopf oscillator) and for $\alpha = 0$, $\beta_1 > 0$, and $\beta_2 < 0$. This is because the two parameter settings have topologically identical autonomous amplitude vector fields (i.e., increasing from zero and then monotonically decreasing as in Figure 1B).

We can classify all possible parameter settings for canonical oscillators into four regimes with distinct driven behaviors (Table 1). Oscillators with an autonomous amplitude vector field that monotonically decreases from zero with no local extremum (Figure 1A) have the same set of driven behaviors as a critical Hopf oscillator ($\alpha = 0$, $\beta_1 < 0$, $\beta_2 = 0$). Linear oscillators ($\alpha < 0$, $\beta_1 = 0$, $\beta_2 = 0$) belong to this category. Oscillators whose amplitude vector fields have one local maximum (Figure 1B) have the same set of driven behaviors and bifurcations as a supercritical Hopf oscillator ($\alpha > 0$, $\beta_1 < 0$, $\beta_2 = 0$). Oscillators with $\alpha < 0$, $\beta_1 > 0$, and $\beta_2 < 0$ are divided into three groups depending on whether the local maximum of the amplitude vector field is above zero (Figure 1C, a supercritical DLC oscillator), is below zero (Figure 1D, a subcritical DLC oscillator), or does not exist (with the same driven behaviors as a critical Hopf oscillator).

Table 1. Classification of parameter regimes by driven behavior.

α	β_1	β_2	Local extrema[a]	Discussed as	Bifurcations[b]
–	0	0	None		None
0	–	0		Critical Hopf	
0	0	–			
–	–	0			
–	0	–			
0	–	–			
–	–	–			
–	+	–	(No max)		
+	–	0	One	Supercritical Hopf	SNIC (low *F*); Super-Hopf (high *F*)
+	0	–			
0	+	–			
+	–	–			
+	+	–			
–	+	–	Two (max > 0)	Supercritical DLC	SNIC (low *F*); SNIC, Sub-Hopf (mid *F*); Super-Hopf, Sub-Hopf (high *F*)
–	+	–	Two (max < 0)	Subcritical DLC	None (low *F*); SN (mid *F*); Super-Hopf, Super-Hopf (high *F*)

SNIC, saddle-node bifurcation on an invariant circle; Super-Hopf, supercritical Hopf bifurcation; Sub-Hopf, subcritical Hopf bifurcation; SN, saddle-node bifurcation.
[a] *The number of local extrema in the autonomous amplitude vector field.*
[b] *Bifurcations at phase-locking boundaries.*

Frequency Scaling of Logarithmic Frequency Networks

The analysis so far shows that for a fixed forcing amplitude the driven behavior of a canonical oscillator depends on the frequency difference Ω, not on the natural frequency ω per se, and is symmetrical about $\Omega = 0$ on a linear scale (see Figure 11A, for example). This is to be expected from the fact that Equation (2), which was used for the above analysis, is not scaled by natural frequency (or time constant) as is Equation (1). Thus, the width of phase-locking range for an oscillator described by Equation (2) is constant regardless of its natural frequency if other intrinsic parameters remain the same. For an oscillator network with logarithmically equally spaced natural frequencies, this might not be desirable if we want each oscillator in the network to cover the same portion of logarithmic frequency space. Frequency scaling solves this problem by making the driven behavior depend on both frequency difference and natural frequency.

The frequency-scaled version of Equation (2),

$$\frac{1}{f}\dot{z} = z\left(\alpha + \mathrm{i}2\pi + \beta_1|z|^2 + \frac{\epsilon\beta_2|z|^4}{1-\epsilon|z|^2}\right) + Fe^{\mathrm{i}\omega_0 t},$$

has the polar form

$$\begin{cases} \frac{1}{f}\dot{r} = \alpha r + \beta_1 r^3 + \frac{\epsilon\beta_2 r^5}{1-\epsilon r^2} + F\cos\psi \\ \frac{1}{f}\dot{\psi} = \frac{\Omega}{f} - \frac{F}{r}\sin\psi, \end{cases} \quad (4)$$

where $f = \frac{\omega}{2\pi}$ is the linear natural frequency. There are two main differences between Equation (4) and its non-scaled version, Equation (3). First, the left-hand sides of the scaled equations are multiplied by the inverse of natural frequency, indicating that high-frequency oscillators have faster dynamics and shorter relaxation time than low-frequency oscillators. Second, the right-hand sides are identical for the two versions except that the scaled version has Ω/f in place of Ω in the non-scaled version. This means that the two versions share the same set of steady-state solutions and stability types, but for the frequency-scaled version the distribution of solutions in frequency space is proportional to the natural frequency. Figure 12A shows that frequency-scaled canonical oscillators of different natural frequencies have steady-state amplitude curves of an identical shape and width when plotted on a logarithmic frequency axis. But note that the amplitude curves and locking ranges are asymmetrical on the logarithmic axis because even with frequency scaling they are symmetrical in linear frequency (Figure 12B).

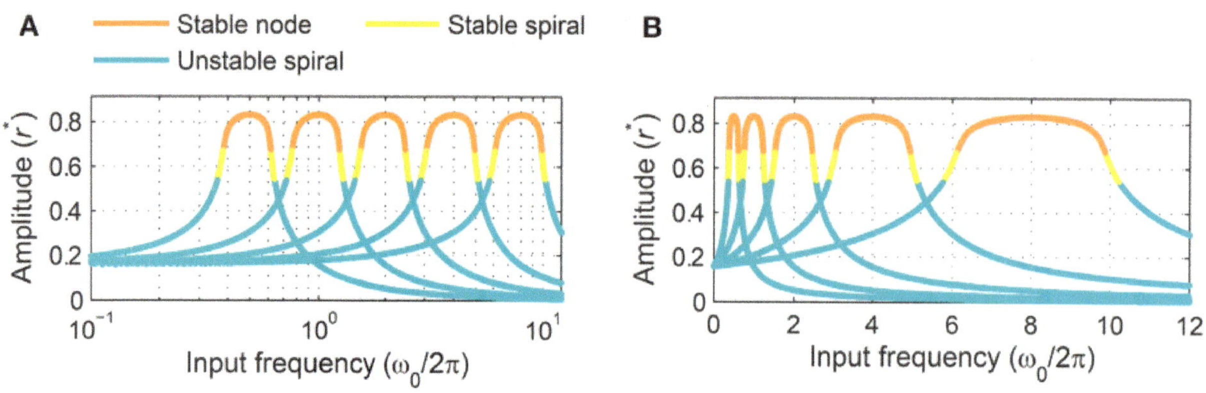

Figure 12. Frequency scaled canonical oscillators in a logarithmic frequency network. The steady-state amplitude of frequency-scaled canonical oscillators ($\alpha = 1$, $\beta_1 = -1$, $\beta_2 = -1$, $\epsilon = 1$, $F = 1$) with logarithmically equally spaced natural frequencies (ω2π=0.5,1,2,4,8ω2π=0.5,1,2,4,8) is plotted as a function of input frequency on **(A)** a logarithmic frequency axis and **(B)** a linear frequency axis.

DISCUSSION

Here we have examined the response of gradient frequency neural networks to periodic forcing by analyzing the driven behavior of a canonical model for such networks. Using a dynamical systems analysis of a canonical oscillator under sinusoidal forcing, we showed that oscillators with distinct autonomous behaviors have different sets of driven behaviors and different types of bifurcation at phase-locking boundaries (summarized in Table 1 and Figure 4). Oscillators that decay to zero without external forcing are found to phase-lock to sinusoidal forcing of any frequency and amplitude, if the autonomous amplitude vector field decreases monotonically (e.g., a critical Hopf oscillator). When the vector field has a below-zero local maximum (i.e., a subcritical DLC oscillator), the oscillator always phase-locks to weak forcing, shows bistability for intermediate forcing, and either phase-locks or frequency-locks to strong forcing. Oscillators with one stable non-zero spontaneous amplitude (e.g., supercritical Hopf oscillators) phase-lock only to forcing frequencies close to their natural frequencies. Just outside the phase-locking range, oscillators of this type frequency-lock to strong forcing. Finally, oscillators with two stable spontaneous amplitudes (i.e., supercritical DLC oscillators) exhibit the most diverse set of behaviors including phase locking, frequency locking, phase slip, bistability between two phase-locked states, and bistability between phase locking and phase slip. We also showed that frequency scaling makes the response of canonical oscillators constant over logarithmic frequency.

The present analysis shows how a gradient frequency network of non-linear oscillators processes a periodic signal. Given the values of intrinsic parameters and the range of natural frequencies, the analysis shows which part of the network would phase-lock to the signal while the other part oscillates near the spontaneous (i.e., autonomous) states. Thus, the response of the network as a whole would include both the signal frequency and the natural frequencies of oscillators not locked to the signal, plus some non-linear combination frequencies in non-locked, fluctuating oscillations. Here we analyzed the phase-locking behavior of an oscillator with a linear input term (i.e., the signal itself), but the canonical model also includes non-linear terms that govern mode locking (i.e., synchronization in integer ratios other than 1:1). For instance, the input term xkz□ m−1xkz□m-1 allows the oscillator, z, to mode-lock to the external signal, x, in a k:m ratio, where k and m are positive integers. With such non-linear input terms, the response of a gradient frequency network contains not only signal frequencies and natural frequencies but also non-linear responses such as harmonics, subharmonics, and combination frequencies including quadratic and cubic difference tones (Cartwright et al., 2001; Large et al., 2010), which are essential for explaining non-linearities found in the neural responses of the auditory system to acoustic signals (Large and Almonte, 2012;Lerud et al., 2014). The analysis of the canonical model with non-linear input terms, which will be given elsewhere, combines with the present analysis of linear forcing to demonstrate the full signal processing capabilities of gradient frequency neural networks.

An understanding of the relationship between auditory neurophysiology, auditory population dynamics and auditory perception remains an elusive goal, due to the intricate circuitry, the many structural levels involved, and the highly non-linear nature of the neural responses. Traditional signal processing approaches employ linear systems almost exclusively, and they approximate human perceptual capabilities only roughly. From the linear systems point of view, the basic job of the auditory system is to decompose signals into orthogonal frequency bands for subsequent pattern analysis. However, cochlear outer hair cells and auditory neurons do not decompose signals into orthogonal bands; instead, each process responds to multiple related frequencies, in a manner that is fundamentally different from linear techniques such as Fourier analysis. It appears that active networks form spatiotemporal patterns that may correspond to the perception of pitch, to the recognition of specific auditory objects, or to the induction of a beat in a musical rhythm. Significant theoretical advances will be necessary to understand signal processing, pattern formation, and plasticity in this complex and highly non-linear system. Here, we have taken the first step, by studying the responses of gradient frequency networks of non-linear oscillators forced with periodic signals. Our aim is to extend our understanding of signal processing to include more biologically realistic elements, such as oscillatory neural networks. This will enable analysis of non-linear auditory physiology from a signal processing point of view, and facilitate the design of artificial auditory networks for performing specific functions. By studying the way in which realistic neurodynamic processes respond to sounds, we hope to shed light on the remarkable capabilities of human perception that arise from non-linear processes in the auditory system.

Conflict of Interest Statement

The authors declare that the research was conducted in the absence of any commercial or financial relationships that could be construed as a potential conflict of interest.

ACKNOWLEDGMENTS

This work was supported by NSF BCS-1027761 and AFOSR FA9550-12-10388 awarded to EL.

REFERENCES

1. Buzsáki, G., and Draguhn, A. (2004). Neuronal oscillations in cortical networks. Science 304, 1926–1929. doi: 10.1126/science.1099745
2. Camalet, S., Duke, T., Jülicher, F., and Prost, J. (2000). Auditory sensitivity provided by self-tuned critical oscillations of hair cells. Proc. Natl. Acad. Sci. U.S.A. 97, 3183–3188. doi: 10.1073/pnas.97.7.3183
3. Cartwright, J. H. E., González, D. L., and Piro, O. (2001). Pitch perception: A dynamical-systems perspective. Proc. Natl. Acad. Sci. U.S.A. 98, 4855–4859. doi: 10.1073/pnas.081070998
4. Childs, L. M., and Strogatz, S. H. (2008). Stability diagram for the forced Kuramoto model. Chaos 18:043128. doi: 10.1063/1.3049136
5. Crawford, A. C., and Fettiplace, R. (1985). The mechanical properties of ciliary bundles of turtle cochlear hair cells. J. Physiol.364, 359–379. doi: 10.1113/jphysiol.1985.sp015750
6. Duke, T., and Jülicher, F. (2003). Active traveling wave in the cochlea. Phys. Rev. Lett. 90:158101. doi: 10.1103/PhysRevLett.90.158101
7. Eguíluz, V. M., Ospeck, M., Choe, Y., Hudspeth, A. J., and Magnasco, M. O. (2000). Essential nonlinearities in hearing. Phys. Rev. Lett. 84, 5232–5235. doi: 10.1103/PhysRevLett.84.5232
8. Fredrickson-Hemsing, L., Ji, S., Bruinsma, R., and Bozovic, D. (2012). Mode-locking dynamics of hair cells of the inner ear.Phys. Rev. E. Stat. Nonlin. Soft. Matter Phys. 86:021915. doi: 10.1103/PhysRevE.86.021915
9. Guckenheimer, J., and Holmes, P. (1983). Nonlinear Oscillations, Dynamical Systems, and Bifurcations of Vector Fields. New York, NY: Springer-Verlag.
10. Hodgkin, A. L., and Huxley, A. F. (1952). A quantitative description of membrane current and its application to conduction and excitation in nerve. J. Physiol. 117, 500–544. doi: 10.1113/jphysiol.1952.sp004764
11. Holmes, P. J., and Rand, D. A. (1978). Bifurcations of the forced van der Pol oscillator. Q. Appl. Math. 35, 495–509.
12. Hoppensteadt, F. C., and Izhikevich, E. M. (1996). Synaptic organizations and dynamical properties of weakly connected neural oscillators. II. Learning phase information. Biol. Cybern. 75, 129–135. doi: 10.1007/s004220050280
13. Hoppensteadt, F. C., and Izhikevich, E. M. (1997). Weakly Connected Neural Networks. New York, NY: Springer-Verlag.
14. Hoppensteadt, F. C., and Izhikevich, E. M. (2001). "Canonical neural models," in The Handbook of Brain Theory and Neural Networks, 2nd Edn. ed M. A. Arbib (Cambridge, MA, MIT Press), 181–186.
15. Hudspeth, A. J., Jülicher, F., and Martin, P. (2010). A critique of the critical cochlea: Hopf–a bifurcation–is better than none. J. Neurophysiol. 104, 1219–1229. doi: 10.1152/jn.00437.2010
16. Izhikevich, E. M. (2000). Neural excitability, spiking and bursting. Int. J. Bifurcat. Chaos 10, 1171–1266. doi: 10.1142/S0218127400000840
17. Izhikevich, E. M. (2001). Synchronization of Elliptic Bursters. SIAM Rev. Soc. Ind. Appl. Math. 43, 315–344. doi: 10.1137/s0036144500382064
18. Jülicher, F., Andor, D., and Duke, T. (2001). Physical basis of two-tone interference in hearing. Proc. Natl. Acad. Sci. U.S.A. 98, 9080–9085. doi: 10.1073/pnas.151257898
19. Kern, A., and Stoop, R. (2003). Essential role of couplings between hearing nonlinearities. Phys. Rev. Lett. 91:128101. doi: 10.1103/PhysRevLett.91.128101
20. Koepsell, K., Wang, X., Hirsch, J. A., and Sommer, F. T. (2010). Exploring the function of neural oscillations in early sensory systems. Front. Neurosci. 4:53. doi: 10.3389/neuro.01.010.2010
21. Large, E. W. (2010). "A dynamical systems approach to musical tonality," in Nonlinear Dynamics in Human Behavior, eds R. Huys, and V. K. Jirsa (Berlin; Heidelberg: Springer), 193–211.
22. Large, E. W. (2011). "Musical tonality, neural resonance and Hebbian learning," in Mathematics and Computation in Music, Vol. 6726 of Lecture Notes in Computer Science, eds C. Agon, M. Andreatta, G. Assayag, E. Amiot, J. Bresson, and J. Mandereau (Berlin; Heidelberg: Springer), 115–125.
23. Large, E. W., and Almonte, F. V. (2012). Neurodynamics, tonality, and the auditory brainstem response. Ann. N.Y. Acad. Sci.1252, E1–E7. doi: 10.1111/j.1749-6632.2012.06594.x
24. Large, E. W., Almonte, F. V., and Velasco, M. J. (2010). A canonical model for gradient frequency neural networks. Phys. D 239, 905–911. doi: 10.1016/j.physd.2009.11.015

25. Large, E. W., Herrera, J. A., and Velasco, M. J. (2015). Neural networks for beat perception in musical rhythm. Front. Syst. Neurosci. 9:159. doi: 10.3389/fnsys.2015.00159

26. Large, E. W., Kim, J. C., Flaig, N. K., Bharucha, J. J., and Krumhansl, C. L. (in press). A neurodynamic account of musical tonality. Music Percept.

27. Laudanski, J., Coombes, S., Palmer, A. R., and Sumner, C. J. (2010). Mode-locked spike trains in responses of ventral cochlear nucleus chopper and onset neurons to periodic stimuli. J. Neurophysiol. 103, 1226–1237. doi: 10.1152/jn.00070.2009

28. Lerud, K. D., Almonte, F. V., Kim, J. C., and Large, E. W. (2014). Mode-locking neurodynamics predict human auditory brainstem responses to musical intervals. Hear. Res. 308, 41–49. doi: 10.1016/j.heares.2013.09.010

29. Lerud, K. D., Kim, J. C., Almonte, F. V., Carney, L. H., and Large, E. W. (2015). A canonical nonlinear cochlear model. Assoc. Res. Otolaryngol. Abs. 38, 211–212.

30. Magnasco, M. O. (2003). A wave traveling over a Hopf instability shapes the cochlear tuning curve. Phys. Rev. Lett. 90:058101. doi: 10.1103/PhysRevLett.90.058101

31. Martin, P., Bozovic, D., Choe, Y., and Hudspeth, A. J. (2003). Spontaneous oscillation by hair bundles of the bullfrog's sacculus.J. Neurosci. 23, 4533–4548.

32. Meddis, R., and O'Mard, L. P. (2006). Virtual pitch in a computational physiological model. J. Acoust. Soc. Am. 120, 3861–3869. doi: 10.1121/1.2372595

33. Ospeck, M., Eguíluz, V. M., and Magnasco, M. O. (2001). Evidence of a Hopf bifurcation in frog hair cells. Biophys. J. 80, 2597–2607. doi: 10.1016/S0006-3495(01)76230-3

34. Pikovsky, A., Rosenblum, M., and Kurths, J. (2000). Phase synchronization in regular and chaotic systems. Int. J. Bifurcat. Chaos 10, 2291–2305. doi: 10.1142/S0218127400001481

35. Pikovsky, A., Rosenblum, M., and Kurths, J. (2001). Synchronization: A Universal Concept in Nonlinear Sciences. Cambridge: Cambridge University Press.

36. Ramunno-Johnson, D., Strimbu, C., Fredrickson, L., Arisaka, K., and Bozovic, D. (2009). Distribution of frequencies of spontaneous oscillations in hair cells of the bullfrog sacculus. Biophys. J. 96, 1159–1168. doi: 10.1016/j.bpj.2008.09.060

37. Schwarz, D. W. F., Dezsö, A., and Neufeld, P. R. (1993). Frequency selectivity of central auditory neurons without inner ear. Acta Otolaryngol. 113, 266–270. doi: 10.3109/00016489309135807

38. Sejnowski, T. J., and Paulsen, O. (2006). Network oscillations: emerging computational principles. J. Neurosci. 26, 1673–1676. doi: 10.1523/JNEUROSCI.3737-05d.2006

39. Stoop, R., Steeb, W.-H., Gallas, J. C., and Kern, A. (2005). Auditory two-tone suppression from a subcritical Hopf cochlea. Phys. A 351, 175–183. doi: 10.1016/j.physa.2004.12.019

40. Strogatz, S. H. (1994). Nonlinear Dynamics and Chaos: With Applications to Physics, Biology, Chemistry, and Engineering. Cambridge, MA: Perseus Books.

CHAPTER 14

THE BANACH FIXED POINT THEOREM APPLICATION TO HOPF BIFURCATION OF A GENERALIZED BOUSSINESQ SYSTEM

Junling Sun and Xuefeng Han

Keywords: Hopf bifurcation, periodic solution, Boussinesq systems

INTRODUCTION

The Banach fixed-point theorem is one of the most celebrated results in fixed-point theory and it states that a contraction on a complete metric space has a unique fixed point; its proof hinges on 'Picard iterations'. This method is applicable to a variety of subjects such as partial differential equations and engineering of image processing. In this paper, we apply the Banach fixed-point theorem to the study of the Hopf bifurcation of the Boussinesq system with the temperature-dependent viscosity and thermal diffusivity under external time-independent forces,

$$U_t + (U\cdot\nabla)U - \nabla\cdot\big(\nu(T)\nabla U\big) + \nabla P = Te_3 + f_\alpha, \tag{1.1}$$

$$T_t + (U\cdot\nabla)T - \nabla\cdot\big(\kappa(T)\nabla T\big) = h_\alpha, \tag{1.2}$$

$$\nabla\cdot U=0, \tag{1.3}$$

where $(t,x)\in R^+\times R^3$, U is the velocity field, T is the temperature function, the viscosity ν and the thermal diffusivity κ depend on the temperature, P denotes the pressure, the vector $e_3=(0,0,1)$, f_α, and h_α are external time-independent forces, which depend smoothly on some parameter α. One may refer to the literature [1] for the derivation of this model and the related parameters.

The Boussinesq system is a very important model in fluid mechanics. It exhibits extremely rich phenomena. For example, Rayleigh-Bénard convection, geophysical fluid dynamics (see [2–6]). One of the key problems in the study of the dynamic behavior for the Boussinesq system is how to understand the bifurcation of solutions. To our knowledge, there are no results concerning the dynamical behavior of a Boussinesq system with ν and κ depending on the temperature.

We assume that $\nu(T)\nu$ and $\kappa(T)$ satisfy

$$\nu^{(k)}(\cdot),\kappa^{(k)}(\cdot)<0 \quad \text{for } k=0,1,$$
$$\nu^{(k)}(\cdot)=\kappa^{(k)}(\cdot)=0 \quad \text{for } k=2,3,\ldots, \tag{1.4}$$

where $\nu(\cdot)$ and $\kappa(\cdot)$ are smooth functions and are analytic at 0.

Due to the technical restriction, two cases are still open. One is the case of $\nu^{(k)}(\cdot),\kappa^{(k)}(\cdot)\neq 0$for k=2,3,…k=2,3,… ; the other case is that $\nu(T)$ and $\kappa(T)$ are not smooth functions. We also assume that external forces f_α and h_α can be chosen suitably such that $(U_\alpha(x)+U_{c1},T_\alpha(x)+T_{c1},P_\alpha(x))$ are a smooth solution of the steady Boussinesq system

$$-\nabla\cdot\big(\nu(T)\nabla U\big) + (U\cdot\nabla)U + \nabla P = Te_3 + f_\alpha,$$
$$-\nabla\cdot\big(\kappa(T)\nabla T\big) + (U\cdot\nabla)T = h_\alpha,$$
$$\nabla\cdot U=0,$$

where $U_{c1}=(c1,0,0)^T$, $T_{c1}=(c1,0,0)$, and

Citation: Junling Sun, Xuefeng Han, The Banach fixed point theorem application to Hopf bifurcation of a generalized Boussinesq system, DOI: 10.1186/s13662-014-0352-z.

$$\lim_{|x|\to\infty} U_\alpha(x) = \mathbf{0}, \qquad \lim_{|x|\to\infty} T_\alpha(x) = \mathbf{0}.$$

Linearizing system (1.1)-(1.2) around the steady state $(U_\alpha(x)+U_{c_1}, T_\alpha(x)+T_{c_1}, P_\alpha(x))$, we obtain

$$u_t + c_1\partial_{x_1}u + u_\alpha\cdot\nabla u + u\cdot\nabla u_\alpha + u\cdot\nabla u - \nabla\cdot\left(\nu(v+T_\alpha+T_{c_1})\nabla u\right) - \nabla p$$
$$-\nabla\cdot\left(\left(\nu(v+T_\alpha+T_{c_1}) - \nu(T_\alpha+T_{c_1})\right)\nabla u_\alpha\right) - ve_3 = 0, \tag{1.5}$$
$$v_t + c_1\partial_{x_1}v + T_\alpha\cdot\nabla v + u\cdot\nabla T_\alpha + u\cdot\nabla v - \nabla\cdot\left(\kappa(v+T_\alpha+T_{c_1})\nabla v\right)$$
$$-\nabla\cdot\left(\left(\kappa(v+T_\alpha+T_{c_1}) - \kappa(T_\alpha+T_{c_1})\right)\nabla T_\alpha\right) = 0, \tag{1.6}$$

with the incompressibility condition $\nabla\cdot u=0$. This condition leads to $\nabla\cdot(uu^T)=u\cdot\nabla u+u\nabla\cdot u=u\cdot\nabla u$.

Thus it follows from (1.4)-(1.6) that

$$u_t + c_1\partial_{x_1}u + \nu_0\Delta u + \nu_0' c_1\partial_{x_1x_1}u + \nabla\cdot\left(u_\alpha u^T\right) + \nabla\cdot\left(uu_\alpha^T\right) + \nabla\cdot\left(uv^T\right) - \nabla p$$
$$-\nu_0'\nabla\cdot\left((v+T_\alpha)\nabla u\right) - \nu_0'\nabla\cdot(v\nabla u_\alpha) - ve_3 = 0, \tag{1.7}$$
$$v_t + c_1\partial_{x_1}v + \kappa_0\Delta v + \kappa_0' c_1\partial_{x_1x_1}v + T_\alpha\cdot\nabla v + u\cdot\nabla T_\alpha + u\cdot\nabla v$$
$$+\kappa_0'\nabla\cdot\left((v+T_\alpha)\nabla v\right) - \kappa_0'\nabla\cdot(v\nabla T_\alpha) = 0. \tag{1.8}$$

The vorticity associated with the velocity field u of the fluid is defined by $\omega=\nabla\times u$. Note that $\nabla\times\nabla\cdot(uu^T) = \nabla\cdot(\omega u^T - u\omega^T)$. We can obtain the vorticity equation from (1.7) as

$$\omega_t + \nu_0\Delta\omega + c_1\partial_{x_1}\omega + \nu_0' c_1\partial_{x_1x_1}\omega + \nabla\cdot\left(\omega_\alpha u^T - u_\alpha\omega^T\right) + \nabla\cdot\left(\omega u_\alpha^T - u\omega_\alpha^T\right)$$
$$+\nabla\cdot\left(\omega u^T - u\omega^T\right) - \nu_0'\nabla\cdot\left(\nabla\times(v+T_\alpha)\nabla u^T + (v+T_\alpha)\nabla\omega^T\right)$$
$$-\nu_0'\nabla\cdot\left((\nabla\times v)\nabla u_\alpha^T + v\nabla\omega_\alpha^T\right) - \partial_{x_3}v = 0. \tag{1.9}$$

Note that the space of divergence free vector fields is invariant under the evolution (1.9). We can assume that $\nabla\cdot\omega=0$. Moreover, we can reconstruct the velocity u from the vorticity ω by solving the equation

$\nabla\times u=\omega, \nabla\cdot\omega=0$

On the other hand, the Biot-Savart law shows that the velocity field u is defined in terms of the vorticity

$$u(x) = -\frac{1}{4\pi}\int_{\mathbf{R}^3}\frac{(x-y)^\perp\times\omega(y)}{|x-y|^3}dy, \quad x\in\mathbf{R}^3. \tag{1.10}$$

Let $\varphi=(\omega,v)^T$. Then we can rewrite system (1.8)-(1.9) as the evolution equation

$$\frac{d\varphi}{dt} + \mathcal{N}\varphi + G(\varphi) = F(\varphi), \tag{1.11}$$

where

$$\mathcal{N} = \begin{pmatrix} \nu_0\Delta + c_1\partial_{x_1} + \nu_0' c_1\partial_{x_1x_1} & -\partial_{x_3} \\ 0 & \kappa_0\Delta + c_1\partial_{x_1} + \kappa_0' c_1\partial_{x_1x_1} \end{pmatrix}$$

and

$$G(\varphi) = \begin{pmatrix} g_1 \\ g_2 \end{pmatrix}, \qquad F(\varphi) = \begin{pmatrix} g_3 \\ g_4 \end{pmatrix}$$

with

$$g_1 = \nabla \cdot \left(\omega_\alpha u^T - u_\alpha \omega^T - \omega u_\alpha^T + u\omega_\alpha^T - \nu_0'(\nabla \times T_\alpha)\nabla u^T + \nu_0' T_\alpha \nabla \omega^T - \nu_0'(\nabla \times \nu)\nabla u_\alpha^T + \nu_0' \nu \nabla \omega_\alpha^T\right),$$

$$g_2 = T_\alpha \cdot \nabla \nu + u \cdot \nabla T_\alpha - \kappa_0' \nabla \cdot (\nu \nabla T_\alpha - T_\alpha \nabla \nu),$$

$$g_3 = -\nabla \cdot \left(\omega u^T - u\omega^T - \nu_0'(\nabla \times \nu)\nabla u^T + \nu \nabla \omega^T\right),$$

$$g_4 = -u \cdot \nabla \nu - \kappa_0' \nabla \cdot (\nu \nabla \nu).$$

To reach the essential spectrum of the operator $-(\widehat{\mathcal{N}} + \widehat{G})$ up to the imaginary axis, we need to make a similar assumption to [7]:

(H1)For any $\alpha \in [\alpha_c - \alpha_0, \alpha_c + \alpha_0]$, 0 is not an eigenvalue of $\widehat{\mathcal{N}} + \widehat{G}$.

(H2)For $\alpha = \alpha c$, the operator $-(\widehat{\mathcal{N}} + \widehat{G})$ has two pairs of eigenvalues (λ_0^+, μ_0^+) and (λ_0^-, μ_0^-) satisfying

$$\lambda_0^\pm(\alpha_c) = \mu_0^\pm(\alpha_c) = \pm i\xi_0 \neq 0, \quad \text{for } \xi_0 > 0,$$

$$\frac{d}{d\alpha}\mathrm{Re}\left(\lambda_0^\pm(\alpha)\right)|_{\alpha=\alpha_c}, \qquad \frac{d}{d\alpha}\mathrm{Re}\left(\mu_0^\pm(\alpha)\right)|_{\alpha=\alpha_c} > 0.$$

(H3) The remaining eigenvalue of $-(\widehat{\mathcal{N}} + \widehat{G})$ is strictly bounded away from the imaginary axis in the left half plane for all $\alpha \in [\alpha_c - \alpha_0, \alpha_c + \alpha_0]$.

Under the generic assumption that the cubic coefficient terms a1, a2≠0 in (2.40)-(2.41), our main result is stated thus.

Theorem 1.1 *Assume that* (H1)-(H3) *hold. Then system* (1.1)-(1.3) *has a one-dimensional family of small time-periodic solutions, i.e.*

$$\mathcal{U}(x,t) = \mathcal{U}(x, t + 2\pi/\xi_1), \qquad T(x,t) = T(x, t + 2\pi/\xi_2)$$

with $\alpha = \alpha_c + \epsilon, \epsilon \in (0, \alpha_0)$. *Moreover,* $\xi_1 = \xi_0 + \mathcal{O}(\epsilon)$, $\xi_2 = \xi_0 + \mathcal{O}(\epsilon)$ *and*

$$\left\|\mathcal{U}(x,t)\right\|_{\mathbf{C}_b^0(\mathbf{R}^3 \times [0, 2\pi/\xi_1])} = \mathcal{O}(\epsilon), \qquad \left\|T(x,t)\right\|_{\mathbf{C}_b^0(\mathbf{R}^3 \times [0, 2\pi/\xi_2])} = \mathcal{O}(\epsilon).$$

PROOF OF THEOREM

We consider the following standard Sobolev space and spatially weighted Lebesgue space: $\mathbf{W}_p^q := \{u : \|u\|_p^q := \sum_{|\beta| \le p} \|D^\beta u\|_{\mathbf{L}^q}^q < \infty\}$, $\mathbf{L}_m^n := \{u : \|u\|_m^n := \int_{\mathbf{R}^3} \rho^m(x) u^n(x) dx < \infty\}$ with the weighted function $\rho(x) = \sqrt{1 + |x|^2}$. We denote $\mathbf{W}_p^2$ by $\mathbf{H}^p$ with the norm $\|u\|_{\mathbf{H}^p} = \sum_{|\beta| \le p} \|D^\beta u\|_{\mathbf{L}^2}^2$. When $q = 2$. When q=2, the Fourier transform is an isomorphism between H^p and $\mathbf{L}_p^2$ with $\|u\|_{\mathbf{L}_p^2} = \|\rho^p u\|_{\mathbf{L}^2}$. We also introduce the space $\mathbf{X} := \{u = (u_n)_{n \in \mathbf{Z}} : \|u\|_{\mathbf{X}} < \infty\}$ and the weighted spaces $\mathcal{L}_s^p = \mathbf{L}_s^p \times \mathbf{L}_s^p$, $\mathcal{H}^m = \mathbf{H}^m \times \mathbf{H}^m$, $\mathcal{X} = \mathbf{X} \times \mathbf{X}$, which are equipped with the norms $\|u\|_{\mathbf{X}} = \sum_{n \in \mathbf{Z}} \|u_n\|_{\mathbf{H}^m}$, $\|\varphi\|_{\mathcal{X}} := \|u\|_{\mathbf{X}} + \|v\|_{\mathbf{X}}$, $\|\varphi\|_{\mathcal{L}_m^p} := \|u\|_{\mathbf{L}_m^p} + \|v\|_{\mathbf{L}_m^p}$, $\|\varphi\|_{\mathcal{H}^m} := \|u\|_{\mathbf{H}^m} + \|v\|_{\mathbf{H}^m}$ for $\forall \varphi = (u, v)^T \in \mathcal{L}_s^p$ or $\mathcal{X}$.

Next, we look for 2π time-periodic solutions of

$$\Xi \frac{d\varphi}{dt} + \mathcal{N}\varphi + G(\varphi) = F(\varphi), \tag{2.1}$$

where

$$\Xi = \begin{pmatrix} \xi_1 & 0 \\ 0 & \xi_2 \end{pmatrix}, \qquad \mathcal{N} = \begin{pmatrix} \nu_0 \Delta + c_1 \partial_{x_1} + \nu_0' c_1 \partial_{x_1 x_1} & -\partial_{x_3} \\ 0 & \kappa_0 \Delta + c_1 \partial_{x_1} + \kappa_0' c_1 \partial_{x_1 x_1} \end{pmatrix},$$

and

$$G(\varphi) = \begin{pmatrix} g^1 \\ g^2 \end{pmatrix}, \qquad F(\varphi) = \begin{pmatrix} g^3 \\ g^4 \end{pmatrix}$$

with

$$g^1 = \nabla \cdot (\omega_\alpha u^T - u_\alpha \omega^T - \omega u_\alpha^T + u\omega_\alpha^T - \nu_0'(\nabla \times T_\alpha)\nabla u^T + \nu_0' T_\alpha \nabla \omega^T$$
$$- \nu_0'(\nabla \times \nu)\nabla u_\alpha^T + \nu_0' \nu \nabla \omega_\alpha^T),$$
$$g^2 = T_\alpha \cdot \nabla \nu + u \cdot \nabla T_\alpha - \kappa_0' \nabla \cdot (\nu \nabla T_\alpha - T_\alpha \nabla \nu),$$
$$g^3 = -\nabla \cdot (\omega u^T - u\omega^T - \nu_0'(\nabla \times \nu)\nabla u^T + \nu \nabla \omega^T), \tag{2.2}$$
$$g^4 = -u \cdot \nabla \nu - \kappa_0' \nabla \cdot (\nu \nabla \nu). \tag{2.3}$$

According to the classical result in [8], we know that the essential spectrum of the operator N+G is a relatively compact perturbation of N. It has the essential spectrum

$$\text{essspec}(\widehat{\mathcal{N}}) = \{\lambda \in \mathcal{C}^2 : \lambda = (-\nu_0|y|^2 - \nu_0' c_1 |y_1|^2 + icy_1, -\kappa_0|y|^2 - \kappa_0' c_1 |y_1|^2 + icy_1), y \in \mathbf{R}^3\}.$$

Moreover, the spectra of N+G and N only differ by isolated eigenvalues of finite multiplicity. The above spectrum properties are critical to prove our main result.

For convenience, we rewrite (2.1) as

$$\xi_1 \omega_t = M_1 \omega + g^3(\omega, u, \nu), \tag{2.4}$$
$$\xi_2 \nu_t = M_2 \nu + g^4(\omega, u, \nu), \tag{2.5}$$

where g_3 and g_4 defined in (2.2)-(2.3),

$$M_1 \omega = \overline{M_1} \omega + g^1 = \nu_0 \Delta \omega + \nu_0' c_1 \partial_{x_1 x_1} \omega + c_1 \partial_{x_1} \omega - \partial_{x_3} \nu + g^1, \tag{2.6}$$
$$M_2 \nu = \overline{M_2} \nu + g^2 = \kappa_0 \Delta \nu + \kappa_0' c_1 \partial_{x_1 x_1} \nu + c_1 \partial_{x_1} \nu + g^2. \tag{2.7}$$

We make the ansatz

$\omega(x,t) = \sum_{n\in\mathbf{Z}} \omega_n(x) e^{int}$ and $\nu(x,t) = \sum_{n\in\mathbf{Z}} \nu_n(x) e^{int}$ to (2.4)-(2.5). Then we obtain

$$(in\xi_1 - M_1)\omega_n = g_n^3(\omega, u, \nu), \tag{2.8}$$
$$(in\xi_2 - M_2)\nu_n = g_n^4(\omega, u, \nu) \tag{2.9}$$

where

$$g^3(\omega, u, \nu)(x,t) = \sum_{n\in\mathbf{Z}} g_n^3(\omega, u, \nu) e^{int}, \quad g^4(\omega, u, \nu)(x,t) = \sum_{n\in\mathbf{Z}} g_n^4(\omega, u, \nu) e^{int}.$$

Note that we are interested in a real valued solution only. So we will always suppose that $(\omega_n, \nu_n) = (\omega_{-n}, \nu_{-n})$ for $n \in \mathbf{Z}$. These series are uniformly convergent on $R^3\times[0,2\pi]$ in the spaces which we have chosen. More precisely, we have the following three results, which are taken from [7, 9].

Lemma 2.1 *A linear operator* $\mathbf{J} : \mathcal{X} \longrightarrow \mathbf{C}_b^0(\mathbf{R}^3 \times [0,\pi], \mathbf{C}^2)$ *is defined by*

$$(\mathbf{J}u)(x,t) = \tilde{u}(x,t) := \sum_{n\in\mathbf{Z}} u_n(x) e^{int}, \quad u = (u_n)_{n\in\mathbf{Z}} \in \mathcal{X}.$$

Then **J** *is bounded. Here* C^2 *denotes the twice continuous differentiable function space.*

Lemma 2.2 *For* $u = (u_n)_{n\in\mathbf{Z}}, \nu = (\nu_n)_{n\in\mathbf{Z}} \in \mathbf{X}$, *the convolution* $u*\nu \in X$ *is defined by*

$$(u * \nu)_n = \sum_{k\in\mathbf{Z}} u_{n-k} \nu_k, \quad n \in \mathbf{Z}.$$

Then there exists $C>0$ *such that* $\|u*\nu\|X \leq C\|u\|X\|\nu\|X$.

Lemma 2.3 *Let a linear operator* $M: X \longrightarrow X$ *be defined component-wise as* $(Mu)_n = M_n u_n$ *for* $u = (u_n)_{n\in\mathbf{Z}}$. *Then* $\|Mu\|_{\mathbf{X}} = (\|M_0\|_{\mathbf{H}^m \longrightarrow \mathbf{H}^m} + \sup_{n\in\mathbf{Z}\setminus\{0\}} \|M\|_{\mathbf{H}^m \longrightarrow \mathbf{H}^m})\|u\|_{\mathbf{X}}$.

By (H2) and (H3), we know that the operators M_1 and M_2 have two eigenvalues $\lambda_0^{\pm}(\alpha)$ and $\mu_0^{\pm}(\alpha)$, respectively, and all other eigenvalues of M_1 and M_2 are strictly bounded away from the imaginary axis in the left half plane. Thus we construct the M_j-invariant (j=1,2) projections $P\pm1,c$ by

$$P_{1,c}\omega = (\psi^{+,*}, \omega)_{\mathbf{L}^2} \psi^+, \qquad P_{-1,c}\omega = (\psi^{-,*}, \omega)_{\mathbf{L}^2} \psi^-, \tag{2.10}$$

$$P_{1,c}v = \left(\psi^{+,*}, v\right)_{\mathbf{L}^2}\psi^{+}, \qquad P_{-1,c}v = \left(\psi^{-,*}, v\right)_{\mathbf{L}^2}\psi^{-}, \tag{2.11}$$

where $\psi^{\pm}$ denotes the associated normalized eigenfunctions, $\psi^{\pm1,*}$ denotes the associated normalized eigenfunctions of the adjoint operator M_j^*. The bounded 'stable' part of the projection is $\mathbf{P}_{\pm1,s} = I - \mathbf{P}_{\pm1,c}$, and we also know that $\mathbf{P}_{\pm,c}M_j = M_j\mathbf{P}_{\pm,c}$ and $\mathbf{P}_{\pm,s}M_j = M_j\mathbf{P}_{\pm,s}$ Thus we can split $\omega_{\pm1}$ and $v_{\pm1}$ as $\omega_1 = \omega_{1,c} + \omega_{1,s}$, $\omega_{-1} = \omega_{-1,c} + \omega_{-1,s}$, $v_1 = v_{1,c} + v_{1,s}$ and $v_{-1} = v_{-1,c} + v_{-1,s}$, where $\omega_{\pm1,c} = \mathbf{P}_{\pm1,c}\omega_1$, $\omega_{\pm1,s} = \mathbf{P}_{\pm1,s}\omega_1$, $v_{\pm1,c} = \mathbf{P}_{\pm1,c}v_1$, and $v_{\pm1,s} = \mathbf{P}_{\pm1,s}v_1$. Applying the above decompositions to (2.8)-(2.9), we have

$$(in\xi_1 - M_1)\omega_n = g_n^3(\omega, u, v), \quad n = \pm2, \pm3, \ldots, \tag{2.12}$$

$$(in\xi_2 - M_2)v_n = g_n^4(\omega, u, v), \quad n = \pm2, \pm3, \ldots, \tag{2.13}$$

$$M_1\omega_0 = g_0^3(\omega, u, v), \quad n = 0, \tag{2.14}$$

$$M_2v_0 = g_0^4(\omega, u, v), \quad n = 0, \tag{2.15}$$

$$(\pm i\xi_1 - M_1)\omega_{\pm1,s} = \mathbf{P}_{\pm1,s}g_{\pm1}^3(\omega, u, v), \tag{2.16}$$

$$(\pm i\xi_2 - M_2)v_{\pm1,s} = \mathbf{P}_{\pm1,s}g_{\pm1}^4(\omega, u, v), \tag{2.17}$$

$$(\pm i\xi_1 - M_1)\omega_{\pm1,c} = \mathbf{P}_{\pm1,c}g_{\pm1}^3(\omega, u, v), \tag{2.18}$$

$$(\pm i\xi_2 - M_2)v_{\pm1,c} = \mathbf{P}_{\pm1,c}g_{\pm1}^4(\omega, u, v). \tag{2.19}$$

The organization of the proof of Theorem 1.1 is as follows: we first solve (2.14)-(2.15), then we use the fixed point theorem to solve (2.12)-(2.13) and (2.16)-(2.17). This is a nontrivial process due to the nonlinear terms $g_n^3(\omega, u, v)$ and $g_n^4(\omega, u, v)$. Finally, we employ the implicit function theorem to solve (2.18)-(2.19). The process of solving (2.18)-(2.19) is inspired by the classical Hopf-Bifurcation result [10].

Equations (2.12)-(2.19) can be rewritten as

$$(in\Xi + \mathcal{N} + G)\varphi_n = F_n(\varphi, u), \quad n = \pm2, \pm3, \ldots, \tag{2.20}$$

$$(\mathcal{N} + G)\varphi_0 = F_0(\varphi, u), \quad n = 0, \tag{2.21}$$

$$(\pm i\Xi + \mathcal{N} + G)\varphi_{\pm1,s} = \mathbf{P}_{\pm1,s}F_{\pm1}(\varphi, u), \tag{2.22}$$

$$(\pm i\Xi + \mathcal{N} + G)\varphi_{\pm1,c} = \mathbf{P}_{\pm1,c}F_{\pm1}(\varphi, u). \tag{2.23}$$

Next we solve (2.21). The linear operator N has an essential spectrum up to the imaginary axis, and it is an invertible operator in the following sense. One may refer to [9] for the details of the proof.

Lemma 2.4 *For* j=1,2 *and* $f = (f^1, f^2)^T \in \mathcal{H}^{m-1} \cap \mathcal{L}^1$, *the equation* $N\varphi = \partial_j f$ *has a unique solution* $\varphi = \mathcal{N}^{-1}\partial_j f \in \mathcal{H}^m$. *Moreover,* $\|\varphi\|_{\mathcal{H}^m} \leq C\|f\|_{\mathcal{H}^{m-1}\cap\mathcal{L}^1}$.

This lemma tells us that $\widehat{\mathcal{N}}(iy_i, iy_i)^T$ is a bounded compact operator from $\mathcal{L}_m^2$ to itself. Furthermore, the spectra of $\widehat{\mathcal{N}} + \widehat{G}$ and $\widehat{\mathcal{N}}$ only differ by isolated eigenvalues of finite multiplicity (see the book of Henry [8, p.136]). The following three lemmas give the solvability of (2.21); they are similar to Lemma 6-8 in [9]. Here we omit their proofs.

Lemma 2.5 *Assume that* (H1)-(H3) *hold. Then* (2.21) *has a unique solution* $\varphi_0 = (\mathcal{N} + G)^{-1}F_0(\varphi, u)$. *Moreover,* $\|\varphi_0\|_{\mathcal{H}^m} \leq C\|y_j^{-1}I_{2\times2}\widehat{F}_0(\hat{\varphi}, \hat{u})\|_{\mathcal{L}_m^2}$, *where* $I_{2\times2}$ *and* $\widehat{F}_0$ *denote the* 2×2 *unit matrix and the application of Fourier transform to* F_0, *respectively.*

Lemma 2.6 *There exists a constant* C>0 *such that*

$$\|u\|_{\mathbf{H}^m} \leq C\|\omega\|_{\mathbf{H}^m}, \qquad \|\partial_{x_i}u\|_{\mathbf{H}^m} \leq C\|\omega\|_{\mathbf{H}^m}.$$

Lemma 2.7 *For* $m > \frac{3}{2}$, *there exists a positive constant C such that*

$$\|\hat{\omega} * \hat{u}\|_{\mathbf{L}_m^2} \leq C\|\hat{\omega}\|_{\mathbf{L}_m^2}\|\hat{u}\|_{\mathbf{L}_m^2}.$$

Applying the Fourier transform to g^3 and g^4 in (2.2)-(2.3), we get

$$\widehat{g^3} = -iy(\hat{\omega} * \hat{u}^T - \hat{u} * \hat{\omega}^T + v_0'|y|^2\hat{v} * \hat{u}^T + iy\hat{v} * \hat{\omega}), \tag{2.24}$$

$$\widehat{g^4} = -iy\hat{u} * \hat{v} - 2\kappa_0'|y|^2\hat{v} * \hat{v}. \tag{2.25}$$

From the form of the nonlinear terms g^3 and g^4, it is critical to estimate the term as uv and u^2. For convenience, we derive some estimates as regards the nonlinear terms $N^1(\varphi)=\varphi^2$ and $N^2(\varphi, \psi)=\varphi\psi$. Since the proof of the next lemma is similar to Lemma

4 in [7], we omit it.

Lemma 2.8 *Define* $N^1 : \mathcal{X} \longrightarrow \mathcal{X}$ *by* $N^1(\varphi)_n = N^1_n(\mathbf{J}\varphi)$ *and* $N^2 : \mathcal{X} \times \mathcal{X} \longrightarrow \mathcal{X}$ *by* $N^2(\varphi)_n = N^2_n(\mathbf{J}\varphi, \mathbf{J}\psi)$ *for* $\varphi, \psi \in \mathcal{X}$. *Then we have*

$$\|N^1(\varphi)\|_{\mathcal{X}} \le C\|\varphi\|^2_{\mathcal{X}}, \qquad \|N^2(\varphi,\psi)\|_{\mathcal{X}} \le C\|\psi\|_{\mathcal{X}}\|\varphi\|_{\mathcal{X}}$$

for $\varphi, \psi \in \mathcal{X}$ *with* $\|\varphi\|_{\mathcal{X}} \le 1$ *and* $\|\psi\|_{\mathcal{X}} \le 1$. *Moreover, we have*

$$\|N^1(\varphi^1) - N^1(\varphi^2)\|_{\mathcal{X}} \le C(\|\varphi^1\|_{\mathcal{X}} + \|\varphi^2\|_{\mathcal{X}})\|\varphi^1 - \varphi^2\|_{\mathcal{X}},$$

$$\|N^2(\varphi^1,\psi^1) - N^2(\varphi^2,\psi^2)\|_{\mathcal{X}} \le C(\|\varphi^1\|_{\mathcal{X}} + \|\varphi^2\|_{\mathcal{X}} + \|\psi^1\|_{\mathcal{X}} + \|\psi^2\|_{\mathcal{X}}) \times (\|\varphi^1 - \varphi^2\|_{\mathcal{X}} + \|\psi^1 - \psi^2\|_{\mathcal{X}}),$$

for $\varphi^1, \varphi^2, \psi^1, \psi^2 \in \mathcal{X}$ *with* $\|\varphi^1\|_{\mathcal{X}}, \|\varphi^2\|_{\mathcal{X}}, \|\psi^1\|_{\mathcal{X}}, \|\psi^2\|_{\mathcal{X}} \le 1$.

By a small modification of the proof of Lemma 7 in [7], we have the following result.

Lemma 2.9 *Assume that* ξ *is close enough to* ξ_0. *Then there exists a constant* C>0 *such that*

$$\|(in\xi_i - \overline{M_i})^{-1}\|_{\mathbf{H}^m \longrightarrow \mathbf{H}^m} \le C, \qquad \|(in\xi_i - \overline{M_i})^{-1}\nabla^j\cdot\|_{\mathbf{H}^m \longrightarrow \mathbf{H}^m} \le C,$$

$$\|(in\xi_i - M_i)^{-1}\|_{\mathbf{H}^m \longrightarrow \mathbf{H}^m} \le C, \qquad \|(in\xi_i - M_i)^{-1}\nabla^j\cdot\|_{\mathbf{H}^m \longrightarrow \mathbf{H}^m} \le C,$$

$$\|(in\xi_i - M_i)^{-1}\mathbf{P}_{\pm 1,s}\|_{\mathbf{H}^m \longrightarrow \mathbf{H}^m} \le C, \qquad \|(in\xi_i - M_i)^{-1}\nabla^j \cdot \mathbf{P}_{\pm 1,s}\|_{\mathbf{H}^m \longrightarrow \mathbf{H}^m} \le C,$$

For n≠0 *and* j=1,2. *Here* $\overline{M_i}$ *and* M_i *are defined in* (2.6)-(2.7), *respectively.*

Thus by Lemma 2.9, we can rewrite (2.20) and (2.22) as

$$\varphi_n = (in\Xi + \mathcal{N} + G)^{-1}F_n(\varphi, u), \quad n = \pm 2, \pm 3, \ldots,$$

$$\varphi_{\pm 1,s} = (\pm i\Xi + \mathcal{N} + G)^{-1}\mathbf{P}_{\pm 1,s}F_{\pm 1}(\varphi, u),$$

i.e.

$$\omega_n = (in\xi_1 - M_1)^{-1}g^3_n(\omega, u, v), \quad n = \pm 2, \pm 3, \ldots, \tag{2.26}$$

$$v_n = (in\xi_2 - M_2)^{-1}g^4_n(\omega, u, v), \quad n = \pm 2, \pm 3, \ldots, \tag{2.27}$$

$$\omega_{\pm 1,s} = (\pm i\xi_1 - M_1)^{-1}\mathbf{P}_{\pm 1,s}g^3_{\pm 1}(\omega, u, v), \tag{2.28}$$

$$v_{\pm 1,s} = (\pm i\xi_2 - M_2)^{-1}\mathbf{P}_{\pm 1,s}g^4_{\pm 1}(\omega, u, v). \tag{2.29}$$

Using Lemmas 2.8 and 2.9, we obtain the solvability of (2.26)-(2.29). Since the proof is similar to [9], we omit it.

Lemma 2.10 *Assume that there exist* $\sigma_1,\sigma_2>0$ *such that for all* ξ_1, $\xi_2>0$ *with* $|\xi_1-\xi_0|,|\xi_2-\xi_0|\le\sigma_1$ *and al* $\omega_{\pm 1,c}, v_{\pm 1,c} \in \mathbf{H}^m$ *with* $\|\omega_{\pm 1,c}\|_{\mathbf{H}^m}, \|v_{\pm 1,c}\|_{\mathbf{H}^m} \le \sigma_2$. *Then* (2.26)-(2.29) *has a unique solution* $(\tilde{\omega}, \tilde{v}) = \Phi(\omega_c, v_c) \in \mathcal{X}$, *where* $\omega_c = (\omega_{-1,c}, \omega_{1,c})$, $v_c = (v_{-1,c}, v_{1,c})$, $\tilde{\omega} = (\ldots, \omega_{-2}, \omega_{-1,c} + \omega_{-1,s}, \omega_0, \omega_{1,c} + \omega_{1,s}, \omega_2, \ldots)$, $\tilde{v} = (\ldots, v_{-2}, v_{-1,c} + v_{-1,s}, v_0, v_{1,c} + v_{1,s}, v_2, \ldots)$. *Moreover, there exists a positive C such that*

$$\Phi(0,0) = (0,0), \qquad \|\tilde{\omega} - \omega_c\|_{\mathbf{X}} \le C(\|\omega_{-1,c}\|^2_{\mathbf{H}^m} + \|\omega_{1,c}\|^2_{\mathbf{H}^m}), \tag{2.30}$$

$$\|\tilde{v} - v_c\|_{\mathbf{X}} \le C(\|v_{-1,c}\|^2_{\mathbf{H}^m} + \|v_{1,c}\|^2_{\mathbf{H}^m}), \tag{2.31}$$

With $\tilde{\omega} - \omega_c := (\ldots, 0, \omega_{-1,c}, 0, \omega_{1,c}, 0, \ldots)$ *and* $\tilde{v} - v_c := (\ldots, 0, v_{-1,c}, 0, v_{1,c}, 0, \ldots)$.

Proof

For fixed ξ_1, $\xi_2>0$ so close to ξ_0 and given $\omega_{\pm 1,c}, v_{\pm 1,c} \in \mathbf{H}^m$ with $\|\omega_{\pm 1,c}\|_{\mathbf{H}^m}, \|v_{\pm 1,c}\|_{\mathbf{H}^m} \le \sigma_2$. Define the operator

$$\begin{aligned}\Gamma : (\tilde{\omega}^*, \tilde{v}^*) &\longmapsto (\tilde{\omega}, \tilde{v}) \\ &= (\tilde{\omega}^* + (\ldots, 0, \omega_{-1,c}, 0, \omega_{1,c}, 0, \ldots), \tilde{v}^* + (\ldots, 0, v_{-1,c}, 0, v_{1,c}, 0, \ldots)) \\ &\longmapsto (\omega, v) \longmapsto (\tilde{\omega}^{**}, \tilde{v}^{**}) = \text{the right hand side of (2.26)-(2.29)},\end{aligned}$$

where $(\omega, v) = (\mathbf{J}\tilde{\omega}, \mathbf{J}\tilde{v})$ are defined in Lemma 2.1 and

$$(\tilde{\omega}^*, \tilde{v}^*) = ((\ldots, \omega_{-2}, \omega_{-1,s}, \omega_0, \omega_{1,s}, \omega_2, \ldots), (\ldots, v_{-2}, v_{-1,s}, v_0, v_{1,s}, v_2, \ldots)),$$

$$(\tilde{\omega}, \tilde{v}) = (\tilde{\omega}^* + \omega_c, \tilde{v}^* + v_c)$$
$$= (\tilde{\omega}^* + (\ldots, 0, \omega_{-1,c}, 0, \omega_{1,c}, 0, \ldots), \tilde{v}^* + (\ldots, 0, v_{-1,c}, 0, v_{1,c}, 0, \ldots)).$$

Now we prove the operator Γ is a self-map of a sufficiently small ball in X. By Lemmas 2.6-2.9and the form of the nonlinear terms $\widehat{g^3}, \widehat{g^4}$ in (2.24)-(2.25), we derive

$$\begin{aligned}
\|\tilde{\omega}^{**}\|_{\mathbf{X}} &\le C\sup\{\|(in\xi_1 - M_1)^{-1}\|_{\mathbf{H}^m \to \mathbf{H}^m}, \|(\pm i\xi_1 - M_1)^{-1}\mathbf{P}_{\pm 1,s}\|_{\mathbf{H}^m \to \mathbf{H}^m}, \\
&\quad \|(in\xi_1 - M_1)^{-1}\nabla^j\|_{\mathbf{H}^m \to \mathbf{H}^m}, \|(\pm i\xi_1 - M_1)^{-1}\nabla^j\mathbf{P}_{\pm 1,s}\|_{\mathbf{H}^m \to \mathbf{H}^m} \\
&\quad : n \in \mathbf{Z}\setminus\{-1,1\}\} \times \|(\tilde{g}_n^3(\omega, u, v))_{n\in\mathbf{Z}}\|_{\mathbf{X}} \\
&\le C\|\tilde{g}^3(\tilde{\omega}, u, \tilde{v})\|_{\mathbf{X}} \\
&\le C(\|\tilde{\omega}\|_{\mathbf{X}}\|u\|_{\mathbf{X}} + \|\tilde{\omega}\|_{\mathbf{X}}\|\tilde{v}\|_{\mathbf{X}} + \|u\|_{\mathbf{X}}\|\tilde{v}\|_{\mathbf{X}}) \\
&\le C(\|\tilde{\omega}\|_{\mathbf{X}}^2 + \|\tilde{\omega}\|_{\mathbf{X}}\|\tilde{v}\|_{\mathbf{X}}) \\
&\le C(\|\tilde{\omega}^*\|_{\mathbf{X}}^2 + \|\omega_{-1,c}\|_{\mathbf{H}^m}^2 + \|\omega_{1,c}\|_{\mathbf{H}^m}^2 + \|\tilde{v}^*\|_{\mathbf{X}}^2 + \|v_{-1,c}\|_{\mathbf{H}^m}^2 + \|v_{1,c}\|_{\mathbf{H}^m}^2) \\
&\le C(\|\tilde{\omega}^*\|_{\mathbf{X}}^2 + \|\tilde{v}^*\|_{\mathbf{X}}^2 + \sigma_2^2),
\end{aligned} \tag{2.32}$$

$$\begin{aligned}
\|\tilde{v}^{**}\|_{\mathbf{X}} &\le C\sup\{\|(in\xi_2 - M_2)^{-1}\|_{\mathbf{H}^m \to \mathbf{H}^m}, \|(\pm i\xi_2 - M_2)^{-1}\mathbf{P}_{\pm 1,s}\|_{\mathbf{H}^m \to \mathbf{H}^m}, \\
&\quad \|(in\xi_2 - M_2)^{-1}\nabla^j\|_{\mathbf{H}^m \to \mathbf{H}^m}, \|(\pm i\xi_2 - M_2)^{-1}\nabla^j\mathbf{P}_{\pm 1,s}\|_{\mathbf{H}^m \to \mathbf{H}^m} \\
&\quad : n \in \mathbf{Z}\setminus\{-1,1\}\} \times \|(\tilde{g}_n^4(\omega, u, v))_{n\in\mathbf{Z}}\|_{\mathbf{X}} \\
&\le C\|\tilde{g}^4(\tilde{\omega}, u, \tilde{v})\|_{\mathbf{X}} \\
&\le C(\|\tilde{v}\|_{\mathbf{X}}^2 + \|\tilde{\omega}\|_{\mathbf{X}}\|\tilde{v}\|_{\mathbf{X}}) \\
&\le C(\|\tilde{\omega}^*\|_{\mathbf{X}}^2 + \|\omega_{-1,c}\|_{\mathbf{H}^m}^2 + \|\omega_{1,c}\|_{\mathbf{H}^m}^2 + \|\tilde{v}^*\|_{\mathbf{X}}^2 + \|v_{-1,c}\|_{\mathbf{H}^m}^2 + \|v_{1,c}\|_{\mathbf{H}^m}^2) \\
&\le C(\|\tilde{\omega}^*\|_{\mathbf{X}}^2 + \|\tilde{v}^*\|_{\mathbf{X}}^2 + \sigma_2^2),
\end{aligned} \tag{2.33}$$

where

$$\tilde{g}^3 = 2\omega \cdot u^T + v_0' v \cdot u^T + v \cdot \omega^T,$$

$$\tilde{g}^4 = u \cdot v^T + 2\kappa_0' v \cdot v^T.$$

Thus, for $\sigma_2 \le \frac{1}{\sqrt{2C}}$ and $(\tilde{\omega}^*, \tilde{v}^*) \in \mathcal{X}$ with $\|(\tilde{\omega}^*, \tilde{v}^*)\|_{\mathcal{X}} \le \frac{1}{\sqrt{2C}}$, we have

$$\begin{aligned}
\|\Gamma(\tilde{\omega}^*, \tilde{v}^*)\|_{\mathcal{X}} &= \|\tilde{\omega}^{**}\|_{\mathcal{X}} + \|\tilde{v}^{**}\|_{\mathbf{X}} \\
&\le C((\|\tilde{\omega}^*\|_{\mathbf{X}} + \|\tilde{v}^*\|_{\mathbf{X}})^2 + \sigma_2^2) \le 1,
\end{aligned}$$

which implies that for sufficient small $\sigma_2 > 0$. Hence, by the Banach fixed point theorem, we obtain a unique fixed point $(\tilde{\theta}^*, \tilde{v}^*) \in \mathcal{X}$ of Γ, which means that (2.26)-(2.29) have a solution of $(\tilde{\omega}, \tilde{v}) = (\tilde{\omega}^* + \omega_c, \tilde{v}^* + v_c)$. Moreover, if $(\omega_{\pm 1,c}, v_{\pm 1,c}) = (0,0)$, then $\Phi(0,0)=(0,0)$. Next we prove the second inequality in (2.30) and (2.31). Note that $(\tilde{\omega}^*, \tilde{v}^*) = \Gamma(\tilde{\omega}^*, \tilde{v}^*) = (\tilde{\theta}^{**}, \tilde{v}^{**})$. This, combined with (2.32)-(2.33), allows us to derive

$$\begin{aligned}
\|\tilde{\omega} - \omega_c\|_{\mathbf{X}} &= \|\tilde{\omega}^*\|_{\mathbf{X}} = \|\tilde{\omega}^{**}\|_{\mathbf{X}} \\
&\le C(\|\tilde{\omega}^*\|_{\mathbf{X}}^2 + \|\omega_{-1,c}\|_{\mathcal{H}^m}^2 + \|\omega_{1,c}\|_{\mathcal{H}^m}^2),
\end{aligned}$$

$$\begin{aligned}
\|\tilde{v} - v_c\|_{\mathbf{X}} &= \|\tilde{v}^*\|_{\mathbf{X}} = \|\tilde{v}^{**}\|_{\mathbf{X}} \\
&\le C(\|\tilde{v}^*\|_{\mathbf{X}}^2 + \|v_{-1,c}\|_{\mathcal{H}^m}^2 + \|v_{-1,c}\|_{\mathcal{H}^m}^2).
\end{aligned}$$

Thus we deduce that for a sufficient small ball $B_r(0) \subset B_1(0)$,

$$\|\tilde{\omega} - \omega_c\|_{\mathbf{X}} \le C(\|\omega_{-1,c}\|_{\mathbf{H}^m}^2 + \|\omega_{1,c}\|_{\mathbf{H}^m}^2),$$

$$\|\tilde{v} - v_c\|_{\mathbf{X}} \le C(\|v_{-1,c}\|_{\mathbf{H}^m}^2 + \|v_{1,c}\|_{\mathbf{H}^m}^2),$$

where

$\tilde{\omega} - \omega_c := (\ldots, 0, \omega_{-1,c}, 0, \omega_{1,c}, 0, \ldots),$

$\tilde{v} - v_c := (\ldots, 0, v_{-1,c}, 0, v_{1,c}, 0, \ldots).$

This completes the proof.

It remains to analyze (2.18)-(2.19). We restate the equations:

$(\pm i\xi_1 - M_1)\omega_{\pm 1,c} = \mathbf{P}_{\pm 1,c} g^3_{\pm 1}(\omega, u, v),$

$(\pm i\xi_2 - M_2)v_{\pm 1,c} = \mathbf{P}_{\pm 1,c} g^4_{\pm 1}(\omega, u, v).$ It follows from $(\omega_{-1}, v_{-1}) = (\overline{\omega_1}, \overline{v_1})$ and $(g^3_{-1}, g^4_{-1}) = (\overline{g^3_1}, \overline{g^4_1})$ that the '−' equation is the complex conjugate of the '+' equation. By Lemma 2.1, we can denote $(\omega, v) = (\mathbf{J}\tilde{\omega}, \mathbf{J}\tilde{v})$ by means of $(\tilde{\omega}, \tilde{v}) = \Phi(\omega_c, v_c) = \Phi((\overline{\omega_{1,c}}, \omega_{1,c}), (\overline{v_{1,c}}, v_{1,c}))$. Our target is to find (ξ_1,α) and (ξ_2,α) close to (ξ_0,α_c)(and a nontrivial solution $(\omega_{1,c}, v_{1,c}) = (\omega_{1,c}, v_{1,c})(x)$ of

$$-i\xi_1\omega_{1,c} + M_1\omega_{1,c} + \mathbf{P}_{1,c} g^3_1\left(\mathbf{J}\Phi(\overline{\omega_{1,c}}, \omega_{1,c}, \overline{v_{1,c}}, v_{1,c})\right) = 0, \tag{2.34}$$

$$-i\xi_2 v_{1,c} + M_2 v_{1,c} + \mathbf{P}_{1,c} g^4_1\left(\mathbf{J}\Phi(\overline{\omega_{1,c}}, \omega_{1,c}, \overline{v_{1,c}}, v_{1,c})\right) = 0. \tag{2.35}$$

Since $\omega_{1,c}, v_{1,c} \in \mathbb{C}\psi^+$ and $(M_1\psi^+, M_2\psi^+) = (\lambda_0^+(\alpha)\psi^+, \mu_0^+(\alpha)\psi^+)$, we can write $\omega_{1,c}=\eta\psi^+$ and $v_{1,c}=\delta\psi^+$. Then by (2.34)-(2.35), for some $\eta,\delta \in C\backslash\{0\}$, we obtain

$$-i\xi_1\eta\psi^+ + \lambda_0^+(\alpha)\eta\psi^+ + \mathbf{P}_{1,c} g^3_1\left(\mathbf{J}\Phi(\overline{\eta\psi^+}, \eta\psi^+, \overline{\delta\psi^+}, \delta\psi^+)\right) = 0, \tag{2.36}$$

$$-i\xi_2\delta\psi^+ + \mu_0^+(\alpha)\delta\psi^+ + \mathbf{P}_{1,c} g^4_1\left(\mathbf{J}\Phi(\overline{\eta\psi^+}, \eta\psi^+, \overline{\delta\psi^+}, \delta\psi^+)\right) = 0. \tag{2.37}$$

Now we introduce $(p_{1,c}, \theta_{1,c})$ by $(\mathbf{P}_{1,c}\omega, \mathbf{P}_{1,c}v) = (p_{1,c}(\omega)\psi^+, \theta_{1,c}(v)\psi^+)$. Then (2.36)-(2.37) can be simplified to

$$-i\xi_1\eta + \lambda_0^+(\alpha)\eta + g^3(\alpha, \eta, \delta) = 0, \quad \text{for some } \eta \in C, \tag{2.38}$$

$$-i\xi_2\delta + \mu_0^+(\alpha)\delta + g^4(\alpha, \eta, \delta) = 0, \quad \text{for some } \delta \in C, \tag{2.39}$$

where the cubic coefficient $a_1 \neq 0$ and $a_2 \neq 0$ in

$$g^3(\alpha, \eta, \delta) := p_{1,c}\left(g^3_1\left(\mathbf{J}\Phi(\overline{\eta\psi^+}, \eta\psi^+, \overline{\delta\psi^+}, \delta\psi^+)\right)\right), \tag{2.40}$$

$$g^4(\alpha, \eta, \delta) := \theta_{1,c}\left(g^4_1\left(\mathbf{J}\Phi(\overline{\eta\psi^+}, \eta\psi^+, \overline{\delta\psi^+}, \delta\psi^+)\right)\right). \tag{2.41}$$

Note that $|p_{1,c}(\omega)| \le C\|\mathbf{P}_{1,c}\omega\|_{\mathbf{H}^m} \le C\|\omega\|_{\mathbf{H}^m}$ and $|\theta_{1,c}(v)| \le C\|\mathbf{P}_{1,c}v\|_{\mathbf{H}^m} \le C\|v\|_{\mathbf{H}^m}$. Thus by (2.30)-(2.31) and (2.40)-(2.41), we derive

$$\left|p_{1,c}\left(g^3_1\left(\mathbf{J}\Phi(\overline{\eta\psi^+}, \eta\psi^+, \overline{\delta\psi^+}, \delta\psi^+)\right)\right)\right| \le C(|\eta|^2 + |\delta|^2),$$

$$\left|\theta_{1,c}\left(g^4_1\left(\mathbf{J}\Phi(\overline{\eta\psi^+}, \eta\psi^+, \overline{\delta\psi^+}, \delta\psi^+)\right)\right)\right| \le C(|\eta|^2 + |\delta|^2),$$

where we use the following relation: $(\tilde{\omega}, \tilde{v}) = \Phi(\omega_c, v_c) = \Phi(\overline{\eta\psi^+}, \eta\psi^+, \overline{\delta\psi^+}, \delta\psi^+)$.

Analogously to the case of the classical Hopf-bifurcation result [10], the Banach fixed point theorem can be applied to (2.38)-(2.39), as soon as the zero solution is divided out. It is sufficient to find the real value solutions (*i.e.*$(\gamma_1, \gamma_2)=(\eta, \delta) \in R^2$ of (2.38)-(2.39). Hence we define the complex-valued smooth function

$$\Upsilon^1(\gamma_1, \gamma_2; \varrho, \epsilon) := \begin{cases} -i(\xi_0 + \varrho) + \lambda_0^+(\alpha_c + \epsilon) + \gamma_1^{-1} g^3(\alpha_c + \epsilon, \gamma_1, \gamma_2), & \gamma_1 \neq 0, \\ -i(\xi_0 + \varrho) + \lambda_0^+(\alpha_c + \epsilon), & \gamma_1 = 0, \end{cases}$$

$$\Upsilon^2(\gamma_1, \gamma_2; \varrho, \epsilon) := \begin{cases} -i(\xi_0 + \varrho) + \mu_0^+(\alpha_c + \epsilon) + \gamma_2^{-1} g^4(\alpha_c + \epsilon, \gamma_1, \gamma_2), & \gamma_2 \neq 0, \\ -i(\xi_0 + \varrho) + \mu_0^+(\alpha_c + \epsilon), & \gamma_2 = 0. \end{cases}$$

It follows from

$(\lambda_0^+(\alpha_c), \mu_0^+(\alpha_c)) = (i\xi_0, i\xi_0)$ that $(\Upsilon^1(0,0,0,0), \Upsilon^2(0,0,0,0)) = (0,0)$. Moreover, by assumption (H2) for the Jacobi matrix

$$\mathbf{D}_{\rho,\epsilon}\Upsilon^1(\gamma_1, \gamma_2; \varrho, \epsilon)|_{\gamma_1=\gamma_2=\varrho=\epsilon=0} = \begin{pmatrix} 0 & \frac{d}{d\beta}\operatorname{Re}\lambda_0^+(\alpha)|_{\alpha=\alpha_c} \\ -1 & \frac{d}{d\beta}\operatorname{Im}\lambda_0^+(\alpha)|_{\alpha=\alpha_c} \end{pmatrix},$$

$$\mathbf{D}_{\rho,\epsilon}\Upsilon^2(\gamma_1, \gamma_2; \varrho, \epsilon)|_{\gamma_1=\gamma_2=\varrho=\epsilon=0} = \begin{pmatrix} 0 & \frac{d}{d\beta}\operatorname{Re}\mu_0^+(\alpha)|_{\alpha=\alpha_c} \\ -1 & \frac{d}{d\beta}\operatorname{Im}\mu_0^+(\alpha)|_{\alpha=\alpha_c} \end{pmatrix}$$

with respect to ρ, ϵ one has det

$\mathbf{D}_{\rho,\epsilon}\Upsilon^1(\gamma_1, \gamma_2; \varrho, \epsilon)|_{\gamma_1=\gamma_2=\varrho=\epsilon=0} = \frac{d}{d\beta}\operatorname{Re}\lambda_0^+(\alpha)|_{\alpha=\alpha_c} > 0$ and det $\mathbf{D}_{\rho,\epsilon}\Upsilon^2(\gamma, \gamma_2; \varrho, \epsilon)|_{\gamma_1=\gamma_2=\varrho=\epsilon=0} = \frac{d}{d\beta}\operatorname{Re}\mu_0^+(\beta)|_{\alpha=\alpha_c} > 0$. Thus for sufficient small $\gamma_1,\gamma_2>0$, we can find a function $\gamma_1 \mapsto (\varrho(\gamma_1),\epsilon(\gamma_1))$ and $\gamma_2 \mapsto (\varrho(\gamma_2),\epsilon(\gamma_2))$ with $\varrho(0)=\epsilon(0)=0$ such that

$$-i(\xi_0+\varrho(\gamma_1))+\lambda_0^+(\alpha_c+\epsilon(\gamma_1))-\gamma_1^{-1}g^3(\alpha_c+\epsilon(\gamma_1),\gamma_1,\alpha_c+\epsilon(\gamma_2),\gamma_2)=0,$$

$$-i(\xi_0+\varrho(\gamma_2))+\mu_0^+(\alpha_c+\epsilon(\gamma_2))-\gamma_2^{-1}g^4(\alpha_c+\epsilon(\gamma_1),\gamma_1,\alpha_c+\epsilon(\gamma_2),\gamma_2)=0.$$

Note the degree of nonlinearity. Then it follows from differentiating this equation that $\epsilon^{(i)}\neq 0$ for some first i. Hence the function $\gamma_1\mapsto\epsilon(\gamma_1)$ and $\gamma_1\mapsto\epsilon(\gamma_2)$ can locally be inverted to yield a function $\epsilon\mapsto\gamma_1(\epsilon)$ and $\epsilon\mapsto\gamma_2(\epsilon)$. This implies that the following equation holds:

$$-i(\xi_0+\varrho(\gamma_1(\epsilon)))\gamma_1(\epsilon)+\lambda_0^+(\alpha_c+\epsilon)\gamma_1(\epsilon)-g^3(\alpha_c+\epsilon,\gamma_1(\epsilon),\gamma_2(\epsilon))=0,$$

$$-i(\xi_0+\varrho(\gamma_2(\epsilon)))\gamma_2(\epsilon)+\mu_0^+(\alpha_c+\epsilon)\gamma_2(\epsilon)-g^4(\alpha_c+\epsilon,\gamma_1(\epsilon),\gamma_2(\epsilon))=0,$$

for sufficient small $\epsilon>0$.

Therefore we obtain the desired solutions of (2.34)-(2.35) by setting $(\xi_1, \xi_2)=(\xi_0+\varrho(\gamma_1(\epsilon)),\xi_0+\varrho(\gamma_2(\epsilon)))$, $\alpha=\alpha_c+\epsilon$, and $(\omega_{1,c},\nu_{1,c})=(\gamma_1(\epsilon)\psi^+_{\alpha_c+\epsilon},\gamma_2(\epsilon)\psi^+_{\alpha_c+\epsilon})(x)$. This result combined with Lemmas 3.2, 3.8, and (1.10) gives the proof of Theorem 1.1.

ACKNOWLEDGEMENTS

The authors expresses his sincere thanks to the anonymous referees for a very careful reading and for providing many valuable comments and suggestions which led to an improvement of this paper.

Competing Interests

The authors declare that they have no competing interests.

REFERENCES

1. Turcotte, DL, Schubert, G: Geodynamics: Applications of Continuum Physics to Geological Problems. Wiley, New York (1982)
2. Bodenschatz, E, Pesch, W, Ahlers, G: Recent developments in Rayleigh-Bénard convection. In: Annual Review of Fluid Mechanics. Annu. Rev. Fluid Mech., vol. 32, pp. 709-778. Annual Reviews, Palo Alto (2000)
3. Getling, AV: Rayleigh-Bénard Convection. Structures and Dynamics. Advanced Series in Nonlinear Dynamics, vol. 11. World Scientific, River Edge (1998)
4. Lions, JL, Temam, R, Wang, S: On the equations of large-scale ocean. Nonlinearity **5**, 1007-1053 (1992)
5. Pedlosky, J: Geophysical Fluid Dynamics, 2nd edn. Springer, New York (1987)
6. Tritton, DJ: Physical Fluid Dynamics, 2nd edn. Oxford Science Publications. Clarendon, New York (1988)
7. Brand, T, Kunze, M, Schneider, G, Seelbach, T: Hopf bifurcation and exchange of stability in diffusive media. Arch. Ration. Mech. Anal. **171**, 263-296 (2004)
8. Henry, D: Geometric Theory of Semilinear Parabolic Equations. Springer, New York (1981)
9. Yan, WP: Hopf-Bifurcation theorem and stability for the Magneto-Hydrodynamics equations. Topol. Methods Nonlinear Anal. (to appear)
10. Marsden, JE, Mccracken, M: The Hopf Bifurcation and Its Applications. Springer, Berlin (1976)

CHAPTER 15

BIFURCATION ANALYSIS AND SOLUTIONS OF A HIGHER-ORDER NONLINEAR SCHRÖDINGER EQUATION

Yi Li, Wen-rui Shan, Tianping Shuai and Ke Rao

INTRODUCTION

In the past decades, communication systems have scored a great growth of the transmission capacity. Due to the undamped and unchanged characteristics in a far distance, optical solitons are the focus of many research groups during the past decades and stand a good chance to be the main information carriers in telecommunications in the future. Ignoring optical losses, the wave dynamics of nonlinear pulse propagation in a monomode fiber is described by the nonlinear Schrödinger equation [1, 2], which accounts for the group velocity dispersion and self-phase modulation. To increase the bit rate, it is often desirable to use shorter femtosecond pulses. However, when short pulses are considered, the equation can no longer represent the propagations of light pulses in fibers because higher-order dispersion terms and the non-Kerr nonlinearity effects cannot be neglected. This phenomenon can be expressed by a higher-order nonlinear Schrödinger equation [3]:

$$iE_z - \frac{\beta_2}{2}E_{tt} + \gamma_1 |E|^2 E = i\frac{\beta_3}{6}E_{ttt} + i\alpha_1 \left(|E|^2 E\right)_t + i\alpha_2 E\left(|E|^2\right)_t, \quad (1)$$

where α_1, α_2, β_2, β_3, and γ_1 are real constants. (t, x) is a slowly varying envelope amplitude, t represents the normalized retarded time (in the group velocity frame), and z represents the normalized distance along the direction of propagation. β_2 comes from the group velocity dispersion (GVD). γ_1 is proportional to the nonlinear index which originates from the Kerr effect. β_3 is the coefficients with the relevant work of the third-order dispersion. α_1 is related to self-steepening due to stimulated Raman scattering. The coefficient of the last term that is proportional to $\alpha_2 = \gamma_1 T_R$ has its origin in the delayed Raman response T_R. Generally speaking, T_R can be estimated from the slope of the Raman gain and is defined as the first moment of the nonlinear response function [4]. In fact, α_2 should be an imaginary number, but many analytical studies have been done when α_2 is real, such as Painleve property [5], inverse scattering transform [6], Hirota direct method, and conservation laws [7]. These researches verify its integrable nature and have obtained many exact wave solutions.

Laser spectroscopic techniques have been widely used in all fields of science. It can help us observe the physical processes in materials and molecules which occur on a femtosecond time scale by using ultrashort lasers. The pulses can also be applied in telecommunication and ultrafast signal routing systems. Research indicates that non-Kerr nonlinear effects begin to have some effects when the pulse width becomes narrower and the intensity of the incident light field becomes stronger. The influence is described by the NLS family of equations with nonlinear terms [8]. The nonlinearity due to fifth-order susceptibility can be obtained in many optical materials such as semiconductors and some transparent organic materials. Actually, it is also important to include some additional higher-order perturbation effects into the HNLS equation to analyze

Citation: Yi Li, Wen-rui Shan, Tianping Shuai, and Ke Rao, "Bifurcation Analysis and Solutions of a Higher-Order Nonlinear Schrödinger Equation," Mathematical Problems in Engineering, vol. 2015, Article ID 408586, 10 pages, 2015. doi:10.1155/2015/408586.

the solitary wave solution in a non-Kerr nonlinear medium. In this paper, with the aid of Mathematica, we study the new traveling wave solutions for a higher-order NLS equation that contains the non-Kerr nonlinear terms, which describes propagation of very short pulses in highly nonlinear optical fibers by using different elliptic functions. The bifurcation theory method is widely used to solve differential equations [9–12]. By using this method of dynamical systems, we obtain the explicit expressions of the bounded traveling wave solutions for the equation and investigate the relation between the bounded orbit of the traveling wave system and the energy level h. The new solutions correspond to the orbits on phase portraits and they include solitary, kink, and periodic wave solutions. Note that the existence of solitary wave solutions depends essentially on the model coefficients and therefore on the specific nonlinear features of the medium.

BIFURCATION AND PHASE PORTRAITS

We consider the higher-order NLS equation with non-Kerr term [13]:

$$iE_z - \frac{\beta_2}{2}E_{tt} + \gamma_1 |E|^2 E = i\frac{\beta_3}{6}E_{ttt} + i\alpha_1\left(|E|^2 E\right)_t + i\alpha_2 E\left(|E|^2\right)_t - \gamma_2 |E|^4 E + i\alpha_3\left(|E|^4 E\right)_t + i\alpha_4 E\left(|E|^4\right)_t, \quad (2)$$

where $\alpha_1, \alpha_2, \alpha_3, \alpha_4, \beta_2, \beta_3, \gamma_1$, and γ_2 are real constants. γ_2, α_3, and α_4 represent the coefficients of quintic nonKerr nonlinearities. The quintic nonlinearities arise from the expansion of the refractive index of the light pulse. The polarizations induced through the susceptibilities give the cubic and quintic (non-Kerr) terms in a nonlinear Schrodinger ¨ equation. When $\gamma_2 = \alpha_3 = \alpha_4 = 0$, (2) reduces to (1).

Assume that (2) has the form of the exact solution

$$E(z,t) = u(\xi)e^{i(kz-ct)}, \quad \xi = pz - qt, \quad (3)$$

where $u(\xi)$ is the real-valued function and the parameters of p, q, k, and c are real constants to be determined later. Substituting (3) into (2) and removing the exponential term, we change (2) into the form

$$\frac{1}{6}\beta_3 c^3 u(\xi) - \frac{1}{2}i\beta_3 c^2 q u'(\xi) + \frac{1}{2}\beta_2 c^2 u(\xi) - \frac{1}{2}\beta_3 c q^2 u''(\xi) - i\beta_2 c q u'(\xi) + \alpha_3(-c)u(\xi)^5 - \alpha_1 c u(\xi)^3 - k u(\xi) + ipu'(\xi) + \frac{1}{6}i\beta_3 q^3 u^{(3)}(\xi) - \frac{1}{2}\beta_2 q^2 u''(\xi) + 5i\alpha_3 q u(\xi)^4 u'(\xi) + 4i\alpha_4 q u(\xi)^4 u'(\xi) + 3i\alpha_1 q u(\xi)^2 u'(\xi) + 2i\alpha_2 q u(\xi)^2 u'(\xi) + \gamma_2 u(\xi)^5 + \gamma_1 u(\xi)^3 = 0. \quad (4)$$

The real and imaginary parts of (4), respectively, are

$$3q^2(\beta_2 + \beta_3 c)u''(\xi) - \left(\beta_3 c^3 + 3\beta_2 c^2 - 6k\right)u(\xi) - (6\gamma_1 - 6\alpha_1 c)u(\xi)^3 - (6\gamma_2 - 6\alpha_3 c)u(\xi)^5 = 0, \quad (5)$$

$$-\frac{1}{2}\beta_3 c^2 q u'(\xi) - \beta_2 c q u'(\xi) + pu'(\xi) + \frac{1}{6}\beta_3 q^3 u^{(3)}(\xi) + 5\alpha_3 q u(\xi)^4 u'(\xi) + 4\alpha_4 q u(\xi)^4 u'(\xi) + 3\alpha_1 q u(\xi)^2 u'(\xi) + 2\alpha_2 q u(\xi)^2 u'(\xi) = 0. \quad (6)$$

Integrating (6), we can get

$$\frac{1}{6}\beta_3 q^3 u''(\xi) + \left(-\frac{1}{2}\beta_3 c^2 q - \beta_2 c q + p\right)u(\xi) + \left(\alpha_1 q + \frac{2\alpha_2 q}{3}\right)u(\xi)^3 + \left(\alpha_3 q + \frac{4\alpha_4 q}{5}\right)u(\xi)^5 + n = 0, \quad (7)$$

where n is a constant. Equations (5) and (7) can be reduced to an equation if n=0 and

$$-\frac{3q^2(\beta_2 + \beta_3 c)}{\beta_3 q^3/6} = \frac{\beta_3 c^3 + 3\beta_2 c^2 - 6k}{-(1/2)\beta_3 c^2 q - \beta_2 c q + p} = \frac{6\gamma_1 - 6\alpha_1 c}{\alpha_1 q + (2\alpha_2 q/3)} = \frac{6\gamma_2 - 6\alpha_3 c}{\alpha_3 q + (4\alpha_4 q/5)}. \quad (8)$$

By solving (8), we get

$$c = \frac{-15\alpha_3\gamma_1 - 12\alpha_4\gamma_1 + 15\alpha_1\gamma_2 + 10\alpha_2\gamma_2}{2(5\alpha_2\alpha_3 - 6\alpha_1\alpha_4)} \neq -\frac{\beta_2}{\beta_3},$$

$$p = \frac{-q}{45(5\alpha_2\alpha_3 - 6\alpha_1\alpha_4)^2(\alpha_3\gamma_1 - \alpha_1\gamma_2)} \times \left(-540\alpha_4^2\alpha_3\beta_3\gamma_1^3 - 375\alpha_2^2\alpha_3\beta_3\gamma_1\gamma_2^2 + 900\alpha_2\alpha_4\alpha_3\beta_3\gamma_1^2\gamma_2 - 432\alpha_4^3\beta_3\gamma_1^3 + 250\alpha_2^3\beta_3\gamma_2^3 + 375\alpha_1\alpha_2^2\beta_3\gamma_2^3 - 900\alpha_2^2\alpha_4\beta_3\gamma_1\gamma_2^2 - 900\alpha_1\alpha_2\alpha_4\beta_3\gamma_1\gamma_2^2 + 540\alpha_1\alpha_4^2\beta_3\gamma_1^2\gamma_2 + 1080\alpha_2\alpha_4^2\beta_3\gamma_1^2\gamma_2 + 750\alpha_2^3\alpha_3^3 k - 2700\alpha_1\alpha_2^2\alpha_4\alpha_3^2 k + 3240\alpha_1^2\alpha_2\alpha_4^2\alpha_3 k - 1296\alpha_1^3\alpha_4^3 k\right),$$

$$\beta_2 = -\frac{\beta_3(-5\alpha_3\gamma_1 - 6\alpha_4\gamma_1 + 5\alpha_1\gamma_2 + 5\alpha_2\gamma_2)}{5\alpha_2\alpha_3 - 6\alpha_1\alpha_4}, \quad (9)$$

or

$$k = \frac{\beta_2^3}{3\beta_3^2}, \quad c = -\frac{\beta_2}{\beta_3}, \quad \gamma_1 = -\frac{\alpha_1\beta_2}{\beta_3}, \quad \gamma_2 = -\frac{\alpha_3\beta_2}{\beta_3}. \quad (10)$$

Then (5) and (7) reduce to the following planar dynamic system:

$$\frac{du}{d\xi} = y, \qquad \frac{dy}{d\xi} = a_1 u\left(u^4 + a_2 u^2 + a_3\right), \quad (11)$$

where $a_1 = (-3\beta_3 c^2 q - 6\beta_2 cq + 6p)/\beta_3 q^3$, $a_2 = (6\alpha_1 q + 4\alpha_2 q)/\beta_3 q^3$, and $a_3 = (6\alpha_3 q + (24\alpha_4 q/5))/\beta_3 q^3$.

Obviously, the above system (11) has the first integral

$$H(u,y) = \frac{1}{2}y^2 - \frac{a_1}{6}u^6 - \frac{a_1 a_2}{4}u^4 - \frac{a_1 a_3}{2}u^2 = h. \quad (12)$$

We suppose that

$$M(u_0, y_0) = \begin{pmatrix} 0 & 1 \\ a_1 a_3 & 0 \end{pmatrix} \quad (13)$$

is the coefficient matrix of the linearized system (11) at an equilibrium point (u_0, y_0) and

$$J(u_0, y_0) = \begin{bmatrix} 0 & 1 \\ a_1\left(4u_0^4 + 2a_2u_0^2 + a_3\right) & 0 \end{bmatrix} \quad (14)$$

is the Jacobian determinant. By the bifurcation theory of planar dynamical system, we know that if $J<0$ and Trace $(M) = 0$, then it is a center point; if $J>0$ and $(\mathrm{Trace}(M))^2 - 4J = 0$, then it is a node; if $J=0$ and Poincare index of the equilibrium point is 0, then it is a cusp point. By using the above facts to do qualitative analysis, we have the following.

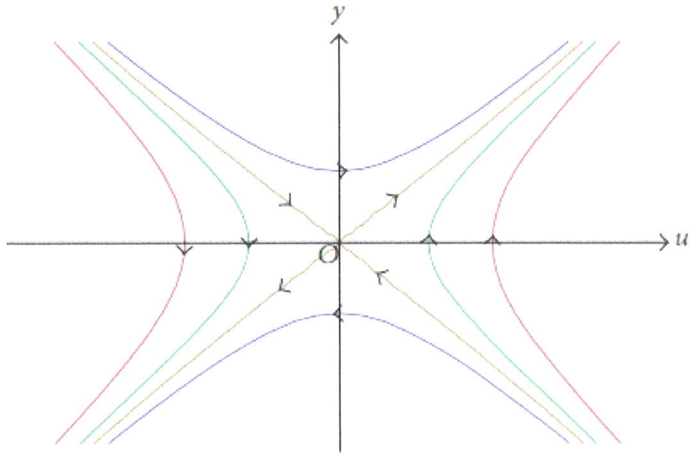

Figure 1: The phase portraits of (11) when $a_1 > 0$ and $a_2^2 < 4a_3$ or $a_1 > 0$, $a_2 > 0$, $a_3 > 0$, and $a_2^2 > 4a_3$.

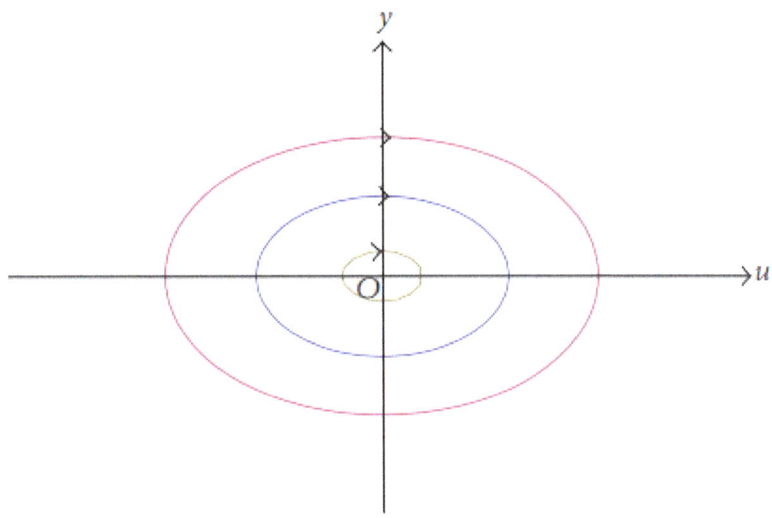

Figure 2: The phase portraits of (11) when $a_1 < 0$, $a_2 > 0$, $a_3 > 0$, and $a_2^2 > 4a_3$.

(1) If $a_1 > 0$ and $a_2^2 < 4a_3$ or $a_1 > 0$, $a_2 > 0$, $a_3 > 0$, and $a_2^2 > 4a_3$, then system (11) has only one equilibrium point (0, 0). It is easy to find $(O) < 0$. So it is a saddle point (see Figure 1).

(2) If $a_1 < 0$, $a_2 > 0$, $a_3 > 0$, and $a_2^2 > 4a_3$, then system (11) has only one equilibrium point (0, 0). It is easy to find that $(O) > 0$ and $\mathrm{Trace}(M(O)) = 0$. So it is a center point (see Figure 2).

(3) If $a_1 > 0$, $a_2 < 0$, $a_3 > 0$, and $a_2^2 = 4a_3$, then system (11) has three equilibrium points (0, 0) and $P_{1\pm}(\pm\sqrt{-a_2/2}, 0)$. $(O) < 0$, $J(P_{1\pm})=0$, and Poincare′ index of $P_{1\pm}$ is equal to zero. So O is a saddle point; $P_{1\pm}$ are cusp points (see Figure 3).

(4) If $a_1 < 0$, $a_2 < 0$, $a_3 > 0$, and $a_2^2 = 4a_3$, then system (11) has three equilibrium points (0, 0) and $P_{2\pm}(\pm\sqrt{-a_2/2}, 0)$. $(O) > 0$, $\mathrm{Trace}(M(O)) = 0$, $J(P_{1\pm}) = 0$, and Poincare index of ′ $P1\pm$ is equal to zero. So O is a center point; $P_{2\pm}$ are cusp points (see Figure 4).

(5) If $a_1 > 0$ and $a_3 < 0$, then system (11) has three equilibrium points (0, 0) and $Q_{1\pm}\left(\pm\sqrt{\left(-a_2 + \sqrt{a_2^2 - 4a_3}\right)/2}, 0\right)$. $(O) > 0$, $\mathrm{Trace}(M(O)) = 0$, and $J(Q_{1\pm})$

(6) If $a_1 < 0$ and $a_3 < 0$, then system (11) has three equilibrium points (0, 0) and $Q_{2\pm}(\pm Q_{2\pm}(\pm\sqrt{(-a_2 + \sqrt{a_2^2 - 4a_3})/2}$, , 0). $(O) < 0$, $J(Q_{1\pm}) > 0$, and Trace$(M(Q_{1\pm})) = 0$. So O is a saddle point; $Q_{2\pm}$ are center points (see Figure 6).

(7) If $a_1 > 0$, $a_2 < 0$, $a_3 > 0$, and $a_2^2 > 4a_3$, then system (11) has five equilibrium points $O(0, 0)$, $R_{1\pm}(\pm\sqrt{(-a_2 - \sqrt{a_2^2 - 4a_3})/2}, 0)$, and $S_{1\pm}(\pm\sqrt{(-a_2 + \sqrt{a_2^2 - 4a_3})/2}, 0)$. $J(O) < 0$, $J(S_{1\pm}) < 0$, $J(S_{1\pm})0$, and Trace$(M(R_{1\pm})) = 0$. So O and $S_{1\pm}$ are saddle points; $R_{1\pm}$ are center points. Furthermore, if $a_2^2 < (16/3)a_3$, then $H(R_{1\pm}) < H(S_{1\pm}) < H(O)$ (see Figure 7); if $a_2^2 = (16/3)a_3$, then $H(R_{1\pm}) < H(S_{1\pm}) = H(O)$ (see Figure 8); if $a_2^2 > (16/3)a_3$, then $H(R_{1\pm}) < H(O) < H(S_{1\pm})$ (see Figure 9).

(8) If $a_1 < 0$, $a_2 < 0$, $a_3 > 0$, and $a_2^2 > 4a_3$, then system (11) has five equilibrium points $O(0, 0)$, $R_{2\pm}(\pm\sqrt{(-a_2 - \sqrt{a_2^2 - 4a_3})/2}, 0)$, , and $S_{2\pm}(\pm\sqrt{(-a_2 + \sqrt{a_2^2 - 4a_3})/2}, 0)$. $J(O)$ $J(O) >$ 0, Trace$(M(O)) = 0$, $J(S_{2\pm}) > 0$, Trace$(M(S_{2\pm})) = 0$, and $J(R_{1\pm}) < 0$. So O and $S_{2\pm}$ are center points; $R_{2\pm}$ are saddle points (see Figure 10).

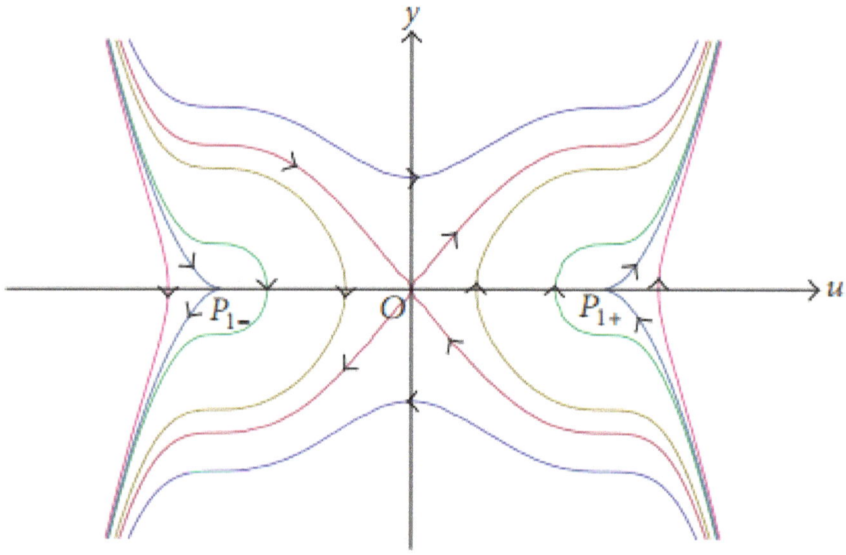

Figure 3: The phase portraits of (11) when $a_1 > 0$, $a_2 < 0$, $a_3 > 0$, and $a_2^2 = 4a_3$.

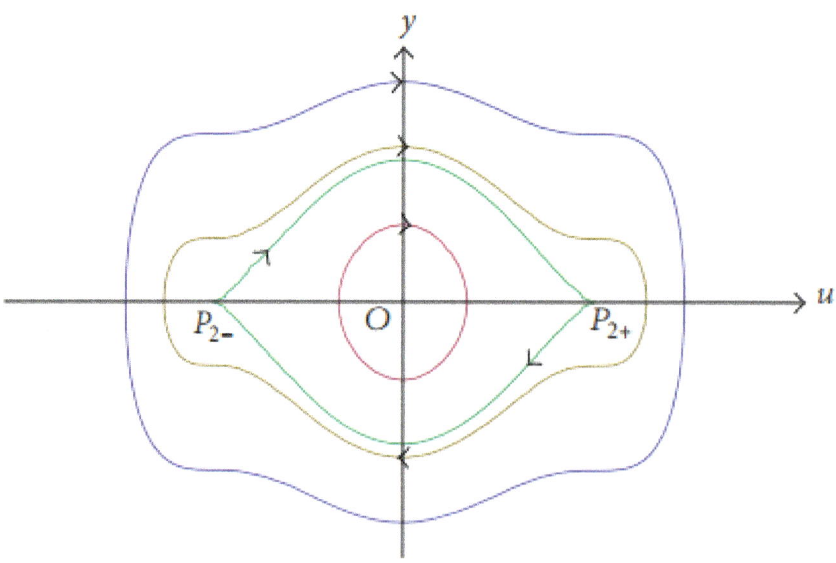

Figure 4: The phase portraits of (11) when $a_1 < 0$, $a_2 < 0$, $a_3 > 0$, and $a_2^2 = 4a_3$.

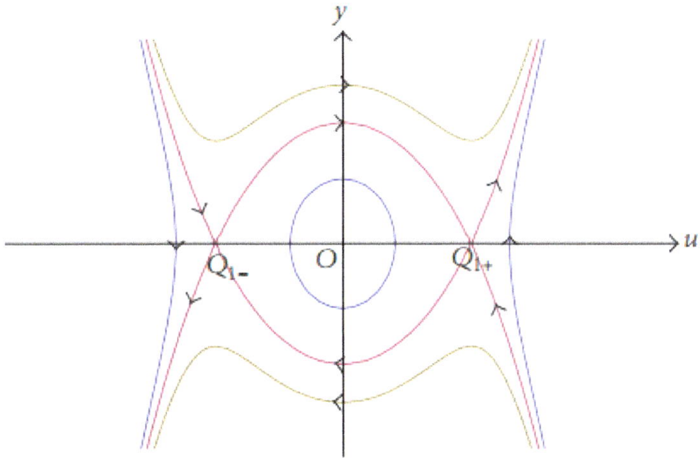

Figure 5: The phase portraits of (11) when $a_1 > 0$ and $a_3 < 0$.

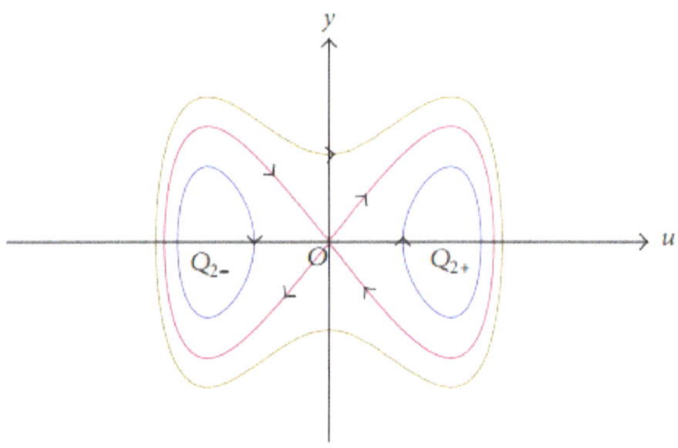

Figure 6: The phase portraits of (11) when $a_1 < 0$ and $a_3 < 0$.

For a fixed $h \in R$, the curve

$$\overline{C}_h = \{(u, y) \in R \times R : H(u, y) = h\} \tag{15}$$

is called a level curve with energy level h [14]. Obviously, each orbit of (12) is a branch of certain energy curve. For convenience, we name the orbit as the orbit with energy level h.

To facilitate further analysis, we investigate the relation between the bounded orbit of (12) and the energy level h. Put

$$F_h(u) = h + \frac{a_1}{6}u^6 + \frac{a_1 a_2}{4}u^4 + \frac{a_1 a_3}{2}u^2. \tag{16}$$

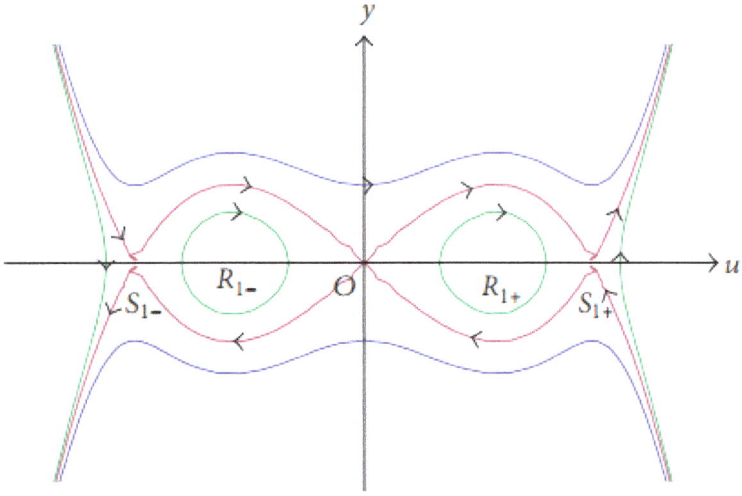

Figure 7: The phase portraits of (11) when $a_1 > 0$, $a_2 < 0$, $a_3 > 0$, and $a_2^2 > (16/3)_3$.

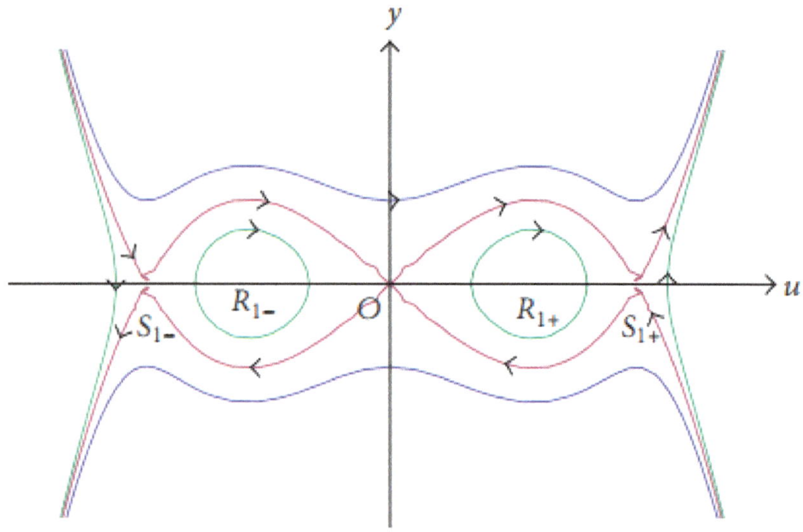

Figure 8: The phase portraits of (11) when $a_1 > 0$, $a_2 < 0$, $a_3 > 0$, and $a_2^2 = (16/3)_3$.

If $a_1 > 0$, $a_2 < 0$, and $a_3 > 0$, it is easy to obtain the five extreme points of $Fh(u)$ as follows:

$$u_0 = 0,$$

$$u_{1\pm} = \pm\sqrt{\frac{-a_2 - \sqrt{a_2^2 - 4a_3}}{2}},$$

$$u_{2\pm} = \pm\sqrt{\frac{-a_2 + \sqrt{a_2^2 - 4a_3}}{2}}. \tag{17}$$

Let

$$h_1 = -F_0(u_{1\pm}), \qquad h_2 = -F_0(u_{2\pm}). \tag{18}$$

Therefore we can easily draw the graphics of the function $F_h(u)$ in Figures 11, 12, and 13. Observe that the energy curve $\overline{C}_h$ is equivalent to the curve defined by $(1/2)^2 = F_h(u)$. According to the above analysis, we have the following.

Case 1. Assume that $a_1 > 0$, $a_2 < 0$, $a_3 > 0$, and $a_2^2 > (16/3)_3$; system (11) has three saddle points O, $S_{1\pm}$, and two center points $R_{1\pm}$. There exist two homoclinic orbits with energy level 0 to the saddle point O (corresponding to red closed lines $L(O, O)$ in the left half-plane and the right half-plane in Figure 7) and two families of periodic orbits with energy level h ($h \in (h_1, 0)$) around the centers $R_{1\pm}$ (corresponding to orange closed lines in the left half-plane and the right half-plane in Figure 7) which lie inside of above two homoclinic orbits, respectively. Besides, system (11) has a singular closed orbit with energy level h_2 to the closed curve $L(S_{1-}, S_{1+}, S_{1-})$ (corresponding to the green closed lines in Figure 7) and a family of periodic orbits with energy level h ($h \in (0, h_2)$) (corresponding to the family of periodic orbits enclosing the equilibrium points O and $S_{1\pm}$) which lie inside of above singular closed orbit. It implies that for (11) there exist three families of periodic wave solutions, two solitary wave solutions, and two kink wave solutions.

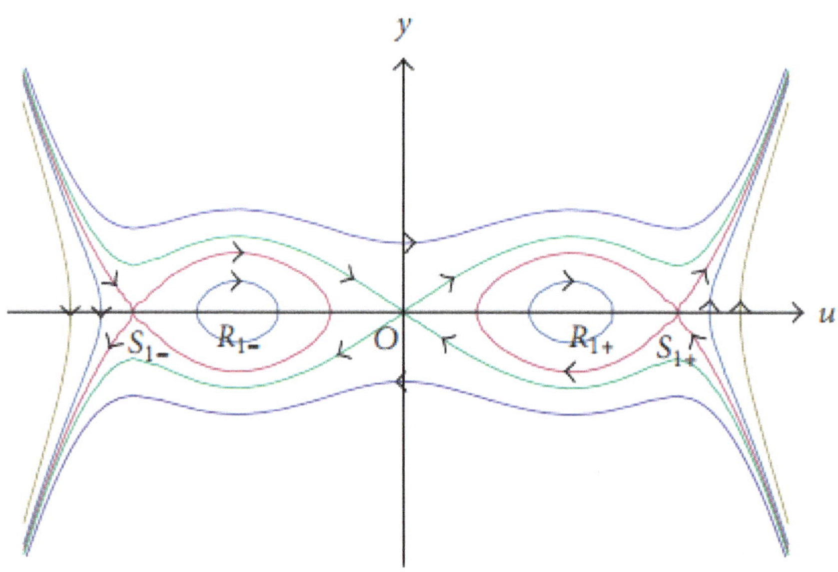

Figure 9: The phase portraits of (11) when $a_1 > 0$, $a_2 < 0$, $a_3 > 0$, and $4a_3 < a_2^2 < (16/3)_3$.

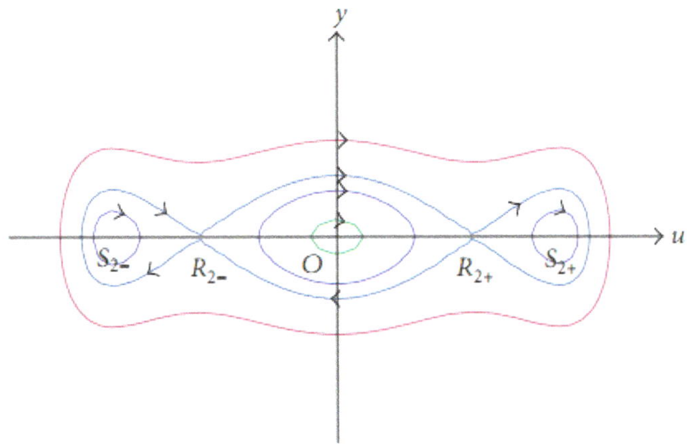

Figure 10: The phase portraits of (11) when $a_1 < 0$, $a_2 < 0$, $a_3 > 0$, and $a_2^2 > 4a_3$.

Case 2. Assume that $a_1 > 0$, $a_2 < 0$, $a_3 > 0$, and $a_2^2 = (16/3)_3$; system (11) has three saddle points O, $S_{1\pm}$, and two center points $R_{1\pm}$. There exist a singular closed orbit with energy level 0 to the closed curve (S_{1-}, 0, S_{1+}, 0, S_{1-}) (corresponding to the red closed lines in Figure 8) and two families of periodic orbits with energy level h ($h \in (h_1, 0)$) around the centers $R_{1\pm}$ (corresponding to green closed lines in the left half-plane and the right half-plane in Figure 8) which lie inside of above singular closed orbit. It means that there exist two families of periodic wave solutions and four kink wave solutions.

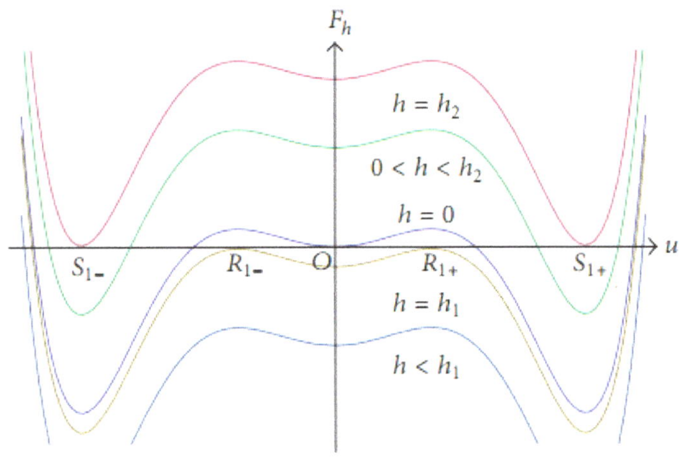

Figure 11: The phase portraits of (11) when $a_1 > 0$, $a_2 < 0$, $a_3 > 0$, and $a_2^2 > (16/3)_3$.

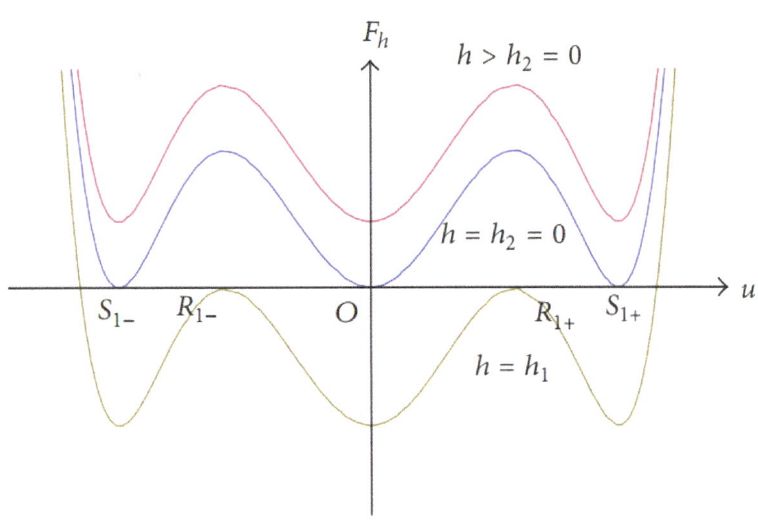

Figure 12: The phase portraits of (11) when $a_1 > 0$, $a_2 < 0$, $a_3 > 0$, and $a_2^2 = (16/3)_3$.

Case 3. Assume that $a_1 > 0$, $a_2 < 0$, $a_3 > 0$, and $4a_3 < a_2^2 < (16/3)_3$; system (11) has three saddle points O, $S_{1\pm}$, and two center points $R_{1\pm}$. There exist two homoclinic orbits with energy level h_2 at the saddle points $S_{1\pm}$ (corresponding to red closed lines $L(S_{1-}, S_{1-})$ in the left half-plane and $L(S_{1+}, S_{1+})$ in the right half-plane in Figure 9) and two families of periodic orbits with energy level h ($h \in (h_1, 0)$) around the centers $R_{1\pm}$ (corresponding to blue closed lines in the left half-plane and the right half-plane in Figure 9) which lie inside of above two homoclinic orbits, respectively.

TRAVELING WAVE SOLUTIONS OF THE HNLS EQUATION

(1) Solitary, Kink, and Periodic Wave Solutions When $a_1 > 0$, $a_2 < 0$, $a_3 > 0$, and $a_2^2 > (16/3)_3$. (i) If h=0, as is seen in Figure 7, there are two symmetric homoclinic orbits connected at the saddle point O. In (u, y)-plane the expressions of the homoclinic orbits are given as

$$y = \pm\sqrt{\frac{a_1}{3}u^2 (m_1 - u^2)(m_2 - u^2)},$$

(19)

where

$$m_{1,2} = \frac{-3a_2 \mp \sqrt{9a_2^2 - 48a_3}}{4}.$$

(20)

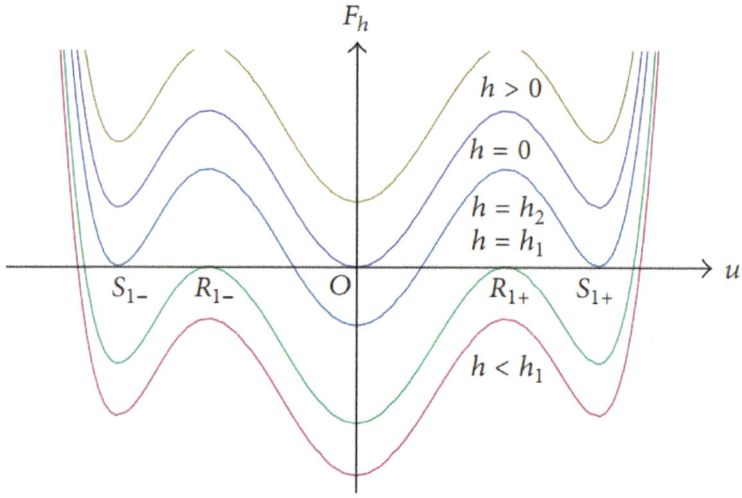

Figure 13: The phase portraits of (11) when $a_1 > 0$, $a_2 < 0$, $a_3 > 0$, and $4a3 < a_2^2 < (16/3)_3$.

Substituting (19) into $du/d\xi = y$ and integrating them along the homoclinic orbits, noticing that $E(z, t) = u(\xi)e^{i(kz-ct)}$ and $\xi = pz - qt$, we get two solitary wave solutions:

$$E_{1\pm}(z,t) = \pm\sqrt{\frac{m_1 m_2 \operatorname{sech}^2 \sqrt{a_1 m_1 m_2/3}\,(pz - qt)}{m_2 - m_1 \tanh^2 \sqrt{a_1 m_1 m_2/3}\,(pz - qt)}}\,e^{i(kz-ct)},$$

(21)

which correspond to the family of homoclinic orbits $L(O, O)$ in the left half-plane and $L(O, O)$ in the right half-plane shown in Figure 7.

(ii) If h=h_2, as is seen in Figure 7, there are two heteroclinic orbits connected at the saddle points S_{1-} and S_{1+}. In (u, y)-plane the expressions of the heteroclinic orbits are given as

$$y = \pm\sqrt{\frac{a_1}{3}(u^2 - n_1)(n_2 - u^2)^2},$$

(22)

where

$$n_1 = \frac{-a_2 - 2\sqrt{a_2^2 - 4a_3}}{2},$$

$$n_2 = \frac{-a_2 + \sqrt{a_2^2 - 4a_3}}{2}.$$

(23)

Substituting (22) into $du/d\xi = y$ and integrating them along the heteroclinic orbits, noticing that $(z, t) = u(\xi)e^{i(kz-ct)}$ and $\xi = pz - qt$, we get two kink wave solutions:

$$E_{2\pm}(z,t) = \pm\frac{\sqrt{-n_1 n_2}\tanh\sqrt{a_1 n_2 (n_2 - n_1)/3}\,(pz - qt)}{\sqrt{n_2 \operatorname{sech}^2 \sqrt{a_1 n_2 (n_2 - n_1)/3}\,(pz - qt) - n_1}} \times e^{i(kz-ct)},$$

(24)

which correspond to the family of heteroclinic orbits $L(S_{1-}, S_{1+})$ and $L(S_{1+}, S_{1-})$ shown in Figure 7.

(iii) If $h \in (h_1, 0)$, as is seen in Figure 7, there are two families of periodic orbits inside the homoclinic orbits. In (u, y)-plane the expressions of the periodic orbits are given as $(\varphi_1 \in ((-a_2 - \sqrt{a_2^2 - 4a_3})/2, n_1))$:

$$y = \pm\sqrt{\frac{a_1}{3}(\varphi_1 - u^2)(u^2 - \varphi_2)(\varphi_3 - u^2)}, \tag{25}$$

Where

$$\varphi_2 = \frac{-2\varphi_1 - 3a_2 - \sqrt{(2\varphi_1 + 3a_2)^2 - 8(2\varphi_1^2 + 3a_2\varphi_1 + 6a_3)}}{4},$$
$$\varphi_3 = \frac{-2\varphi_1 - 3a_2 + \sqrt{(2\varphi_1 + 3a_2)^2 - 8(2\varphi_1^2 + 3a_2\varphi_1 + 6a_3)}}{4}. \tag{26}$$

Substituting (25) into $du/d\xi = y$ and integrating them along the periodic orbits, noticing that $(z, t) = u(\xi)e^{i(kz-ct)}$ and $\xi = pz - qt$, we get two families of periodic wave solutions:

$$E_{3\pm}(z,t) = \pm\left((\varphi_1\varphi_2) \times \left(\varphi_1 - (\varphi_1 - \varphi_2) \times \mathrm{sn}^2\left(\sqrt{\frac{a_1\varphi_1(\varphi_3 - \varphi_2)}{3}}(pz - qt), \sqrt{\frac{\varphi_3(\varphi_1 - \varphi_2)}{\varphi_1(\varphi_3 - \varphi_2)}}\right)\right)^{-1}\right)^{1/2} \times e^{i(kz-ct)}, \tag{27}$$

which correspond to two families of periodic orbits inside the homoclinic orbits in the right half-plane and in the left halfplane shown in Figure 7.

When the energy level $h \to 0$, the above periodic solutions tend to the solitary wave solutions (21). (iv) If $h \in (0, h_2)$, as is seen in Figure 7, there is a family of periodic orbits enclosing the equilibrium points R_{1-}, O and R_{1+}. In (u, y)-plane the expressions of the periodic orbits are given as $(\psi_1 \in (n_1, m_2))$:

$$y = \pm\sqrt{\frac{a_1}{3}(\psi_1 - u^2)(u^2 + \psi_2)(\psi_3 - u^2)}, \tag{28}$$

Where

$$\psi_2 = \frac{2\psi_1 + 3a_2 + \sqrt{(2\psi_1 + 3a_2)^2 - 8(2\psi_1^2 + 3a_2\psi_1 + 6a_3)}}{4},$$
$$\psi_3 = \frac{-2\psi_1 - 3a_2 + \sqrt{(2\psi_1 + 3a_2)^2 - 8(2\psi_1^2 + 3a_2\psi_1 + 6a_3)}}{4}. \tag{29}$$

Substituting (28) into $du/d\xi = y$ and integrating them along the periodic orbits, noticing that $(z, t) = u(\xi)e^{i(kz-ct)}$ and $\xi = pz - qt$, we get a family of periodic wave solutions:

$$E_{4\pm}(z,t) = \pm\left(\left(\psi_1\psi_2 \times \mathrm{sn}^2\left(\sqrt{\frac{a_1\psi_3(\psi_1 + \psi_2)}{3}}(pz - qt), \sqrt{\frac{\psi_1(\psi_2 + \psi_3)}{\psi_3(\psi_1 - \psi_2)}}\right)\right) \times \left(\psi_2 + \psi_1 \times \mathrm{cn}^2\left(\sqrt{\frac{a_1\psi_3(\psi_1 + \psi_2)}{3}}(pz - qt), \sqrt{\frac{\psi_1(\psi_2 + \psi_3)}{\psi_3(\psi_1 - \psi_2)}}\right)\right)^{-1}\right)^{1/2} \times e^{i(kz-ct)}, \tag{30}$$

which correspond to a family of periodic orbits enclosing three equilibrium points shown in Figure 7. When the energy level $h \to 0$, the above periodic solutions also tend to the solitary wave solutions (21).

(2) Kink and Periodic Wave Solutions When $a_1 > 0$, $a_2 < 0$, $a_3 > 0$, and $a_2^2 = (16/3)_3$. (i) If $h = h_2$, as is seen in Figure 8, there are four heteroclinic orbits. In (u, y)-plane the expressions of the heteroclinic orbits are given as

$$y = \pm\sqrt{\frac{a_1}{3}}u\left(u^2 + \frac{3a_2}{4}\right). \tag{31}$$

Substituting (31) into $du/d\xi = y$ and integrating them along the heteroclinic orbits, noticing that $(z, t) = u(\xi)e^{i(kz-ct)}$ and $\xi = pz - qt$, we get four kink wave solutions:

$$E_{5\pm}(z,t) = \pm\sqrt{\frac{\sqrt{3a_3}}{2}\left[1 + \tanh\sqrt{a_1a_2}\left((pz - qt) - \ln\frac{1}{\sqrt{a_1a_2}}\right)\right]} \times e^{i(kz-ct)},$$

$$E_{6\pm}(z,t) = \pm\sqrt{\frac{\sqrt{3a_3}}{2}\left[1 - \tanh\sqrt{a_1a_2}\left((pz - qt) + \ln\frac{1}{\sqrt{a_1a_2}}\right)\right]} \times e^{i(kz-ct)}, \tag{32}$$

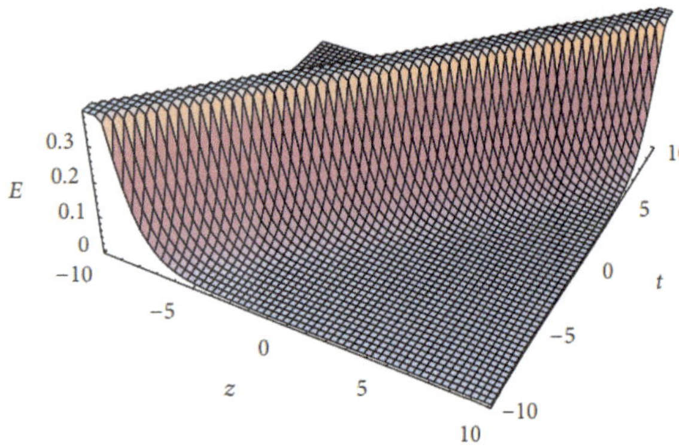

Figure 14: $u = |E_{1+}|$ when $a_1 = 1$, $a_2 = -6$, $a_3 = 1$, $p=1$, and $q=1$.

which correspond to $L(O.B_+)$, $L(O, B_-)$, $L(B_+, O)$, and $L(B_-, O)$ in Figure 8.

(ii) If $h_1 < h < h_2$, as is seen in Figure 8, there are two families of periodic orbits inside the homoclinic orbits. Thus we have the same solutions as (27).

(3) Solitary and Periodic Wave Solutions When $a_1 > 0$, $a_2 < 0$, $a_3 > 0$, and $4a_3 < a_2^2 < (16/3)_3$. (i) If $h=h_2$, as is seen in Figure 9, there are two symmetric homoclinic orbits connected at the saddle points S_{1-} and S_{1+}. In (u, y)-plane the expressions of the homoclinic orbits are given as

$$y = \pm\sqrt{\frac{a_1}{3}\left(u^2 - n_1\right)\left(n_2 - u^2\right)^2}, \tag{33}$$

where n_1 and n_2 satisfy (23).

Substituting (33) into $du/d\xi = y$ and integrating them along the homoclinic orbits, noticing that $E(z, t) = u(\xi)e^{i(kz-ct)}$ and $\xi = pz - qt$, we get two solitary wave solutions:

$$E_{7\pm}(z,t) = \pm\sqrt{\frac{n_1 n_2}{n_2 + \left(n_1 - n_2\right)\tanh^2\sqrt{a_1 n_2\left(n_2 - n_1\right)/3}\left(pz - qt\right)}} \times e^{i(kz-ct)}. \tag{34}$$

(ii) If $h \in (h_1, h_2)$, as is seen in Figure 9, there are two families of periodic orbits inside the homoclinic orbits. Thus we have the same solutions as (27).

We will draw the figures of some solutions under the special conditions. Figure 14 is a bell solitary wave solution of the field $|E_{1+}|$ given by (21) when the energy level $h=0$. Figure 15 shows the kink wave solution of the field $|E_{2+}|$ given by (21) when the energy level $h=h_2$. Figure 16 is a plot for the periodic wave solution of the field $|E_{3+}|$ given by (27) when the energy level $h = -0.0390$. Figure 17 is a periodic wave solution of the field $|E_{4+}|$ given by (30) when the energy level $h = 0.0030$. Figure 18 shows periodic wave solutions of the field $|E_{3+}|$ given by (27) when $\varphi_1 \in ((-a_2 - \sqrt{a_2^2 - 4a_3})/2, n_1)$. We can see that when $\varphi_1 \to (-a_2 - \sqrt{a_2^2 - 4a_3})/2$, or we say $h \to h_1$, these periodic wave solutions will be reduced back to zero. When $\varphi_1 \to n_1$, or we say $h \to 0$, the periodic solutions tend to the solitary wave solutions (21). Figure 19 is a periodic wave solution of the field $|E_{4+}|$ when ϕ_1 changes. When the energy level $h \to 0$, the above periodic solutions also tend to the solitary wave solutions (21). Figures 20, 21, 22, 23, 24, and 25 show the relationship near critical points between $E_{1+}(\xi)$, $E_{2+}(\xi)$ and a_1, a_2, and a_3, respectively.

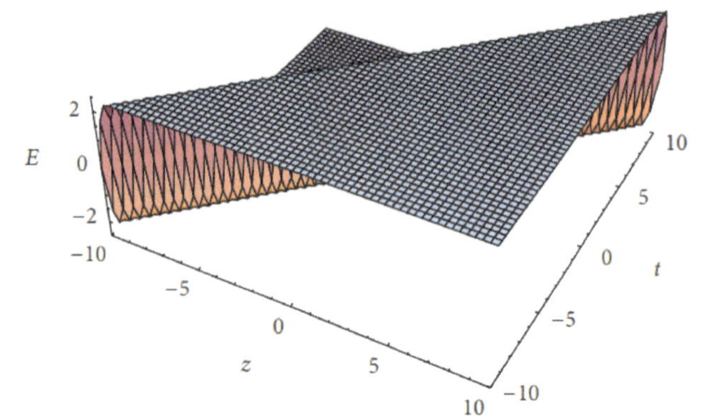

Figure 15: $u = |E_{2+}|$ when $a_1 = 1$, $a_2 = -6$, $a_3 = 1$, $p=1$, and $q=1$.

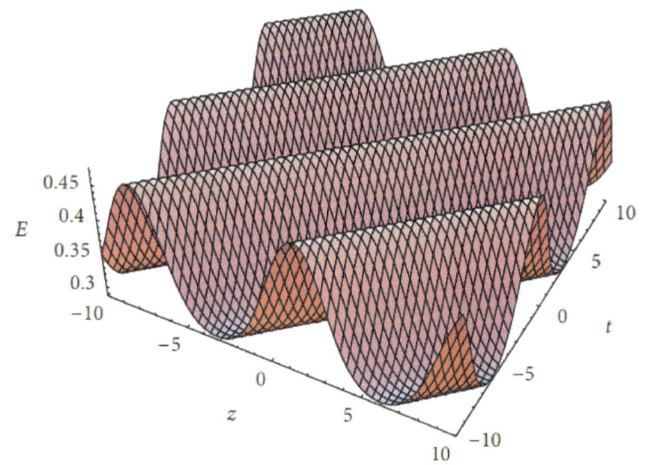

Figure 16: $u = |E_{3+}|$ when $a_1 = 1$, $a_2 = -6$, $a_3 = 1$, $p=1$, $q=1$, and $h = -0.0390$.

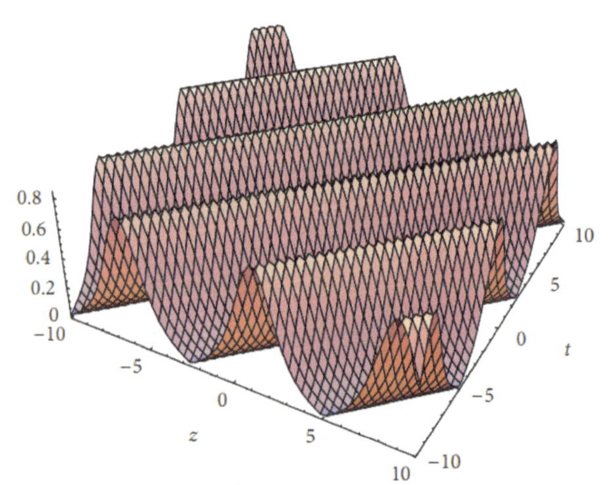

Figure 17: $u = |E_{4+}|$ when $a_1 = 1$, $a_2 = -6$, $a_3 = 1$, $p=1$, $q=1$, and $h = 0.0030$.

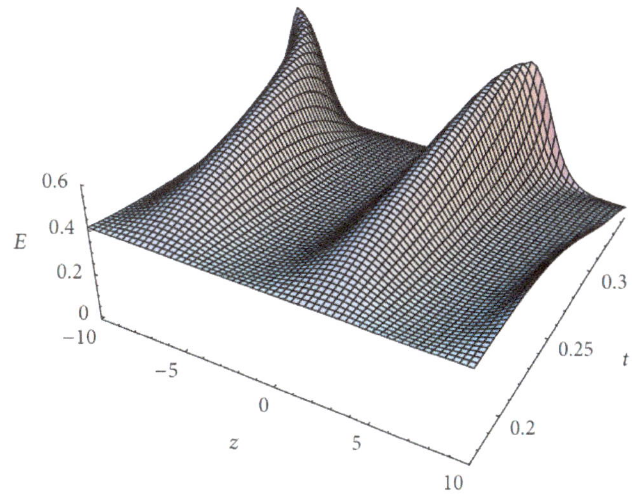

Figure 18: $u = |E_{3+}|$ when $a_1 = 1$, $a_2 = -6$, and $a_3 = 1$.

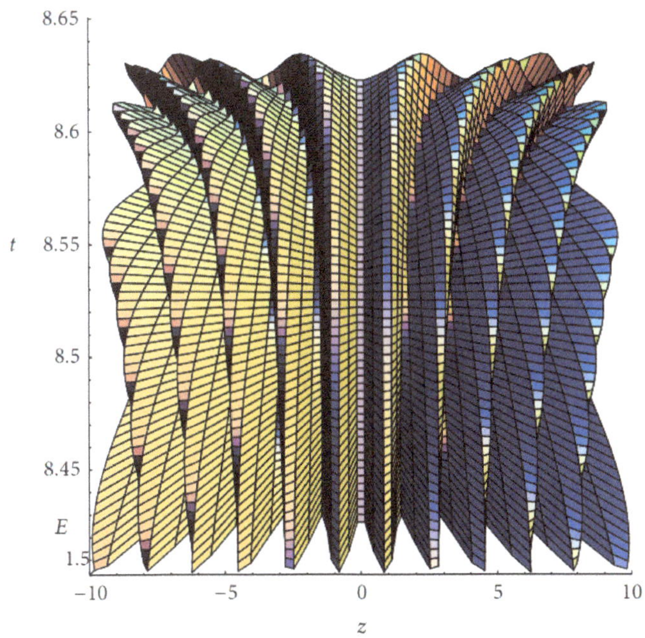

Figure 19: $u = |E_{4+}|$ when $a_1 = 1$, $a_2 = -6$, and $a3 = 1$.

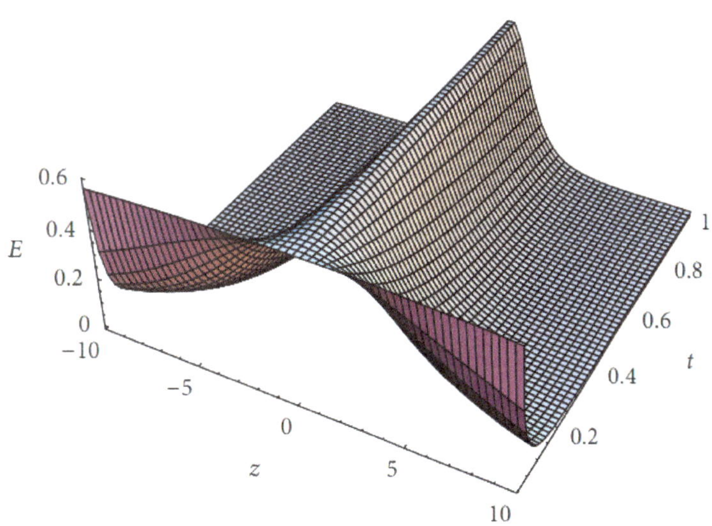

Figure 20: $u = |E_{1+}|$ when $a_2 = -6$ and $a_3 = 1$.

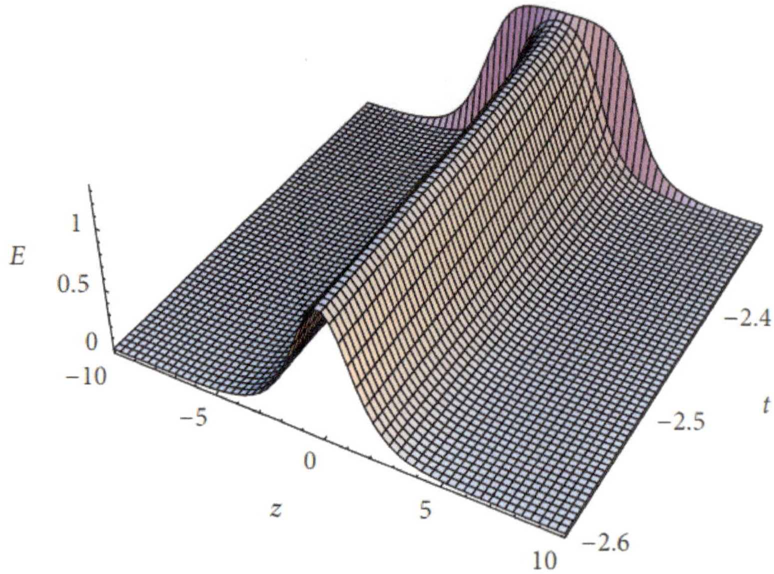

Figure 21: $u = |E_{1+}|$ when $a_1 = 1$ and $a_3 = 1$.

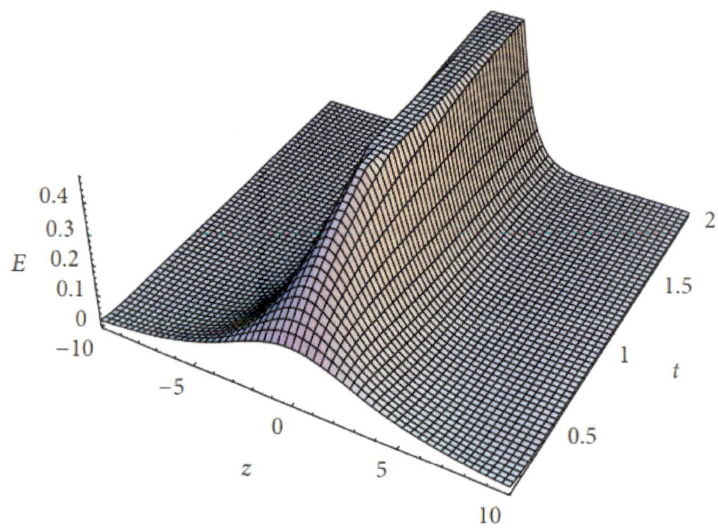

Figure 22: $u = |E_{2+}|$ when $a_1 = 1$ and $a_2 = -6$.

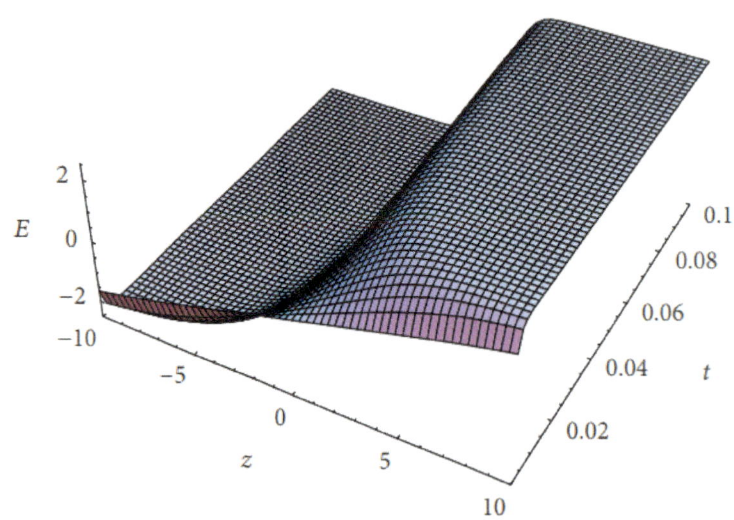

Figure 23: $u = |E_{2+}|$ when $a_2 = -6$ and $a_3 = 1$.

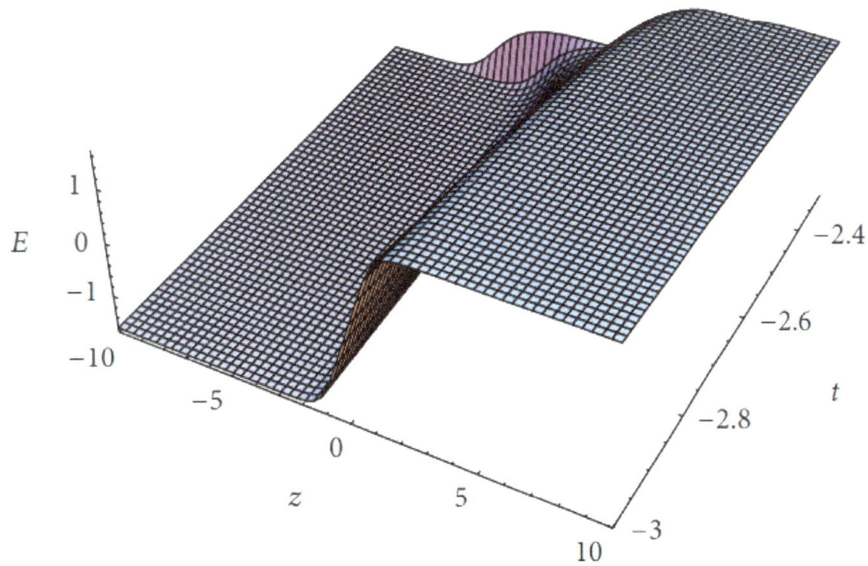

Figure 24: $u = |E_{2+}|$ when $a_1 = 1$ and $a_3 = 1$.

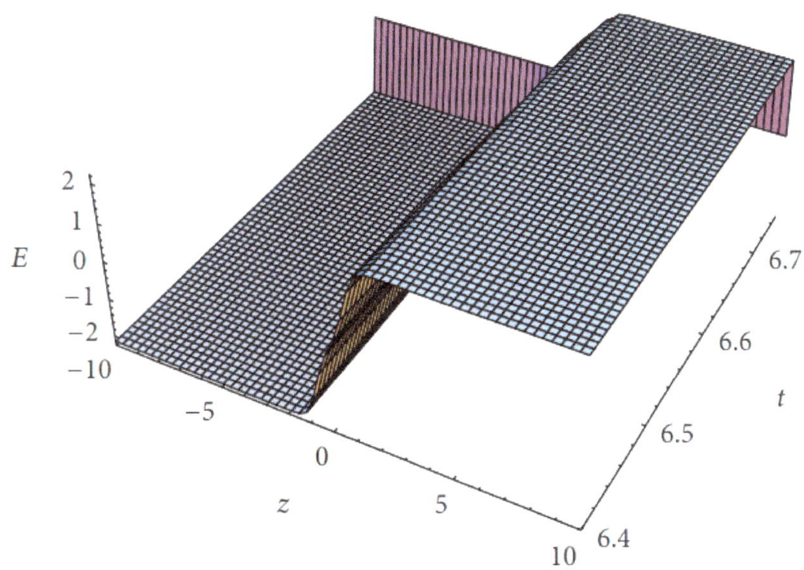

Figure 25: $u = |E_{1+}|$ when $a_1 = 1$ and $a_2 = -6$.

CONCLUSION

By using the bifurcation theory method of dynamical systems, we successfully obtain 10 phase portraits for the corresponding dynamic system of (2). Through analysing three of these phase portraits, we get new traveling wave solutions, including solitary wave solutions, kink wave solutions, and periodic wave solutions. The solutions are new and have not been investigated. By the dependent variable transformations, four linear forms for (2) can also be obtained. We will continue to consider the problem by using Hirota method in the future.

Conflict of Interests

The authors declare that there is no conflict of interests regarding the publication of this paper.

ACKNOWLEDGMENTS

The authors express their sincere thanks to the editor and reviewers for their valuable comments. This work has been supported by the National Natural Science Foundation of China under Grant no. 11272023, by the Open Fund of State Key Laboratory of Information Photonics and Optical Communications (Beijing University of Posts and Telecommunications) under Grant no. IPOC2013B008, and by the Fundamental Research Funds for the Central Universities of China under Grant no. 2011BUPTYB02.

REFERENCES

1. G. P. Agrawal, Nonlinear Fiber Optics, Academic Press, San Diego, Calif, USA, 1995.
2. A. Hasegawa and Y. Kodama, Solitons in Optical Communications, Oxford University Press, Oxford, UK, 1995.

3. Y. Kodama and A. Hasegawa, "Nonlinear pulse propagation in a monomode dielectric guide," IEEE Journal of Quantum Electronics, vol. 23, no. 5, pp. 510–524, 1987.

4. G. P. Agrawal, Nonlinear Fiber Optics, Academic Press, New York, NY, USA, 2006.

5. K. Porsezian and K. Nakkeeran, "Optical solitons in presence of Kerr dispersion and self-frequency shift,"Physical Review Letters, vol. 76, no. 21, pp. 3955–3958, 1996. ·

6. S. Palacios, A. Guinea, J. Fernández-Díaz, and R. D. Crespo, "Dark solitary waves in the nonlinear Schrödinger equation with third order dispersion, self-steepening, and self-frequency shift," Physical Review E, vol. 60, no. 1, pp. R45–R47, 1999.

7. W.-J. Liu, "Soliton interaction in the higher-order nonlinear Schrödinger equation investigated with Hirota's bilinear method," Physical Review E, vol. 77, Article ID 066605, 2008.

8. R. Radhakrishnan, A. Kundu, and M. Lakshmanan, "Coupled nonlinear Schrödinger equations with cubic-quintic nonlinearity: integrability and soliton interaction in non-Kerr media," Physical Review E—Statistical Physics, Plasmas, Fluids, and Related Interdisciplinary Topics, vol. 60, no. 3, pp. 3314–3323, 1999.

9. Z. Zhang, F.-L. Xia, and X.-P. Li, "Bifurcation analysis and the travelling wave solutions of the Klein-Gordon-Zakharov equations," Pramana, vol. 80, no. 1, pp. 41–59, 2013.

10. M. Song, B. S. Ahmed, and A. Biswas, "Topological soliton solution and bifurcation analysis of the Klein-Gordon-Zakharov equation in (1+1)-dimensions with power law nonlinearity," Journal of Applied Mathematics, vol. 2013, Article ID 972416, 7 pages, 2013.

11. M. Song, Z. Liu, E. Zerrad, and A. Biswas, "Singular solitons and bifurcation analysis of quadratic nonlinear Klein-Gordon equation," Applied Mathematics & Information Sciences, vol. 7, no. 4, pp. 1333–1340, 2013.

12. Z. S. Wen, "Bifurcation of solitons, peakons, and periodic cusp waves for theta-equation," Nonlinear Dynamics, vol. 77, no. 1-2, pp. 247–253, 2014.

13. A. Choudhuri and K. Porsezian, "Higher-order nonlinear Schrödinger equation with derivative non-Kerr nonlinear terms: a model for sub-10-fs-pulse propagation," Physical Review A, vol. 88, Article ID 033808, 2013.

14. J. B. Li and H. H. Dai, On the Study of Singular Nonlinear Traveling Wave Equations: Dynamica System Approach, Science Press, Beijing, China, 2007.

CHAPTER 16

FFT BIFURCATION ANALYSIS OF ROUTES TO CHAOS VIA QUASI-PERIODIC SOLUTIONS

L. Borkowski and A. Stefanski

INTRODUCTION

The spectral analysis of a signal using the fast Fourier transform (FFT) is a widespread method for investigation and diagnostics of dynamical systems in science and engineering. The bibliography concerning the FFT algorithms and their application is very huge. Therefore, this paper concentrates on selected application associated with the researched problem only. The FFT is an algorithm for computing the discrete Fourier transform (DFT) and its inverse [1]. For $x_0,\ldots,_{N-1}$, which are complex numbers, the DFT is defined by the following formula:

$$X_k = \sum_{n=0}^{N-1} x_n e^{-(2\pi i/N)nk}, \qquad (1)$$

where $k = 0, \ldots, N-1$. Using the FFT analysis, the frequency components included in the time waveform can be presented. Calculation of the sum by formula (1) would take (N^2) operations. Using the Cooley-Tukey algorithm [2], which is based on the divide and conquer algorithm, the fast Fourier transform is calculated recursively dividing the transform of size $N = N1N2$ into transform of size $N1$ and $N2$ with the use of $O(N)$ multiplications. The computational complexity of the fast Fourier transform is $(N\log_2 N)$, instead of $O(N^2)$ algorithm which follows from the formula determining the DFT. There are other algorithms for calculating the DFT, for example, the Prime-factor algorithm also called the Good-Thomas algorithm [3, 4], Bruun's algorithm [5], Rader's algorithm [6], and Bluestein's algorithm [7]. As a result of the existence of the above-mentioned algorithms, it became possible to apply digital signal processing (DSP) [8, 9] and the use of discrete cosine transform (DCT) to data compression [10, 11] (e.g., JPEG or MP3 files).

Nowadays, in many areas of science and technology, we can observe the use of the fast Fourier transform in order to present the results of research and calculations. The use of the FFT analysis to study nonlinear dynamical systems is present in works of many scientists and researchers. A few selected examples of such applications are mentioned in [12–14]. In a series of three articles, Krysko et al. used the FFT analysis to study

(i) dynamics of continuous dynamical systems such as flexible plate and shallow shells [15],

(ii) classical and novel scenarios of transition from periodic to chaotic solutions of dissipative continuous mechanical systems [16],

(iii) dynamic loss of stability and different routes of transition to chaos of flexible curvilinear beam using Lyapunov exponents [17].

One can also mention a few examples of the FFT application in cases similar to the system analyzed in this paper. In 2006, Sánchez et al. [18] studied in their works a ring of unidirectionally coupled Lorenz oscillators. They observed occurrence of so-called rotating wave between oscillators and the transition from periodic rotating wave through quasiperiodic solutions to chaotic rotating wave. Numerical investigations were confirmed by experimental research. They

Citation: L. Borkowski and A. Stefanski, "FFT Bifurcation Analysis of Routes to Chaos via Quasiperiodic Solutions,"Mathematical Problems in Engineering, vol. 2015, Article ID 367036, 9 pages, 2015. doi:10.1155/2015/367036.

used the FFT analysis as a tool for the presentation of results. Also in the electrical systems the FFT analysis is widely used. For example, Hajimiri and Lee used the FFT analysis to study phase noise in nonlinear electrical oscillators [19, 20]. Also in the article of Razavi we can observe the use of the FFT analysis test phase noise in a ring of CMOS oscillators [21].

In this paper, the FFT analysis is applied to study dynamics and bifurcations of the ring of unidirectionally coupled nonlinear Duffing oscillators. In this system a route to chaos via 2-frequency and 3-frequency quasiperiodicity can be observed. The FFT investigation accompanies classical qualitative and quantitative tools for dynamical systems research as Poincaré maps, bifurcation diagrams, and Lyapunov exponents. The paper is organized as follows. Section 2 contains a brief description of analyzed ring of Duffing oscillators. In Section 3, the results of numerical investigation of the system under consideration are demonstrated. Classical bifurcation diagrams and values of Lyapunov exponents are summarized with results of the FFT analysis. Finally, Section 4presents a discussion of our results and conclusions.

ANALYZED SYSTEM

The system under consideration is a closed ring of N=7 unidirectionally coupled identical oscillators shown in Figure 1. As a node system we took autonomous single-well Duffing oscillator given by

$$\ddot{x}_j + d\dot{x}_j + ax_j + bx_j^3 = 0, \tag{2}$$

where a, b, and d are real positive parameters. Introducing the substitution $y_j = \dot{x}_j$ and assuming diffusive coupling between the oscillators, we can describe the dynamics of each jth ring node by the following pair of 1st-order ODEs:

$$\dot{x}_j = y_j,$$

$$\dot{y}_j = -ax_j - bx_j^3 - dy_j + \sigma\left(x_{j-1} - x_j\right), \tag{3}$$

where $j = 1, 2, \ldots, 7$ and σ is an overall coupling coefficient [22].

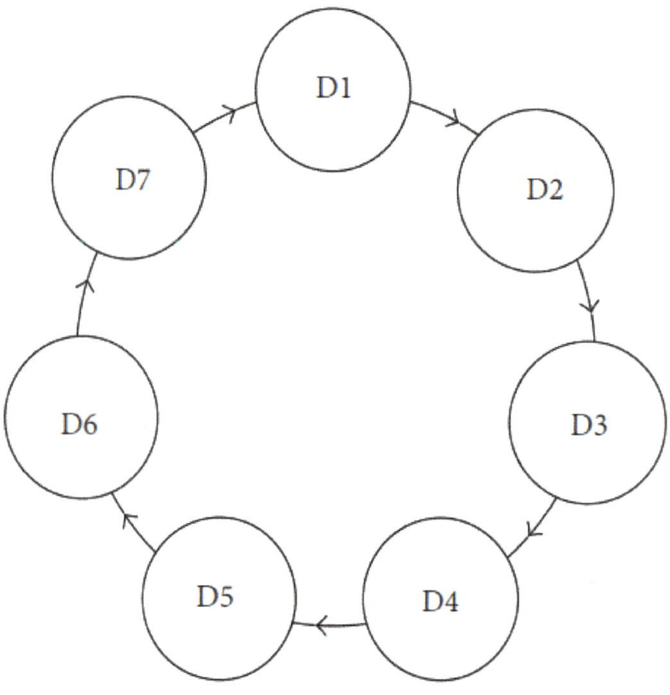

Figure 1: Scheme representing a ring of seven unidirectionally coupled oscillators.

NUMERICAL INVESTIGATIONS

The results of numerical analysis of system (3) are demonstrated in Figures 2(a)–2(c) and Figures 3–7. The number of coupled oscillators (N=7) was associated with a range of the 3-frequency quasiperiodic solution. For the ring of seven coupled oscillators, the range of occurrence in the 3-frequency quasiperiodic solution was the biggest. Numerical modeling and calculations for this work were done in MATLAB R2009b and Borland-Delphi 6 software, while the graphical presentation of bifurcation diagrams, time series, Poincaré maps, Lyapunov exponents, and the FFT analysis was drawn by means of Origin Pro 8.0 program. Assumed parameters of system (3) are $a = 1.0$, $b = 10.0$, and $d = 0.03162$ and coupling coefficient σ is considered as the control parameter.

In Figures 2(a)–2(c), the bifurcation diagram of individual node variable (Figure 2(a)), the parallel course of five largest Lyapunov exponents (Figure 2(b)), and the corresponding bifurcation diagram of the frequency spectrum (Figure 2(c)), depicting how to vary frequency distribution and density with increasing control parameter, are shown. In order to improve the readability of the FFT analysis and elimination of the noise power, the range of the amplitude from which the results were presented started at level 10^{-2} [dB]. Then, we can see the emergence of new peaks along with the emergence of the next Hopf bifurcation. On the other hand in Figures 3–7 time series, Poincaré maps, and related FFT spectra for selected values of coupling strength σ are presented.

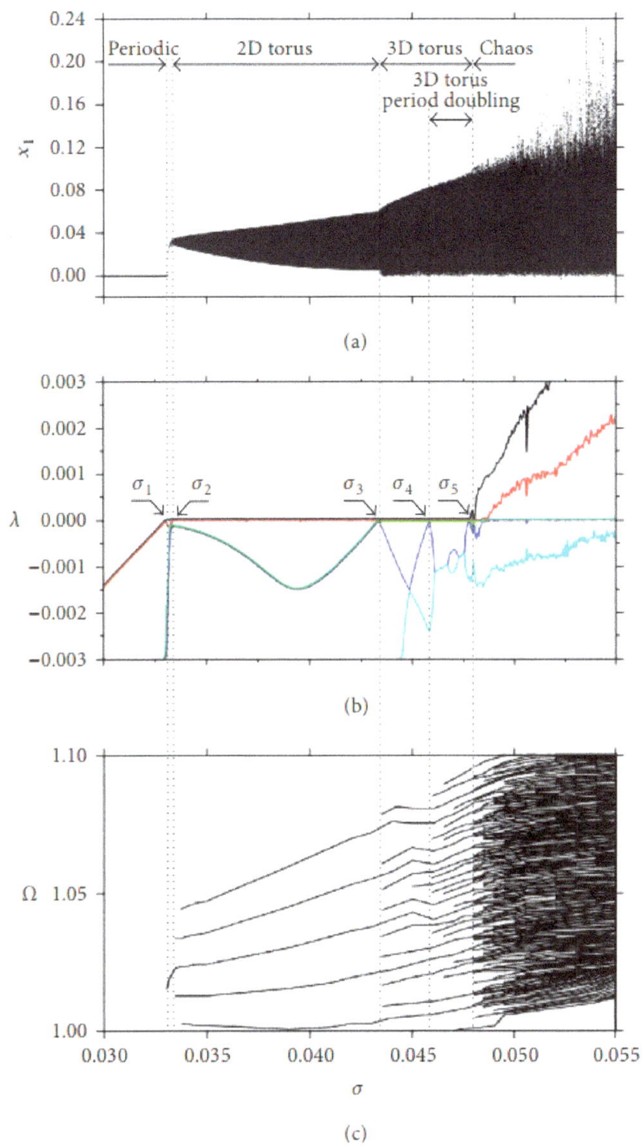

Figure 2: Bifurcation diagram of individual node variable x_1 (a), graph of five largest Lyapunov exponents (b), and parallel profile of frequency spectrum (c) for the ring of seven unidirectionally coupled nonlinear Duffing oscillators versus coupling coefficient σ.

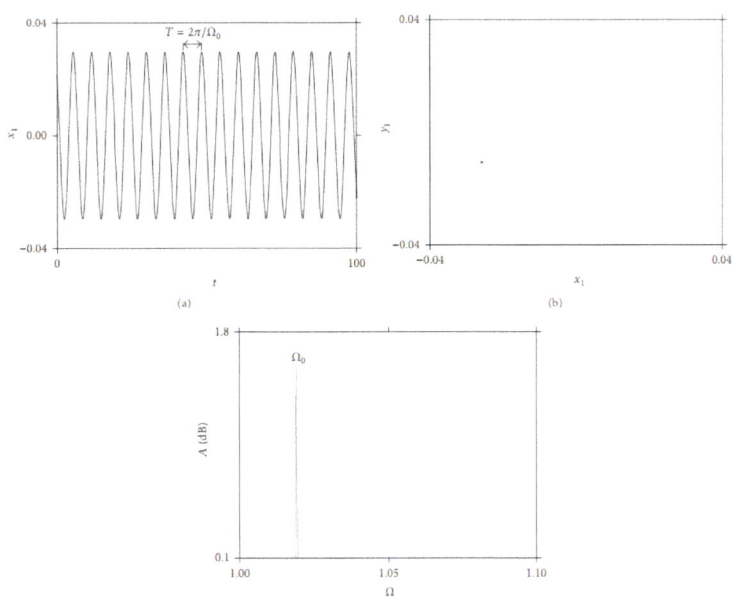

Figure 3: Time series (a), Poincare map (b), and FFT spectrum (c) of system (3) for ′ $\sigma = 0.0332$.

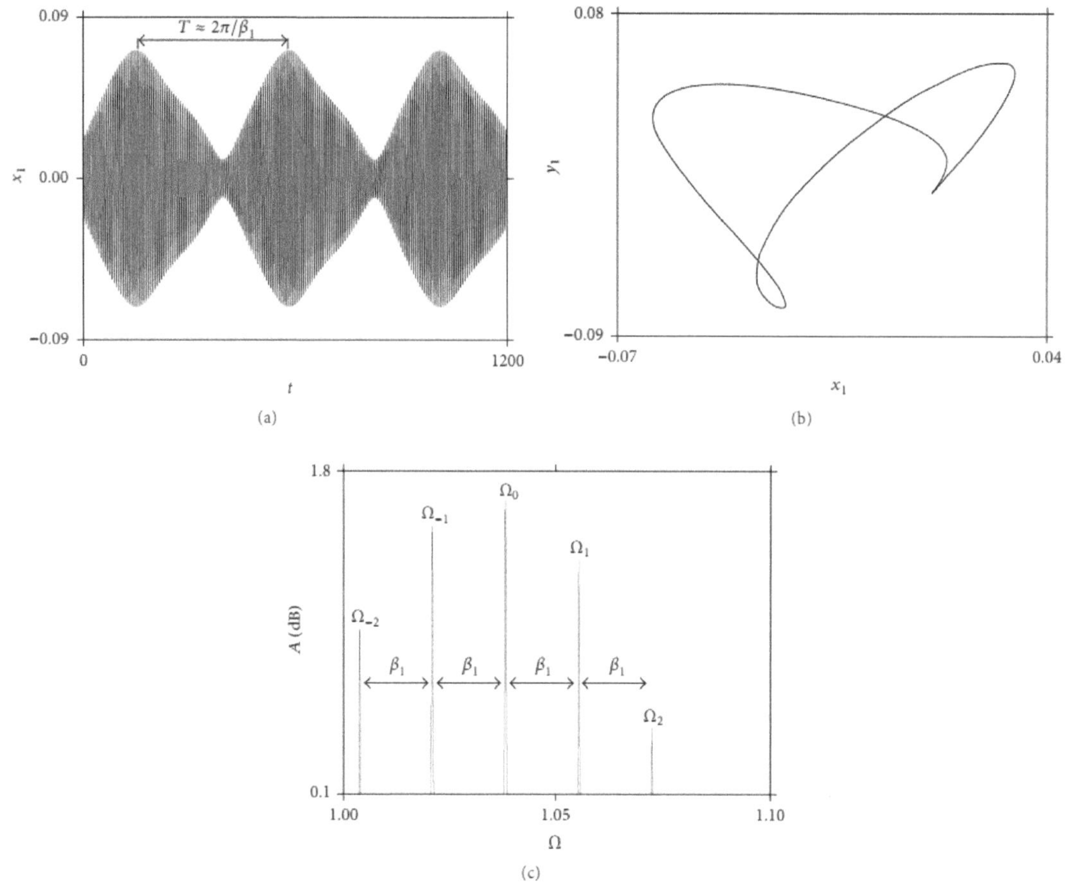

Figure 4: Time series (a), Poincare map (b), and FFT spectrum (c) of system (3) for ´ $\sigma = 0.0430$.

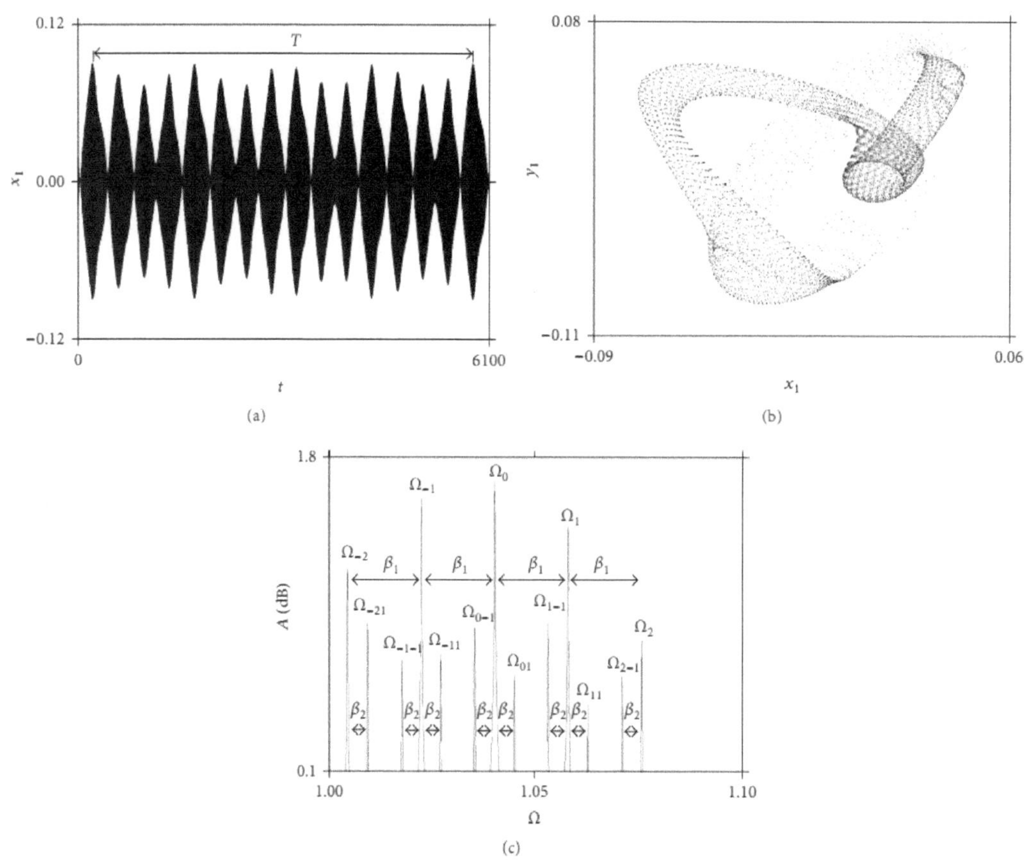

Figure 5: Time series (a), Poincare map (b), and FFT spectrum (c) of system (3) for ´ $\sigma = 0.0440$.

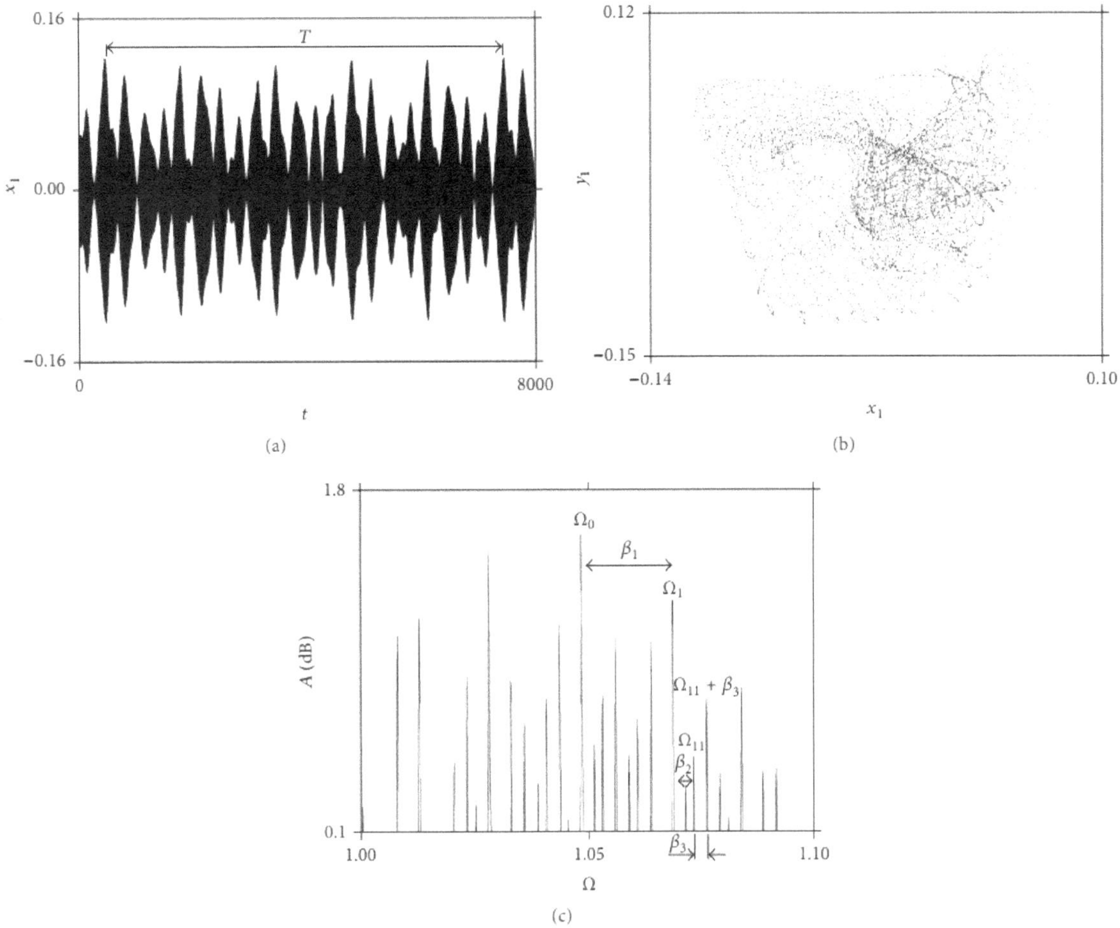

Figure 6: Time series (a), Poincare map (b), and FFT spectrum (c) of system (3) for ′ $\sigma = 0.0471$.

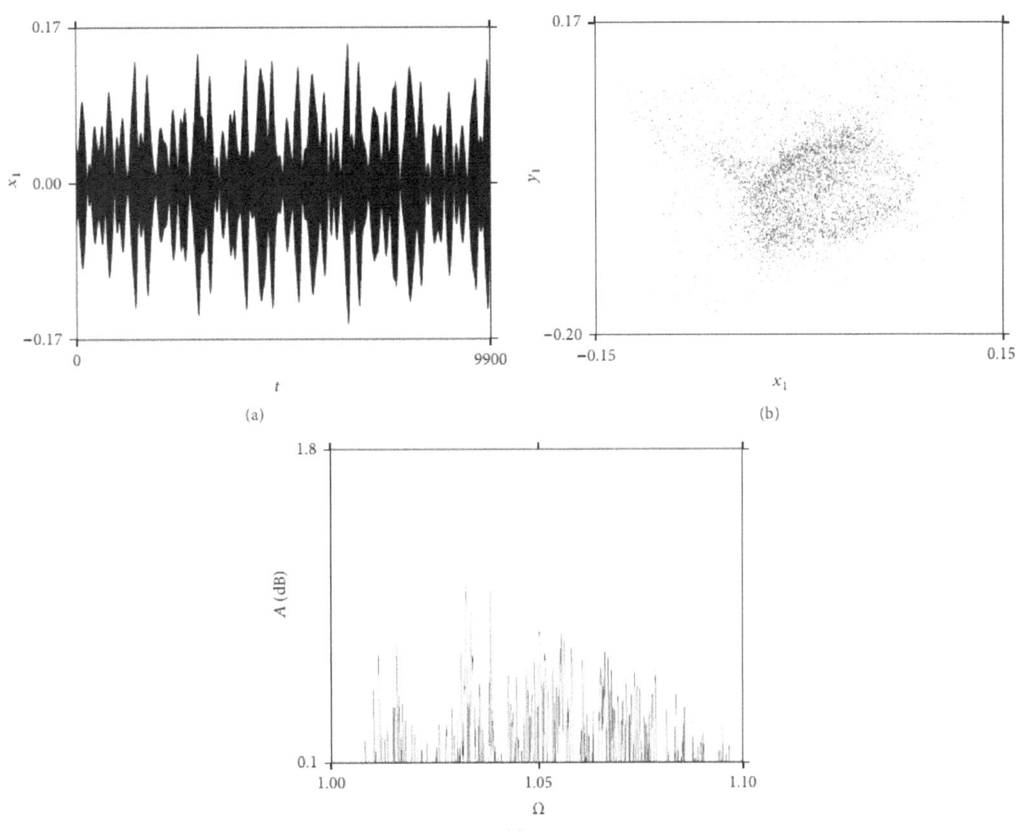

Figure 7: Time series (a), Poincare map (b), and FFT spectrum (c) of system (3) for ′ $\sigma = 0.0500$.

The bifurcation analysis in Figures 2(a)–2(c) depicts a transition from stable critical point (equilibrium position) to chaos via sequence of four consecutive Hopf-type bifurcations. The first time such a route to chaos was postulated was by Landau and Hopf—the Landau-Hopf transition to the turbulence after a series of infinite number of Hopf-type bifurcations [23, 24]. On the other hand, Newhouse, Ruelle, and Takens had formulated the theorem that just after third successive Hopf bifurcation the 3D torus decays into strange chaotic attractor in effect of arbitrarily small perturbation, a NRT scenario [25, 26]. Thus, in the researched case, we can observe an intermediary scenario of transition to chaos after fourth Hopf bifurcation. The first Hopf bifurcation occurs for $\sigma=\sigma_1$ where the system response passes from stationary to periodic solutions (Figures 3(a)-3(b)) and the first frequency of oscillation Ω_0 appears which is represented by a single peak Ω_0 in Figure 3(c). A small increase of coupling strength leads to the second Hopf bifurcation at $\sigma=\sigma_2$. The limit cycle is converted into quasiperiodic solution (Figures 4(a)- 4(b)), characterized by two incommensurate frequencies Ω_0 and Ω_1 (Figure 4(c)) and two largest Lyapunov exponents equal to zero (in black and red in Figure 2(b)). Analyzing the newly formed peaks (Figure 4(c)), the constant difference (offset) β_1 can be seen, which represent a beat frequency defining the approximate period of the torus cycle $T \approx 2\pi/\beta_1$; see Figure 4(a). Hence, the remaining incommensurate frequency peaks are distributed in the FFT spectrum according to the following formula:

$$\Omega_n = \Omega_0 + n\beta_1, \tag{4}$$

where n is a number of torus frequency. The 2D torus exists until the next Hopf-type bifurcation at where the transition from the 2D torus to the 3-frequency quasiperiodic solution (the 3D torus) takes place (Figures 5(a)-5(b)). As such, the 3D torus is distinguished with three largest Lyapunov exponents of zero value in spectrum (black, red, and green in Figure 2(b)) and third independent frequency Ω_2 in the FFT spectrum (Figure 5(c)). We can also observe new frequency peaks which are characterized by a constant offset β_2; see Figure 5(c). In the analogy to (4) the third disproportionate frequency can be described as

$$\Omega_{nm} = \Omega_n + m\beta_2, \tag{5}$$

where n, m are number of frequency peaks. For instance, using formula (5), the frequency Ω_{-21} from FFT spectrum depicted in Figure 5(c) can be calculated as follows:

$$\begin{aligned}\Omega_{-21} &= \Omega_{-2} + \beta_2 = \Omega_0 - 2\beta_1 + \beta_2 \\ &= 1,04286 - 2 * 0,01670 + 0,00446 = 1,01392.\end{aligned} \tag{6}$$

The 3D torus dominates in the interval $\sigma_3 < \sigma < \sigma_5$, but in the middle of this range the period-doubling bifurcation of the 3D torus takes place at $\sigma=\sigma_4$ (see Figures 6(a)-6(b)). This phenomenon is also reflected on the FFT spectrum (Figure 6(c)) where additional frequency peaks, shifted by frequency interval β_3, appear after the period-doubling of the 3D torus. Further increase of the coupling strength causes destruction of the 3D torus, direct transition to chaos, and next, after small increase of σ to hyperchaos on T_2 (the 2D torus), two positive and two Lyapunov exponents equal to zero $(+, +, 0, 0, -, -, \ldots)$ in the spectrum (see Figures 7(a)- 7(b)). The chaotic response manifests with the FFT spectrum with randomly distributed huge number of frequency peaks of various amplitudes but frequencies indicating the presence of the 2D torus in the skeleton of hyperchaotic attractor are still dominant (see Figure 7(c)).

CONCLUSIONS AND REMARKS

In this paper the dynamical system composed of the ring of seven unidirectionally coupled nonlinear Duffing oscillators is examined using the FFT bifurcation analysis. In addition, in order to confirm the results obtained by the FFT method, corresponding study with use of other classical tools for dynamical systems research (bifurcation diagrams, Poincaré maps, and Lyapunov exponents) was carried out. The considered system (3) was selected due to observed scenario of the transition to chaos via stable 3-frequency quasiperiodicity and its period doubling. Such an original route to chaos allows exhibiting advantages of the FFT spectrum analysis.

In general, the FFT method is a tool commonly known and used in engineering, diagnostics, and also science. Presented results show that the FFT analysis can be precise and useful instrument to nonlinear systems research. Obviously, for systems simulated numerically, as demonstrated here, calculation of the spectrum of Lyapunov exponents seems to be sufficiently accurate research approach. However, comparing Figures 2(b) and2(c), we can see that the bifurcation analysis of the FFT spectrum can be treated as a valuable complement of quantitative tools, for example, Lyapunov exponents, and support for more detailed identification of the system motion character. Then attractor of the system is additionally characterized by frequency peaks distribution and their signal strength in dB which is determined by peak height in logarithmic scale. It is clearly visible that bifurcations indicated in the spectrum of Lyapunov exponents (Figure 2(b)) are reflected in the FFT bifurcation graph (Figure 2(c)) and they manifest with newly emerging frequency lines at bifurcation values of the control parameter σ (i.e. σ_{1-5},); side peaks shifted by constant frequency intervals β_{1-3} (see Figures 4(c)–6(c)). A detailed identification of the bifurcation type (Hopf-type or period-doubling) only on the basis of the FFT graph from Figure 2(c) requires its juxtaposition with time series, Poincaré maps, and so forth reconstructed from investigated signal. The transition to chaos (for $\sigma=\sigma_5$ in Figures 2(a)–2(c)) manifests with a transition from the discrete FFT spectrum which is characteristic for regular solution (periodic, multiperiodic, and quasiperiodic; see Figures 3(c)–6(c)) to the FFT spectrum typical for chaos which is continuous in some frequency ranges (Figure 7(c)).

Calculation or estimation of Lyapunov exponents can be in many cases not straightforward (systems with discontinuities [27–32]) or very complex (experimental data), even in spite of existence of some algorithms allowing the estimation of these exponents from time series [33–35]. In such cases demonstrated approach of the FFT bifurcation analysis of the complex dynamical system as well as the nonclassical approach to the estimation of Lyapunov exponents [36–39] can turn out to be especially noteworthy.

Conflict of Interests

The authors declare that there is no conflict of interests regarding the publication of this paper.

ACKNOWLEDGMENT

This work has been supported by the Polish National Centre of Science (NCN) under project PRELUDIUM nr UMO-2012/07/N/ST8/03248.

REFERENCES

1. W. T. Cochran, J. W. Cooley, D. L. Favin et al., "What is the fast Fourier transform?" Proceedings of the IEEE, vol. 55, no. 10, pp. 1664–1674, 1967.
2. J. W. Cooley and J. W. Tukey, "An algorithm for the machine calculation of complex Fourier series,"Mathematics of Computation, vol. 19, pp. 297–301, 1965.
3. I. J. Good, "The interaction algorithm and practical Fourier analysis," Journal of the Royal Statistical Society, Series B: Methodological, vol. 20, pp. 361–372, 1958.
4. L. H. Thomas, "Using a computer to solve problems in physics," in Applications of Digital Computers, pp. 44–45, Ginn, 1963.
5. G. Bruun, "Z-transform DFT filters and FFT's," IEEE Transactions on Acoustics, Speech, and Signal Processing, vol. 26, no. 1, pp. 56–63, 1978.
6. C. M. Rader, "Discrete Fourier transforms when the number of data samples is prime," Proceedings of the IEEE, vol. 56, no. 6, pp. 1107–1108, 1968.
7. L. I. Bluestein, "A linear filtering approach to the computation of discrete Fourier transform," IEEE Transactions on Audio and Electroacoustics, vol. 18, no. 4, pp. 451–455, 1970. ·
8. S. M. Kuo and D. Morgan, Active Noise Control Systems: Algorithms and DSP Implementations, John Wiley & Sons, 1995.
9. L. R. Rabiner and B. Gold, Theory and Application of Digital Signal Processing, Prentice-Hall, Englewood Cliffs, NJ, USA, 1975.
10. N. Ahmed, T. Natarajan, and K. R. Rao, "Discrete cosine transform," IEEE Transactions on Computers, vol. 100, pp. 90–93, 1974.
11. Y. Yang, N. P. Galatsanos, and A. K. Katsaggelos, "Regularized reconstruction to reduce blocking artifacts of block discrete cosine transform compressed images," IEEE Transactions on Circuits and Systems for Video Technology, vol. 3, no. 6, pp. 421–432, 1993.
12. K. Hadjiyannakis, R. D. Ogilvie, C. E. D. Alloway, and C. Shapiro, "FFT analysis of EEG during stage 2-to-REM transitions in narcoleptic patients and normal sleepers," Electroencephalography and Clinical Neurophysiology, vol. 103, no. 5, pp. 543–553, 1997.
13. G. S. Spagnolo, G. Guattari, C. Sapia, D. Ambrosini, D. Paoletti, and G. Accardo, "Contouring of artwork surface by fringe projection and FFT analysis," Optics and Lasers in Engineering, vol. 33, no. 2, pp. 141–156, 2000.
14. I. Steller, R. Bolotovsky, and M. G. Rossmann, "An algorithm for automatic indexing of oscillation images using Fourier analysis," Journal of Applied Crystallography, vol. 30, no. 6, pp. 1036–1040, 1997.
15. J. Awrejcewicz, V. A. Krysko, I. V. Papkova, and A. V. Krysko, "Routes to chaos in continuous mechanical systems. Part 1. Mathematical models and solution methods," Chaos, Solitons and Fractals, vol. 45, no. 6, pp. 687–708, 2012.
16. A. V. Krysko, J. Awrejcewicz, I. V. Papkova, and V. A. Krysko, "Routes to chaos in continuous mechanical systems: part 2. Modelling transitions from regular to chaotic dynamics," Chaos, Solitons & Fractals, vol. 45, no. 6, pp. 709–720, 2012.
17. J. Awrejcewicz, A. V. Krysko, I. V. Papkova, and V. A. Krysko, "Routes to chaos in continuous mechanical systems. Part 3. The Lyapunov exponents, hyper, hyper-hyper and spatial-temporal chaos," Chaos, Solitons & Fractals, vol. 45, no. 6, pp. 721–736, 2012.
18. E. Sánchez, D. Pazó, and M. A. Matías, "Experimental study of the transitions between synchronous chaos and a periodic rotating wave," Chaos, vol. 16, no. 3, Article ID 033122, 2006. ·
19. A. Hajimiri and T. H. Lee, "A general theory of phase noise in electrical oscillators," IEEE Journal of Solid-State Circuits, vol. 33, no. 2, pp. 179–194, 1998.
20. A. Hajimiri and T. H. Lee, "Design issues in CMOS differential LC oscillators," IEEE Journal of Solid-State Circuits, vol. 34, no. 5, pp. 717–724, 1999.
21. B. Razavi, "A study of phase noise in CMOS oscillators," IEEE Journal of Solid-State Circuits, vol. 31, no. 3, pp. 331–343, 1996.
22. P. Perlikowski, S. Yanchuk, M. Wolfrum, A. Stefanski, P. Mosiolek, and T. Kapitaniak, "Routes to complex dynamics in a ring of unidirectionally coupled systems," Chaos, vol. 20, no. 1, Article ID 013111, 2010.
23. L. D. Landau, "On the problem of turbulence," in Proceedings of the USSR Academy of Sciences, vol. 44, pp. 1–314, 1944.
24. E. Hopf, "A mathematical example displaying features of turbulence," Communications on Pure and Applied Mathematics, vol. 1, pp. 303–322, 1948.
25. D. Ruelle and F. Takens, "On the nature of turbulence," Communications in Mathematical Physics, vol. 20, no. 3, pp. 167–192, 1971.
26. S. Newhouse, D. Ruelle, and F. Takens, "Occurrence of strange Axiom A attractors near quasi periodic flows on Tm, m≧3," Communications in Mathematical Physics, vol. 64, no. 1, pp. 35–40, 1978.
27. B. Blazejczyk-Okolewska and T. Kapitaniak, "Co-existing attractors of impact oscillator," Chaos, Solitons & Fractals, vol. 9, no. 8, pp. 1439–1443, 1998.
28. B. Blazejczyk-Okolewska and T. Kapitaniak, "Dynamics of impact oscillator with dry friction," Chaos, Solitons & Fractals, vol. 7, no. 9, pp. 1455–1459, 1996.

29. T. Kapitaniak, "Generating strange nonchaotic trajectories," Physical Review E, vol. 47, no. 2, pp. 1408–1410, 1993.

30. J. Wojewoda, A. Stefanski, M. Wiercigroch, and T. Kapitaniak, "Hysteretic effects of dry friction: modelling and experimental studies," Philosophical Transactions of the Royal Society of London A: Mathematical, Physical and Engineering Sciences, vol. 366, pp. 747–765, 2008.

31. M. Kapitaniak, J. Strzalko, J. Grabski, and T. Kapitaniak, "The three-dimensional dynamics of the die throw," Chaos, vol. 22, no. 4, Article ID 047504, 2012.

32. A. Stefanski, J. Wojewoda, M. Wiercigroch, and T. Kapitaniak, "Chaos caused by non-reversible dry friction," Chaos, Solitons & Fractals, vol. 16, no. 5, pp. 661–664, 2003.

33. S. Oberst and J. C. S. Lai, "A statistical approach to estimate the Lyapunov spectrum in disc brake squeal,"Journal of Sound and Vibration, vol. 334, pp. 120–135, 2015.

34. H. Abarbanel, Analysis of Observed Chaotic Data, Springer Science & Business Media, 2012.

35. H. Kantz and T. Schreiber, Nonlinear Time Series Analysis, Cambridge University Press, 2004.

36. A. Stefanski, "Estimation of the largest Lyapunov exponent in systems with impacts," Chaos, Solitons & Fractals, vol. 11, no. 13, pp. 2443–2451, 2000.

37. A. Stefanski and T. Kapitaniak, "Estimation of the dominant Lyapunov exponent of non-smooth systems on the basis of maps synchronization," Chaos, Solitons and Fractals, vol. 15, no. 2, pp. 233–244, 2003.

38. A. Stefanski, A. Dabrowski, and T. Kapitaniak, "Evaluation of the largest Lyapunov exponent in dynamical systems with time delay," Chaos, Solitons & Fractals, vol. 23, no. 5, pp. 1651–1659, 2005.

39. W. Serweta, A. Okolewski, B. Blazejczyk-Okolewska, K. Czolczynski, and T. Kapitaniak, "Lyapunov exponents of impact oscillators with Hertz's and Newton's contact models," International Journal of Mechanical Sciences, vol. 89, pp. 194–206, 2014.

CHAPTER 17

GENERALIZED SYNCHRONIZATION WITH UNCERTAIN PARAMETERS OF NONLINEAR DYNAMIC SYSTEM VIA ADAPTIVE CONTROL

Cheng-Hsiung Yang and Cheng-Lin Wu

INTRODUCTION

The chaos synchronization phenomenon has the following feature: the trajectories of the master and the slave chaotic system are identical in spite of starting from different initial conditions or different nonlinear dynamic system. However, slight differentiations of initial conditions, for chaotic dynamical systems, will lead to completely different trajectories [1–14]. The issue may be treated as the control law design for observer of slave chaotic system using the master chaotic system so as to ensure that the controlled receiver synchronizes with the master chaotic system. Hence, the slave chaotic system completely traces the dynamics of the master chaotic system in the course of time [15– 19]. The key technique of chaos synchronization for secret communication has been widely investigated. Until now, a wide variety of approaches have been proposed for control and synchronization of chaotic systems, such as adaptive control [20, 21], backstepping control [22–25], sliding mode control [26–28], and fuzzy control [29–31], just to name a few. The forenamed strategies and many other existing skills of synchronization mainly concern the chaos synchronization of two identical chaotic systems with known parameters or identical unknown parameters [32–38].

Among many kinds of chaos synchronizations, the generalized synchronization is widely studied. This means that there exists a given functional relationship between the states of the master system and that of the slave system $y = (x)$. In this paper, a new generalized synchronization with uncertain parameters,

$$\dot{x} = f(t, x, A(t)),$$
$$y = x + F(t), \tag{1}$$

is studied, where x, y are the state vectors of the master and slave system, respectively, and the $A(t)$ is uncertain chaotic parameters in f. The (t) may be given a regular/chaotic dynamical system. The rest of the paper is organized as follows. In Section 2, by the Lyapunov asymptotic stability theorem, the generalized synchronization with uncertain chaotic parameters by adaptive control scheme is given. In Section 3, various adaptive controllers and update laws are designed for the generalized synchronization with uncertain parameters of the identical Lorenz-Stenflo systems. The numerical simulation of three examples is also given in Section 3. Finally, some concluding remarks are given in Section 4.

GENERALIZED ADAPTIVE SYNCHRONIZATION WITH UNCERTAIN PARAMETERS SCHEME

Consider the master system

$$\dot{x} = f(t, x, A(t)) \tag{2}$$

and the slave system

$$\dot{y} = f\left(t, y, \widehat{A}(t)\right) + u, \tag{3}$$

where $x = [x_1, x_2, \ldots, x_n]^T \in R^n$, $y = [y_1, y_2, \ldots, y_n]^T \in R^n$ denote the master states vector and slave states vector, respectively, the f is nonlinear vector functions, the $A(t)$

Citation: Cheng-Hsiung Yang and Cheng-Lin Wu, "Generalized Synchronization with Uncertain Parameters of Nonlinear Dynamic System via Adaptive Control," The Scientific World Journal, vol. 2014, Article ID 152485, 9 pages, 2014. doi:10.1155/2014/152485.

is uncertain chaotic parameters in f, the $\widehat{A}(t)$ is estimates of uncertain chaotic parameters in f, and the $u = [u_1, u_2, \ldots, u_n]^T \in R^n$ is adaptive control vector

Our goal is to design a controller $u(t)$ and an adaptive law $\dot{\widetilde{A}}$ so that the state vector of the slave system equation (3) asymptotically approaches the state vector of the master system equation (2) plus a given vector regular/chaotic function $F(t) = [F_1(t), F_2(t), \ldots, F_n(t)]^T$, and finally the generalized adaptive synchronization with uncertain parameters will be accomplished in the sense that the limit of the states error vector $e(t) = [e_1, e_2, \ldots, e_n]^T$ and parameters error vector $\widetilde{A}(t) = [\widetilde{A}_1, \widetilde{A}_2, \ldots, \widetilde{A}_m]^T$ approaches zero:

$$\lim_{t \to \infty} e(t) = 0, \tag{4a}$$

$$\lim_{t \to \infty} \widetilde{A}(t) = 0, \tag{4b}$$

where $e_i = x_i - y_i + F_i(t), (i = 1, 2, \ldots, n)$ and $\widetilde{A}_j = A_j(t) - \widehat{A}_j(t), (j = 1, 2, \ldots, m)$.

From (4a), we have

$$\dot{e}_i = \dot{x}_i - \dot{y}_i + \dot{F}_i(t), \quad i = 1, 2, \ldots, n. \tag{5}$$

Introduce (2) and (3) in (5) as

$$\dot{e} = f(t, x, A(t)) - f\left(t, y, \widehat{A}(t)\right) + \dot{F}(t) - u(t). \tag{6}$$

A Lyapunov function candidate $V(e, \widetilde{A})$ is chosen as a positive definite function as

$$V\left(e, \widetilde{A}\right) = \frac{1}{2}e^T e + \frac{1}{2}\widetilde{A}^T \widetilde{A}. \tag{7}$$

Its derivative along the solution of (7) is

$$\dot{V}\left(e, \widetilde{A}\right) = e^T\left[f(t, x, A(t)) - f\left(t, y, \widehat{A}(t)\right) + \dot{F}(t) - u(t)\right] + \widetilde{A}^T \dot{\widetilde{A}}, \tag{8}$$

where $u(t)$ and $\dot{\widetilde{A}}$ are chosen so that $\dot{V} = e^T C_1 e + \widetilde{A}^T C_2 \widetilde{A}$, C_1 and C_2 are negative constants, and $\dot{V}$ is a negative definite function of $e_1, e_2, \ldots, e_n$ and $\widetilde{A}_1, \widetilde{A}_2, \ldots, \widetilde{A}_m$. When

$$\lim_{t \to \infty} e = 0, \qquad \lim_{t \to \infty} \widetilde{A} = 0, \tag{9}$$

the generalized adaptive synchronization with uncertain parameters is obtained.

RESULTS OF NUMERICAL SIMULATION

In this section, a mathematical proof is provided for the three cases' results of numerical, adaptive synchronization, generalized adaptive synchronization, and generalized adaptive synchronization with uncertain parameters.

Case I Adaptive Synchronization

The master system is new Lorenz-Stenflo system [39]:

$$\begin{aligned} \dot{x}_1 &= a(x_2 - x_1), \\ \dot{x}_2 &= cx_1 - x_1 x_3 - x_2, \\ \dot{x}_3 &= bx_4 - dx_3 + x_1 x_2, \\ \dot{x}_4 &= -x_1 - ax_4, \end{aligned} \tag{10}$$

where $a = 3.7$, $b = 1.5$, $c = 26$, and $d = 0.7$. The initial conditions are $x_1(0) = 30$, $x_2(0) = 30$, $x_3(0) = 30$, and $x_4(0) = 30$. The phase portrait is shown in Figure 1. The slave system is

$$\begin{aligned} \dot{y}_1 &= \widehat{a}y_2 - \widehat{a}y_1 + u_1, \\ \dot{y}_2 &= \widehat{c}y_1 - y_1 y_3 - y_2 + u_2, \\ \dot{y}_3 &= y_1 y_2 - \widehat{d}y_3 + \widehat{b}y_4 + u_3, \\ \dot{y}_4 &= -y_1 - \widehat{a}y_4 + u_4, \end{aligned} \tag{11}$$

Where $\widehat{a}, \widehat{b}, \widehat{c}, \widehat{d}$, and $\widehat{r}$ are estimates of uncertain parameters a, b, c, d, and r, respectively? The initial conditions of salve system are $y_1(0) = -50$, $y_2(0) = -50$, $y_3(0) = -50$, and $y_4(0) = -50$.

Our objective is to design the controllers such that the trajectories, (t) and $y(t)$, of the master system and slave system satisfy

$$\lim_{t \to \infty} \|x(t) - y(t)\| = 0. \tag{12a}$$

Our objective is to design the controllers parameters estimation update laws $\dot{\widetilde{A}} = [\dot{\widetilde{a}}, \dot{\widetilde{b}}, \dot{\widetilde{c}}, \dot{\widetilde{d}}]^T$ such that the trajectories, (t) and $\widehat{A}(t)$, of the uncertain chaotic parameters and estimates of uncertain chaotic parameters satisfy

$$\lim_{t \to \infty} \left\|A(t) - \widehat{A}(t)\right\| = 0, \tag{12b}$$

where $\|\cdot\|$ denotes the Euclidean norm. Define an error vector function

$$e_i = x_i - y_i, \quad (i = 1, 2, 3, 4). \tag{13}$$

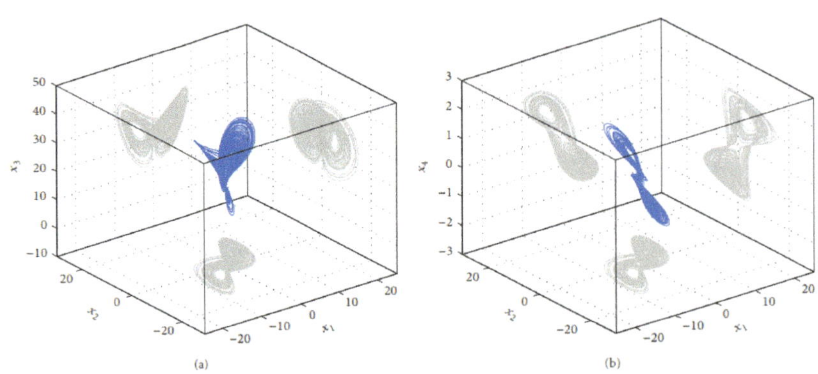

Figure 1: Three-dimension phase portrait of the four-dimensional Lorenz-Stenflo system and its projection.

From the error functions, we get the error dynamics

$$\begin{aligned}\dot{e}_1 &= \hat{a}e_2 + \tilde{a}x_2 - \hat{a}e_1 - \tilde{a}x_1 - u_1,\\ \dot{e}_2 &= \hat{c}e_1 + \tilde{c}x_1 + e_1e_3 - x_1e_3 - x_3e_1 - e_2 - u_2,\\ \dot{e}_3 &= -e_1e_2 + x_1e_2 + x_2e_1 - \hat{d}e_3 - \tilde{d}x_3 + \hat{b}e_4 + \tilde{b}x_4 - u_3,\\ \dot{e}_4 &= -e_1 - \hat{a}e_4 - \tilde{a}x_4 - u_4,\end{aligned} \quad (14)$$

where $e_1 = x_1 - y_1, e_2 = x_2 - y_2, e_3 = x_3 - y_3, e_4 = x_4 - y_4$, $\tilde{a} = a - \hat{a}, \tilde{b} = b - \hat{b}, \tilde{c} = c - \hat{c}$, and $\tilde{d} = d - \hat{d}$.

Choose a Lyapunov function candidate in the form of a positive definite function

$$V(e_1, e_2, e_3, e_4, \tilde{a}, \tilde{b}, \tilde{c}, \tilde{d}) = \frac{1}{2}\left(e_1^2 + e_2^2 + e_3^2 + e_4^2 + \tilde{a}^2 + \tilde{b}^2 + \tilde{c}^2 + \tilde{d}^2\right), \quad (15)$$

and its time derivative is

$$\begin{aligned}\dot{V}&(e_1, e_2, e_3, e_4, \tilde{a}, \tilde{b}, \tilde{c}, \tilde{d})\\ &= e_1\dot{e}_1 + e_2\dot{e}_2 + e_3\dot{e}_3 + e_4\dot{e}_4 + \tilde{a}\dot{\tilde{a}} + \tilde{b}\dot{\tilde{b}} + \tilde{c}\dot{\tilde{c}} + \tilde{d}\dot{\tilde{d}}\\ &= e_1(\hat{a}e_2 + \tilde{a}x_2 - \hat{a}e_1 - \tilde{a}x_1 - u_1)\\ &\quad + e_2(\hat{c}e_1 + \tilde{c}x_1 + e_1e_3 - x_1e_3 - x_3e_1 - e_2 - u_2)\\ &\quad + e_3(-e_1e_2 + x_1e_2 + x_2e_1 - \hat{d}e_3 - \tilde{d}x_3\\ &\qquad + \hat{b}e_4 + \tilde{b}x_4 - u_3)\\ &\quad + e_4(-e_1 - \hat{a}e_4 - \tilde{a}x_4 - u_4) + \tilde{a}(-\dot{\hat{a}}) + \tilde{b}(-\dot{\hat{b}})\\ &\quad + \tilde{c}(-\dot{\hat{c}}) + \tilde{d}(-\dot{\hat{d}}) + \tilde{r}(-\dot{\hat{r}}).\end{aligned} \quad (16)$$

Choose the parameters estimation update laws as follows:

$$\begin{aligned}\dot{\tilde{a}} &= -\dot{\hat{a}} = -e_1(x_2 - x_1) + e_4x_4 - \tilde{a},\\ \dot{\tilde{b}} &= -\dot{\hat{b}} = -e_3x_4 - \tilde{b},\\ \dot{\tilde{c}} &= -\dot{\hat{c}} = -e_2x_1 - \tilde{c},\\ \dot{\tilde{d}} &= -\dot{\hat{d}} = e_3x_3 - \tilde{d}.\end{aligned} \quad (17)$$

The initial values of estimates for uncertain parameters are $\hat{a}(0) = 0, \hat{b}(0) = 0, \hat{c}(0) = 0$, and $\hat{d}(0) = 0$. Through (16) and (17), the appropriate controllers can be designed as

$$\begin{aligned}u_1 &= \hat{a}e_2 - \hat{a}e_1 + e_1,\\ u_2 &= \hat{c}e_1 + e_1e_3 - x_1e_3 - x_3e_1 - e_2 + e_2,\\ u_3 &= -e_1e_2 + x_1e_2 + x_2e_1 - \hat{d}e_3 + \hat{b}e_4 + e_3,\\ u_4 &= -e_1 - \hat{a}e_4 + e_4.\end{aligned} \quad (18)$$

Substituting (18) and (17) into (16), we obtain

$$\dot{V}(e_1, e_2, e_3, e_4, \tilde{a}, \tilde{b}, \tilde{c}, \tilde{d}) = -e_1^{\,2} - e_2^{\,2} - e_3^{\,2} - e_4^{\,2} - \tilde{a}^2 - \tilde{b}^2 - \tilde{c}^2 - \tilde{d}^2 < 0. \quad (19)$$

Since the Lyapunov function $V(e_1, e_2, e_3, e_4, \tilde{a}, \tilde{b}, \tilde{c}, \tilde{d})$ is positive definite and its derivative $\dot{V}(e_1, e_2, e_3, e_4, \tilde{a}, \tilde{b}, \tilde{c}, \tilde{d})$ is negative definite in the neighborhood of the zero solutions for (12a) and (12b), according to the Lyapunov stability theory, the zero solutions of error states dynamic and parameters error vector are asymptotically stable; namely, the slave system equation (11) can asymptotically converge to its master system equation (10) with the adaptive control law equation (18) and the estimation parameter update law equation (17). The adaptive synchronization concept proof had to be completed. The numerical simulation results are shown in Figures 2, 3, and 4.

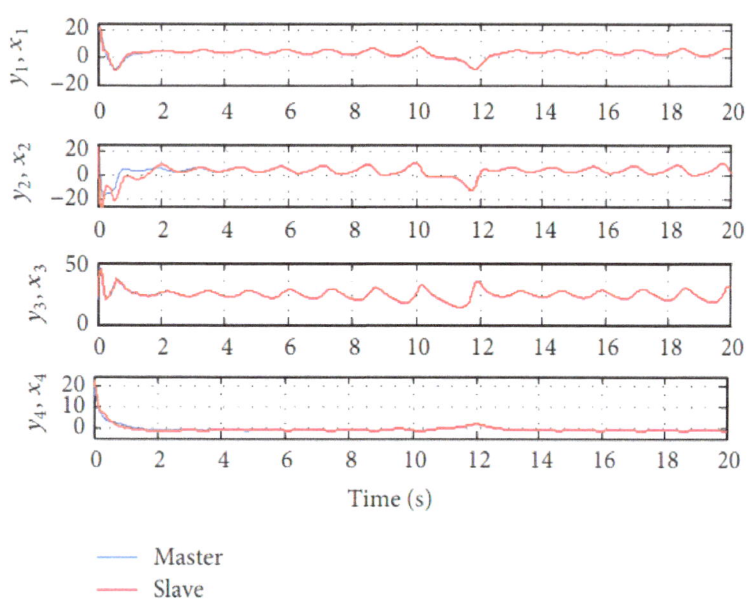

Figure 2: Time histories of the master system and slave system for Case I.

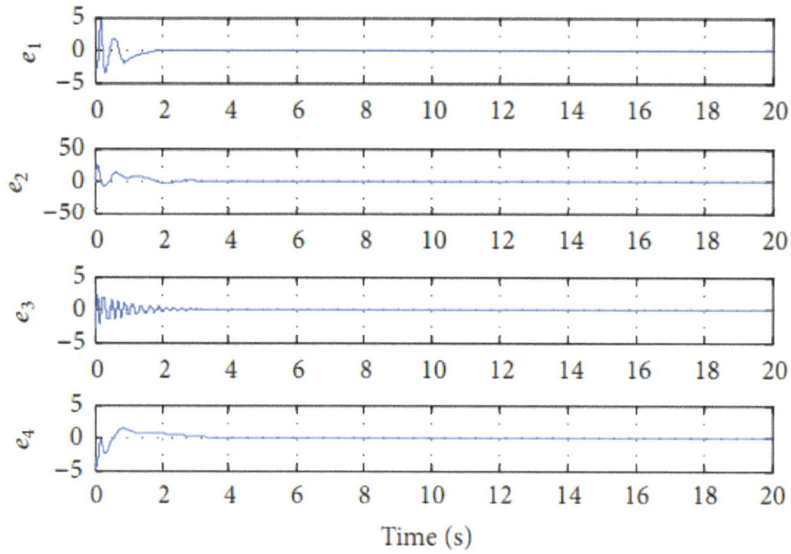

Figure 3: Time histories of error states for Case I.

Case II Generalized Adaptive Synchronization

The given functional system for generalized synchronization is also a new Lorenz-Stenflo system but with different initial conditions: $w_1(0) = 25$, $w_2(0) = 25$, $w_3(0) = 25$, and $w_4(0) = 25$:

$$\begin{aligned} \dot{w}_1 &= a(w_2 - w_1), \\ \dot{w}_2 &= cw_1 - w_1 w_3 - w_2, \\ \dot{w}_3 &= bw_4 - dw_3 + w_1 w_2, \\ \dot{w}_4 &= -aw_4 - w_1. \end{aligned} \tag{20}$$

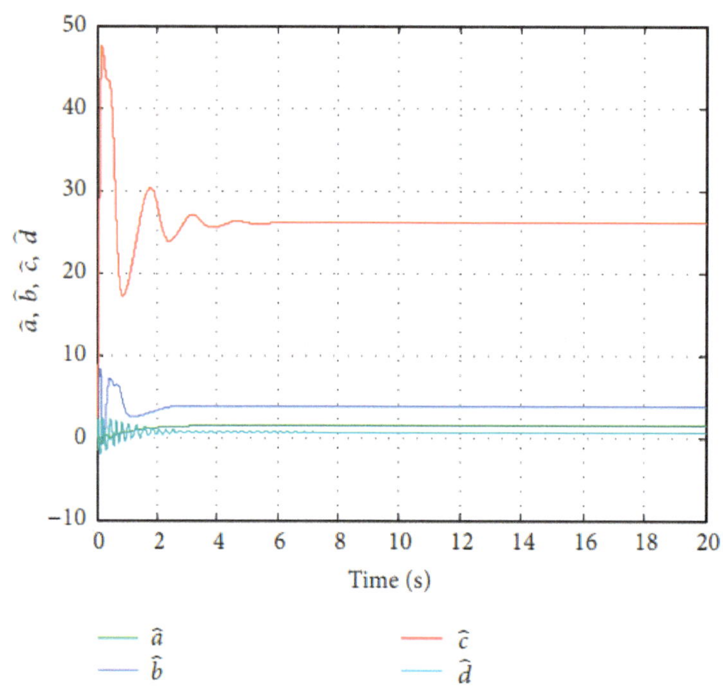

Figure 4: Time histories of estimated parameters for Case I

When the time approaches infinite, the error functions approach zero. The generalized adaptive synchronization can be accomplished as

$$\lim_{t \to \infty} e = 0, \tag{21}$$

where the error functions here can be defined as

$$e_i = x_i + w_i - y_i, \quad (i = 1, 2, 3, 4). \tag{22}$$

$$
\begin{aligned}
\dot{e}_1 &= \hat{a}(e_2 - e_1) + \bar{a}(x_2 - x_1 + w_2 - w_1) - u_1, \\
\dot{e}_2 &= \hat{c}e_1 + \bar{c}(x_1 + w_1) - e_2 + x_1(w_3 - e_3) + w_1(x_3 - e_3) \\
&\quad - e_1(x_3 + w_3 - e_3) - u_2, \\
\dot{e}_3 &= -\hat{d}e_3 - \bar{d}(x_3 + w_3) + \hat{b}e_4 + \bar{b}(x_4 + w_4) \\
&\quad - x_1(w_2 - e_2) - w_1(x_2 - e_2) + (x_2 + w_2 - e_2) - u_3, \\
\dot{e}_4 &= -e_1 - \hat{a}e_4 - \bar{a}(x_4 + w_4) - u_4,
\end{aligned} \tag{23}
$$

Where $\bar{a} = a - \hat{a}, \bar{b} = b - \hat{b}, \bar{c} = c - \hat{c}$, and $\bar{d} = d - \hat{d}$.

Choose a Lyapunov function in the form of a positive definite function

$$
\begin{aligned}
&V(e_1, e_2, e_3, e_4, \bar{a}, \bar{b}, \bar{c}, \bar{d}) \\
&= \frac{1}{2}(e_1^2 + e_2^2 + e_3^2 + e_4^2 + \bar{a}^2 + \bar{b}^2 + \bar{c}^2 + \bar{d}^2),
\end{aligned} \tag{24}
$$

and its time derivative is

$$
\begin{aligned}
&\dot{V}(e_1, e_2, e_3, e_4, \bar{a}, \bar{b}, \bar{c}, \bar{d}) \\
&= e_1\dot{e}_1 + e_2\dot{e}_2 + e_3\dot{e}_3 + e_4\dot{e}_4 + \bar{a}\dot{\bar{a}} + \bar{b}\dot{\bar{b}} + \bar{c}\dot{\bar{c}} + \bar{d}\dot{\bar{d}} \\
&= e_1(\hat{a}(e_2 - e_1) + \bar{a}(x_2 - x_1 + w_2 - w_1) - u_1) \\
&\quad + e_2(\hat{c}e_1 + \bar{c}(x_1 + w_1) - e_2 + x_1(w_3 - e_3) \\
&\qquad + w_1(x_3 - e_3) - e_1(x_3 + w_3 - e_3) - u_2) \\
&\quad + e_3(-\hat{d}e_3 - \bar{d}(x_3 + w_3) + \hat{b}e_4 + \bar{b}(x_4 + w_4) \\
&\qquad - x_1(w_2 - e_2) - w_1(x_2 - e_2) \\
&\qquad + e_1(x_2 + w_2 - e_2) - u_3) \\
&\quad + e_4(-e_1 - \hat{a}e_4 - \bar{a}(x_4 + w_4) - u_4) + \bar{a}(-\dot{\hat{a}}) + \bar{b}(-\dot{\hat{b}}) \\
&\quad + \bar{c}(-\dot{\hat{c}}) + \bar{d}(-\dot{\hat{d}}) + \bar{r}(-\dot{\hat{r}}).
\end{aligned} \tag{25}
$$

Choose the parameters estimation update laws as follows:

$$
\begin{aligned}
\dot{\bar{a}} &= -\dot{\hat{a}} = -e_1(x_2 - x_1 + w_2 - w_1) + e_4(x_4 + w_4) - \bar{a}, \\
\dot{\bar{b}} &= -\dot{\hat{b}} = -e_3(x_4 + w_4) - \bar{b}, \\
\dot{\bar{c}} &= -\dot{\hat{c}} = -e_2(x_1 + w_1) - \bar{c}, \\
\dot{\bar{d}} &= -\dot{\hat{d}} = e_3(x_3 + w_3) - \bar{d}.
\end{aligned} \tag{26}
$$

The initial values of estimates for uncertain parameters are $\hat{a}(0) = 0, \hat{b}(0) = 0, \hat{c}(0) = 0$, and $\hat{d}(0) = 0$. Through (25) and (26), the appropriate controllers can be designed as

$$
\begin{aligned}
u_1 &= \hat{a}e_2 - \hat{a}e_1 + e_1, \\
u_2 &= \hat{c}e_1 - e_2 + x_1(w_3 - e_3) + w_1(x_3 - e_3) \\
&\quad - e_1(x_3 + w_3 - e_3) + e_2, \\
u_3 &= -\hat{d}e_3 + \hat{b}e_4 - x_1(w_2 - e_2) - w_1(x_2 - w_2) \\
&\quad + e_1(x_2 + w_2 - e_2) + e_3, \\
u_4 &= -e_1 - \hat{a}e_4 + e_4.
\end{aligned} \tag{27}
$$

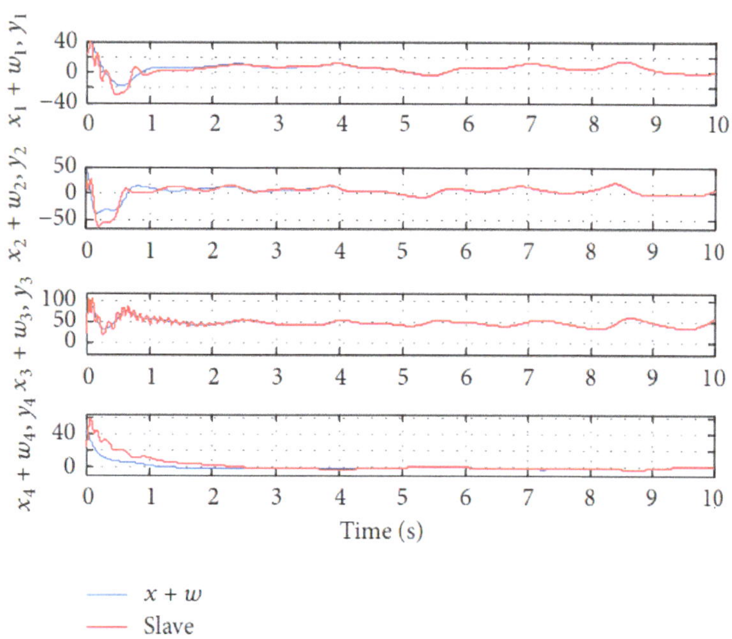

Figure 5: Time histories of the generalized signal (master x system plus given function w) and slave system for Case II.

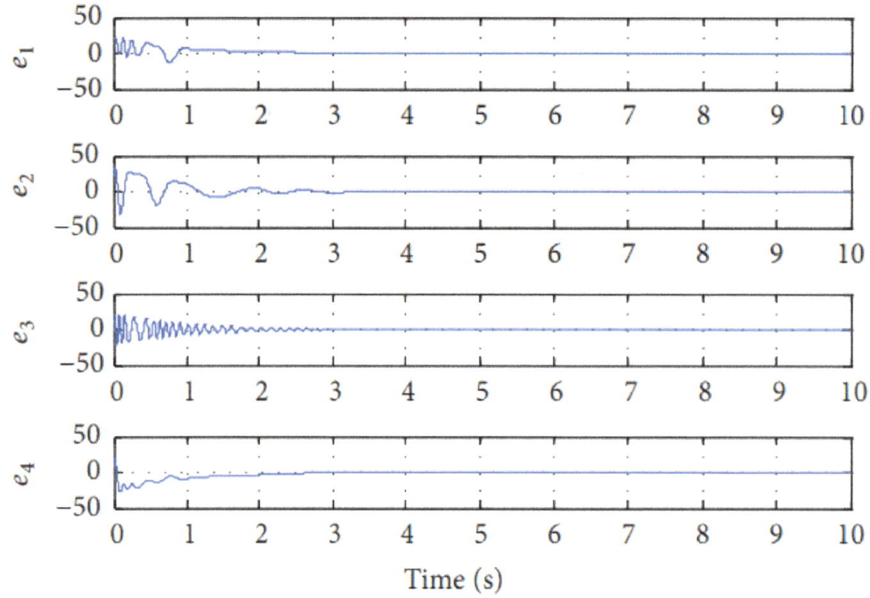

Figure 6: Time histories of error states for Case II.

Substituting (26) and (27) into (25), we obtain

$$\dot{V}\left(e_1, e_2, e_3, e_4, \tilde{a}, \tilde{b}, \tilde{c}, \tilde{d}\right)$$
$$= -e_1^{\ 2} - e_2^{\ 2} - e_3^{\ 2} - e_4^{\ 2} - \tilde{a}^2 - \tilde{b}^2 - \tilde{c}^2 - \tilde{d}^2 < 0. \tag{28}$$

Since the Lyapunov function $V(e_1, e_2, e_3, e_4, \tilde{a}, \tilde{b}, \tilde{c}, \tilde{d})$positive definite and its derivative $\dot{V}(e_1, e_2, e_3, e_4, \tilde{a}, \tilde{b}, \tilde{c}, \tilde{d})$ is negative definite in the neighborhood of the zero solutions for (12a) and (12b), according to the Lyapunov stability theory, the zero solutions of error states dynamic and parameters error vector are asymptotically stable; namely, the slave system equation (11) can asymptotically converge to its master system equation (10) with the adaptive control law equation (27) and the estimation parameter update law equation (26). The generalized adaptive synchronization concept proof had to be completed. The numerical simulation results are shown in Figures 5, 6, and 7.

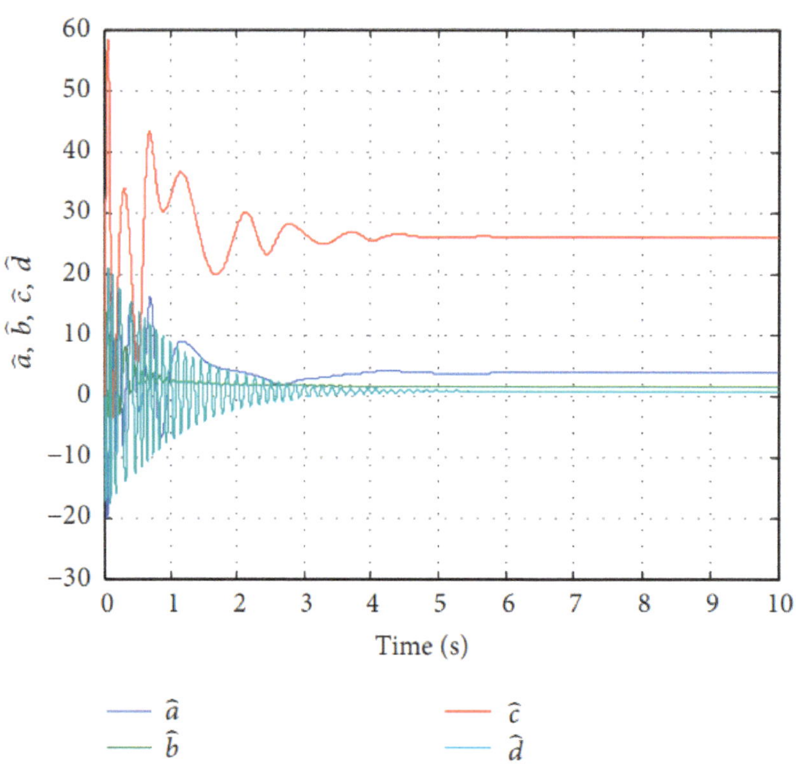

Figure 7: Time histories of estimated parameters for Case II.

Case III Generalized Adaptive Synchronization with Uncertain Parameters

Consider that the master system is the new Lorenz-Stenflo system with uncertain chaotic parameters

$$\begin{aligned}\dot{x}_1 &= A_1(t)(x_2 - x_1),\\ \dot{x}_2 &= A_3(t)x_1 - x_1x_3 - x_2,\\ \dot{x}_3 &= A_2(t)x_3 - A_4(t)x_3 + x_1x_2,\\ \dot{x}_4 &= -A_1(t)x_4 - x_1,\end{aligned} \tag{29}$$

where $A_1(t)$, $A_2(t)$, $A_3(t)$, and $A_4(t)$ are uncertain chaotic parameters. The uncertain parameters are given as

$$\begin{aligned}A_1(t) &= a(1 + f_1z_1),\\ A_2(t) &= b(1 + f_2z_2),\\ A_3(t) &= c(1 + f_3z_3),\\ A_4(t) &= d(1 + f_4z_4),\end{aligned} \tag{30}$$

where f_1, f_2, f_3, and f_4 are arbitrary positive constants. Positive constants are $f_1 = f_2 = f_3 = f_4 = 0.005$. The chaotic signals z_1, z_2, z_3, and z_4 are given as the states of system as follows:

$$\begin{aligned}\dot{z}_1 &= a(z_2 - z_1),\\ \dot{z}_2 &= cz_1 - z_1z_3 - z_2,\\ \dot{z}_3 &= bz_4 + -dz_3 + z_1z_2,\\ \dot{z}_4 &= -az_4 - z_1.\end{aligned} \tag{31}$$

The initial constants of the chaotic signals are $z_1(0) = 0.7$, $z_2(0) = 0.7$, $z_3(0) = 0.7$, and $z_4(0) = 0.7$. The new LorenzStenflo system with uncertain chaotic parameters of master system will exhibit a more complex dynamic behavior since the parameters of the system change over time. The generalized synchronization error functions can be defined as

$$e_i = x_i + w_i - y_i, \quad (i = 1, 2, 3, 4). \tag{32}$$

From the error functions equation (32), the error dynamics becomes

$$\begin{aligned}\dot{e}_1 &= \widehat{A}_1(e_2 - e_1 - w_2 + w_1) + \widetilde{A}_1(x_2 - x_1)\\ &\quad + a(w_2 - w_1) - u_1,\\ \dot{e}_2 &= \widehat{A}_3(e_1 - w_1) + \widetilde{A}_3x_1 + cw_1 + x_1(w_3 - e_3)\\ &\quad + w_1(x_3 - e_3) - e_1(x_3 + w_3 - e_3) - e_2 - u_2,\\ \dot{e}_3 &= \widehat{A}_2(e_4 - w_4) + \widetilde{A}_2x_4 + bw_4 - \widehat{A}_4(e_3 - w_3) - \widetilde{A}_4x_3\\ &\quad - dw_3 - x_1(w_2 - e_2) - w_1(x_2 - e_2)\\ &\quad + e_1(x_2 + w_2 - e_2) - u_3,\\ \dot{e}_4 &= -e_1 - \widehat{A}_1(e_4 - w_4) - \widetilde{A}_1x_4 - aw_4 - u_4,\end{aligned} \tag{33}$$

Where $\widetilde{A}_i(t) = A_i(t) - \widehat{A}_i(t)$ $(i = 1, 2, 3, 4)$.

Choose a Lyapunov function in the form of a positive definite function:

$$\begin{aligned}&V(e_1, e_2, e_3, e_4, \widetilde{A}_1, \widetilde{A}_2, \widetilde{A}_3, \widetilde{A}_4)\\ &= \frac{1}{2}\left(e_1^2 + e_2^2 + e_3^2 + e_4^2 + \widetilde{A}_1^2(t) + \widetilde{A}_2^2(t)\right.\\ &\quad \left.+ \widetilde{A}_3^2(t) + \widetilde{A}_4^2(t)\right),\end{aligned} \tag{34}$$

and its time derivative is

$$\begin{aligned}&\dot{V}(e_1, e_2, e_3, e_4, \widetilde{A}_1, \widetilde{A}_2, \widetilde{A}_3, \widetilde{A}_4)\\ &= e_1\dot{e}_1 + e_2\dot{e}_2 + e_3\dot{e}_3 + e_4\dot{e}_4 + \widetilde{A}_1\dot{\widetilde{A}}_1 + \widetilde{A}_2\dot{\widetilde{A}}_2\\ &\quad + \widetilde{A}_3\dot{\widetilde{A}}_3 + \widetilde{A}_4\dot{\widetilde{A}}_4\\ &= e_1\left(\widehat{A}_1(e_2 - e_1 - w_2 + w_1) + \widetilde{A}_1(x_2 - x_1)\right.\\ &\quad \left.+ a(w_2 - w_1) - u_1\right)\\ &\quad + e_2\left(\widehat{A}_3(e_1 - w_1) + \widetilde{A}_3x_1 + cw_1 + x_1(w_3 - e_3)\right.\\ &\quad \left.+ w_1(x_3 - e_3) - e_1(x_3 + w_3 - e_3) - e_2 - u_2\right)\\ &\quad + e_3\left(\widehat{A}_2(e_4 - w_4) + \widetilde{A}_2x_4 + bw_4 - \widehat{A}_4(e_3 - w_3)\right.\\ &\quad - \widetilde{A}_4x_3 - dw_3 - x_1(w_2 - e_2) - w_1(x_2 - e_2)\\ &\quad \left.+ e_1(x_2 + w_2 - e_2) - u_3\right)\\ &\quad + e_4\left(-e_1 - \widehat{A}_1(e_4 - w_4) - \widetilde{A}_1x_4 - aw_4 - u_4\right)\\ &\quad + \widetilde{A}_1\left(-\dot{\widehat{A}}_1\right) + \widetilde{A}_2\left(-\dot{\widehat{A}}_2\right) + \widetilde{A}_3\left(-\dot{\widehat{A}}_3\right) + \widetilde{A}_4\left(-\dot{\widehat{A}}_4\right).\end{aligned} \tag{35}$$

Choose the parameters estimation update laws for those uncertain parameters as follows:

$$\begin{aligned}\dot{\widetilde{A}}_1 &= \dot{A}_1 - \dot{\widehat{A}}_1 = -e_1(x_2 - x_1) + e_4x_4 - \widetilde{A}_1,\\ \dot{\widetilde{A}}_2 &= \dot{A}_2 - \dot{\widehat{A}}_2 = -e_3x_4 - \widetilde{A}_2,\\ \dot{\widetilde{A}}_3 &= \dot{A}_3 - \dot{\widehat{A}}_3 = -e_2x_1 - \widetilde{A}_3,\\ \dot{\widetilde{A}}_4 &= \dot{A}_4 - \dot{\widehat{A}}_4 = e_3x_3 - \widetilde{A}_4.\end{aligned} \tag{36}$$

The initial values of estimates for uncertain parameters are $\widehat{A}_1(0) = 0, \widehat{A}_2(0) = 0, \widehat{A}_3(0) = 0$, and $\widehat{A}_4(0) = 0$. Through (35) and (36), the appropriate controllers can be designed as

$$
\begin{aligned}
u_1 &= \widehat{A}_1 \left(e_2 - e_1 - w_2 + w_1\right) + a\left(w_2 - w_1\right) + e_1, \\
u_2 &= \widehat{A}_3 \left(e_1 - w_1\right) + cw_1 + x_1\left(w_3 - e_3\right) + w_1\left(x_3 - e_3\right) \\
&\quad - e_1\left(x_3 + w_3 - e_3\right) - e_2 + e_2, \\
u_3 &= \widehat{A}_2 \left(e_4 - w_4\right) + bw_4 - \widehat{A}_4 \left(e_3 - w_3\right) - dw_3 \\
&\quad - x_1\left(w_2 - e_2\right) - w_1\left(x_2 - w_2\right) \\
&\quad + e_1\left(x_2 + w_2 - e_2\right) + e_3, \\
u_4 &= -\widehat{A}_1 \left(e_4 - w_4\right) - aw_4 - e_1 + e_4.
\end{aligned} \tag{37}
$$

Substituting (36) and (37) into (35), we obtain

$$
\begin{aligned}
&\dot{V}\left(e_1, e_2, e_3, e_4, \widetilde{A}_1, \widetilde{A}_2, \widetilde{A}_3, \widetilde{A}_4\right) \\
&\quad = -e_1^2 - e_2^2 - e_3^2 - e_4^2 - \widetilde{A}_1^2(t) - \widetilde{A}_2^2(t) - \widetilde{A}_3^2(t) - \widetilde{A}_4^2(t) \\
&\quad < 0.
\end{aligned} \tag{38}
$$

Since the Lyapunov function $V(e_1, e_2, e_3, e_4, \tilde{a}, \tilde{b}, \tilde{c}, \tilde{d})$ is positive definite and its derivative $\dot{V}(e_1, e_2, e_3, e_4, \tilde{a}, \tilde{b}, \tilde{c}, \tilde{d})$ is negative definite in the neighborhood of the zero solutions for (12a) and (12b), according to the Lyapunov stability theory, the zero solutions of error states dynamic and parameters error vector are asymptotically stable; namely, the slave system equation (11) can asymptotically converge to its master system equation (10) with the adaptive control law equation (36) and the estimation parameter update law equation (37). The generalized adaptive synchronization with uncertain parameters concept proof had to be completed.The numerical simulation results are shown in Figures 8, 9,10, and 11.

CONCLUSION

A generalized adaptive synchronization with uncertain chaotic parameters is new chaos synchronization concept.

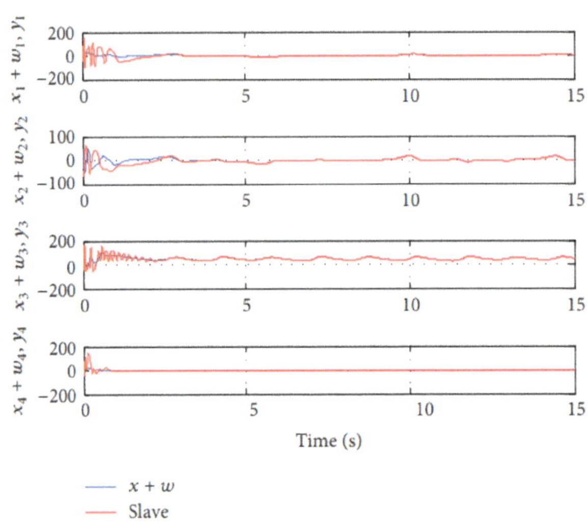

Figure 8: Time histories of the generalized signal (master x system plus given function w) and slave system for Case III.

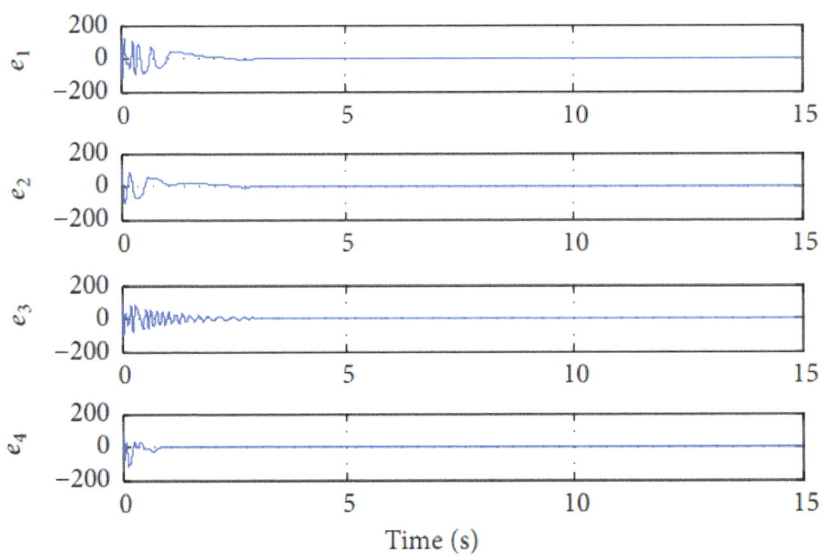

Figure 9: Time histories of error states for Case III.

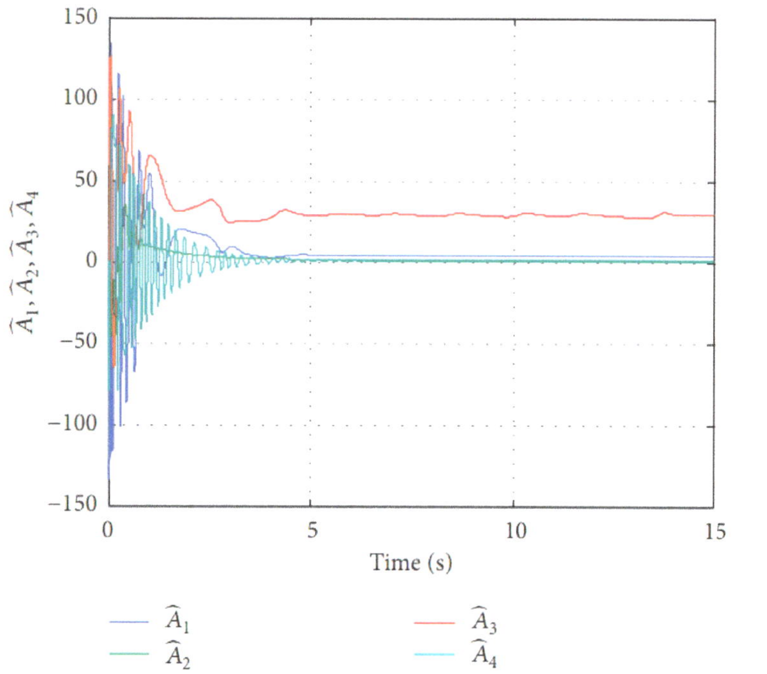

Figure 10: Time histories of estimated parameters for Case III.

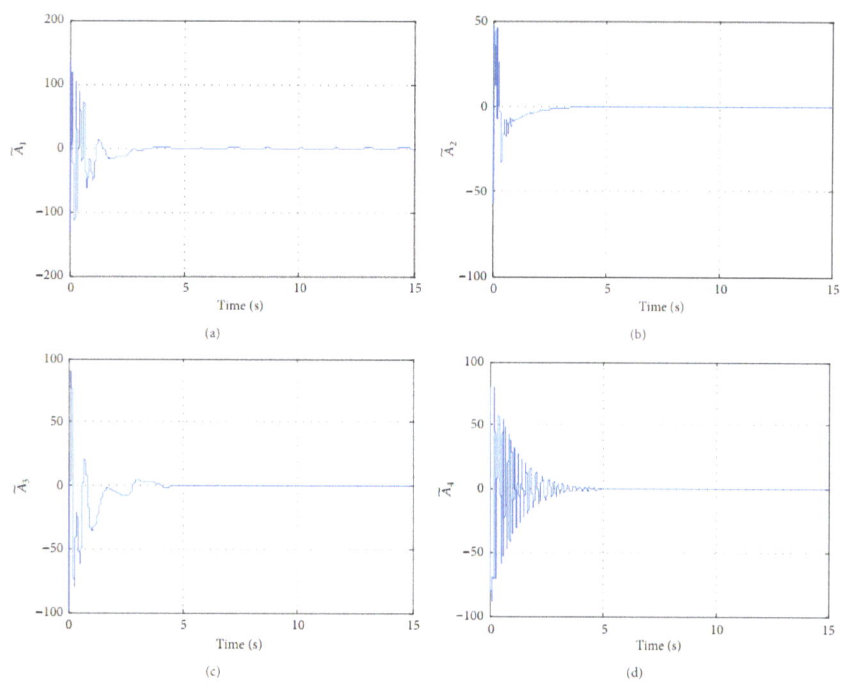

Figure 11: Time histories of different parameters for Case III.

The theoretical analysis and numerical simulation results of three cases, adaptive synchronization, generalized adaptive synchronization, and generalized adaptive synchronization with uncertain parameters, are shown in the corresponding figures which imply that the adaptive controllers and update laws we designed are feasible and effective. In this paper, the three examples can be used to increase the security of secret communication system.

Conflict of Interests

The authors declare that there is no conflict of interests regarding the publication of this paper.

ACKNOWLEDGMENT

This research was supported by the National Science Council, China, under Grant no. NSC 102-2221-E-011-034.

REFERENCES

1. L. M. Pecora and T. L. Carroll, "Synchronization in chaotic systems," Physical Review Letters, vol. 64, no. 8, pp. 821–824, 1990.
2. J. R. Terry and G. D. Vanwiggeren, "Chaotic communication using generalized synchronization," Chaos, Solitons and Fractals, vol. 12, no. 1, pp. 145–152, 2001.

3. W. Guo Xu and Q. Shu Li, “Chemical chaotic schemes derived from NSG system,” Chaos, Solitons and Fractals, vol. 15, no. 4, pp. 663–671, 2003.
4. Z.-M. Ge and C.-H. Yang, “Symplectic synchronization of different chaotic systems,” Chaos, Solitons and Fractals, vol. 40, no. 5, pp. 2532–2543, 2009.
5. Z.-M. Ge and C.-H. Yang, “The generalized synchronization of a Quantum-CNN chaotic oscillator with different order systems,” Chaos, Solitons and Fractals, vol. 35, no. 5, pp. 980–990, 2008.
6. Z.-M. Ge and C.-H. Yang, “Synchronization of complex chaotic systems in series expansion form,” Chaos, Solitons & Fractals, vol. 34, no. 5, pp. 1649–1658, 2007.
7. U. E. Vincent, “Synchronization of identical and non-identical 4-D chaotic systems using active control,” Chaos, Solitons and Fractals, vol. 37, no. 4, pp. 1065–1075, 2008.
8. L. Stenflo, “Generalized Lorenz equations for acoustic-gravity waves in the atmosphere,” Physica Scripta, vol. 53, no. 1, pp. 83– 84, 1996.
9. C.-H. Yang, “Enhanced symplectic synchronization between two different complex chaotic systems with uncertain parameters,” Abstract and Applied Analysis, vol. 2013, Article ID 193138, 12 pages, 2013.
10. S.-Y. Li, C.-H. Yang, and C.-T. Lin, “Chaotic motions in the real fuzzy electronic circuits,”Abstract and Applied Analysis, vol. 2013, Article ID 875965, 8 pages, 2013.
11. C.-H. Yang, “Symplectic synchronization of Lorenz-Stenflo system with uncertain chaotic parameters via adaptive control,” Abstract and Applied Analysis, vol. 2013, Article ID 528325, 14 pages, 2013.
12. S.-Y. Li, C.-H. Yang, L.-W. Ko, C.-T. Lin, and Z.-M. Ge, “Implementation on electronic circuits and RTR pragmatical adaptive synchronization: time-reversed uncertain dynamical systems’ analysis and applications,” Abstract and Applied Analysis, vol. 2013, Article ID 909721, 10 pages, 2013.
13. C.-H. Yang, “Chaos hybrid generalized synchronization of liuchen system by GYC partial region stability theory,” Journal of Computational and Theoretical Nanoscience, vol. 10, no. 4, pp. 825–831, 2013.
14. C. H. Yang, P. C. Tsen, S. Y. Li, and Z. M. Ge, “Pragmatical adaptive synchronization by variable strength linear coupling,” Journal of Computational and Theoretical Nanoscience, vol. 10, no. 4, pp. 1007–1013, 2013.
15. C.-H. Yang, “Chaos control of the Quantum-CNN systems,” Journal of Computational and Theoretical Nanoscience, vol. 10, no. 1, pp. 171–176, 2013.
16. Z. Liu, “The first integrals of nonlinear acoustic gravity wave equations,” Physica Scripta, vol. 61, no. 5, article 526, 2000.
17. C. Zhou, C. H. Lai, and M. Y. Yu, “Bifurcation behavior of the generalized Lorenz equations at large rotation numbers,” Journal of Mathematical Physics, vol. 38, no. 10, pp. 5225–5239, 1997.
18. S. Banerjee, P. Saha, and A. R. Chowdhury, “Chaotic scenario in the Stenflo equations,” Physica Scripta, vol. 63, no. 3, pp. 177–180, 2001.
19. Y. Chen, X. Wu, and Z. Gui, “Global synchronization criteria for two Lorenz-Stenflo systems via single-variable substitution control,” Nonlinear Dynamics, vol. 62, no. 1-2, pp. 361–369, 2010.
20. L. Yang and J. Jiang, “Adaptive synchronization of driveresponse fractional-order complex dynamical networks with uncertain parameters,” Communications in Nonlinear Science and Numerical Simulation, vol. 19, no. 5, pp. 1496–1506, 2014.
21. D. Zhu, C. Liu, and B. Yan, “Modeling and adaptive pinning synchronization control for a chaotic-motion motor in complex network,” Physics Letters A, vol. 378, no. 5-6, pp. 514–518, 2014.
22. W. Xiao-Qun and L. Jun-An, “Parameter identification and backstepping control of uncertain Lu system,” ¨ Chaos, Solitons and Fractals, vol. 18, no. 4, pp. 721–729, 2003.
23. Y. Yu and S. Zhang, “Controlling uncertain Lu system using ¨ backstepping design,” Chaos, Solitons and Fractals, vol. 15, no. 5, pp. 897–902, 2003.
24. J. A. Laoye, U. E. Vincent, and S. O. Kareem, “Chaos control of 4D chaotic systems using recursive backstepping nonlinear controller,” Chaos, Solitons and Fractals, vol. 39, no. 1, pp. 356– 362, 2009.
25. M. T. Yassen, “Chaos control of chaotic dynamical systems using backstepping design,” Chaos, Solitons & Fractals, vol. 27, no. 2, pp. 537–548, 2006.
26. W. M. Bessa, A. S. de Paula, and M. A. Savi, “Chaos control using an adaptive fuzzy sliding mode controller with application to a nonlinear pendulum,” Chaos, Solitons and Fractals, vol. 42, no. 2, pp. 784–791, 2009.
27. J.-F. Chang, M.-L. Hung, Y.-S. Yang, T.-L. Liao, and J.-J. Yan, “Controlling chaos of the family of Rossler systems using sliding ¨ mode control,” Chaos, Solitons & Fractals, vol. 37, no. 2, pp. 609– 622, 2008.
28. S. Zhankui and K. Sun, “Nonlinear and chaos control of a microelectro-mechanical system by using second-order fast terminal sliding mode control,” Communications in Nonlinear Science and Numerical Simulation, vol. 18, no. 9, pp. 2540–2548, 2013.
29. D. Chen, W. Zhao, J. C. Sprott, and X. Ma, “Application of Takagi-Sugeno fuzzy model to a class of chaotic synchronization and anti-synchronization,” Nonlinear Dynamics, vol. 73, no. 3, pp. 1495–1505, 2013.
30. Y. Nian and Y. Zheng, “Controlling discrete time T-S fuzzy chaotic systems via adaptive adjustment,” Physics Procedia, vol. 24, pp. 1915–1921, 2012.
31. H. Hua, Y. Liu, J. Lu, and J. Zhu, “A new impulsive synchronization criterion for T-S fuzzy model and its applications,” Applied Mathematical Modelling, vol. 37, pp. 8826–8835, 2013.
32. S. Yanchuka and T. Kapitaniak, “Chaos—hyperchaos transition in coupled Rossler systems,” ¨ Physics Letters A, vol. 290, pp. 139– 144, 2001.

33. P. Perlikowski, A. Stefanski, and T. Kapitaniak, "Ragged synchronizability and clustering in a network of coupled oscillators," in Advances in Nonlinear Dynamics and Synchronization, pp. 49–75, 2009.

34. M. Chadlia, A. Abdob, and S. X. Dingb, "H−/H∞ fault detection filter design for discrete-time Takagi–Sugeno fuzzy system," Automatica, vol. 49, no. 7, pp. 1996–2005, 2013.

35. M. Chadli and H. R. Karimi, "Robust observer design for unknown inputs takagi-sugeno models," IEEE Transactions on Fuzzy Systems, vol. 21, no. 1, pp. 158–164, 2013.

36. S. Aouaouda, M. Chadli, V. Cocquempot, and M. Tarek Khadir, "Multi-objective $H-H\infty$ fault detection observer design for Takagi-Sugeno fuzzy systems with unmeasurable premise variables: descriptor approach," International Journal of Adaptive Control and Signal Processing, vol. 27, no. 12, pp. 1031–1047, 2013.

37. M. Chadlia, I. Zelinkab, and T. Youssef, "Unknown inputs observer design for fuzzy systems with application to chaotic system reconstruction," Computers & Mathematics with Applications, vol. 66, no. 2, pp. 147–154, 2013.

38. I. Zelinka, M. Chadli, D. Davendra, R. Senkerik, and R. Jasek, "An investigation on evolutionary reconstruction of continuous chaotic systems," Mathematical and Computer Modelling, vol. 57, no. 1-2, pp. 2–15, 2013.

39. C.-H. Yang and C.-L. Wu, "Nonlinear dynamic analysis and synchronization of four-dimensional Lorenz-Stenflo system and its circuit experimental implementation," Abstract and Applied Analysis, vol. 2014, Article ID 213694, 17 pages, 2014.

CHAPTER 18

NONLINEAR DYNAMIC ANALYSIS AND SYNCHRONIZATION OF FOUR-DIMENSIONAL LORENZ-STENFLO SYSTEM AND ITS CIRCUIT EXPERIMENTAL IMPLEMENTATION

Cheng-Hsiung Yang and Cheng-Lin Wu

INTRODUCTION

Chaotic systems provide a rich mechanism for signal design and generation, with potential applications to communications and signal processing. Because chaotic signals are typically broadband, noise-like, and difficult to predict, they can be used in various contexts for masking information bearing waveforms. They can also be used as modulating waveforms in spread spectrum systems.

This can be useful in many practical circumstances like securing communication channels [1], masking signals [2], spreading data sequence, or for generating random signals [3]. Of course, chaos is an unwanted phenomenon in many situations. The essential problem which is faced here is how to distinguish such motion from long transient behavior. A spectrum called Lyapunov exponents (LE) is often used as a quantifier [4] of chaotic motion or regular motion. These real numbers measure the average ratio of the exponential separation of the two neighborhood trajectories. For chaos it is necessary to have at least one positive of Lyapunov exponents; the second one represents a direction of the flow and must converge to zero. The last one must be negative with the largest absolute value, since the flow is dissipative. Circuits of nonlinear dynamic system provide an excellent tool for the study of chaotic behavior. Some of these circuits treat time as a discrete variable, employing sample-and-hold subcircuits and analog multipliers to model iterated maps such as the logistic map [5, 6].

In this paper, an adaptive generalized synchronization of the four-dimensional Lorenz-Stenflo system [7–12] with uncertain chaotic parameters strategy by GYC partial region stability theory is proposed [13–16]. By using the GYC partial region stability theory the Lyapunov function is a simple linear homogeneous function of states and the structure of controllers are simpler and have less simulation error because they are in lower order than that of traditional controllers design.

Citation: Cheng-Hsiung Yang and Cheng-Lin Wu, "Nonlinear Dynamic Analysis and Synchronization of Four-Dimensional Lorenz-Stenflo System and Its Circuit Experimental Implementation," Abstract and Applied Analysis, vol. 2014, Article ID 213694, 17 pages, 2014. doi:10.1155/2014/213694.

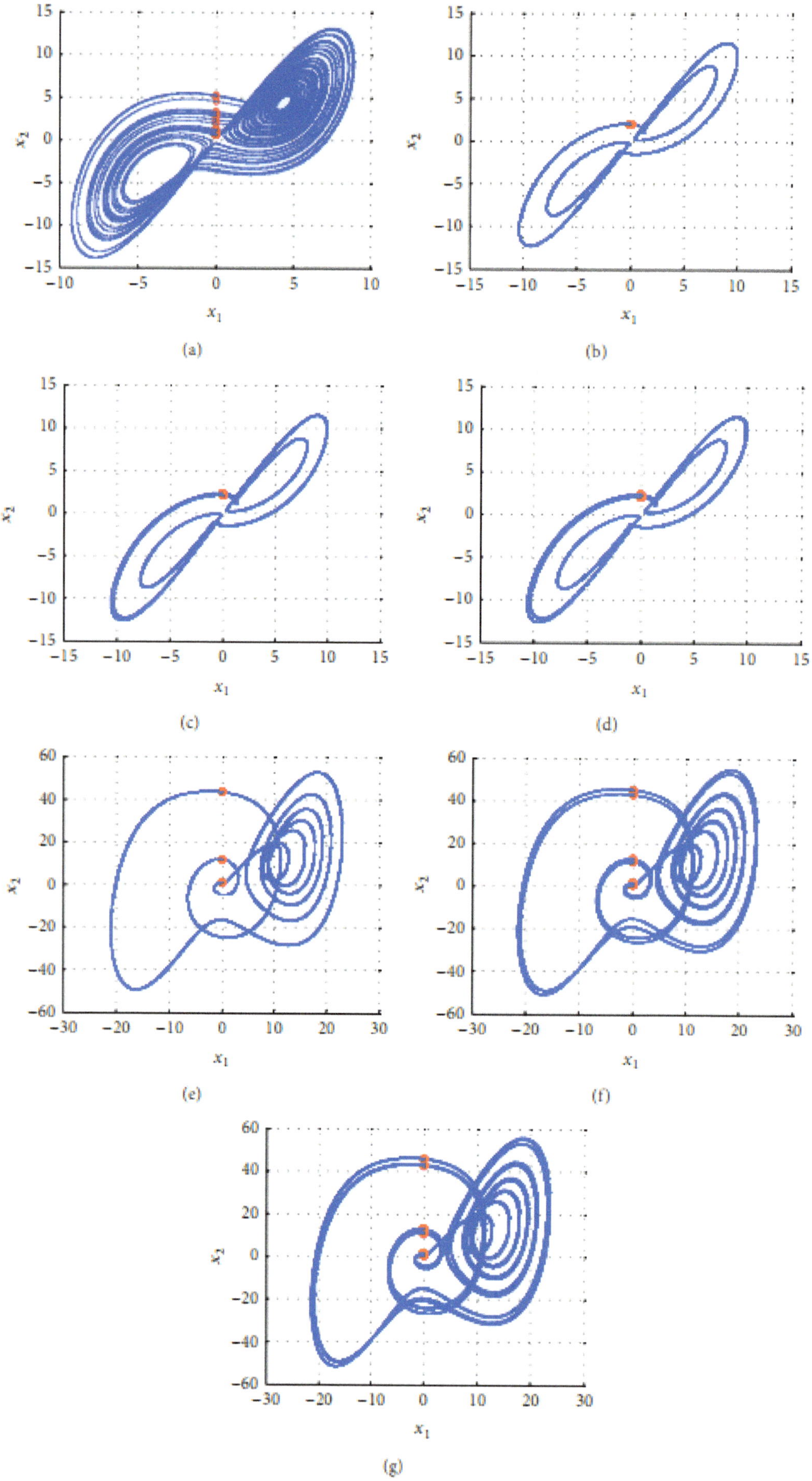

Figure 1: Phase portraits and Poincare maps for (a) chaotic, (b) period-1, (c) period-2, (d) period-4, (e) period-3, (f) period-6, and (g) ′ period-12, respectively, of the four-dimensional Lorenz-Stenflo system.

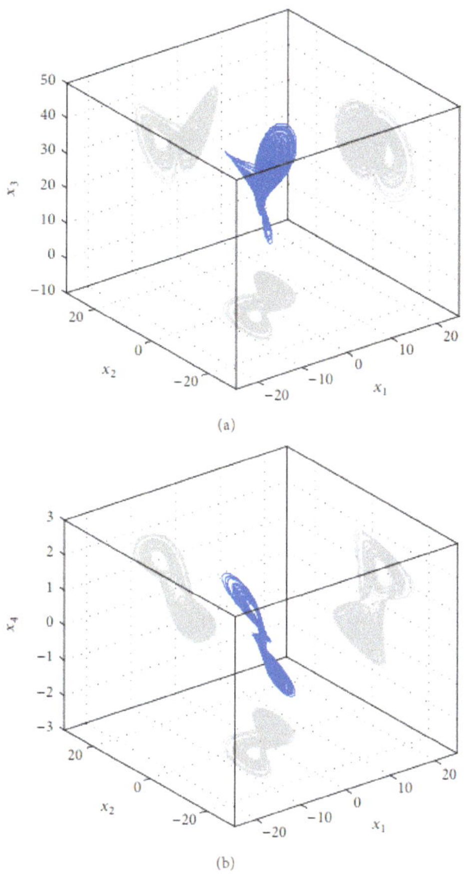

Figure 2: Three dimensions phase portrait of the four-dimensional Lorenz-Stenflo system and its projection.

This paper is organized as follows. In Section 2, the four-dimensional Lorenz-Stenflo system analyzes characteristics of nonlinear dynamic behavior, construction realization circuit of the four-dimensional Lorenz-Stenflo system. In Section 3, adaptive generalized chaos synchronization with uncertain parameters strategy by GCY partial region stability theory is proposed. In Section 4, numerical results of adaptive generalized synchronization of the four-dimensional Lorenz-Stenflo system with uncertain chaotic parameters using GYC partial region stability. Finally, some concluding remarks are given in Section 5.

CHAOTIC DYNAMIC ANALYSIS OF A FOUR-DIMENSIONAL LORENZ-STENFLO SYSTEM

Phase Portraits and Poincaré Map

The Lorenz-Stenflo system is described as [7, 8]

$$\begin{aligned} \dot{x}_1 &= ax_2 - ax_1 + bx_4 \\ \dot{x}_2 &= cx_1 - x_1x_3 - x_2 \\ \dot{x}_3 &= x_1x_2 - dx_3 \\ \dot{x}_4 &= -x_1 - ax_4, \end{aligned} \tag{4}$$

where a is Prandtl number, b is rotation number, c is Rayleigh number, and d is geometric parameter.

The bx_4 of Lorenz-Stenflo system from first item changes to third item, new four-dimensional Lorenz-Stenflo system. Consider

$$\begin{aligned} \dot{x}_1 &= ax_2 - ax_1 \\ \dot{x}_2 &= cx_1 - x_1x_3 - x_2 \\ \dot{x}_3 &= x_1x_2 - dx_3 + bx_4 \\ \dot{x}_4 &= -x_1 - ax_4. \end{aligned} \tag{2}$$

The Lorenz-Stenflo system has many dynamical behaviors that are one chaotic motion and six different periodic motions as shown in Figure 1. To have a better understanding of the four-dimensional Lorenz-Stenflo system, we use an innovative technique to present the three-dimensional phase portraits shown in Figure 2. These figures plot three dimensions phase portraits and their projection simultaneously.

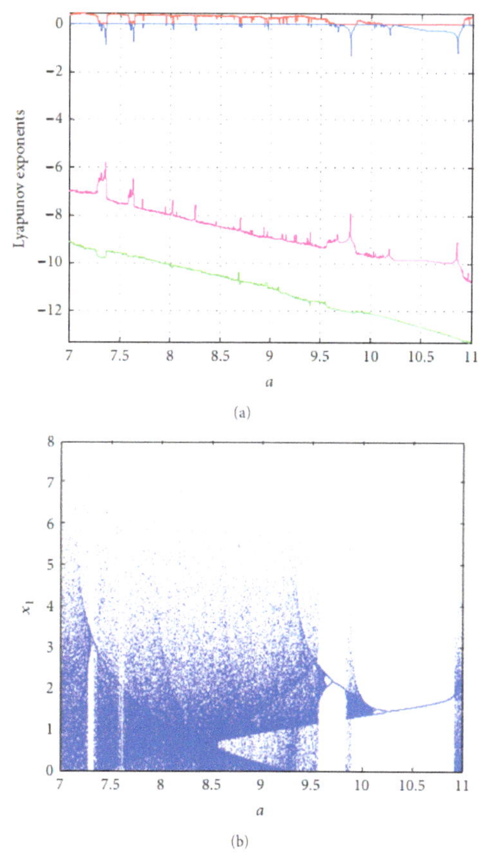

Figure 3: Lyapunov exponents and bifurcation diagrams of the four-dimensional Lorenz-Stenflo system with parameters $b = 1.5$, $c = 26$, and $d = 0.7$.

Thus, we can comprehend the four-dimensional LorenzStenflo system much easier.

Equilibrium Analysis

The equilibrium points of the system are shown as follows:

$$E_0\,(0,0,0,0)$$

$$E_1\left(\begin{array}{l}\dfrac{b-\sqrt{b^2-4a^2d+4a^2cd}}{2a},\dfrac{b-\sqrt{b^2-4a^2d+4a^2cd}}{2a},c-1,\\ \dfrac{-b+\sqrt{b^2-4a^2d+4a^2cd}}{2a^2}\end{array}\right)$$

$$E_2\left(\begin{array}{l}\dfrac{b+\sqrt{b^2-4a^2d+4a^2cd}}{2a},\dfrac{b+\sqrt{b^2-4a^2d+4a^2cd}}{2a},c-1,\\ \dfrac{b+\sqrt{b^2-4a^2d+4a^2cd}}{2a^2}\end{array}\right). \tag{3}$$

The parameters of four-dimensional Lorenz-Stenflo system are $a = 3.7$, $b = 1.5$, $c = 26$, and $d = 0.7$, so the equilibrium points become

$$E_0\,(0,0,0,0)$$
$$E_1\,(-3.98551,-3.98551,25,1.07716)$$
$$E_2\,(4.39091,4.39091,25,-1.18673)\,. \tag{4}$$

For the equilibrium point E_0, the Jacobian matrix of the system is

$$J_0 = \begin{bmatrix} -a & a & 0 & 0 \\ c & -1 & 0 & 0 \\ 0 & 0 & -d & b \\ -1 & 0 & 0 & -a \end{bmatrix} = \begin{bmatrix} -3.7 & 3.7 & 0 & 0 \\ 26 & -1 & 0 & 0 \\ 0 & 0 & -0.7 & 1.5 \\ -1 & 0 & 0 & -3.7 \end{bmatrix}. \tag{5}$$

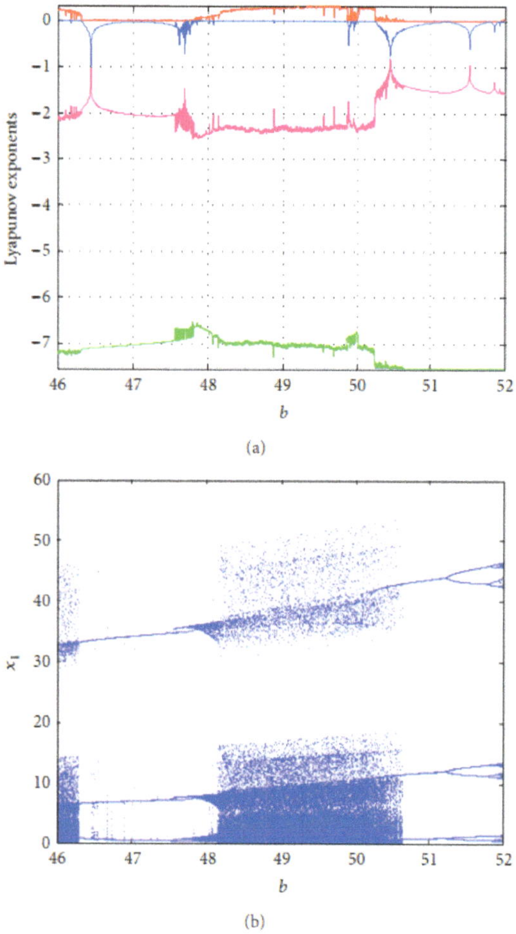

Figure 4: Lyapunov exponents and bifurcation diagrams of the four-dimensional Lorenz-Stenflo system with parameters $a = 3.7$, $c = 26$, and $d = 0.7$.

The eigenvalues of the Jacobian matrix below are obvious:

$$\lambda_1 = 3.7, \qquad \lambda_2 = -1,$$
$$\lambda_3 = -0.7, \qquad \lambda_4 = -3.7. \tag{7}$$

Then we linearize the system about the nonzero equilibrium points and obtaine the Jacobian matrix:

$$J_{nz} = \begin{bmatrix} -a & a & 0 & 0 \\ c - x_3 & -1 & -x_1 & 0 \\ x_2 & x_1 & -d & b \\ -1 & 0 & 0 & -a \end{bmatrix}. \tag{7}$$

The J_{nz} eigenvalues are

$$\det\left|J_{nz} - \lambda I\right| = \begin{bmatrix} -a-\lambda & a & 0 & 0 \\ c - x_3 & -1-\lambda & -x_1 & 0 \\ x_2 & x_1 & -d-\lambda & b \\ -1 & 0 & 0 & -a-\lambda \end{bmatrix}$$
$$= \lambda^4 + (2a + d + 1)\,\lambda^3$$
$$+ \left(a\,(2 + 2d + x_3 - c) + a^2 + d + x_1^2\right)\lambda^2$$
$$+ \left(a^2 + 2ad - a^2c + a^2d - acd\right.$$
$$\left. + 2ax_1^2 + a^2x_3 + abx_3 + ax_1x_2\right)\lambda$$
$$+ a^2\left(x_1^2 + x_1x_2 + dx_3\right) - abx_1$$
$$+ a^2d - a^2cd. \tag{8}$$

By letting $\det|J_{nz} - \lambda I| = 0$, the eigenvalues of the matrix at each equilibrium point can be obtained as shown in Table 1. The eigenvalues of each equilibrium point are the same. And all the equilibrium points are unstable, since at least one eigenvalue has positive real part for each equilibrium point.

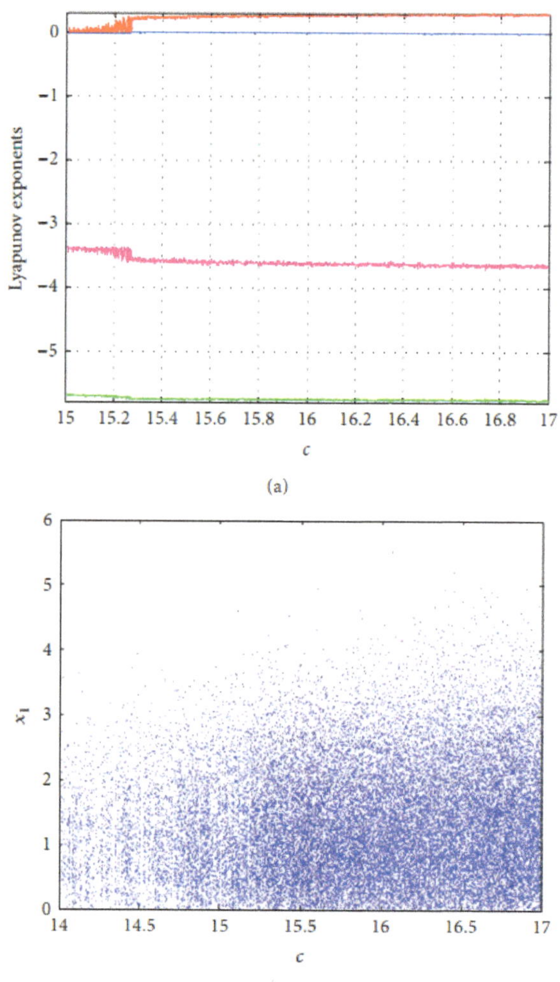

Figure 5: Lyapunov exponents and bifurcation diagrams of the four-dimensional Lorenz-Stenflo system with parameters $a = 3.7$, $b = 1.5$, and $d = 0.7$.

Table 1: Eigenvalues and stability of equilibrium points.

Equilibrium point	Eigenvalues	Stable/Unstable
E_0	$\lambda_1 = 3.7, \lambda_2 = -1, \lambda_3 = -0.7, \lambda_4 = -3.7$	Unstable
E_1	$\lambda_1 = 2.152 - 6.27i, \lambda_2 = -12.5777, \lambda_3 = 2.152 + 6.27i, \lambda_4 = -0.826$	Unstable
E_2	$\lambda_1 = 2.1489 + 6.50198i, \lambda_2 = -12.5416, \lambda_3 = 2.1489 - 6.50198i, \lambda_4 = -0.8561$	Unstable

Divergence Analysis

The divergence represents the volume density of the outward flux of a vector field. The divergence of a system implies whether this system dissipative or not. For the four-dimensional Lorenz-Stenflo system, its divergence is

$$\operatorname{div}\Phi = \nabla \cdot \Phi = \frac{\partial \Phi}{\partial x_1} + \frac{\partial \Phi}{\partial x_2} + \frac{\partial \Phi}{\partial x_3} + \frac{\partial \Phi}{\partial x_4} = -a - 1 - d - a, \tag{9}$$

Where $\Phi = (\dot{x}_1, \dot{x}_2, \dot{x}_3, \dot{x}_4)$.

The parameters are taken as $a = 3.7$, $b = 1.5$, $c = 26$, and $d = 0.7$; hence, the divergence of the system becomes

$$\operatorname{div}\Phi = 3.7 - 1 + 0.7 - 3.7 = -9.1. \tag{10}$$

So, (2) is dissipative, with an exponential contraction rate:

$$\frac{d\Phi}{dt} = e^{-(2a+d+1)}. \tag{11}$$

So, a volume element Π_0 is contracted by the flow into a volume element $\Phi_0 e^{-(2a+d+1)}$ in time. All orbits of

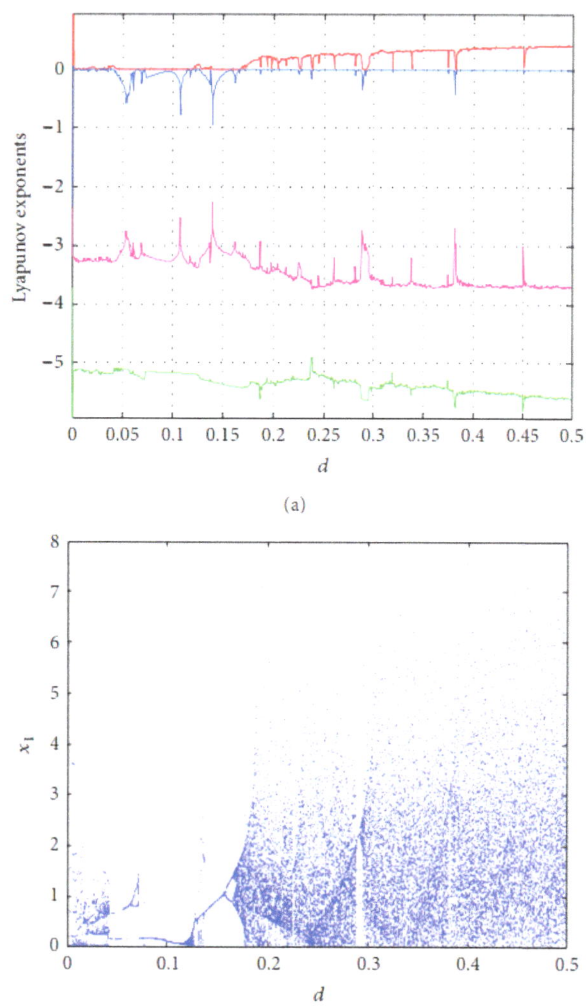

Figure 6: Lyapunov exponents and bifurcation diagrams of the four-dimensional Lorenz-Stenflo system with parameters $a = 3.7$, $b = 1.5$, and $c = 26$.

(2) are ultimately confined to a set having zero volume and asymptotic motion settles onto an attractor. The fourdimensional Lorenz-Stenflo system has bund system.

Bifurcation Diagram

For numerical analysis, all of the initial conditions are $x_1(0) = 20$, $x_2(0) = 20$, $x_3(0) = 20$, and $x_4(0) = 20$, in this and next sections. The bifurcation diagrams of the four-dimensional Lorenz-Stenflo system are plotted in Figure 3(a) to Figure 6(a). The parameters a, b, c, and d in (2) will be varied, respectively, for plotting the bifurcation diagrams. From the bifurcation diagrams, we can learn more about the behaviors of the four-dimensional Lorenz-Stenflo system.

Lyapunov Exponent and Lyapunov Dimension

The Lyapunov exponent may be used to measure the sensitive dependence on initial conditions. Once the Lyapunov exponent is greater than zero at specific value of parameters, it is said to be chaos, otherwise, periodic solution. The Lyapunov exponents diagrams of the four-dimensional Lorenz-Stenflo system are shown in Figure 3(b) to Figure 6(b). In order to verify the chaotic behavior of the system, Lyapunov exponents diagrams and bifurcation diagrams can be compared together.

The Lyapunov dimension has been proposed by Frederickson et al. [17] to measure the complexity of the attractor. The relation between Lyapunov dimension and Lyapunov exponent is

$$d_L = j + \frac{\sum_{i=1}^{j} \lambda_i}{\left|\lambda_{j+1}\right|}, \tag{12}$$

where λ_i is the largest Lyapunov exponent and j is defined by the condition as

$$\sum_{i=1}^{j} \lambda_i \geq 0, \quad \sum_{i=1}^{j+1} \lambda_i < 0. \tag{13}$$

The Lyapunov dimension diagrams are plotted in Figure 7.

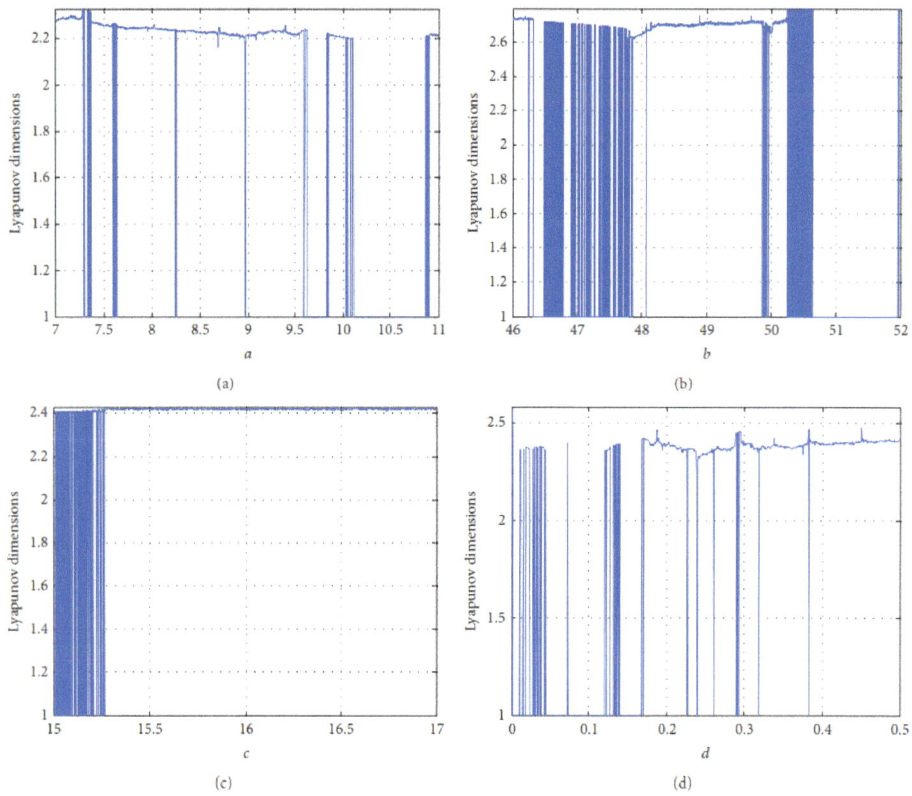

Figure 7: Lyapunov dimension diagrams of the four-dimensional Lorenz-Stenflo system.

Electronic Circuit Implementation

The four-dimensional Lorenz-Stenflo system can be modeled by an electronic circuit [18–21]. The main electronic components in the circuit are integrated amplifiers and inverting amplifiers. In this circuit simulation, we use electronic circuit simulation software named Multisim. All the parameters and initial conditions are the same as numerical simulation. But here we modify the value of capacitors to decrease the output voltage within a range from −12.0 to 12.0, and all the state values reduced tenfold in circuit simulation. The chaotic circuit configuration diagram is shown in Figure 8, and the governing integral equation of the circuit can be written as

$$\begin{aligned} \dot{x}_1 &= \frac{1}{R_1C_1}x_2 - \frac{1}{R_2C_1}x_1, \\ \dot{x}_2 &= \frac{1}{R_3C_2}x_1 - \frac{1}{R_4C_2}x_1x_3 - \frac{1}{R_5C_2}x_2, \\ \dot{x}_3 &= \frac{1}{R_7C_3}x_1x_2 + \frac{1}{R_6C_3}x_4 - \frac{1}{R_8C_3}x_3, \\ \dot{x}_4 &= -\frac{1}{R_9C_4}x_1 - \frac{1}{R_{10}C_4}x_4. \end{aligned} \tag{14}$$

Components of the chaotic system circuit are chosen to be $R_1 = R_2 = R_{10} = 270$ kΩ, $R_3 = 38.461$ kΩ, $R_6 = 667$ kΩ, $R_4 = R_5 = R_7 = R_9 = 1$ MΩ, $R_8 = 1429$ kΩ, $R_{11} = R_{12} = R_{13} = R_{14} = R_{15} = R_{16} = R_{17} = R_{18} = 100$ kΩ, and $C_1 = C_2 = C_3 = C_4 = 1\mu$F. The voltage output signals of the chaotic circuit simulated in Multisim software are shown in Figure 9. In addition, the real circuit of the four-dimensional Lorenz-Stenflo system was constructed on a breadboard shown in Figure 13. The apparatus and the components used in the circuit are listed in Table 2. The experimental setup is shown in Figure 11 and the experimental results are shown in Figure 12.

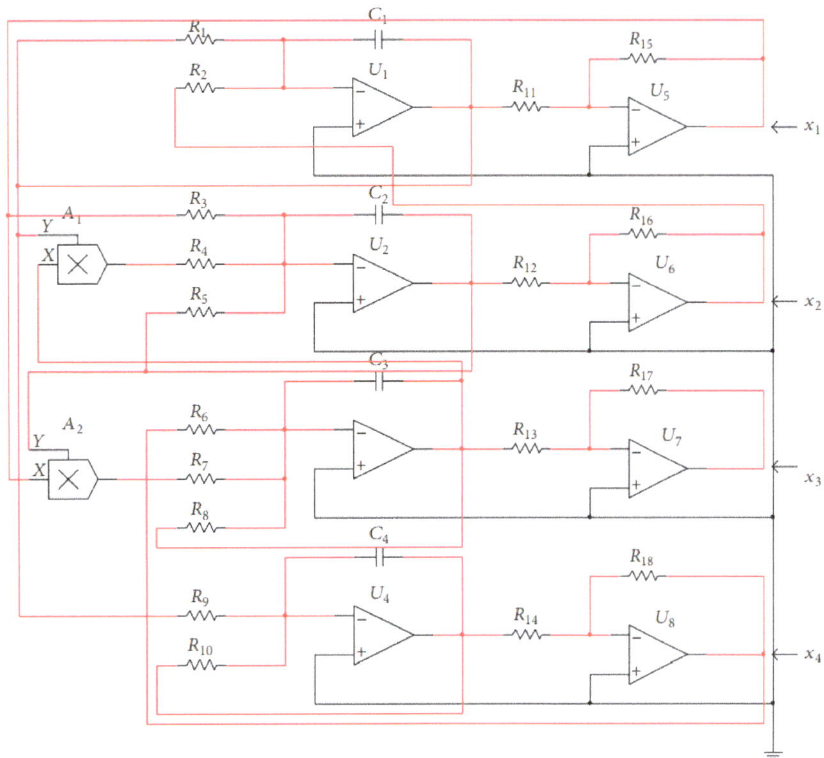

Figure 8: Chaotic circuit of the four-dimensional Lorenz-Stenflo system.

In order to have a better concept about chaotic motions and periodic motions, we present a four-dimensional diagram, which is a combination of the phase portraits of the circuits simulation and the bifurcation diagram of variable parameter a, plotted in Figure 10.

ADAPTIVE GENERALIZED SYNCHRONIZATION OF THE FOUR-DIMENSIONAL LORENZ-STENFLO SYSTEM WITH UNCERTAIN CHAOTIC PARAMETERS VIA GYC PARTIAL REGION STABILITY THEORY STRATEGY

Consider the master system

$$\dot{x} = f(t, x, A(t)), \quad (15)$$

and the slave system

$$\dot{y} = f\left(t, y, \widehat{A}(t)\right) + u, \quad (16)$$

where $x = [x_1, x_2,...,x_n]^T \in R^n$ and $y = [y_1, y_2,...,y_n]^T \in R^n$ denote the master state vector and slave state vector, respectively, f is nonlinear vector functions, $A(t)$is uncertain chaotic parameter in f, $\widehat{A}(t)$is estimates of uncertain chaotic parameter in f, and $u = [u_1, u_2,...,u_n]^T \in R^n$ is a control input vector.

Our goal is to design a controller $u(t)$ so that the state vector of the slave system (16) asymptotically approaches the state vector of the master system (15) plus a given vector function $F(t) = [F_1(t), F_2(t), \ldots, F_n(t)]^T$ which is either a regular or a chaotic function of time, and finally the synchronization will be accomplished in the sense that the limit of the error vector $e(t) = [e_1, e_2,...,e_n]^T$ approaches zero:

$$\lim_{t\to\infty} e(t) = 0, \quad (17)$$

where $e_i = x_i - y_i + F_i(t) + Ke_i$, $i = 1, 2, \ldots, n$. The Kei are positive constants which make the error dynamics always happens in the first quadrant.

From (17) we have

$$\dot{e}_i = \dot{x}_i - \dot{y}_i + \dot{F}_i(t), \quad i = 1, 2, \ldots, n. \quad (18)$$

Introducing (15) and (16) in (18), we get

$$\dot{e} = f(t, x, A(t)) - f\left(t, y, \widehat{A}(t)\right) + \dot{F}(t) - u(t). \quad (19)$$

A Lyapunov function $V(e, \widetilde{A})$ is chosen as a positive definite function in the first quadrant.

Consider

$$V\left(e, \widetilde{A}\right) = e + \widetilde{A}, \quad (20)$$

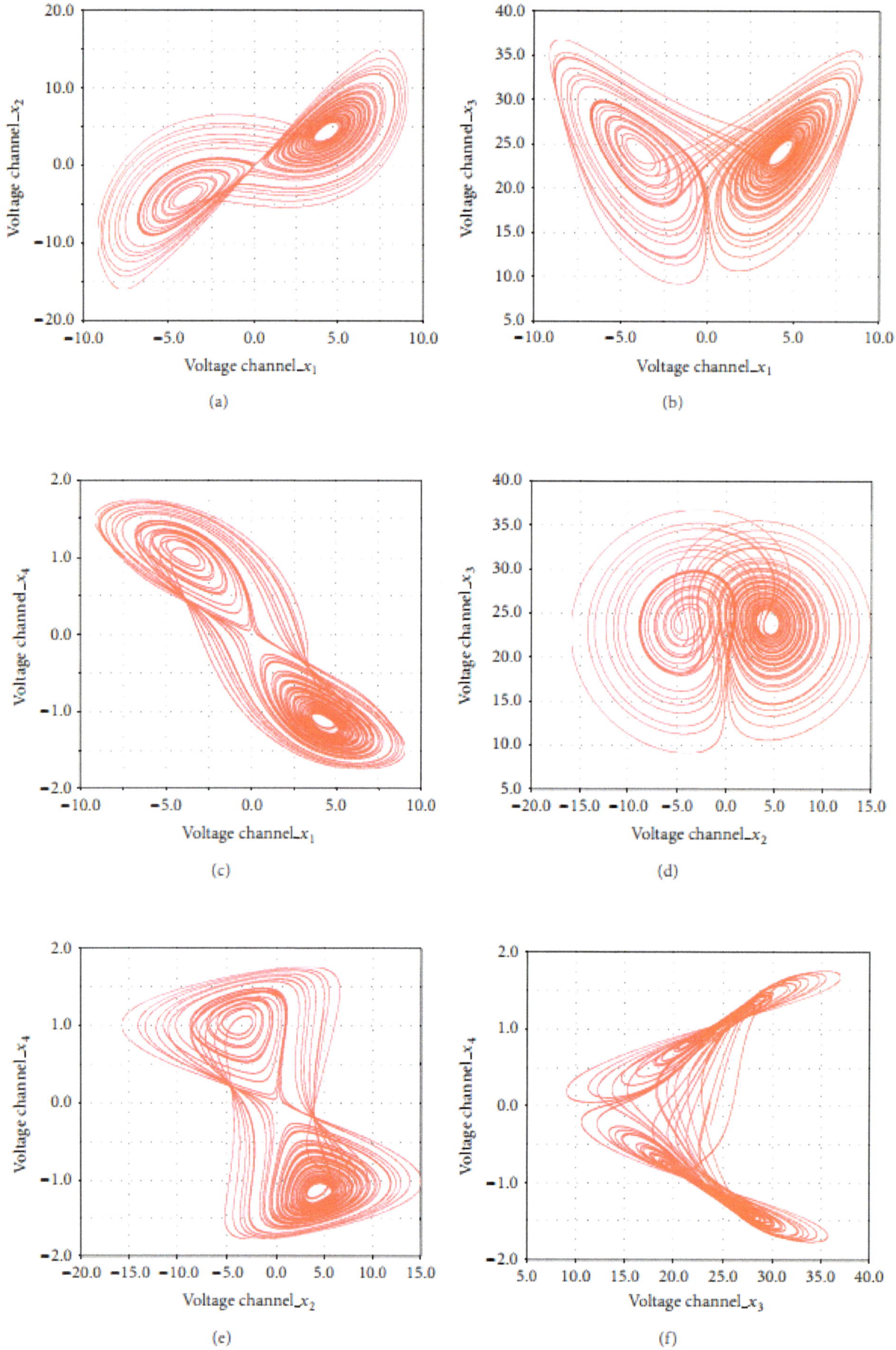

Figure 9: Oscilloscope graphics for voltage signals outputs of the four-dimensional Lorenz-Stenflo system: (a) V_{x_1}-V_{x_2}, (b) V_{x_1}-V_{x_3}, (c) V_{x_1}-V_{x_4}, (d) V_{x_2}-V_{x_3}, (e) V_{x_2}-V_{x_4}, and (f) V_{x_3}-V_{x_4}.

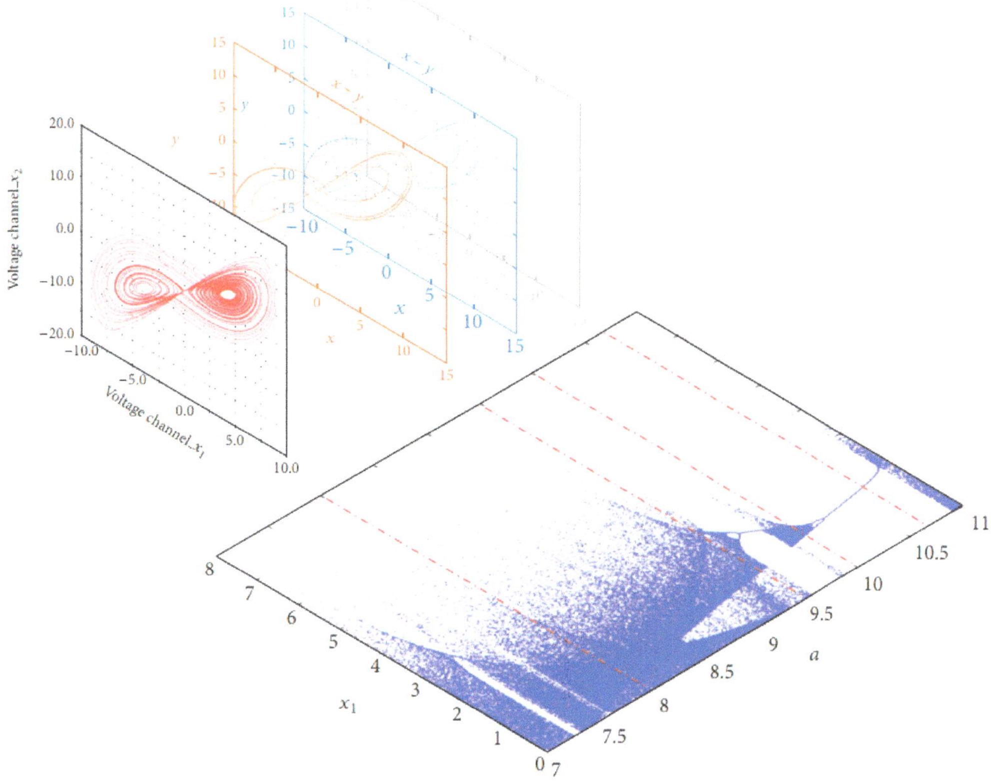

Figure 10: Bifurcation diagram with phase portraits of the four-dimensional Lorenz-Stenflo system.

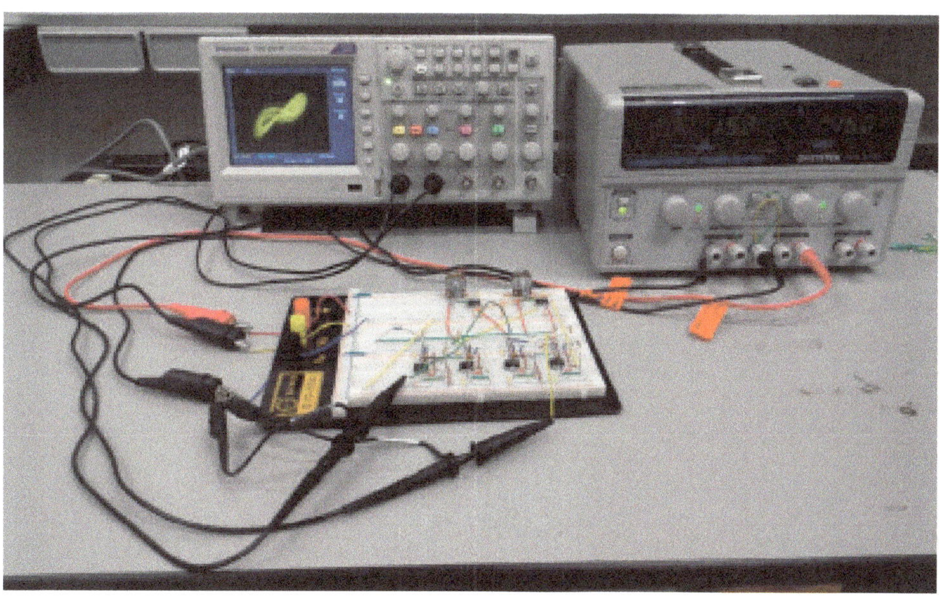

Figure 11: Experimental setup for realization of the fourdimensional Lorenz-Stenflo system.

where $\widetilde{A}_j = A_j(t) - \widehat{A}_j(t) + Kp_i, (j = 1, 2, \ldots, m)$. The Kpi are positive constants which make the update laws always happen in the first quadrant. Its derivative along the solution of (20) is

$$\dot{V}\left(e, \widetilde{A}\right) = f\left(t, x, A(t)\right) - f\left(t, y, \widehat{A}(t)\right) + \dot{F}(t) - u(t) + \dot{\widetilde{A}}, \tag{21}$$

where $u(t)$ and $\dot{\widehat{A}}$ are chosen so that $\dot{V} = Ce + D\widetilde{A}$, C and D are negative constants, and $\dot{V}$ is a negative definite function of e_1, e_2,..., e_n and $\tilde{A}_1$, $\tilde{A}_2$,..., $\tilde{A}_m$. When

$$\lim_{t\to\infty} e = 0, \qquad \lim_{t\to\infty} \widetilde{A} = 0, \tag{22}$$

the generalized synchronization is obtained.

By using the GYC partial region stability theory, the Lyapunov function is easier to find, since the terms of first degree can be used to construct the definite Lyapunov function and the controller can be designed in lower order.

NUMERICAL RESULTS OF ADAPTIVE GENERALIZED SYNCHRONIZATION OF THE FOUR-DIMENSIONAL LORENZ-STENFLO SYSTEM WITH UNCERTAIN CHAOTIC PARAMETERS VIA GYC PARTIAL REGION STABILITY THEORY

The master system is

$$\dot{x}_1 = A_1(t)(x_2 - x_1)$$

$$\dot{x}_2 = A_3(t)x_1 - x_1x_3 - x_2$$

$$\dot{x}_3 = A_2(t)x_3 - A_4(t)x_3 + x_1x_2$$

$$\dot{x}_4 = -A_1(t)x_4 - x_1, \tag{23}$$

where $A_1(t)$, $A_2(t)$, $A_3(t)$, and $A_4(t)$ are uncertain chaotic parameters. Equation (23) is called four-dimensional LorenzStenflo system with uncertain chaotic parameters.

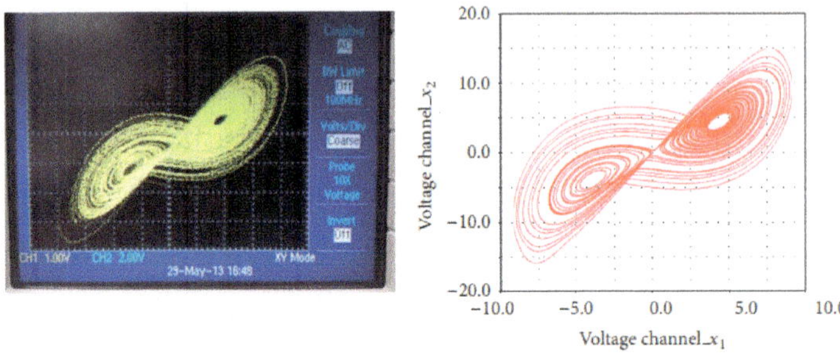

(a)

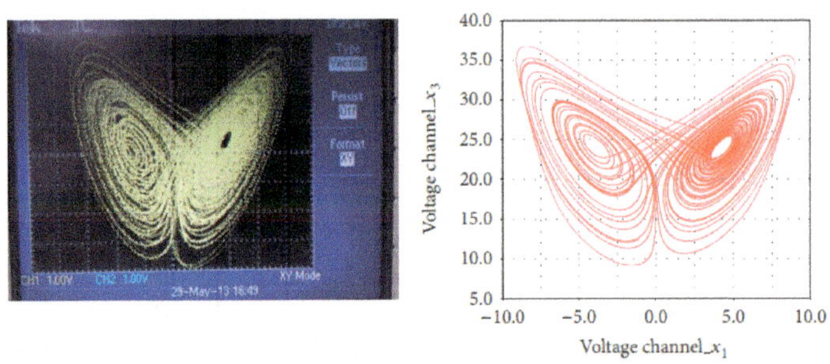

(b)

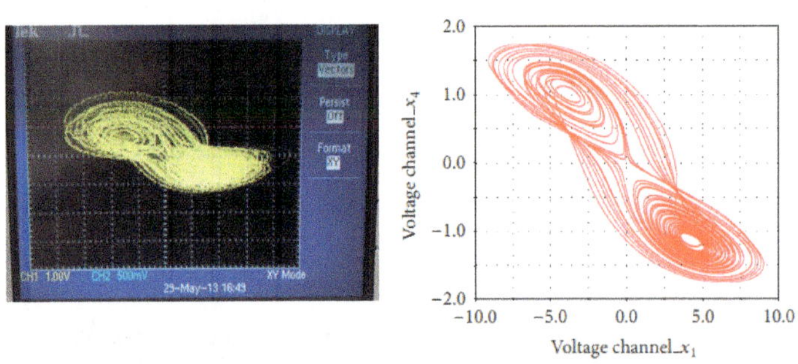

(c)

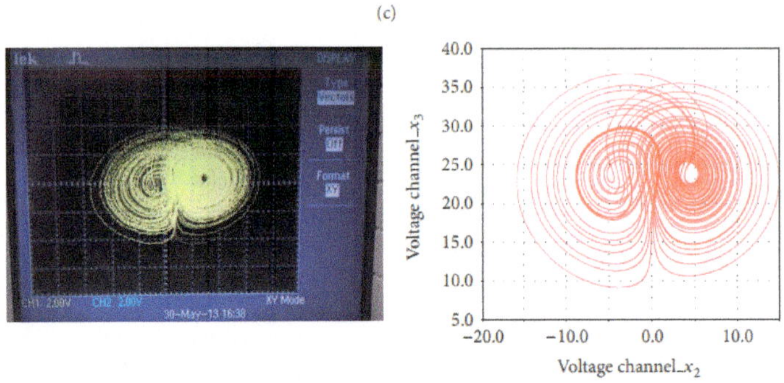

(d)

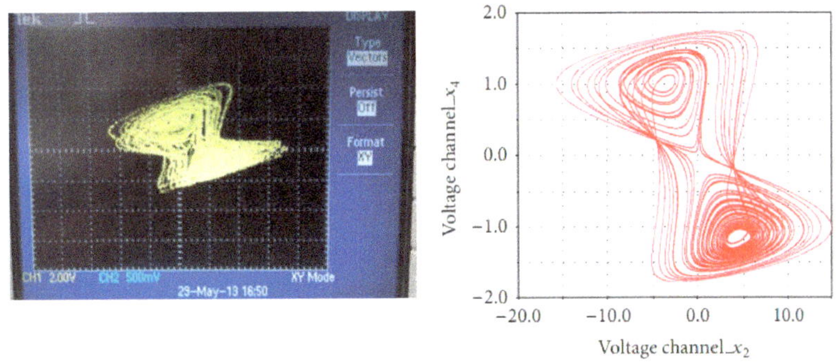

(e)

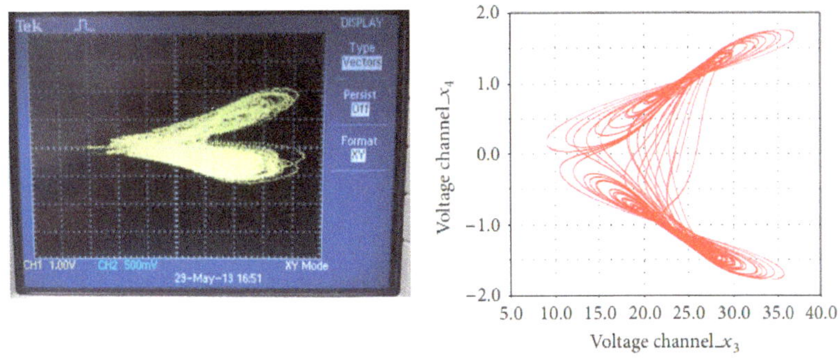

(f)

Figure 12: Realization and simulation circuit compare results for voltage signals outputs of the four-dimensional Lorenz-Stenflo system: (a) V_{x_1}-V_{x_2}, (b) V_{x_1}-V_{x_3}, (c) V_{x_1}-V_{x_4}, (d) V_{x_2}-V_{x_3}, (e) V_{x_2}-V_{x_4}, and (f) V_{x_3}-V_{x_4}.

Consider

$$\begin{aligned}
A_1(t) &= a(1 + f_1 z_1)\\
A_2(t) &= b(1 + f_2 z_2)\\
A_3(t) &= c(1 + f_3 z_3)\\
A_4(t) &= d(1 + f_4 z_4),
\end{aligned} \tag{24}$$

where f_1, f_2, f_3, and f_4 are arbitrary positive constant and $f_1 = f_2 = f_3 = f_4 = 0.005$. The chaotic signals system is

$$\begin{aligned}
\dot{z}_1 &= -z_2 z_3 + a z_1 + b z_4\\
\dot{z}_2 &= z_1 z_3 + c z_2\\
\dot{z}_3 &= \frac{1}{3} z_1 z_2 + d z_3\\
\dot{z}_4 &= z_2 z_3 + r z_4,
\end{aligned} \tag{25}$$

where $a = 3.7$, $b = 1.5$, $c = 26$, and $d = 0.7$.

The slave system is

$$\begin{aligned}
\dot{y}_1 &= \widehat{A}_1 (y_2 - y_1)\\
\dot{y}_2 &= \widehat{A}_3 y_1 - y_1 y_3 - y_2\\
\dot{y}_3 &= \widehat{A}_2 y_4 - \widehat{A}_4 y_3 + y_1 y_2\\
\dot{y}_4 &= -\widehat{A}_1 y_4 - y_1,
\end{aligned} \tag{26}$$

where $\widehat{A}_1(t), \widehat{A}_2(t), \widehat{A}_3(t)$, and $\widehat{A}_4(t)$ are estimates of uncertain chaotic parameters, respectively. The nonlinear controllers, u_1, u_2, u_3, and u_4, are added to each equation in (26), respectively. Consider

$$\begin{aligned}
\dot{y}_1 &= \widehat{A}_1 (y_2 - y_1) + u_1\\
\dot{y}_2 &= \widehat{A}_3 y_1 - y_1 y_3 - y_2 + u_2\\
\dot{y}_3 &= \widehat{A}_2 y_4 - \widehat{A}_4 y_3 + y_1 y_2 + u_3\\
\dot{y}_4 &= -\widehat{A}_1 y_4 - y_1 + u_4.
\end{aligned} \tag{27}$$

The given functional system for generalized synchronization is also the four-dimensional Lorenz-Stenflo system but with different initial conditions.

Figure 13: Realization circuit of the four-dimensional LorenzStenflo system.

Table 2: List of apparatus and components in the circuit experiment.

Apparatus/Component	Model	Quantity
Multiplier	AD633	2
OP-Amplifier	LF412	4
Capacitor	2000 pF~0.47 uF	4
	1 k~1.2 M	19
	270 kΩ	3
	38.461 kΩ	1
Resister	667 kΩ	1
	1429 kΩ	1
	100 kΩ	8
	1 MΩ	4
Variable resistor	100 k	2
DC power supply	GPS-3303	1

Consider

$$\begin{aligned}\dot{w}_1 &= a(w_2 - w_1)\\ \dot{w}_2 &= cw_1 - w_1 w_3 - w_2\\ \dot{w}_3 &= bw_4 - dw_3 + w_1 w_2\\ \dot{w}_4 &= -aw_4 - w_1.\end{aligned} \tag{28}$$

The initial value of the states of the master system, of the slave system, of the chaotic signals system, and of the given functional system are taken as $x_1(0) = x_2(0) = x_3(0) = x_4(0) = 0.2$, $y_1(0) = y_2(0) = y_3(0) = y_4(0) = 0.3$, $z_1(0) = z_2(0) = z_3(0) = z_4(0) = 0.7$, and $w_1(0) = w_2(0) = w_3(0) = w_4(0) = 0.5$, respectively.

The generalized synchronization error functions are

$$e_i = x_i - y_i + w_i + Ke_i, \quad (i = 1, 2, 3, 4). \tag{29}$$

The addition of constant Ke_i makes the error states always greater than zero. As Figure 15 shows, we can confirm that the error states are always greater than zero. From (29), we get the following error dynamics:

$$\begin{aligned}\dot{e}_1 &= \widehat{A}_1(e_2 - e_1 - w_2 + w_1)\\ &\quad + \widetilde{A}_1(x_2 - x_1) + a(w_2 - w_1) - u_1\\ \dot{e}_2 &= \widehat{A}_3(e_1 - w_1 - K) + \widetilde{A}_3 x_1\\ &\quad + cw_1 - x_1 x_3 - w_1 w_3 + (x_1 + w_1 - e_1 + K)\\ &\quad \times (x_3 + w_3 - e_3 + K) - (e_2 - K) - u_2\\ \dot{e}_3 &= \widehat{A}_2(e_4 - w_4 - K) + \widetilde{A}_2 x_4 + bw_4\\ &\quad - \widehat{A}_4(e_3 - w_3 - K) - \widetilde{A}_4 x_3 - dw_3\\ &\quad + x_1 x_3 + w_1 w_3 - (x_1 + w_1 - e_1 + K)\\ &\quad \times (x_2 + w_2 - e_2 + K) - u_3\\ \dot{e}_4 &= -\widehat{A}_1(e_4 - w_4 - K) - \widetilde{A}_1 x_4 - aw_4 - (e_1 - K) - u_4,\end{aligned} \tag{30}$$

where $e_1 = x_1 - y_1 + w_1$, $e_2 = x_2 - y_2 + w_2$, $e_3 = x_3 - y_3 + w_3$, $e_4 = x_4 - y_4 + w_4$, $\widetilde{A}_1 = A_1 - \widehat{A}_1 + Kp_1$, $\widetilde{A}_2 = A_2 - \widehat{A}_2 + Kp_2$, $\widetilde{A}_3 = A_3 - \widehat{A}_3 + Kp_3$, and $\widetilde{A}_4 = A_4 - \widehat{A}_4 + Kp_4$. The addition of constant $K(i = 1, 2, 3, 4)$ makes the update laws always greater than zero. As Figure 17 shows, we can confirm that the update laws are always greater than zero.

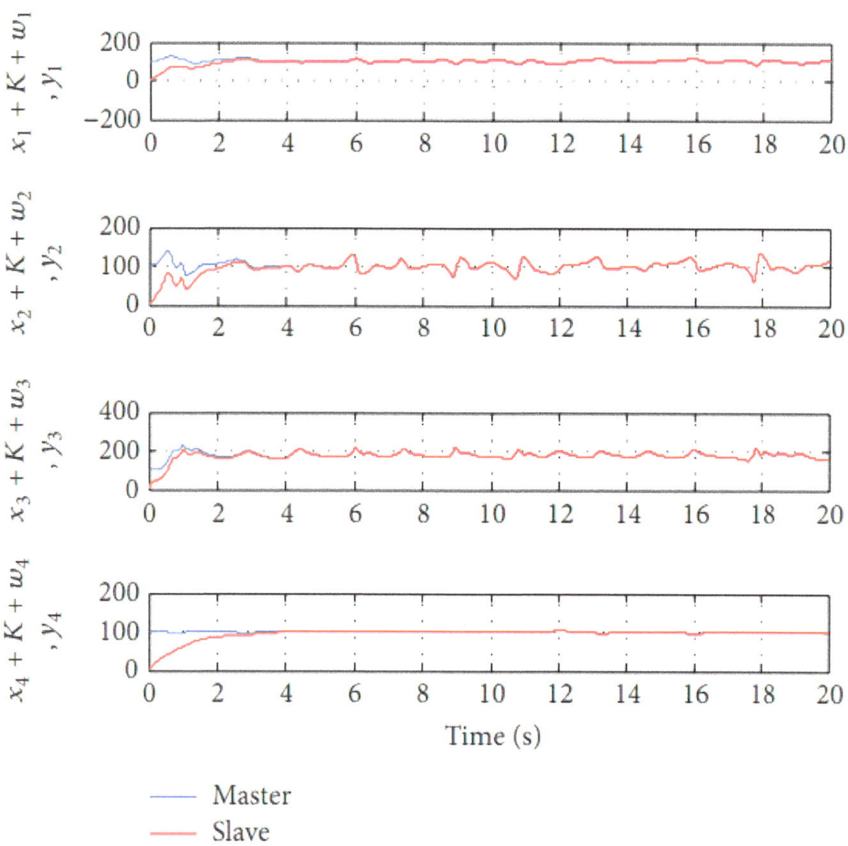

Figure 14: Time histories of the master system and slave system.

Choose a Lyapunov function in the form of a positive definite function:

$$V\left(e_1, e_2, e_3, e_4, \widetilde{A}_1, \widetilde{A}_2, \widetilde{A}_3, \widetilde{A}_4\right) = \left(e_1 + e_2 + e_3 + e_4 + \widetilde{A}_1 + \widetilde{A}_2 + \widetilde{A}_3 + \widetilde{A}_4\right) \quad (31)$$

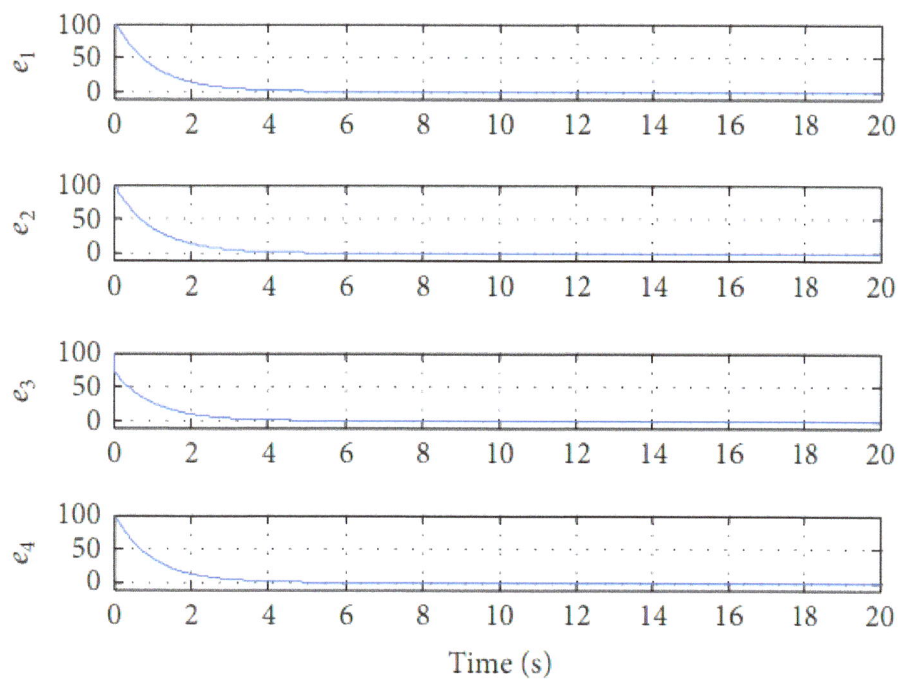

Figure 15: Time histories of error states.

and its time derivative is

$$
\begin{aligned}
&\dot{V}\left(e_1, e_2, e_3, e_4, \widetilde{A}_1, \widetilde{A}_2, \widetilde{A}_3, \widetilde{A}_4\right) \\
&\quad = \dot{e}_1 + \dot{e}_2 + \dot{e}_3 + \dot{e}_4 + \dot{\tilde{a}} + \dot{\tilde{b}} + \dot{\tilde{c}} + \dot{\tilde{d}} \\
&\quad = \left(\widehat{A}_1 \left(e_2 - e_1 - w_2 + w_1\right) + \widetilde{A}_1 \left(x_2 - x_1\right)\right. \\
&\qquad\quad \left. + a \left(w_2 - w_1\right) - u_1\right) \\
&\qquad + \left(\widehat{A}_3 \left(e_1 - w_1 - K\right) + \widetilde{A}_3 x_1 + c w_1 - x_1 x_3 - w_1 w_3\right. \\
&\qquad\quad + \left(x_1 + w_1 - e_1 + K\right)\left(x_3 + w_3 - e_3 + K\right) \\
&\qquad\quad \left. - \left(e_2 - K\right) - u_2\right) \\
&\qquad + \left(\widehat{A}_2 \left(e_4 - w_4 - K\right) + \widetilde{A}_2 x_4 + b w_4 - \widehat{A}_4 \left(e_3 - w_3 - K\right)\right. \\
&\qquad\quad - \widetilde{A}_4 x_3 - d w_3 + x_1 x_3 + w_1 w_3 \\
&\qquad\quad \left. - \left(x_1 + w_1 - e_1 + K\right)\left(x_2 + w_2 - e_2 + K\right) - u_3\right) \\
&\qquad + \left(-\widehat{A}_1 \left(e_4 - w_4 - K\right) - \widetilde{A}_1 x_4 - a w_4 - \left(e_1 - K\right) - u_4\right) \\
&\qquad + \left(-\dot{\hat{a}}\right) + \left(-\dot{\hat{b}}\right) + \left(-\dot{\hat{c}}\right) + \left(-\dot{\hat{d}}\right) + \left(-\dot{\hat{r}}\right).
\end{aligned}
\tag{32}
$$

We choose the update laws for those uncertain parameters as

$$
\begin{aligned}
\dot{\widetilde{A}}_1 &= \dot{A}_1 - \dot{\widehat{A}}_1 = -\widetilde{A}_1 e_1 - \widetilde{A}_1 \\
\dot{\widetilde{A}}_2 &= \dot{A}_2 - \dot{\widehat{A}}_2 = -\widetilde{A}_2 e_2 - \widetilde{A}_2 \\
\dot{\widetilde{A}}_3 &= \dot{A}_3 - \dot{\widehat{A}}_3 = -\widetilde{A}_3 e_3 - \widetilde{A}_3 \\
\dot{\widetilde{A}}_4 &= \dot{A}_4 - \dot{\widehat{A}}_4 = -\widetilde{A}_4 e_4 - \widetilde{A}_4 .
\end{aligned}
\tag{33}
$$

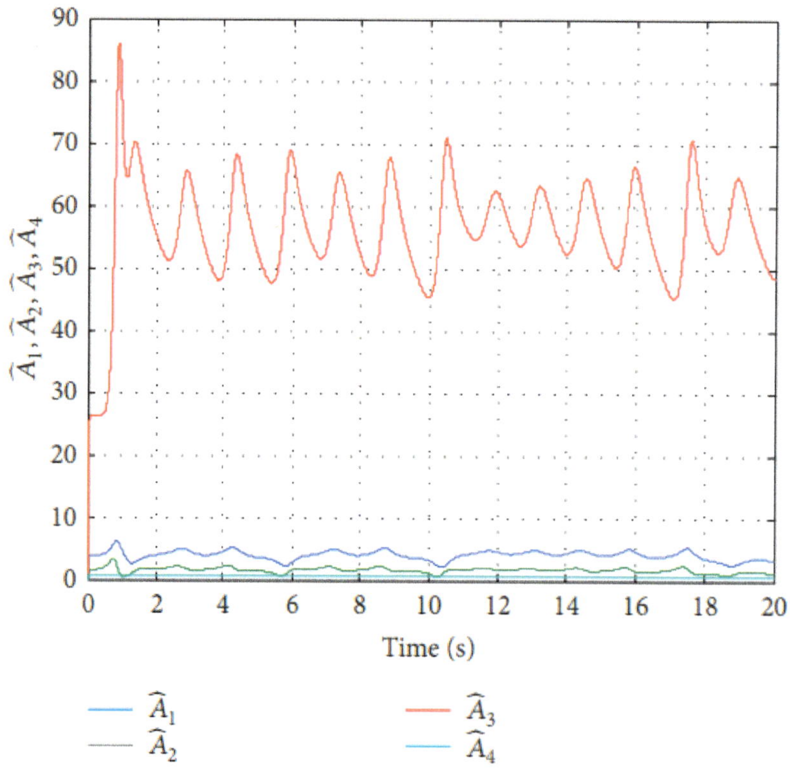

Figure 16: Time histories of estimated parameters.

The initial values of estimates for uncertain parameters are $\widehat{A}_1(0) = \widehat{A}_2(0) = \widehat{A}_3(0) = \widehat{A}_4(0) = 0$. Through (32) and (33), the appropriate controllers can be designed as

$$\begin{aligned}
u_1 &= \widehat{A}_1 (e_2 - e_1 - w_2 + w_1) \\
&\quad + \widetilde{A}_1 (x_2 - x_1 - x_4 - e_1) + a (w_2 - w_1) + e_1 \\
u_2 &= \widehat{A}_3 (e_1 - w_1 - K) + \widetilde{A}_2 (x_4 - e_2) \\
&\quad + cw_1 - x_1 x_3 - w_1 w_3 \\
&\quad + (x_1 - e_1 + w_1 + K)(x_3 - e_3 + w_3 + K) + e_2 \\
u_3 &= \widehat{A}_2 (e_4 - w_4 - K) + bw_4 - \widehat{A}_4 (e_3 - w_3 - K) \\
&\quad - dw_3 + \widetilde{A}_3 (x_1 - e_3) + x_1 x_2 + w_1 w_2 \\
&\quad - (x_1 - e_1 + w_1 + K)(x_2 - e_2 + w_2 + K) + e_3 \\
u_4 &= -\widehat{A}_1 (e_4 - w_4 - K) - aw_4 \\
&\quad + \widetilde{A}_4 (-x_3 - e_4) - (e_1 - K) + e_4.
\end{aligned} \tag{34}$$

Substituting (33) and (34) into (32), we obtain

$$\dot{V} = -e_1 - e_2 - e_3 - e_4 - \widetilde{A}_1 - \widetilde{A}_2 - \widetilde{A}_3 - \widetilde{A}_4 < 0 \tag{35}$$

which is a negative-definite function of e_1, e_2, e_3, e_4, $\tilde{A}_1$, $\tilde{A}_2$, $\tilde{A}_3$, and $\tilde{A}_4$. The numerical simulation results are shown in Figures 14, 15, 16, and 17.

CONCLUSIONS

In summary, the paper has studied the nonlinear dynamical behaviors of the new chaotic attractor, including some basic dynamical properties, bifurcations, Lyapunov exponents, Lyapunov dimension, periodic motion, and routes to chaos motion. Also, the four-dimensional Lorenz-Stenflo system based realization circuit has been constructed to implement various wing butterfly attractors. Furthermore, by electronic circuit implementation of the proposed four-dimensional Lorenz-Stenflo system, it is shown that the chaotic attractors do physically exist. A new strategy, the GYC partial region stability theory, is proposed to achieve adaptive chaos synchronization with uncertain chaotic parameters.

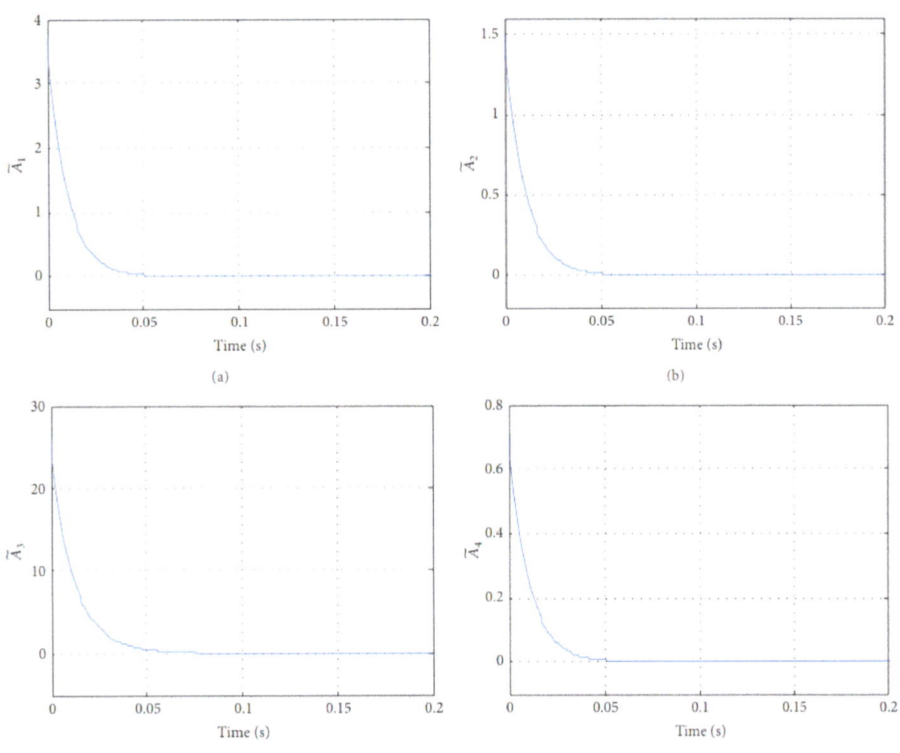

Figure 17: Time histories of parameters differences.

Conflict of Interests

The authors declare that there is no conflict of interests regarding the publication of this paper.

ACKNOWLEDGMENT

This research was supported by the National Science Council, Republic of China, under Grant no. NSC 102-2221-E-011-034.

REFERENCES

1. Z. Li, Y. Soh, and C. Wen, Switched and Impulsive Systems, Analysis, Design, and Applications, vol. 313, Springer, Berlin, Germany, 2005.
2. P. Galajda and D. Kocur, "Chua's circuit in spread spectrum communication systems," Radio Engineering, vol. 11, no. 2, pp. 6–10, 2002.
3. L. Kocarev, Z. Galias, and S. Lian, Eds., Intelligent Computing Based on Chaos, vol. 184 of Studies in Computational Intelligence, Springer, Berlin, Germany, 2009.
4. K. Grygiel and P. Szlachetka, "Lyapunov exponents analysis of autonomous and nonautonomous sets of ordinary differential equations," Acta Physica Polonica B, vol. 26, no. 8, pp. 1321–1331, 1995.
5. S. H. Strogatz, Nonlinear Dynamics and Chaos: With Applications to Physics, Biology, Chemistry, and Engineering, Perseus Books, Cambridge, Mass, USA, 1994.
6. K. Kiers, D. Schmidt, and J. C. Sprott, "Precision measurements of a simple chaotic circuit," American Journal of Physics, vol. 72, no. 4, pp. 503–509, 2004.
7. E. N. Lorenz, "Deterministic nonperiodic flow," Journal of the Atmospheric Sciences, vol. 20, pp. 130–141, 1963.
8. L. Stenflo, "Generalized Lorenz equations for acoustic-gravity waves in the atmosphere," Physica Scripta, vol. 53, no. 1, pp. 83–84, 1996.
9. J. C. Xavier and P. C. Rech, "Regular and chaotic dynamics of the Lorenz-Stenflo system," International Journal of Bifurcation and Chaos in Applied Sciences and Engineering, vol. 20, no. 1, pp. 145–152, 2010.
10. P. Wang, D. Li, and Q. Hu, "Bounds of the hyper-chaotic Lorenz-Stenflo system," Communications in Nonlinear Science and Numerical Simulation, vol. 15, no. 9, pp. 2514–2520, 2010.
11. H. Li, X. Liao, and X. Lei, "Two fuzzy control schemes for Lorenz-Stenflo chaotic system," Journal of Vibration and Control, vol. 18, no. 11, pp. 1675–1682, 2012.
12. W. Xiang, "Equilibrium points and bifurcation control for Lorenz-Stenflo system," ICIC Express Letters, vol. 3, no. 1, pp. 61–66, 2009.
13. C.-H. Yang, "Chaos hybrid generalized synchronization of liu-chen system by GYC partial region stability theory," Journal of Computational and Theoretical Nanoscience, vol. 10, no. 4, pp. 825–831, 2013.
14. Z.-M. Ge and C.-H. Yang, "The generalized synchronization of a Quantum-CNN chaotic oscillator with different order systems," Chaos, Solitons and Fractals, vol. 35, no. 5, pp. 980–990, 2008.
15. Z.-M. Ge, C.-W. Yao, and H.-K. Chen, "Stability on partial region in dynamics," Journal of the Chinese Society of Mechanical Engineers, vol. 15, no. 2, pp. 140–151, 1994.
16. Z.-M. Ge, J.-K. Yu, and H.-K. Chen, "Three asymptotical stability theorems on partial region with applications," Japanse Journal of Applied Physics, vol. 37, article 2762, 1998.
17. P. Frederickson, J.-L. Kaplan, E.-D. Yorke, and J.-A. Yorke, "The Liapunov dimension of strange attractors," Journal of Differential Equations, vol. 49, no. 2, pp. 185–207, 1983.
18. X. Wang and G. A. Chen, "A chaotic system with only one stable equilibrium," Communications in Nonlinear Science and Numerical Simulation, vol. 17, no. 3, pp. 1264–1272, 2012.
19. Y. Liu, "Circuit implementation and finite-time synchronization of the 4D Rabinovich hyperchaotic system," Nonlinear Dynamics, vol. 67, no. 1, pp. 89–96, 2012.
20. A. G. Radwan, A. M. Soliman, and A. L. El-Sedeek, "MOS realization of the double-scroll-like chaotic equation"," IEEE Transactions on Circuits and Systems I: Fundamental Theory and Applications, vol. 50, no. 2, pp. 285–288, 2003.
21. A. Abooee, H. A. Yaghini-Bonabi, and M. R. Jahed-Motlagh, "Analysis and circuitry realization of a novel three-dimensional chaotic system," Communications in Nonlinear Science and Numerical Simulation, vol. 18, no. 5, pp. 1235–1245, 2013.

CHAPTER 19

INTEGER AND FRACTIONAL GENERAL T-SYSTEM AND ITS APPLICATION TO CONTROL CHAOS AND SYNCHRONIZATION

Mihaela Neamţu, Anamaria Liţoiu, and Petru C. Strain

INTRODUCTION

Lorenz found the first canonical chaotic attractor [1]. During the time, it has been proved that chaos can occur in simple three-dimensional autonomous systems with one, two, and three nonlinearities. Tigan and Opris [2] proposed and analyzed a new chaotic three-dimensional nonlinear system, called T system, which is similar to the Lorenz system. Because T system allows a larger possibility in choosing the parameters of the system, it can display more complex dynamics [3–10].

Recently, based on the study of integer order chaos, the fractional order Lorenz system [11] and the fractional order Liu system [12] were introduced. The system with fractional order still shows the chaotic behavior [13,14].

The T system is described by [2]

$$\begin{aligned} \dot{x}(t) &= a\left(y(t)-x(t)\right), \\ \dot{y}(t) &= (b-a)\,x(t)-ax(t)\,z(t), \\ \dot{z}(t) &= -cz(t)+x(t)\,y(t), \end{aligned} \tag{1}$$

which is chaotic when $a = 2.1$, $b = 30$, and $c = 0.6$ [2].

Li et al. [15] have proposed a new Lorenz-like chaotic system derived from (1).The nonlinear differential three-dimensional system is given by

$$\begin{aligned} \dot{x}(t) &= a\left(y(t)-x(t)\right), \\ \dot{y}(t) &= abx(t)-ax(t)\,z(t), \\ \dot{z}(t) &= -cz(t)+x(t)\,y(t), \end{aligned} \tag{2}$$

which is chaotic when $a = 5$, $b = 4$, and $c = 2$ [15].

Chaotic phenomena in electric circuits have been studied with great interest. The electronic circuit for (2) is designed and a chaotic attractor is implemented and verified [15]. Yang [16] proposed another new Lorenz-like system. The nonlinear differential three-dimensional system is

$$\begin{aligned} \dot{x}(t) &= a\left(y(t)-x(t)\right), \\ \dot{y}(t) &= b_1x(t)-b_2x(t)\,z(t), \\ \dot{z}(t) &= -cz(t)+lx(t)^2+ky(t)^2, \end{aligned} \tag{3}$$

which is chaotic when $a = 10$, $b_1 = 40$, $b_2 = 1$, $c = 2.5$, $l = 2$, and $k = 2$ [16].

Citation: Mihaela Neamţu, Anamaria Liţoiu, and Petru C. Strain, "Integer and Fractional General T-System and Its Application to Control Chaos and Synchronization," Abstract and Applied Analysis, vol. 2015, Article ID 413540, 14 pages, 2015. doi:10.1155/2015/413540.

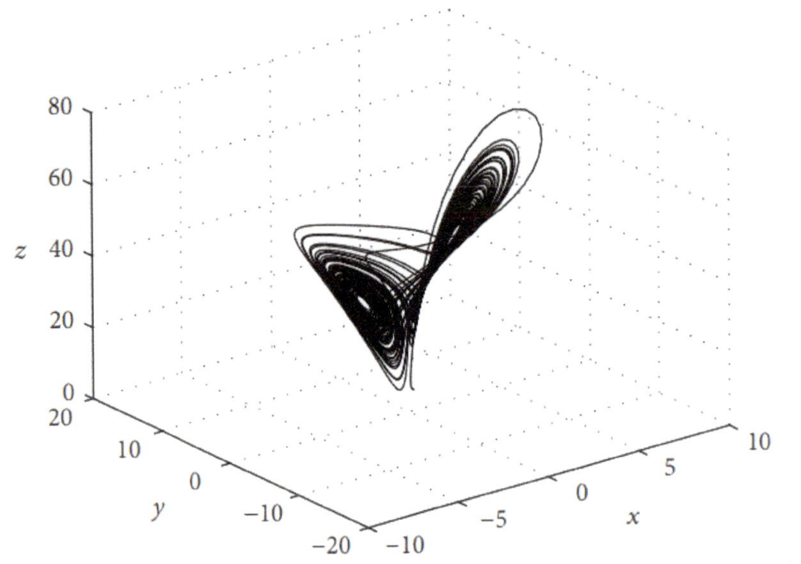

Figure 1: Phase portrait of the chaotic attractor for the general T system (4) with $a = 10$, $b_1 = 40$, $b_2 = 1$, $c = 2.5$, $d = 3$, $l = 0.9$, and $k = 2$.

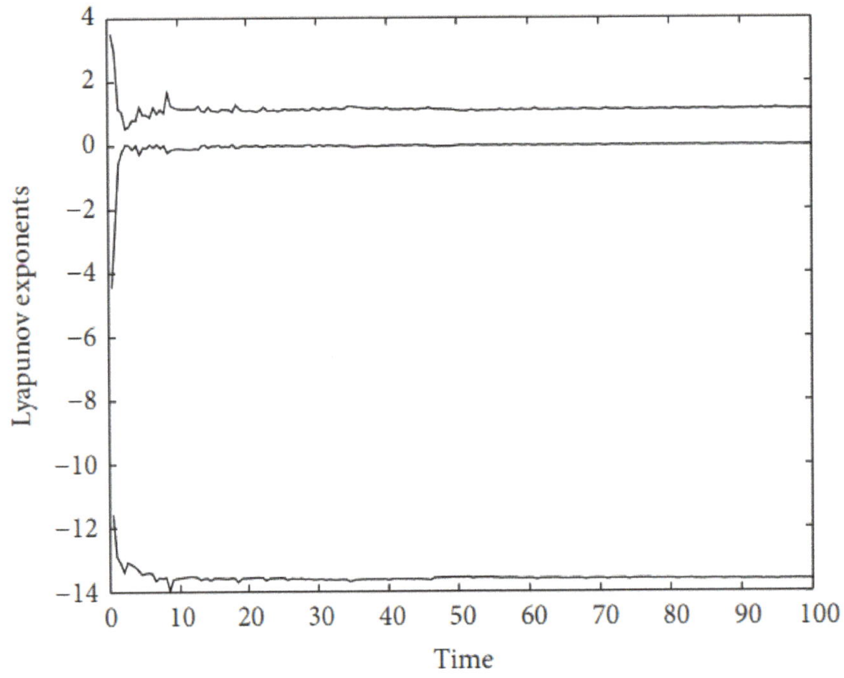

Figure 2: Dynamics of Lyapunov exponents of (4) with $a = 10$, $b_1 = 40$, $b_2 = 1$, $c = 2.5$, $d = 3$, $l = 0.9$, and $k = 2$.

Based on (1), (2), and (3), we propose a general T system described by the following differential equations:

$$\dot{x}(t) = a\left(y(t) - x(t)\right),$$

$$\dot{y}(t) = b_1 x(t) - b_2 x(t) z(t),$$

$$\dot{z}(t) = -cz(t) + dx(t)y(t) + lx(t)^2 + ky(t)^2, \quad (4)$$

where a, b_2, c, d, h, k are positive real parameters, b_1 is a real parameter, and $a \neq 0$.

An experimental electronic circuit for (4) can be described and implemented in a similar way as in [15, 17]. System (4) is chaotic when $a = 10$, $b_1 = 40$, $b_2 = 1$, $c = 2.5$, $d = 3$, $l = 0.9$, and $k = 2$ (see Figure 1). The dynamics of Lyapunov exponents of system (4) is displayed in Figure 2.

For $b_1 = b-a$, $b_2 = a$, $d = 1$, $l = 0$, and $k = 0$, system (4) is given by (1). For $b_1 = ab$, $b_2 = a$, $d = 1$, $l = 0$, and $k = 0$, system (4) becomes (2). For $d = 0$, system (4) is (3). Systems (1), (2), (3), and (4) are chaotic and using the method from [15] we can verify that they are not equivalent to the Lorenz, Chen, and Lu systems. Time-delayed feedback is a powerful tool to control unstable periodic orbits or control unstable steady states [18]. Following the idea of Pyragas [19], as in [9, 18], we add a delayed force $k_1(y(t) - y(t - \tau))$ to the second equation of (4) and we obtain the delayed feedback control system:

$$\dot{x}(t) = a\left(y(t) - x(t)\right),$$

$$\dot{y}(t) = b_1 x(t) - b_2 x(t) z(t) + k_1\left(y(t) - y(t-\tau)\right),$$

$$\dot{z}(t) = -cz(t) + dx(t)y(t) + lx(t)^2 + ky(t)^2, \quad (5)$$

where τ is the time delay

We assume that $\tau > 0$ and $k_1 \in R$, which indicates the strength of the feedback [10]. The fractional-order general T system can be described by

$$D^{q_1} x(t) = a(y(t) - x(t)),$$
$$D^{q_2} y(t) = b_1 x(t) - b_2 x(t) z(t),$$
$$D^{q_3} z(t) = -cz(t) + dx(t) y(t) + lx(t)^2 + ky(t)^2, \quad (6)$$

where ${}_aD_t^q$ is defined by [13]

$$ {}_aD_t^q = \begin{cases} \dfrac{d^q}{dt^q}, & \mathrm{Re}(q) > 0 \\ 1 & \mathrm{Re}(q) = 0 \\ \displaystyle\int_a^t (ds)^{-q} & \mathrm{Re}(q) < 0 \end{cases} \quad (7)$$

and q is the derivative order that can be a complex number with Re(q) the real part of q. The numbers a and t are the limits of the operator. There are many definitions for general fractional derivative. The most frequently used ones are the Grunwald-Letnikov definition, the Riemann-Liouville, and the Caputo definitions.

As in [13], in this paper we use the Caputo definition for the fractional derivative. In the present paper, we focus on (5) and (6).The aim is to provide a new investigation of the Hopf bifurcation and chaos control on the general T system given by (5) and an analysis of the fractional general T-system as well. For system (5), we consider τ as the bifurcation parameter. When it passes through some certain critical values, the equilibrium will lose its stability and Hopf bifurcation will occur. We study the direction of the Hopf bifurcation, as well as the stability and period of the bifurcating periodic solutions. Moreover, with different values for k_1 and τ, we realize the chaos control.

The chaotic dynamics in the general T-system with fractional derivative is taken into account. Some properties are given. Then, synchronization problem of (6) is provided.

The paper is structured as follows. Section 2 provides the stability of the steady states when there is no time delay. For system (5), Section 3 analyzes the local stability, the existence of the Hopf bifurcation, and the direction and the stability of the Hopf bifurcation. The analysis of system (6) is presented in Section 4. The chaos control of (6) is given in Section 5 and the synchronization in Section 6. Finally, conclusions are drawn in Section 7.

EXISTENCE OF STEADY STATES: STABILITY ANALYSIS FOR SYSTEM (4)

The equilibrium of system (4) can be obtained by solving the following algebraic system:

$$a(y - x) = 0,$$
$$b_1 x - b_2 xz = 0,$$
$$-cz + dxy + lx^2 + ky^2 = 0. \quad (8)$$

Then, we have the following.

Proposition 1. Consider the following:

I. If $b_1 < 0$, then system (4) has only one real steady state $S_0(0, 0, 0)$.

II. If $b_1 > 0$, then system (4) has three real steady states: $S_0(0, 0, 0)$, $S-(-x_0, -y_0, z_0)$, $S+(x_0, y_0, z_0)$, where

$$x_0 = y_0 = \sqrt{\frac{b_1 c}{b_2 (d + l + k)}},$$
$$z_0 = \frac{b_1}{b_2}. \quad (9)$$

In order to analyze the local stability of the above steady states, the Jacobian matrix of (4) is given by

$$J(x, y, z) = \begin{pmatrix} -a & a & 0 \\ b_1 - b_2 z & 0 & -b_2 z \\ dy + 2lx & dx + 2ky & -c \end{pmatrix} \quad (10)$$

If $b_1 < 0$, the characteristic equation of (S_0) is given by

$$(\lambda + c)(\lambda^2 + a\lambda - ab_1) = 0. \quad (11)$$

The eigenvalues of (11) are $\lambda_1 = -c$, and $\lambda_{2,3} = (1/2)(-a \pm \sqrt{a_2 + 4ab_1})$.

For $b_1 > 0$, both S_- and S_+ have the same characteristic equation; that is

$$\lambda^3 + (a + c)\lambda^2 + \left(ac + \frac{b_1 c(d + 2k)}{d + l + k}\right)\lambda + 2ab_1 c = 0. \quad (12)$$

The eigenvalues of (12), which are dependent on parameters a, b_1, c, d, l, k, can be obtained by the Cardano formula. Since a, c, d, l, k are all positive real parameters, one can ensure that

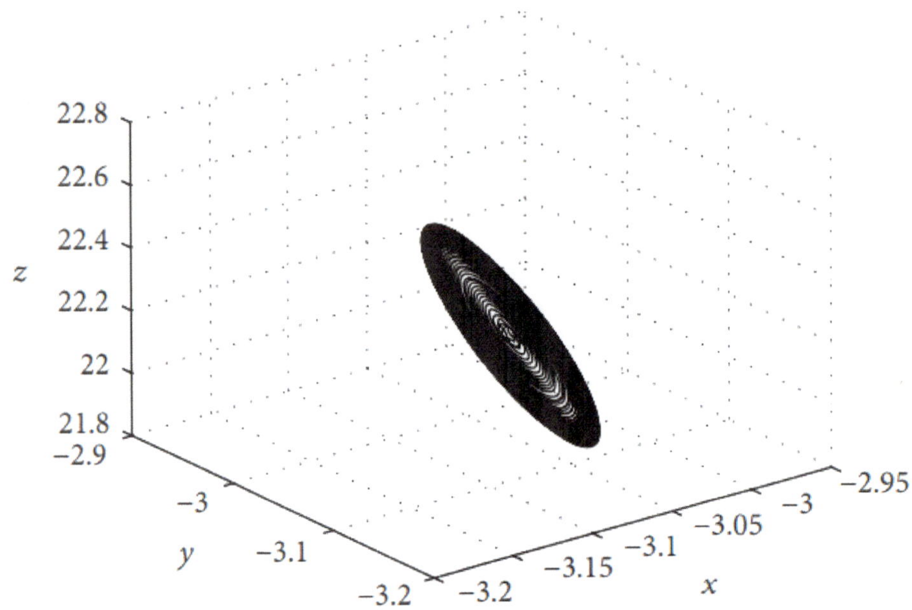

Figure 3: The steady state S_- of system (4) is locally asymptotically stable when $b_1 < b_c$.

(12) has at least one eigenvalue with negative real part as $b_1 > 0$. The other two eigenvalues could be

(a) two negative real roots;

(b) two positive real roots;

(c) two complex-conjugate roots with negative real part;

(d) two complex-conjugate roots with positive real part.

Therefore, analyzing the characteristic equation of(4) and using the Routh-Hurwitz theorem, we obtain the following propositions.

Proposition 2. Consider the following:

(i) If $b_1 < 0$, then the steady state $S_0(0, 0, 0)$ of system (4) is locally asymptotically stable.

(ii) If $b_1 > 0$, then the steady state $S_0(0, 0, 0)$ of system (4) is unstable.

(iii) If $0 < b_1 < b_c$ and $(d + 2l) - c(d + 2k) > 0$, where

$$b_c = \frac{a(a+c)(d+l+k)}{a(d+2l) - c(d+2k)}, \tag{13}$$

then the steady states S_- and S_+ of system (4) are locally asymptotically stable (see Figures 3 and 4).

(iv) If $b_1 > b_c$ and $(d + 2l) - c(d + 2k) > 0$, then the steady states S_- and S_+ of system (4) are unstable (see Figures 5 and 6).

Using the Hopf bifurcation theorem [9, 16], we have the following.

Proposition 3. If $b_1 > 0$ and $b_1 = b_c$, then for the steady state S_+ *(or S_-)* of system (4), the corresponding characteristic equation has three eigenvalues: one negative and one pair of purely imaginary conjugate roots, satisfying $\mathrm{Re}(d\lambda/db_1)|_{b_1=b_c} \neq 0$; that is, system (4) undergoes a Hopf bifurcation at the steady state S_+ *(or S_-)* (see Figures 7 and 8).

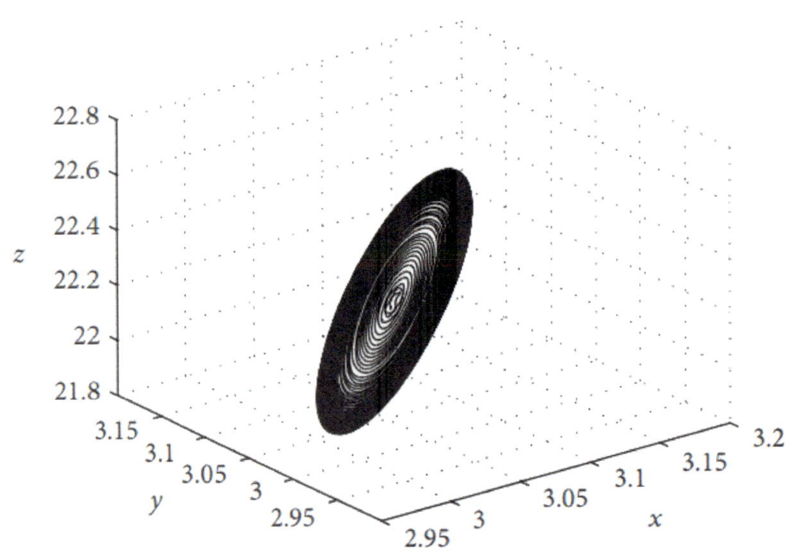

Figure 4: The steady state S_+ of system (4) is locally asymptotically stable when $b_1 < b_c$.

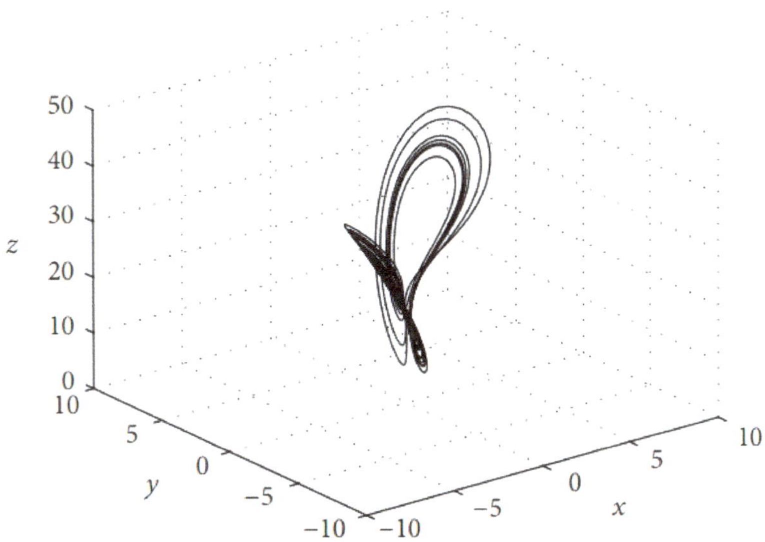

Figure 5: The steady state S_- of system (4) is unstable when $b_1 > b_c$.

LOCAL STABILITY AND HOPF BIFURCATION FOR SYSTEM (5)

Local Stability and the Existence of Hopf Bifurcation for System (5)

Consider system (5). When $\tau = 0$, it becomes (4). Since the time delay does not change, the delayed feedback system (5) has the same equilibria as system (4).

From Proposition 1, under the assumption $b_1 > 0$, system (5) also has three real steady states: $S_0(0, 0, 0)$, $S_-(-x_0, -y_0, z_0)$, and $S+(x_0, y_0, z_0)$. Now, we analyze the effect of the time delay on the stability of these steady states.

The linearization of system (5) at S_0 is

$$\begin{aligned}\dot{u}_1(t) &= a\left(u_2(t) - u_1(t)\right),\\ \dot{u}_2(t) &= b_1 u_1(t) + k_1\left(u_2(t) - u_2(t-\tau)\right),\\ \dot{u}_3(t) &= -cu_3(t).\end{aligned} \qquad (14)$$

The characteristic equation of (14) is

$$(\lambda + c)\left(\lambda^2 + (a - k_1)\lambda - ak_1 - ab_1 + (\lambda + a)k_1 e^{-\lambda\tau}\right) = 0. \qquad (15)$$

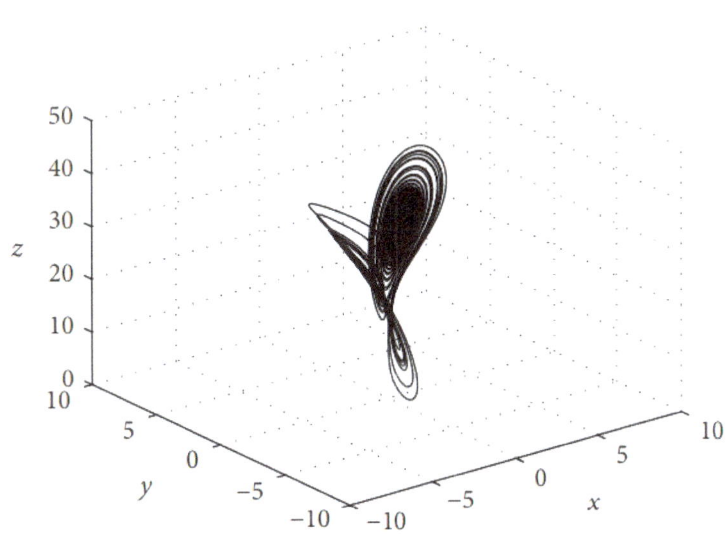

Figure 6: The steady state S_+ of system (4) is unstable when $b_1 > b_c$.

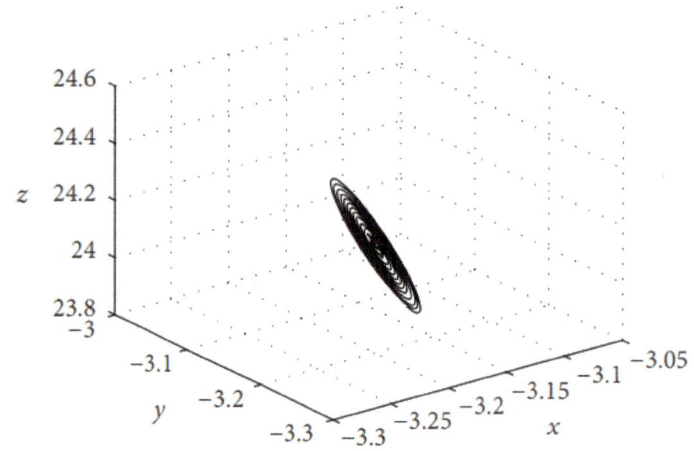

Figure 7: A Hopf bifurcation occurs at the steady state S_- of system (4) when $b_1 = bc$.

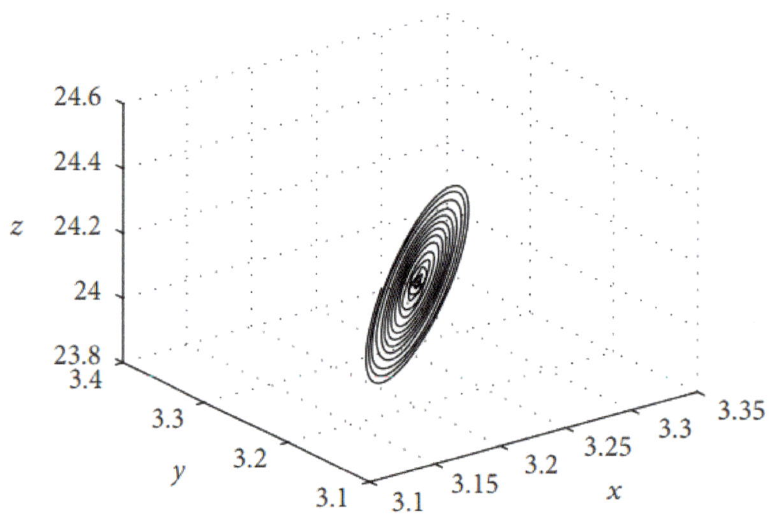

Figure 8: A Hopf bifurcation occurs at the steady state S_+ of system (4) when $b_1 = b_c$.

Equation (15) has a negative root $\lambda = -c$ for all $\tau > 0$; thus, we need to analyse the following equation:

$$\lambda^2 + (a-k_1)\lambda - ak_1 - ab_1 + (\lambda + a)k_1 e^{-\lambda\tau} = 0. \quad (16)$$

If there is no delay, (16) becomes

$$\lambda^2 + a\lambda - ab_1 = 0. \quad (17)$$

Let $i\omega$ ($\omega > 0$) be a root of (16). It follows that

$$\omega^2 + ak_1 + ab_1 = ak_1\cos(\omega\tau) + \omega k_1 \sin(\omega\tau),$$
$$(a-k_1)\omega = ak_1\sin(\omega\tau) - \omega k_1\cos(\omega\tau), \quad (18)$$

which can be written as

$$\omega^4 + A\omega^2 + B = 0, \quad (19)$$

where

$$A = 2ab_1 + a^2,$$
$$B = 2a^2 b_1(2k_1 + b_1). \quad (20)$$

As in [9], we consider the following analysis:

(1) If the conditions

$$(H_1)\ A > 0,\ B > 0 \quad (21)$$

hold, then (19) has no positive roots.

(2) If the condition

$$(H_2)\ B < 0 \quad (22)$$

holds, then (19) has a unique positive root ω_+^2. Using (19), we have

$$\omega_+ = \left(\frac{-A + \sqrt{A^2 - 4B}}{2}\right)^{1/2}. \quad (23)$$

(3) If the conditions

$$(H_3)\ A < 0,\ B > 0,\ A^2 - 4B > 0$$

(24)

hold, then (19) has two positive roots $\omega_\pm^2$. Thus, we have

$$\omega_\pm = \left(\frac{-A + \sqrt{A^2 - 4B}}{2}\right)^{1/2}. \quad (25)$$

A stability switch may occur, through the roots $\pm\omega_\pm i$, where $\omega_\pm$ are given by (25). Therefore, from (18), we have

$$\tau_n^\pm = \frac{1}{\omega_\pm}\arccos\left(\frac{k_1\omega_\pm^2 + a(ak_1 + ab_1)}{k_1(a^2 + \omega_\pm^2)}\right) + \frac{2n\pi}{\omega_\pm}, \quad (26)$$

where $n = 1, 2, \ldots$ at which (16) has a pair of purely imaginary roots $\pm\omega_{\pm}i$. Consider

$$\lambda(\tau) = \mu(\tau) + i\omega(\tau) \tag{27}$$

the root of (16) so that $\mu(\tau_n^{\pm}) = 0$. Using (16) and considering $\lambda = (\tau)$, $(d\lambda(\tau)/d\tau)^{-1}$ is given by

$$\left(\frac{d\lambda(\tau)}{d\tau}\right)^{-1} = \frac{(2\lambda + a - k_1)e^{\lambda\tau}}{\lambda(k_1\lambda + ak_1)} + \frac{k_1}{\lambda(k_1\lambda + ak_1)} - \frac{\tau}{\lambda}. \tag{28}$$

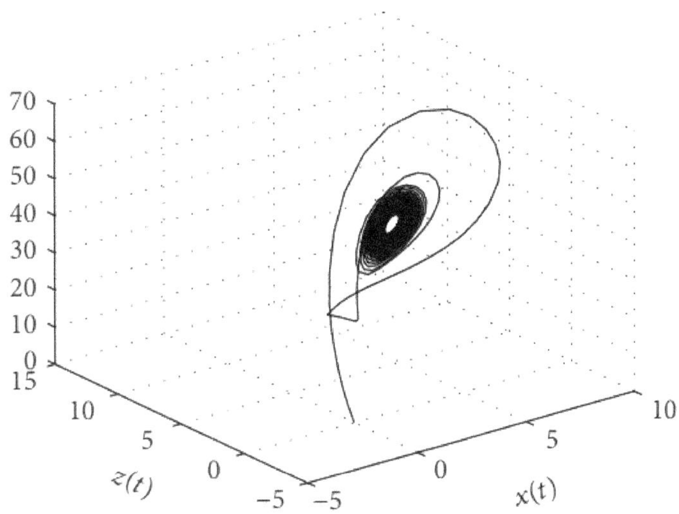

Figure 9: A Hopf bifurcation occurs at the steady state S_0 of system (5), when $k_1 = -20.3$, $\tau_1 = 0.415$.

From (18) and (25), we obtain

$$\operatorname{Re}\left(\frac{d\lambda(\tau)}{d\tau}\right)^{-1}\Bigg|_{\tau=\tau_n^{\pm}} = \frac{1}{k_1^2(\omega_{\pm}^2 + a^2)} \cdot \left(\pm\sqrt{A^2 - 4B}\right). \tag{29}$$

Thus, if $A^2 - 4B > 0$, we have

$$\operatorname{Re}\left(\frac{d\lambda(\tau)}{d\tau}\right)^{-1}\Bigg|_{\tau=\tau_n^{+}} = \frac{1}{k_1^2(\omega_{+}^2 + a^2)}\sqrt{A^2 - 4B} > 0,$$

$$\operatorname{Re}\left(\frac{d\lambda(\tau)}{d\tau}\right)^{-1}\Bigg|_{\tau=\tau_n^{-}} = -\frac{1}{k_1^2(\omega_{-}^2 + a^2)}\sqrt{A^2 - 4B} < 0. \tag{30}$$

From the above findings we have the following

Theorem 4. Let $\tau_n^{\pm}$ $(n = 0, 1, 2, \ldots)$ be defined by (26) and $\tau_0 = \min\{\tau_0^{+}, \tau_0^{-}\}$.

i. If (H_1) holds, (16) and (17) have the same number of roots with positive real part for all $\tau > 0$.
ii. If either (H_2) or (H_3) holds, when $\tau \in [0, \tau_0)$, (16) and (17) have the same number of roots with positive real part. Moreover, if the transversality conditions (30) hold, then a Hopf bifurcation occurs at the steady state S_0 in $\tau = \tau_n^{\pm}$ (see Figure 9).

From the symmetry of S_+ and S_-, it is sufficient to analyze the stability of S_+. By the linear transformation

$$x_1(t) = x(t) - x_0,$$
$$y_1(t) = y(t) - y_0,$$
$$z_1(t) = z(t) - z_0, \tag{31}$$

system (5) becomes

$$\begin{aligned}\dot{x}_1(t) &= a\left(y_1(t) - x_1(t)\right),\\ \dot{y}_1(t) &= -b_2 x_0 z_1(t) - b_2 x_1(t) z_1(t)\\ &\quad + k_1\left(y_1(t) - y_1(t-\tau)\right),\\ \dot{z}_1(t) &= (d+2l)\, x_0 x_1(t) + (d+2k)\, y_0 y_1(t) - c z_1(t)\\ &\quad + d x_1(t) y_1(t) + l x_1(t)^2 + k y_1(t)^2.\end{aligned} \tag{32}$$

The characteristic equation of system (32) in (0, 0, 0) is

$$\lambda^3 + a_2\lambda^2 + a_1\lambda + a_0 + \left(c_2\lambda^2 + c_1\lambda + c_0\right) e^{-\lambda\tau} = 0, \tag{33}$$

where

$$\begin{aligned} a_2 &= a + c - k_1,\\ a_1 &= ac - (a+c)\,k_1 + \frac{b_1 c\,(d+2k)}{d+l+k},\\ a_0 &= -ack_1 + 2ab_1 c,\\ c_2 &= k_1,\\ c_1 &= k_1(a+c),\\ c_0 &= k_1 ac.\end{aligned} \tag{34}$$

If there is no delay, (33) becomes

$$\lambda^3 + (a_2 + c_2)\lambda^2 + (a_1 + c_1)\lambda + a_0 + c_0 = 0. \tag{35}$$

Let $i\omega$ ($\omega > 0$) be a root of (33). Then, ω satisfies the equation

$$\omega^6 + \left(a_2^2 - c_2^2 - 2a_1\right)\omega^4 + \left(a_1^2 - 2a_0 a_2 - c_1^2 + 2c_0 c_2\right)\omega^2 + a_0^2 - c_0^2 = 0. \tag{36}$$

If $z=\omega^2$, then (36) becomes

$$h(z) := z^3 + pz^2 + qz + r = 0, \tag{37}$$

where

$$\begin{aligned} p &= a_2^2 - c_2^2 - 2a_1,\\ q &= a_1^2 - 2a_0 a_2 - c_1^2 + 2c_0 c_2,\\ r &= a_0^2 - c_0^2.\end{aligned} \tag{38}$$

Therefore, applying the findings from [20] we obtain the following.

Proposition 5. For the polynomial equation (37), we have the following:

(i) If $r < 0$, then (37) has at least one positive root.

(ii) If $r \geq 0$ and $\Delta = p^2 - 3q \leq 0$, then (37) has no positive roots.

(iii) If $r \geq 0$ and $\Delta > 0$, then (37) has positive roots if and only if $z_1^* = (1/2)(-p + \sqrt{\Delta}) > 0$ and $h(z_1^*) \leq 0$.

Suppose that (37) has positive roots. Without loss of generality, we assume that it has three positive roots, defined by z_1, z_2, and z_3.Then,(36) has three positive roots $\omega_h = \sqrt{zh}$, $h = 1, 2, 3$

By direct computation, we have the following.

Theorem 6. For the simple pairs of conjugate purely imaginary roots of (36), $\pm i\omega_h$, $h = 1, 2, 3$, we have

$$\tau_h^{(j)} = \frac{1}{\omega_h}\left[\arccos\frac{c_1\omega_h^2\left(\omega_h^2 - a_1\right) - \left(a_2\omega_h^2 - a_0\right)\left(c_2\omega_h^2 - c_0\right)}{\left(c_2\omega_h^2 - c_0\right)^2 + c_1^2\omega_h^2} + 2j\pi\right]. \tag{39}$$

We have $h = 1, 2, 3$, $j = 0, 1, 2, \ldots$, with

$$\begin{aligned}\tau_0 &:= \tau_{h_0}^{(0)} = \min\left\{\tau_1^{(0)}, \tau_2^{(0)}, \tau_3^{(0)}\right\},\\ \omega_0 &= \omega_{h_0}.\end{aligned} \tag{40}$$

For (33), using Proposition 5 and [21] to (33), we have the following.

Proposition 7 (see [10]). For (33) we have the following:

i. If $r \geq 0$ and $\Delta = p^2 - 3q \leq 0$, then all roots with positive real part of (33) have the same sum to those of the polynomial equation (35) for all $\tau \geq 0$.

ii. If either $r < 0$ or $r \geq 0$, $\Delta = p^2 - 3q > 0$, $z_1^* = (1/3)(-p+\sqrt{\Delta}) > 0$, and $h(z_1^*) \leq 0$, then all roots with positive real parts of (33) have the same sum to those of the polynomial equation (35) for $\tau \in [0, \tau_0)$.

Let us denote the root of (33) by $(\tau) = \mu(\tau) + i\omega(\tau)$ with $\mu(\tau_h^{(j)}) = 0,\ \omega(\tau_h^{(j)}) = \omega_h$.

Proposition 8. If $z_h = \omega_h^2$ and $h'(z_h) \neq 0$, then the transversality condition

$$\operatorname{Re}\left(\frac{d\lambda(\tau)}{d\tau}\right)\bigg|_{\tau=\tau_h^{(j)}} \neq 0 \tag{41}$$

is satisfied and $\operatorname{Re}(d\lambda(\tau)/d\tau)|_{\tau=\tau_h^{(j)}}$ and $h'(z_h)$ have the same signs.

Applying Propositions 7 and 8 to (33), we have the following theorems.

Theorem 9. Let $\tau_h^{(j)}$ and $\tau 0$ be defined by (39) and (40). Suppose that condition

$$a(d+2l) - c(d+2k) > 0 \tag{42}$$

holds.

(1) If $0 < b_1 < b_c$, then we have the following:

(i) If $k_1 \leq b_1$ and $\Delta \leq 0$, then (33) has all the roots with negative real part for all $\tau \geq 0$ and the steady state S_+ (or S_-) of system (5) is locally asymptotically stable.

(ii) If $k_1 > b_1$ or $k_1 \leq b_1$ and $\Delta > 0$, $z_1^* > 0$ and $h(z_1^*) \leq 0$, then (33) has all the roots with negative real part for all $\tau \in [0, \tau_0)$ and the steady state S_+ (or S_-) of system (5) is locally asymptotically stable.

(iii) If the conditions of (ii) hold and $h'(z_h) \neq 0$, then a Hopf bifurcation occurs at the steady state S_+ (or S_-) for $\tau = \tau_h^{(j)}$.

(2) If $b_1 > b_c$, then we have the following:

(i) If $k_1 \leq b_1$ and $\Delta \leq 0$, then (33) has two roots with positive real part for all $\tau \geq 0$ and steady state S_+ (or S_-) of system (5) is unstable.

(ii) If $k_1 > b_1$ or $k_1 \leq b_1$ and $\Delta > 0$, $z_1^* > 0$ and $h(z_1^*) \leq 0$, then (33) has two roots with positive real part for all $\tau \in [0, \tau_0)$ and the steady state S_+ (or S_-) of system (5) is unstable.

(iii) If the conditions of (ii) hold and $h'(z_h^*) \neq 0$, then a Hopf bifurcation occurs at the steady state S_+ (or S_-) for $\tau = \tau_h^{(j)}$.

Direction and Stability of the Hopf Bifurcation

In the previous section, for system (5), we have obtained conditions for the Hopf bifurcations to occur for a sequence of values of τ. Using the techniques from normal form theory and center manifold theory introduced by [22], we determine the direction, the stability, and the periodicity of the bifurcating solutions. Let (x_*, y_*, z_*) be the steady state of (5), where system (5) undergoes Hopf bifurcations at $\tau = \tau_h$ and $\pm i\omega_h$ are the corresponding pure imaginary roots of the characteristic equation.

For convenience, let $x_1(t) = x(t) - x_*$, $x_2(t) = y(t) - y_*$, and $x_3(t) = z(t) - z_*$, $\overline{x}_i(t) = x_i(t\tau)$, $\tau = \tau_h + \mu$ and dropping the bars for simplification of notation, system (5) can be written as a FDE (functional differential equation) in $C = C([-1, 0], \mathbb{R}^3)$ as follows:

$$\dot{u}(t) = L_\mu(u_t) + f(\mu, u_t), \tag{43}$$

where $u(t) = (u_1(t), u_2(t), u_3(t))^T \in \mathbb{R}^3$ and $L_\mu : C \to$ □, f : □ $\times C \to$ □ are given, restrictively, by

$$L_\mu(\phi) = (\tau_h + \mu) \cdot \begin{pmatrix} -a & a & 0 \\ 0 & k_1 & -b_2 x_* \\ (d+2l)y_* & (d+2k)x_* & -c \end{pmatrix} \begin{pmatrix} \phi_1(0) \\ \phi_2(0) \\ \phi_3(0) \end{pmatrix} + (\tau_h + \mu) \cdot \begin{pmatrix} 0 & 0 & 0 \\ 0 & -k_1 & 0 \\ 0 & 0 & 0 \end{pmatrix} \begin{pmatrix} 0 \\ -b_2\phi_1(0)\phi_3(0) \\ d\phi_1(0)\phi_2(0) + l\phi_1(0)^2 + k\phi_2(0)^2 \end{pmatrix}. \tag{44}$$

From the previous section, if $\mu = 0$, system (5) undergoes Hopf bifurcations at (x_*, y_*, z_*). By Riesz representation theorem, there exists a function (θ, μ) of bounded variation for $\theta \in [-1, 0]$ so that

$$L_\mu(\phi) = \int_{-1}^{0} d\eta(\theta, 0)\phi(\theta), \quad \phi \in C[-1, 0]. \tag{45}$$

We can choose

$$\eta(\theta, \mu) = (\tau_h + \mu) \begin{pmatrix} -a & a & 0 \\ 0 & k_1 & -b_2 x_* \\ (d+2l)y_* & (d+2k)x_* & -c \end{pmatrix} \delta(\theta) - (\tau_h + \mu) \begin{pmatrix} 0 & 0 & 0 \\ 0 & -k_1 & 0 \\ 0 & 0 & 0 \end{pmatrix} \delta(\theta + 1), \tag{46}$$

where δ is the Dirac delta function.

For $\phi \in ([-1, 0], (\mathbb{R}^3)^*)$ we define

$$A(\mu)\phi = \begin{cases} \dfrac{d\phi(\theta)}{d\theta}, & \theta \in [-1, 0) \\ \displaystyle\int_{-1}^{0} d\eta(\mu, s)\phi(s), & \theta = 0, \end{cases}$$

$$R(\mu)\phi = \begin{cases} 0, & \theta \in [-1, 0) \\ f(\mu, \phi(\theta)), & \theta = 0. \end{cases} \tag{47}$$

Then, system (43) is equivalent to the abstract differential equation:

$$\dot{u}_t = A(\mu)u_t + R(\mu)u_t, \tag{48}$$

where $u_t(\theta) = u(t+\theta)$, for $\theta \in [-1, 0]$.

For $\Psi \in C^1([0, 1], \mathbb{R}^3)$, we define

$$A^*\Psi(s) = \begin{cases} -\dfrac{d\Psi(s)}{ds}, & s \in (0,1] \\ \displaystyle\int_{-1}^{0} d\eta^T(t,\theta)\Psi(-t), & s = 0 \end{cases} \tag{49}$$

and the bilinear form

$$\langle \Psi(s), \phi(\theta)\rangle$$
$$= \overline{\Psi}(0)\phi(0) - \int_{\theta=-1}^{0}\int_{\varepsilon=0}^{\theta} \overline{\Psi}(\varepsilon-\theta)\, d\eta(\theta)\,\phi(\varepsilon)\, d\varepsilon, \tag{50}$$

where $\eta(\theta) = \eta(\theta, 0)$.

Then, $A = (0)$ and $A^* = A^*(0)$ are adjoining operators. From Section 2, $\pm\omega_h\tau_h$ are eigenvalues of A; thus, they are also eigenvalues of A^*.

By direct computation, we obtain that

$$q(\theta) = q_0 e^{i\theta\omega_h\tau_h}, \tag{51}$$

with

$$q_0 = (1, \alpha, \beta)^T$$
$$= \left(1, \frac{b_2 + i\omega_h}{b_2}, \frac{b_2(x_* + y_*) + i\omega_h x_*}{b_2(c + i\omega_h)}\right) \tag{52}$$

the eigenvector of A corresponding to $i\omega_h\tau_h$ and

$$q^*(s) = Dq_0^* e^{is\omega_h\tau_h} \tag{53}$$

with

$$q_0^* = (1, \alpha^*, \beta^*)^T$$
$$= \left(1, -\frac{(a - i\omega_h)(c - i\omega_h)}{a(d+2l)(d+2k)x_* y_*}, \frac{a - i\omega_h}{(d+2l)y_*}\right) \tag{54}$$

the eigenvector of A^* corresponding to $-i\omega_h\tau_h$, where

$$D = \frac{1}{1 + \overline{\alpha}\alpha^* + \overline{\beta}\beta^* - k_1\tau_h\overline{\alpha}\alpha^* e^{i\omega_h\tau_h}}. \tag{55}$$

Using the same notations as in [22], we compute the coordinates to describe the center manifold C_0 at $\mu = 0$.

Let ut be the solution of (43) when $\mu = 0$ and define

$$z(t) = \langle q^*, u_t\rangle,$$
$$W(t,\theta) = u_t(\theta) - 2\mathrm{Re}\{z(t)q(\theta)\}. \tag{56}$$

On the center manifold C_0 we have

$$W(t,\theta) = W(z(t), \overline{z}(\theta), \theta), \tag{57}$$

where

$$W(z,\overline{z},\theta) = W_{20}(\theta)\frac{z^2}{2} + W_{11}(\theta)z\overline{z} + W_{02}(\theta)\frac{\overline{z}^2}{2}$$
$$+ W_{30}(\theta)\frac{z^3}{6} + \cdots \tag{58}$$

and z and $\overline{z}$ are local coordinates for center manifold C_0 in the direction of q^* and $\overline{q}^*$. Note that W is real if ut is real. We consider only real solutions. For the solution $ut \in C_0$ of (43), since $\mu = 0$, we have

$$\dot{z}(t) = \langle q^*, \dot{u}_t\rangle = \langle q^*, A(\mu)u_t + R(\mu)u_t\rangle$$
$$= i\tau_h\omega_h z$$
$$+ \overline{q}^*(0) f(0, W(z,\overline{z},0) + 2\mathrm{Re}\{zq(0)\})$$
$$= i\tau_h\omega_h z + \overline{q}^*(0) f_0(z,\overline{z}). \tag{59}$$

We rewrite this equation as

$$\dot{z}(t) = i\tau_h\omega_h z(t) + g(z(t), \overline{z}(t)), \tag{60}$$

where

$$g(z,\overline{z}) = g_{20}\frac{z^2}{2} + g_{11}z\overline{z} + g_{01}\frac{\overline{z}^2}{2} + g_{21}\frac{z^2\overline{z}}{2} + \cdots. \tag{61}$$

Note that

$$u_t(\theta) = (u_{1t}(\theta), u_{2t}(\theta), u_{3t}(\theta))$$
$$= W(t,\theta) + zq(\theta) + \overline{zq}(\theta) \tag{62}$$

and $q(\theta) = (1 - \alpha, \beta)^T e^{i\theta\omega_h\tau_h}$.

We have

$$u_{1t}(0) = z + \overline{z} + W_{20}^{(1)}(0)\frac{z^2}{2} + W_{11}^{(1)}(0)z\overline{z}$$
$$+ W_{02}^{(1)}(0)\frac{\overline{z}^2}{2} + \cdots,$$
$$u_{2t}(0) = \alpha z + \overline{\alpha}\,\overline{z} + W_{20}^{(2)}(0)\frac{z^2}{2} + W_{11}^{(2)}(0)z\overline{z}$$
$$+ W_{02}^{(2)}(0)\frac{\overline{z}^2}{2} + \cdots,$$
$$u_{3t}(0) = \beta z + \overline{\beta z} + W_{20}^{(3)}(0)\frac{z^2}{2} + W_{11}^{(3)}(0)z\overline{z}$$
$$+ W_{02}^{(3)}(0)\frac{\overline{z}^2}{2} + \cdots. \tag{63}$$

Using (61), we obtain

$$g_{20} = 2\overline{D}\tau_h \left(d\overline{\beta}^* \alpha - b_2\overline{\alpha}^* \beta + \overline{\beta}^* l + \overline{\beta}^* k\alpha^2 \right),$$

$$g_{11} = \overline{D}\tau_h \left(-b_2\alpha^* \left(\beta + \overline{\beta}\right) + \beta^* \left(d\left(\alpha + \overline{\alpha}\right) + 2l + 2k\alpha\overline{\alpha}\right)\right),$$

$$g_{02} = \overline{D}\tau_h \left(d\overline{\beta}^* \overline{\alpha} - b_2\overline{\alpha}^* \overline{\beta} + \overline{\beta}^* l + \beta^* k\overline{\alpha}^2 \right),$$

$$\begin{aligned} g_{21} = &- b_2\overline{D}\tau_h\overline{\alpha}^* \left(2W_{20}^{(3)}(0) + W_{20}^{(3)}(0) + 2\beta W_{11}^{(1)}(0) + \overline{\beta}W_{20}^{(1)}(0)\right) + \overline{D}\tau_h\overline{\beta}^* \left[d\alpha W_{11}^{(2)}(0) + dW_{20}^{(2)}(0) \right.\\ &+ 2\alpha d W_{11}^{(1)}(0) + d\overline{\alpha}W_{20}^{(2)}(0) \\ &+ l\left(2W_{11}^{(1)}(0) + W_{20}^{(1)}(0)\right) \\ &\left. + k\left(2\alpha W_{11}^{(2)}(0) + \overline{\alpha}W_{02}^{(2)}(0)\right)\right]. \end{aligned} \tag{64}$$

Since there are $W_{20}(0)$ and $W_{11}(0)$ in g_{21}, we need to compute them.

From (43) and (56), we have

$$\begin{aligned} \dot{W} &= \dot{x}_t - \dot{z}q - \dot{\overline{z}}\overline{q} \\ &= \begin{cases} AW - 2\mathrm{Re}\{\overline{q}^*(0) f_0 q(\theta)\}, & \theta \in [-1, 0) \\ AW - 2\mathrm{Re}\{\overline{q}^*(0) f_0 q(0)\} + f_0, & \theta = 0 \end{cases} \\ &= AW + H(z, \overline{z}, \theta), \end{aligned} \tag{65}$$

where

$$H(z, \overline{z}, \theta) = H_{20}(\theta)\frac{z^2}{2} + H_{11}(\theta) z\overline{z} + H_{02}(\theta)\frac{\overline{z}^2}{2} + \cdots. \tag{66}$$

Expending the above series and comparing the corresponding coefficients, we obtain

$$\begin{aligned} \left(A - 2i\tau_h\omega_h\right) W_{20}(\theta) &= -H_{20}(\theta), \\ AW_{11}(\theta) &= -H_{11}(\theta). \end{aligned} \tag{67}$$

From (65), we know that, for $\theta \in [-1, 0)$,

$$\begin{aligned} H(z, \overline{z}, \theta) &= -\overline{q}^*(0) f_0 q(\theta) - q^*(0)\overline{f}_0\overline{q}(0) \\ &= -gq(\theta) + \overline{q}q(\theta). \end{aligned} \tag{68}$$

Comparing the coefficients with (66), we obtain

$$\begin{aligned} H_{20}(\theta) &= -g_{20}q(\theta) - \overline{g}_{02}\overline{q}(\theta), \\ H_{11}(\theta) &= -g_{11}q(\theta) - \overline{g}_{11}\overline{q}(\theta). \end{aligned} \tag{69}$$

From (67) and (68) and the definition of A, it follows that

$$\dot{W}_{20}(\theta) = 2i\tau_h\omega_h W_{20}(\theta) + g_{20}q(\theta) + \overline{g}_{02}\overline{q}(\theta). \tag{70}$$

Notice that $q(\theta) = (1, \alpha, \beta)^T e^{i\theta\omega_h\tau_h}$ and

$$W_{20}(\theta) = \frac{ig_{20}}{\omega_h\tau_h} q(0) e^{i\theta\omega_h\tau_h} + \frac{i\overline{g}_{02}}{3\omega_h\tau_h}\overline{q}(0)\overline{e}^{i\theta\omega_h\tau_h} + E_1 e^{2i\theta\omega_h\tau_h}, \tag{71}$$

where $E_1 = (E_1^{(1)}, E_1^{(2)}, E_1^{(3)})^T \in \mathbb{R}^3$ is a constant vector. In a similar way, we can obtain

$$W_{11}(\theta) = -\frac{ig_{11}}{\omega_h\tau_h} q(0) e^{i\theta\omega_h\tau_h} + \frac{i\overline{g}_{11}}{\omega_h\tau_h}\overline{q}(0)\overline{e}^{i\theta\omega_h\tau_h} + E_2, \tag{72}$$

where $E_2 = (E_2^{(1)}, E_2^{(2)}, E_2^{(3)})^T \in \mathbb{R}^3$ is a constant vector. From the definition of A and (67), we obtain

$$\int_{-1}^{0} d\eta(\theta) W_{20}(\theta) = 2i\omega_h\tau_h W_{20}(0) - H_{20}(0), \tag{73}$$

$$\int_{-1}^{0} d\eta(\theta) W_{11}(\theta) = -H_{11}(0), \tag{74}$$

where $\eta(\theta) = \eta(\theta, 0)$. By (65), we have

$$H_{20}(0) = -g_{20}q(0) - \overline{g}_{02}\overline{q}(0) + 2\tau_h \begin{pmatrix} 0 \\ -\beta \\ \alpha \end{pmatrix}, \tag{75}$$

$$H_{11}(0) = -g_{11}q(0) - \overline{g}_{11}\overline{q}(0) + 2\tau_h \begin{pmatrix} 0 \\ -\mathrm{Re}(\beta) \\ \mathrm{Re}(\alpha) \end{pmatrix}. \tag{76}$$

Substituting (71) and (75) into (73), we obtain

$$\left(2i\omega_h\tau_h I - \int_{-1}^{0} e^{2i\theta\omega_h\tau_h} d\eta(\theta)\right) E_1 = 2\tau_h \begin{pmatrix} 0 \\ -\beta \\ \alpha \end{pmatrix} \tag{77}$$

which leads to

$$\begin{pmatrix} 2i\omega_h + a & -a & 0 \\ 0 & 2i\omega_h - k_1 + k_1 e^{-2\omega_h\tau_h} & b_2 x_* \\ -(d + 2l) y_* & -(d + 2k) x_* & 2i\omega_h + c \end{pmatrix} E_1 = 2\begin{pmatrix} 0 \\ -\beta \\ \alpha \end{pmatrix},$$

$$E_1^{(1)} = \frac{-2a\left(c\beta + 2i\omega_h\beta + b_2\alpha x_*\right)}{\Delta_1},$$

$$E_1^{(2)} = -\frac{\beta\left(2i\omega_h + c\right) + \alpha b_2 x_*}{\Delta_1},$$

$$E_1^{(3)} = \frac{\alpha\left(2i\omega_h + a\right)\left(2i\omega_h - k_1 + k_1 e^{-2\omega_h\tau_h}\right)}{\Delta_1} - \frac{\beta a (d + 2l) y_* - \beta (d + 2k) x_* \left(2i\omega_h + a\right)}{\Delta_1},$$

$$\Delta_1 = \det\begin{pmatrix} 2i\omega_h + a & -a & 0 \\ 0 & 2i\omega_h - k_1 + k_1 e^{-2\omega_h\tau_h} & b_2 x_* \\ -(d + 2l) y_* & -(d + 2k) x_* & 2i\omega_h + c \end{pmatrix}. \tag{78}$$

In a similar way, substituting (72) and (76) into (74), we obtain

$$\begin{pmatrix} a & -a & 0 \\ 0 & 0 & b_2 x_* \\ -(d+2l)\,y_* & -(d+2k)\,x_* & c \end{pmatrix} E_2 = 2\begin{pmatrix} 0 \\ -\mathrm{Re}(\beta) \\ \mathrm{Re}(\alpha) \end{pmatrix},$$

$$E_2^{(1)} = E_1^{(2)} = \frac{-2ab_2 x_* \mathrm{Re}(\alpha) - 2ac\mathrm{Re}(\beta)}{\Delta_2},$$

$$E_2^{(3)} = \frac{-2a\mathrm{Re}(\beta)\left((d+2k)\,x_* + (d+2l)\,y_*\right)}{\Delta_2},$$

$$\Delta_2 = ab_2 x_* \left((d+2k)\,x_* + (d+2l)\,y_*\right). \tag{79}$$

Thus, we can determine $W_{20}(0)$ and $W11(0)$ from (71) and (72). Furthermore, we can determine g_{21}. Therefore, each g_{ij} in (64) is determined by the parameters and delay in (43). Thus, we can compute the following values:

$$C_1(0) = \frac{i}{2\tau_h \omega_h}\left(g_{11}g_{20} - 2|g_{11}|^2 - \frac{|g_{02}|^2}{3}\right) + \frac{g_{21}}{2},$$

$$\mu_2 = -\frac{\mathrm{Re}(C_1(0))}{\mathrm{Re}(\lambda'(\tau_h))},$$

$$\beta_2 = 2\mathrm{Re}(C_1(0)),$$

$$T_2 = -\frac{\mathrm{Im}(C_1(0)) + \mu_2 \mathrm{Im}(\lambda'(\tau_h))}{\tau_h \omega_h}. \tag{80}$$

The above quantities characterize the bifurcating periodic solutions in the center manifold at the critical value τ_h [22, 23]:

(i) $\mu = 0$ is the Hopf bifurcation of system (43).

(ii) μ_2 determines the direction of the Hopf bifurcation: if $\mu_2 > 0$ ($\mu_2 < 0$), then the Hopf bifurcation is supercritical (subcritical) and the bifurcating periodic solution exists for $\tau > \tau_h$ ($\tau < \tau_h$).

(iii) β_2 determines the stability of the bifurcating periodic solutions: the bifurcating periodic solutions are stable (unstable) if $\beta_2 < 0$ ($\beta_2 > 0$).

(iv) T_2 determines the period of the bifurcating periodic solutions: the period increases (decreases) if $T_2 > 0$ ($T_2 < 0$).

ANALYSIS OF SYSTEM (6)

Let $q_1 = q_2 = q_3 = q \in (0, 1)$. In this case, the fractional order system is commesurate-order [24].

Proposition 10. The initial value problem of the commensurate order system (6) can be rewritten as follows:

$$D^q X(t) = AX(t) + x(t)BX(t) + y(t)CX(t),$$
$$X(0) = (x_0, y_0, z_0)^T, \tag{81}$$

where $0 < t \leq T$, $X(t) = (x(t), y(t), z(t)) \in \mathbb{R}^3$

$$A = \begin{pmatrix} -a & a & 0 \\ b_1 & 0 & 0 \\ 0 & 0 & -c \end{pmatrix},$$

$$B = \begin{pmatrix} 0 & 0 & 0 \\ 0 & 0 & -b_2 \\ l & d & 0 \end{pmatrix},$$

$$C = \begin{pmatrix} 0 & 0 & 0 \\ 0 & 0 & 0 \\ 0 & k & 0 \end{pmatrix}. \tag{82}$$

Then, for constant $T = \min\{T^*, (\epsilon\Gamma(q+1)/\|f\|_\infty)^{1/q}\}$, $T^* > 0$, fractional order system (6) has a unique solution, where $\Gamma(\cdot)$ is the Gamma function.

Proof. Consider the function $f(X(t)) = AX(t) + x(t)BX(t) + y(t)CX(t)$, which is continuous and bounded on the interval $X \in [0, T^*] \times [X_0 - \epsilon, X_0 + \epsilon]$, for any $T^*, \epsilon > 0$. For $X(t) = (x(t), y(t), z(t))^T$ and $Y(t) = (x_1(t), y_1(t), z_1(t))^T$, where $Y(t) \in [0, T^*] \times [X_0 - \epsilon, X_0 + \epsilon]$, we can obtain

$$|f(X(t)) - f(Y(t))| \leq L(|X(t) - Y(t)|), \tag{83}$$

where $L = \|A\| + (\|B\| + \|C\|)(2|X_0| + 2\epsilon) > 0$.

Therefore, the fractional-order general T system meets the Lipschitz condition. Then, according to the existence and uniqueness theorem of the fractional-order system [25], the initial value problem of the commensurate order system (81) has a unique solution in the interval $T = \min\{T^*, (\epsilon\Gamma(q+1)/\|f\|_\infty)^{1/q}\}$.

System (6) is dissipative for $a+c > 0$, because

$$\nabla V(x, y, z) = -a - c, \tag{84}$$

where $V(x, y, z) = (1/2)(x^2 + y^2 + z^2)$. Meanwhile, it is convergent in an exponential rate

$$\frac{dV(X(t))}{dt} = e^{-(a+c)}. \tag{85}$$

That is to say, an initial volume $V0$ is the volume element at time t contraction of volume element $V_0 e^{-(a+c)t}$. When $t \to \infty$ all the trajectory of systems will eventually be restricted in a volume element for zero point sets, and incremental dynamic behavior of it will be fixed in an attractor.

In order to determine the stability conditions for the steady states S_- and S_+, we first consider the integer-order case. Based on Proposition 2, S_- and S_+ are locally asymptotically stable if and only if $0 < b_1 < b_c$, where

$$b_c = \frac{a(a+c)(d+l+k)}{a(d+2l) - c(d+2k)} \tag{86}$$

and $a(d+2l) - c(d+2k) > 0$. It is known that the fractional-order system is at least as stable as their integer-order counterpart, so we have the following conclusion.

Proposition 11. The steady states S_- and S_+ are stable with $q \in (0, 1)$ for $b_1 < b_c$.

When $b_1 > b_c$, the steady states S_- and S_+ become unstable in an integer-order system but may be stable in the fractionalorder case.

CHAOS CONTROL OF SYSTEM (6)

In this section, we want to control the chaos for the fractionalorder general T system (6) to the steady state denoted by S via feedback control.

An n-dimensional fractional-order system can be described as [13]

$$D^{q}X(t) = f(X(t)), \quad (87)$$

where $X(t) \in \mathbb{R}^n$, $q \in (0, 1)$.

The system with controller is given by [13]

$$D^{q}X(t) = f(X(t)) - K(X(t) - S), \quad (88)$$

where $K = \mathrm{diag}(k_1, k_2, \ldots, k_n)$ is the matrix of positive feedback gains and S is the steady state of system (87).

The controlled fractional-order general T system (6) is

$$\begin{aligned} D^{q_1}x(t) &= a(y(t) - x(t)) - u_1(t), \\ D^{q_2}y(t) &= b_1x(t) - b_2x(t)z(t) - u_2(t), \\ D^{q_3}z(t) &= -cz(t) + dx(t)y(t) + lx(t)^2 + ky(t)^2 \\ &\quad - u_3(t), \end{aligned} \quad (89)$$

where $u_i(t)$, $i = 1, 2, 3$, are the external control inputs. The control law of single state variables feedback has the following form [13]:

$$\begin{aligned} u_1(t) &= k_1(x(t) - x_{01}), \\ u_2(t) &= k_2(y(t) - y_{02}), \\ u_3(t) &= k_3(z(t) - z_{03}), \end{aligned} \quad (90)$$

where $K = \mathrm{diag}(k_1, k_2, k_3)$ is the matrix of positive feedback gains and $S = (x_{01}, y_{02}, z_{03})\,T$ is the steady state of system (6). The characteristic equation of the controlled system (89) evaluated at the steady state S is

$$(\lambda + a + k_1)(\lambda + k_2)(\lambda + c + k_3) = 0. \quad (91)$$

The fractional-order Routh-Hurwitz conditions lead to

$$\begin{aligned} a + k_1 &> 0, \\ k_2 &> 0, \\ c + k_3 &> 0. \end{aligned} \quad (92)$$

SYNCHRONIZATION OF SYSTEM (6)

The nonlinear control method is used to design control in order to make the drive system (6) and response system state synchronization [14]. Two identical systems are introduced, one is the drive system and the other system added a nonlinear control to be the response system. Corresponding to (6), the drive system is

$$\begin{aligned} D^{q_1}x_1(t) &= a(y_1(t) - x_1(t)), \\ D^{q_2}y_1(t) &= b_1x_1(t) - b_2x_1(t)z_1(t), \\ D^{q_3}z_1(t) &= -cz_1(t) + dx_1(t)y_1(t) + lx_1(t)^2 \\ &\quad + ky_1(t)^2, \end{aligned} \quad (93)$$

and the response system is

$$\begin{aligned} D^{q_1}x_2(t) &= a(y_2(t) - x_2(t)) + v_1(t), \\ D^{q_2}y_2(t) &= b_1x_2(t) - b_2x_2(t)z_2(t) + v_2(t), \\ D^{q_3}z_2(t) &= -cz_2(t) + dx_2(t)y_2(t) + lx_2(t)^2 \\ &\quad + ky_2(t)^2 + v_3(t), \end{aligned} \quad (94)$$

where $V(t) = (V_1(t), V_2(t), V_3(t))T$ is the nonlinear synchronization controller. The drive system and the response system achieve synchronization under the driver of (t). From (93) and (94), the error system is obtained:

$$\begin{aligned} D^{q_1}e_1(t) &= a(e_2(t) - e_1(t)) + v_1(t), \\ D^{q_2}e_2(t) &= b_1e_1(t) - b_2x_2(t)e_3 - b_2x_1(t)e_1 + v_2(t), \\ D^{q_3}e_3(t) &= -ce_3(t) + dx_2(t)e_2(t) + dy_1(t)e_1(t) \\ &\quad + l(x_1(t) + x_2(t))e_1(t) \\ &\quad + k(y_1(t) + y_2(t))e_2(t) + v_3(t), \end{aligned} \quad (95)$$

where $e_1(t) = x_2(t) - x_1(t)$, $e_2(t) = y_2(t) - y_1(t)$, $e_3 = z_2(t) - z_1(t)$. We have to find the proper control function $V_i(t)$, $i = 1, 2, 3$, so that the response system (94) globally synchronizes with drive system (93); that is, $\lim_{t \to +\infty}\|e(t)\| = 0$, where $e(t) = (e_1(t), e_2(t), e_3(t))T$. According to [14], we propose the following control law for system (94):

$$\begin{aligned} v_1(t) &= ae_1(t) - k_1e_1(t), \\ v_2(t) &= -b_1e_1(t) - ae_1(t) + b_2z_1(t)e_1(t) \\ &\quad + b_2x_2(t)e_3(t) - k_2e_2(t), \\ v_3(t) &= ce_3(t) - dy_1(t)e_1(t) \\ &\quad - le_1(t)(x_1(t) + x_2(t)) - dx_2(t)e_2(t) \\ &\quad - k(y_2(t) + y_1(t))e_2(t), \end{aligned} \quad (96)$$

where k_1, k_2, and k_3 are the control parameters.

Proposition 12 (see [14]). For any initial conditions, if $k_i > 0$, $i = 1, 2, 3$ then the drive system and response system will synchronize.

NUMERICAL SIMULATIONS

Now, we illustrate the findings from the previous sections. We have proved that at some critical values of the delay, a family of periodic solutions bifurcate from the steady states of system (5) and the stability of the steady states may be changed. The numerical simulations indicate that when the delay passes through certain critical values, chaotic oscillation is converted into a stable steady state or a periodic orbit.

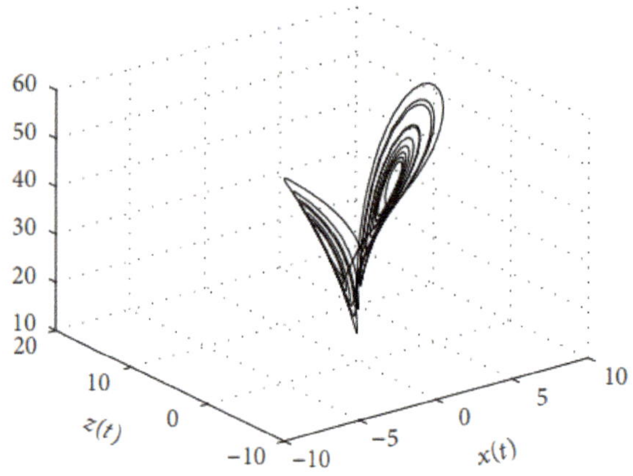

Figure 10: For system (97), there is chaos when $k_1 = -3$, $\tau = 0.01$.

We consider the delayed feedback control system (5) in the following particular form:

$$\dot{x}(t) = 10\left(y(t) - x(t)\right),$$

$$\dot{y}(t) = 40x(t) - x(t)z(t) + k_1\left(y(t) - y(t-\tau)\right),$$

$$\dot{z}(t) = -2.5z(t) + 3x(t)y(t) + 0.9x(t)^2 + 2y(t)^2, \quad (97)$$

which has three steady states $S_0 = (0, 0, 0)$, $S_- = (-4.11, -4.11, 40)$, and $S_+ = (4.11, 4.11, 40)$. When $\tau = 0$ and $k_1 = 0$, system (97) is chaotic.

The characteristic equation of system (97) is given by

$$\lambda^3 + (12.5 - k_1)\lambda^2 + (143.644 - 12.5k_1)\lambda + 2000 - 10k_1 + k_1\left(\lambda^2 + 12.5\lambda + 25\right)e^{-\lambda\tau} = 0. \quad (98)$$

If there is no delay, (98) has a negative root and a pair of complex roots with positive real part.

Using Theorem 9, we have the following:

(i) For $5.0413 < k_1 < 28.403$, $r = 10^5(40 - k_1) > 0$, $\Delta = 1.052 \cdot 10^5 - 3101.694k_1 < 0$, (98) has two roots with positive real part for all $\tau \geq 0$, and the steady state S_+ (or $S-$) of system (97) is unstable.

(ii) For $k_1 < 5.0413$ chaos can occur.

In what follows, we consider $k_1 = -3$. In this case: $\omega_1 = 11.58361053$, $\omega_2 = 13.32108470$,

$$\tau_1^j = 0.473158 + \frac{2j}{\omega_1},$$

$$\tau_2^j = 0.51716 + \frac{2j}{\omega_2},$$

$$j = 0, 1, 2, \ldots. \quad (99)$$

Then, $\tau_1^0 < \tau_2^0 < \tau_2^1 < \tau_1^1$.

i. For $\tau \in [0, \tau_2^0) \cup (\tau_2^1, \infty)$, the steady states S_+ and S_- of system (97) are unstable.

ii. For $\tau \in [\tau_2^0, \tau_2^1)$, the steady states $S+$ and $S-$ of system (97) are locally asymptotically stable.

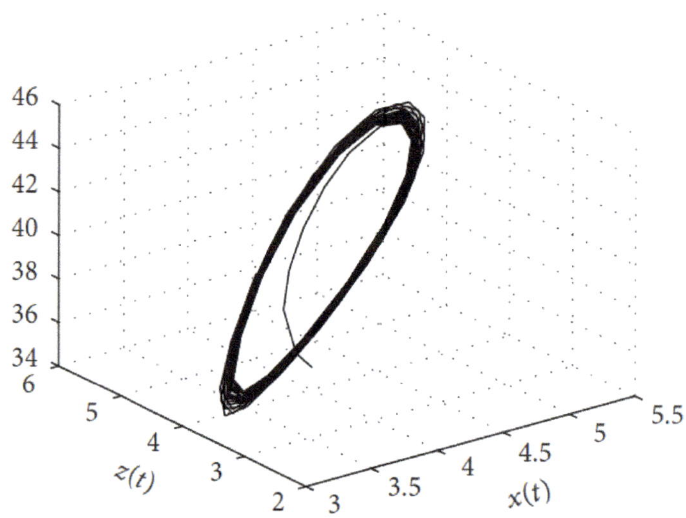

Figure 11: For system (97), there is chaos when $k_1 = -3$, $\tau = 0.051$.

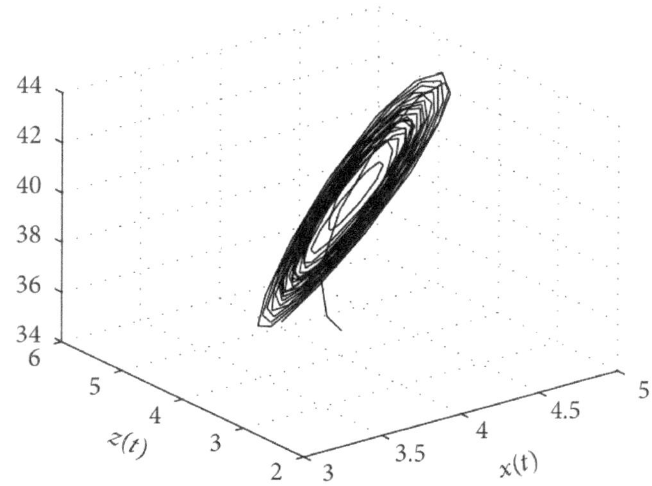

Figure 12: For system (97), an unstable periodic solution bifurcates from the steady state S_+ when $k_1 = -3$, $\tau = 1.5$.

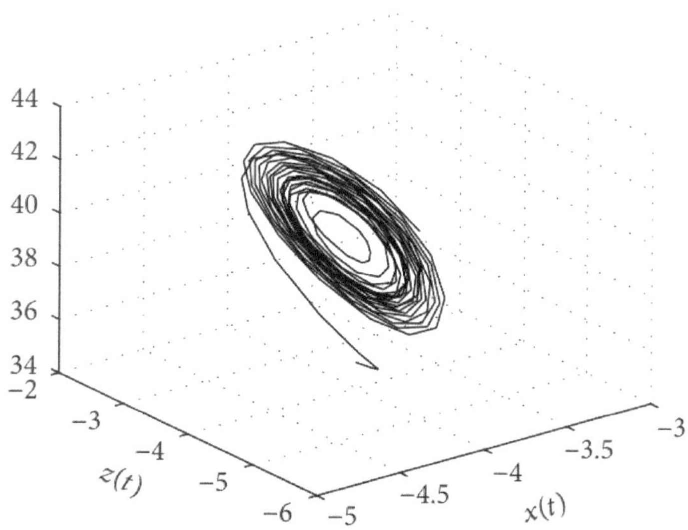

Figure 13: For system (97), an unstable periodic solution bifurcates from the steady state S_- when $k_1 = -3$, $\tau = 1.5$.

iii. For $\tau = \tau_h^j$, for $h = 1, 2$, $j = 0, 1, 2, \ldots$, system (97) undergos a Hopf bifurcation at the steady states S_+ and S_-.

For values of τ, which satisfy the above conditions, we obtain the dynamical behaviors in Figures 10, 11, 12, and 13. For the numerical simulation of the fractional differential equations (FDE) (6), with $a = 10$, $b_2 = 1$, $c = 2.5$, $d = 1$, $l = 2$, $k = 2$, and $q_1 = q_2 = q_3 = 0.99$, we use the method from [26, 27]. We obtain Figures 14, 15, and 16.

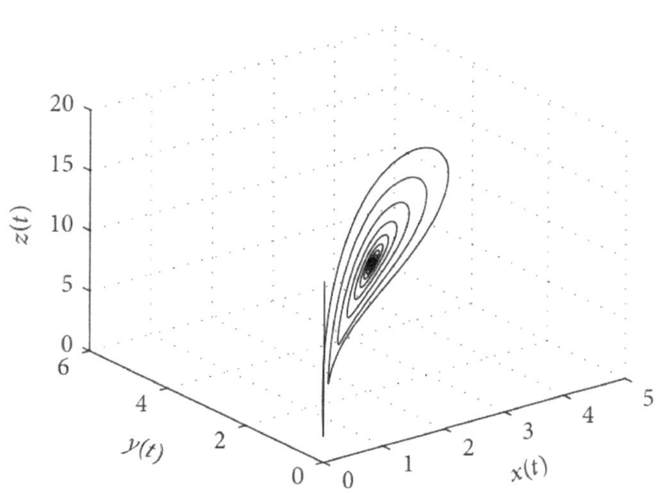

Figure 14: Asymptotic stable solution for the fractional general T system (6) when $b_1 = 10$.

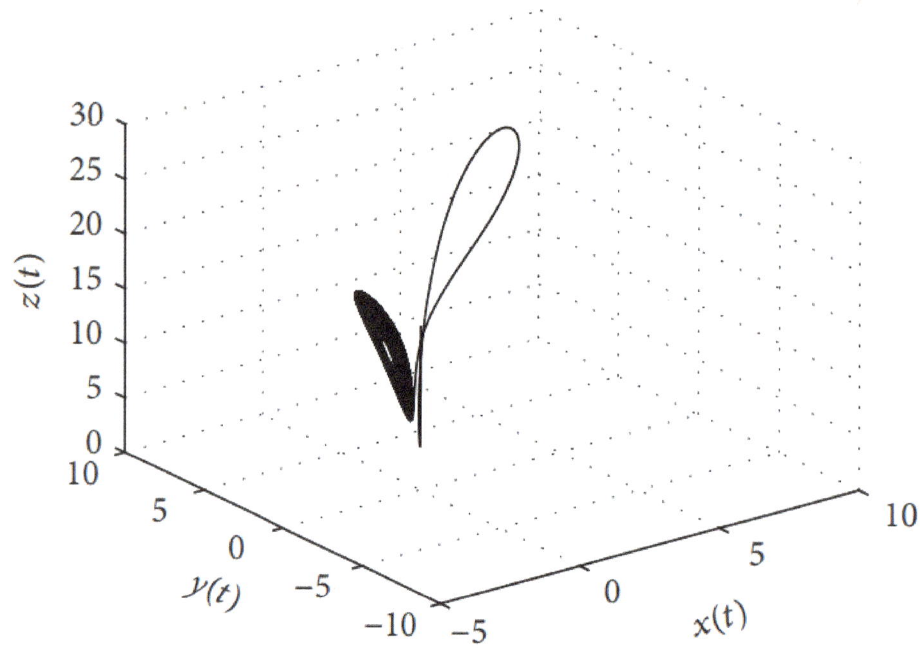

Figure 15: Periodic solution for the fractional general T system (6) when $b_1 = 16.6$.

For the controlled fractional order general T-system (89), using $K = \text{diag}(5, 5, 1)$ the matrix of positive feedback gains, we obtain Figure 17.

CONCLUSION

In the present paper we introduce a generalized T-system where the time delay is present.The linear stability is analyzed by using the Routh Hurwitz criterion. The existence of the Hopf bifurcation is studied. Then, the direction and the stability of the bifurcating periodic solutions are determined by using the normal form theory and the center manifold theorem. Chaotic behavior is also taken into account. The numerical simulations show that when the delay passes through certain critical values, chaotic oscillation is converted into a stable steady state or a periodic orbit. Furthermore, the fractional-order general T system has been proposed. The dynamics, chaos control as well as synchronization have been investigated.

The present study will be continued for the system which describes the financial risk [28]. Also, as in [29], the fractional-order chaotic complex system will be taken into consideration.

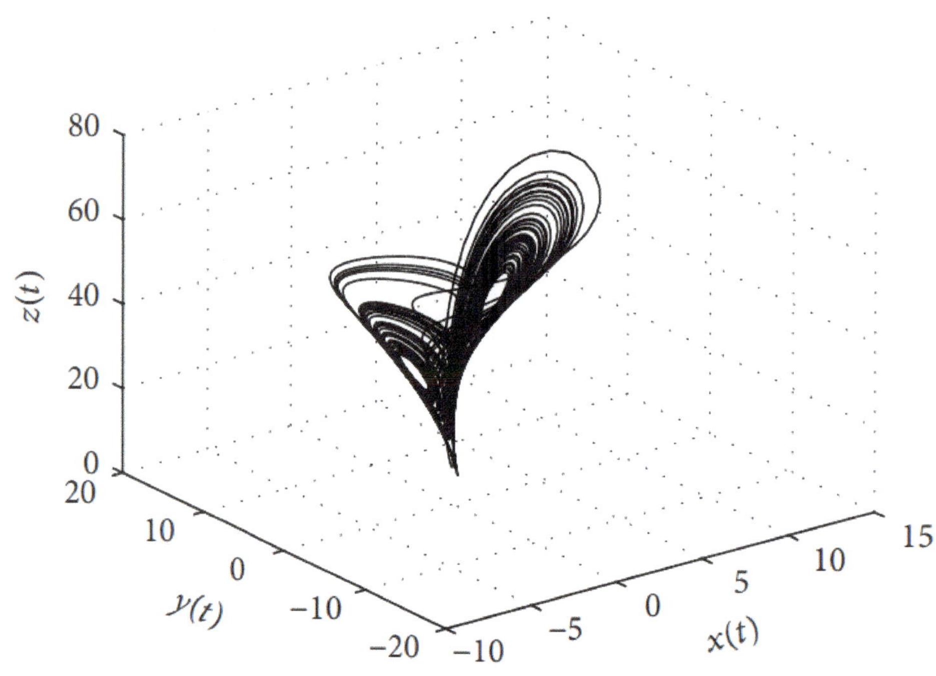

Figure 16: Chaos occurs for the fractional general T system (6) when $b_1 = 40$.

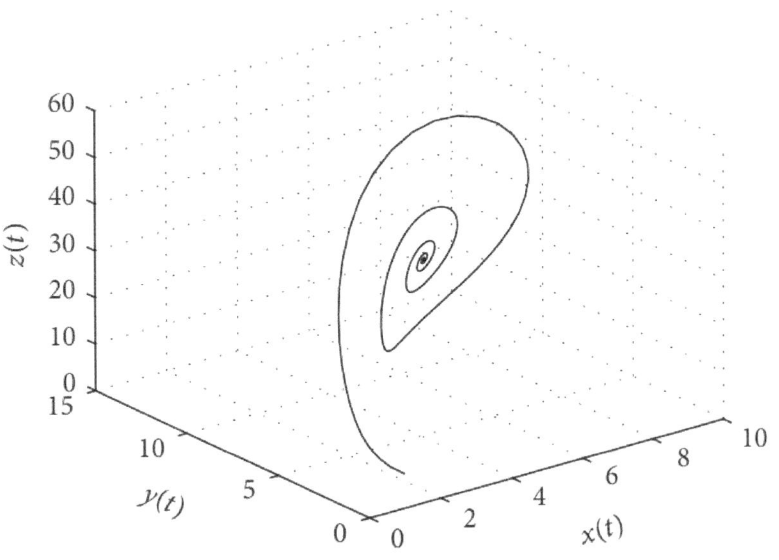

Figure 17: Asymptotic stable solution for the controlled fractional general T system (89), when $b_1 = 10$.

Conflict of Interests

The authors declare that there is no conflict of interests regarding the publication of this paper.

ACKNOWLEDGMENTS

The authors thank professor Dumitru Opris for useful conversations on the topics of this paper. The authors would like to express their gratitude to the editor and the anonymous referees for the comments on this paper.

REFERENCES

1. E. N. Lorenz, "Deterministic non-periodic fows," Journal of the Atmospheric Sciences, vol. 20, pp. 130–141, 1963.
2. G. Tigan and D. Opris, "Analysis of a 3D chaotic system," Chaos, Solitons and Fractals, vol. 36, no. 5, pp. 1315–1319, 2008.
3. A. Algaba, F. Fernandez-Sanchez, and M. Merino, "Existence of heteroclinic and homoclinic orbits in two different chaotic dynamical systems," Applied Mathematics and Computation, vol. 218, pp. 11859–11870, 2012.
4. A. Algaba, F. Fernandez-S ´ anchez, M. Merino, and A. J. ´ Rodr´ıguez-Luis, "Chen's attractor exists if Lorenz repulsor exists: the Chen system is a special case of the Lorenz system," Chaos, vol. 23, no. 3, Article ID 033108, 2013.
5. B. Jiang, X. Han, and Q. Bi, "Hopf bifurcation analysis in the T system," Nonlinear Analysis: Real World Applications, vol. 11, no. 1, pp. 522–527, 2010.
6. G. Tigan and D. Constantinescu, "Heteroclinic orbits in the T and the Lu systems," ¨ Chaos, Solitons & Fractals, vol. 42, no. 1, pp. 20–23, 2009.
7. R. A. van Gorder and S. R. Choudhury, "Classification of chaotic regimes in the T system by use of competitive modes," International Journal of Bifurcation and Chaos, vol. 20, no. 11, pp. 3785– 3793, 2010.
8. R. A. van Gorder and S. R. Choudhury, "Analytical Hopf bifurcation and stability analysis of T system," Communications in Theoretical Physics, vol. 55, no. 4, pp. 609–616, 2011.
9. Y. Chen and Z.-Y. Yan, "Chaos control in a new three-dimensional chaotic T system,"Communications inTheoretical Physics, vol. 49, no. 4, pp. 951–954, 2008.
10. R. Zhang, "Bifurcation analysis for T system with delayed feedback and its application to control of chaos," Nonlinear Dynamics, vol. 72, no. 3, pp. 629–641, 2013.
11. S. Bhalekar, "Chaos control and synchronization in fractionalorder lorenz-like system," International Journal of Differential Equations, vol. 2012, Article ID 623234, 16 pages, 2012.
12. S. Bhalekar and V. Daftardar-Gejji, "Fractional ordered Liu system with time-delay," Communications in Nonlinear Science and Numerical Simulation, vol. 15, no. 8, pp. 2178–2191, 2010.
13. X. Liu, L. Hong, and L. Yang, "Fractional-order complex T system: bifurcations, chaos control, and synchronization," Nonlinear Dynamics, vol. 75, no. 3, pp. 589–602, 2014.
14. S. Ma, J. Zheng, and Y. Li, "Chaos control and synchronization of a new fractional order chaotic system," Journal of Information & Computational Science, vol. 11, no. 10, pp. 3469–3479, 2014.
15. X.-F. Li, Y.-D. Chu, J.-G. Zhang, and Y.-X. Chang, "Nonlinear dynamics and circuit implementation for a new Lorenz-like attractor," Chaos, Solitons & Fractals, vol. 41, no. 5, pp. 2360– 2370, 2009.
16. G. Yang, "Hopf birurcation of Lorenz-like system about parameter h," Modern Applied Science, vol. 4, no. 1, pp. 91–95, 2010.

17. I. Pehlivan and Y. Uyaroglu, "A new chaotic attractor from ˇ general Lorenz system family and its electronic experimental implementation," Turkish Journal of Electrical Engineering & Computer Sciences, vol. 18, no. 2, pp. 171–184, 2010.
18. M. Xiao and J. Cao, "Bifurcation analysis and chaos control for Lu system with delayed feedback," ¨ International Journal of Bifurcation and Chaos, vol. 17, no. 12, pp. 4309–4322, 2007.
19. K. Pyragas, "Continuous control of chaos by self-controlling feedback," Physics Letters A, vol. 170, no. 6, pp. 421–428, 1992.
20. Y. Song and J. Wei, "Bifurcation analysis for Chen's system with delayed feedback and its application to control of chaos," Chaos, Solitons & Fractals, vol. 22, no. 1, pp. 75–91, 2004.
21. S. Ruan and J. Wei, "On the zeros of transcendental functions with applications to stability of delay differential equations with two delays," Dynamics of Continuous, Discrete and Impulsive Systems, vol. 10, pp. 863–874, 2003.
22. B. D. Hassard, N. D. Kazarinoff, and Y. Wan, Theory and Application of Hopf Bifurcation, Cambridge University Press, Cambridge, UK, 1981.
23. S. Wiggins, Introduction to Applied Nonlinear Dynamical, Systems and Chaos, Springer, New York, NY, USA, 2003.
24. M. S. Tavazoei and M. Haeri, "Limitations of frequency domain approximation for detecting chaos in fractional order systems," Nonlinear Analysis: Theory, Methods& Applications, vol. 69, no. 4, pp. 1299–1320, 2008.
25. K. Diethelm and N. J. Ford, "Analysis of fractional differential equations," Journal of Mathematical Analysis and Applications, vol. 265, no. 2, pp. 229–248, 2002.
26. K. Diethelm, N. J. Ford, and A. D. Freed, "A predictor-corrector approach for the numerical solution of fractional differential equations," Nonlinear Dynamics, vol. 29, no. 1–4, pp. 3–22, 2002.
27. I. Petras, Fractional Order Nonlinear Systems: Modeling, Analysis and Simulation, Nonlinear Physical Science, Springer, New York, NY, USA, 2011.
28. Z. Xiao-Dan, L. Xiang-Dong, Z. Yuan, and L. Cheng, "Chaotic dynamic behavior analysis and control for a financial risk system,"Chinese Physics B, vol. 22, no. 3, Article ID 030509, 2013.
29. C. Jiang, S. Liu, and C. Luo, "A new fractional-order chaotic complex system and its antisynchronization," Abstract and Applied Analysis, vol. 2014, Article ID 326354, 12 pages, 2014.

INDEX

O

P

Q

R

S

T

U

V

W